Alkaloids

Nature's Curse or Blessing?

WILEY-
VCH

Alkaloids

Nature's Curse or Blessing?

Manfred Hesse

Verlag Helvetica Chimica Acta · Zürich

Weinheim · New York · Chichester
Brisbane · Singapore · Toronto

Prof. Dr. Manfred Hesse
Organisch-chemisches Institut
der Universität Zürich
Winterthurerstrasse 190
CH-8057 Zürich
Switzerland

Translated by Dr. Andrew Beard

Published jointly by
VHCA, Verlag Helvetica Chimica Acta, Zürich (Switzerland)
WILEY-VCH, Weinheim (Federal Republic of Germany)

Editorial Directors: Dr. Peter M. Wallimann, Dr. M. Volkan Kisakürek
Production Manager: Norbert Wolz
Cover Design: Bettina Bank

Library of Congress Card No. applied for

A CIP catalogue record for this book is available from the British Library

Die Deutsche Bibliothek – CIP-Cataloguing-in-Publication-Data

A catalogue record for this publication is available from Die Deutsche Bibliothek

ISBN 3-906390-24-1

The German edition appeared under the title: Hesse, Manfred: Alkaloide – Fluch oder Segen der Natur?, Verlag Helvetica Chimica Acta, Zürich, 2000

© Verlag Helvetica Chimica Acta, Postfach, CH-8042 Zürich, Switzerland, 2002

Printed on acid-free paper.

Printing: Konrad Triltsch, Print und digitale Medien, D-97199 Ochsenfurt-Hohestadt, Germany
Printed in Germany

For my wife

Preface

We had not expected that '*Alkaloide – Fluch oder Segen der Natur?*', the original German version of this book, would so quickly find such a large and enthusiastic readership: circumstances that would make a new edition necessary. To meet the demand also from non-German speaking countries, we have decided to publish this English translation, which hopefully fulfills both needs. Since my first book entitled '*Alkaloid Chemistry*' appeared some 20 years ago, the world of both alkaloid research and chemistry has altered fundamentally. Even then, though, it was clear that, with the assistance of ever more powerful physical and analytical methods, the elucidation of new alkaloid structures was not going to pose any significant problem for much longer. As well as this, the first methods for identifying compounds in complex mixtures were already appearing. Also, since then, our knowledge of biogenetic processes has been immensely refined, or even altered. The conventional feeding techniques with ^{14}C-labeled precursors has been joined by the use of ^{13}C-enriched compounds. The latter, thanks to ^{13}C-NMR spectroscopy, permit faster identification of labeled preparations.

Alkaloid synthesis, in general, has for a long time known no more real boundaries, with the most structurally complex representatives having been rendered accessible in their correct relative and absolute configurations. In the preparation of physiologically active agents, a great many factors determine whether total syntheses can compete financially with production from natural sources. When alkaloids are being removed as unwanted side products (*e.g.*, caffeine in the decaffeination of *Coffea arabica*), then chemical synthesis is clearly superfluous. Plant extraction, on the other hand, has the advantage that it can be carried out just about anywhere and with the most primitive equipment, while synthesis always requires appreciably greater efforts and specialized knowledge. However, total or partial synthesis is the method of choice if the natural sources are rare or expensive, or if the corresponding extracts do not surrender their desired product without laborious chromatographic separation, examples being vincamine and alloferin.

Especially interesting today is the question in which respect alkaloids are beneficiary or significant for their producers (microorganisms, plants, animals) or for their consumers (animals, people). The small sum of reliable information in this field offers vistas of a fantastic microcosmic world just opening up. Relationships between different plants (suppression or assistance), or between plants and animals, as well as the reciprocal influences between microorganisms and plants, all often depend on certain roles played by alkaloids. The difficulties in exploring this field lie partly in the fact that this kind of studies can profitably be carried out only on living organisms. It is to be expected that findings relating to the symbiotic interdependencies between various organisms will strongly enhance our understanding of natural interrelationships in the next few years.

Long before the chemical structure of any alkaloid was known – may be even before the notion developed that chemical structures could exist – certain alkaloids were already being used by people, or, at least, certain physiological effects (especially of plants) were known and exploited. Although the molecular structures of thousands of alkaloids are known today, the lay public still tend to view strychnine, morphine, or quinine as the quintessential alkaloids, just as they did 50 or 100 years ago. These physiologically and medicinally important compounds have been termed 'drugs' in modern times. However, their cultural histories sometimes date back for millennia. Among early records on alkaloids, particular attention is merited by Chinese, Egyptian, Native American, and, later, also European accounts. They usually bear testimony to the threat of epidemics (*e.g.*, ergotism), medical properties (*e.g.*, use of *Ephedra* or morphine), as well as physical refreshment and benefits (*e.g.*, tea or coffee).

The first isolation of a specific alkaloid – *Sertürner*'s discovery of morphine in 1805 – was not only the beginning of alkaloid chemistry, it was also the birth of organic chemistry, which has proven to be fundamentally more lasting and consequential. The structure elucidation of alkaloids in these early times, particularly with the aid of degradation reactions, gave rise to new compounds, mainly heterocycles, the names of which (*e.g.*, morpholine, piperidine, indole *etc.*) in some cases still reflect their origin. With respect to the structure elucidation of alkaloids, several examples are presented in this book. Thereby, more classical approaches are discussed, as well as the latest analytical techniques (*e.g.*, HPLC-UV(DAD)-APCI-MS/MS).

It has been a matter of concern that this revised version be accessible not only to chemists but also to chemically and biologically interested readers from other fields of science.

Without the invaluable assistance of colleagues, friends, scholars, and co-workers in the form of generous offers concerning illustrations, manuscripts, old books, and advice – this book, in its present form, would not have been possible. In particular, I would like to thank Prof. *Thomas Baumann* (Zürich), *Andreas Baumeler* (Zürich), Prof. *Stefan Bienz* (Zürich), Prof. *Christopher D. K. Cook* (Zürich), Dr. *Konstantin Drandarov* (Zürich and Sofia), Dr. *Max Frenkel* (Zürich), Dr. *Benjamin Frydman* (Madison, Wisconsin), Prof. *N. Fu-*

setani (Tokyo), Profs. *Belkis* and *Tekant Gözler* (Izmir), *Ursula* and *Gerhard Grosskopf* (Berlin), *Rolf Gfeller* (Brissago), *Armin Guggisberg* (Zürich), *Silvia Hase* (Zürich), *Beatrice Häsler* (Zürich), Dr. *Jörg Heerklotz* (Konstanz), Prof. *Edgar Heilbronner* (Herrliberg), *Veronica Herdeg* (Zürich), Dr. *Christian Hesse* (Bern), Dr. *Wenqing Hu* (Zürich), *David Jetzler* (Effretikon), Prof. *Siegfried Johne* (Halle/Saale), *Martha Kalt* (Zürich), *Petra Klein* (Hameln), Dr. *S. Mahboobi* (Regensburg), *Ruth Riemann* (Mühlhausen/ Thüringen), Prof. *G. P. Schiemenz* (Kiel), Prof. *Peter Schönfelder* (Pentling), Prof. *Karlheinz Seifert* (Bayreuth), Prof. *Dieter Sicker* (Leipzig), *Frau Sticher* and Prof. *Otto Sticher* (Zürich), Dr. *Martin Schwela* (Vienna), Prof. *Hans-Jürgen Veith* (Darmstadt), Dr. *Christa Werner* (Zürich), and *Benedikt Zäch* (Winterthur).

I would also like to extend my heartfelt thanks to *Marie-Therese Bohley* for her painstaking transcription of the original, handwritten manuscript. For the excellent formulae and editing of the manuscript, I am deeply grateful to *Pekka Jäckli* and Dr. *Peter M. Wallimann* of *Verlag Helvetica Chimica Acta*. Finally, Dr. *M. Volkan Kisakürek*, thanks to whose encouragement this book exists, deserves my special gratitude for preparing and arranging the manuscript for publication with his own exceptional personal commitment.

Zürich, May 2002   Manfred Hesse

Content

1. Introduction

1.1. The Term 'Alkaloid'

The term 'alkaloid' is made up out of the words 'alkali' and '$\varepsilon\iota\delta o\sigma$' (Greek 'type', 'similarity') or '$\varepsilon\iota\delta\omega$' ('to see', 'to appear'). Essentially, it means a substance with an alkali-like character. The root 'alkali', for its part from the Arabic 'al-qualja', meaning 'plant ashes', is synonymous with the English expression 'potash'. In Europe until the middle of the 18th century, so-called alkali was largely obtained by steeping burnt wood in water and evaporating the lye obtained in pots: hence 'potash'. Potash found its widest application in glassmaking and in the textile industry, where it was required for coloring and bleaching of materials. To meet this need, *ca.* 40 species of plant of the genus *Salsola*, which flourish in salty soil (salt flats, coastal areas), were cultivated in the Mediterranean region especially for alkali production at that time. Of particular significance were *S. kali* L. (Chenopodiaceae, a 30–50 cm high plant; *Fig. 1.1*) and *S. soda* L. The later discovery that potash is particularly rich in potassium carbonate (K_2CO_3) resulted on one hand in the two terms more and more being used synonymously, and on the other hand in the origination of the element name 'potassium', known in other languages as 'kalium'.

The actual word 'alkaloid', though, was first proposed by *W. Meissner*[1] in 1819. During his investigations of *Veratrum*, he wrote (*Fig. 1.2*):

> *'To me it seems wholly appropriate to refer to those plant substances currently known not by the name alkalis, but alkaloids, since in some of their properties they differ from alkalis considerably, and would thus find their place before the plant acids in the field of plant chemistry'.*

Meissner's definition of the term 'alkaloid' is thus very simple: *alkaloids are plant-derived substances that react like alkalis*. This description was put forward at a time, when only very few plant bases (such as morphine) were known, and those only in impure form. Analogous terms for plant acids, such as 'acidoids', did not exist and have never been proposed. Plant acids were later to be classified, like neutral plant substances, according to their structural characteristics rather than to their chemical properties. Structures incorporating, for example, a (C_6–C_3–C_6)-framework are termed 'flavonoids'; small variations (additional C_1- or C_5-moieties *etc.*) hardly affect the character of the main structure. Steroids, saccharides, peptides, or hydrocarbons also represent structurally rather unified substance classes. The use of pre-

Fig. 1.1. *Salsola kali* L., *a Mediterranean plant growing on sandy beaches.* It was formerly used for soda production, its burnt ashes being steeped in water. Tetrahydroisoquinoline alkaloids of the salsoline type have also been isolated from *Salsola* species (photo *P. Schönfelder*).

1

II. Ueber ein neues Pflanzenalkali (A l k a l o i d).

V o m

Dr. W. M e i f s n e r.

Die Reihe leicht zersetzbarer Pflanzenalkalien, zu welcher das Morphium uns den Weg gebahnt hat, scheint sich mit jedem behutsamen Schritt der Pflanzenanalyse zu vermehren, wie diefs noch neuerlich die Auffindung des Strychnin in der faba St. Ignatii und nux vomica durch Pelletier und Caventou bestätigt. Zu den schon bekannten kann ich nun noch ein neues hinzufügen, welches ich zu Anfang dieses Jahres in dem Sabadillsamen fand, und nicht ohne Schwierigkeiten für einen eigenthümlichen alkalischen Pflanzenkörper erkannte.

Man erhält ihn, indem man den Saamen mit mäfsig starken Alkohol ausziehet, diesen bei gelinder Wärme verdampft, oder aus einer Retorte überdestillirt, den harzigen Rückstand mit Wasser behandelt, die braune Auflösung filtrirt, und solange mit kohlenstoffsäuerlichem Kali versetzt, als noch die geringste Trübung entsteht, den Niederschlag so oft mit Wasser auswäscht, bis dieses ungefärbt abläuft, und in gelinder Wärme trocknet.

Der auf diese Art erhaltene Stoff besitzt eine etwas schmutzig weiße Farbe; keinen bemerklichen Geruch, einen sehr brennenden Geschmack, wobei man noch eine sehr unangenehm kratzende Empfindung im Schlunde bemerkt, die auch entsteht, wenn man kaum

Fig. 1.2. *W. Meissner, Journal für Chemie und Physik (Schweiger),* **1819**, *25*, *379. The introduction of the term 'alkaloid': 'To me it seems wholly appropriate to refer to those plant substances known to date not by the name alkalis, but alkaloids, since in some of their properties they differ from alkalis considerably…'.*

fixes such as 'di', 'tri', 'oligo', or 'poly' in front of the class name makes it possible to convey additional structural information. In contrast to this, the basic plant substances had been defined until *ca.* 1890 by completely different criteria. Essential were mainly three characteristics: *1*) their structures were unknown (and in many cases would remain so for a considerable time), *2*) many of them were physiologically active in some way or another, and *3*) they exhibited basic properties.

Circumstances (and no doubt a certain measure of indecision) finally dictated that this large class of unknown substances should be grouped together under the term 'alkaloids'. Classification (*cf. Chapt. 2*) proceeded (among other criteria) according to 'family' occurrence or similarities in chemical and spectroscopic properties, without the overall term 'alkaloids' being narrowed down or replaced in any way. Nonetheless, the term established itself in chemical literature only slowly. *C. J. Löwig* (1803–1890), Professor of Chemistry at the University of Zurich (Switzerland), described *ca.* 50 alkaloids in his textbook[2] *Chemie der Organischen Verbindungen* (1839/41), still referring to them merely as '*nitrogen-containing, nonacidic, organic compounds*'. Nor was *J. J. Berzelius*[3] yet using the term in 1837, speaking of '*vegetable salt bases*'. *W. Königs* suggested in 1880 that only those plant bases known to be pyridine derivatives should be described as alkaloids, but this recommendation was not taken up.

The expression 'alkaloid' finally became established in 1882, when *O. Jacobsen* wrote a detailed review article about alkaloids for *A. Ladenburg*'s *Handwörterbuch der Chemie*[4a]. By then, although a great deal had already been found out about alkaloids, still as good as nothing was known as far as their structures were concerned.

> '*The bodies referred to as alkaloids are nitrogen-containing organic bases. The group, however, is in many respects not sharply defined. Following the earlier practice of commonly terming all carbon-containing bases as alkaloids, and then distinguishing between 'natural alkaloids' such as quinine and morphine and 'artificial alkaloids', such as ethylamine and aniline, the term became further restricted, largely to those bodies found in the plant kingdom to which the name 'alkaloids' has now been reserved. For animal bases, such as choline and creatinine, it is no longer customary or is restricted here to such bodies of which, as in the case of the poisonous bases produced by decaying animal matter, little more than a general similarity with the plant alkaloids is known*'[4b].

In 1896, I. Guareschi[5] gave a broader definition in his *Einführung in das Studium der Alkaloide*:

'The term 'alkaloid' is applicable to all basic, organic compounds – whether obtained either from animal or plant materials, or prepared artificially; that is, the expression 'alkaloid' is synonymous with 'organic base' or 'organic alkali '.

In the book *Die Alkaloide* (1910), by *E. Winterstein* and *G. Trier*[6], there appears:

'By 'alkaloids' in the narrower sense, on the other hand, we mean compounds with nitrogen atoms bound in heterocyclic fashion, with a greater or lesser degree of basic character, marked physiological effects, complicated molecular structure, which are found in plants, and, with a few exceptions, are characteristic for particular plant families, genera, or species'.

A. Stoll[7] put it much more succinctly 43 years later:

'Nitrogen-containing bases of vegetable origin were grouped together under the label 'alkaloids''.

And the great Swiss chemist *P. Karrer*, in his 1959 *Lehrbuch der Organischen Chemie*[8], defined alkaloids in almost the same way:

'Today, by 'alkaloids' we mean nitrogen-containing, basic compounds found in plants in general'.

Lastly, *S. W. Pelletier*[9] supplied a 'modern definition' in 1983, in the first volume of the series *Alkaloids*, which he edited:

'An alkaloid is a cyclic organic compound containing nitrogen in a negative oxidation state which is of limited distribution among living organisms'.

All of the above definitions (among many others at that time) allow a preliminary view into the nature of the substance class. However, none of them really appears satisfactory. For instance, it is clear that not all alkaloids are pyridine bases, and also that they do not necessarily have to be heterocyclic – or even cyclic. Some spider toxins, for example, are open-chain alkaloids. Consequently, it is easier not to incorporate any specific structural prescriptions into the definition at all. Expressions like 'complicated molecular construction' also seem to be of limited meaningfulness, since they are subject to regular qualification from new investigations. Even less must alkaloids necessarily be physiologically active, although many are to some degree

or another. Moreover, it is also accepted now that the occurrence of alkaloids is not restricted to the plant kingdom. Certainly, they are most commonly represented there, but they are also found in 'lower' organisms at
many stages of development (such as agroclavine from the endophytic fungus *Claviceps purpurea*, which grows on *Elymus mollis*), and in animals
(such as samandarine from *Salamandra salamandra* (Salamandridae), adaline
from both *Adalia bipunctata* and *A. decimpunctata*, or nitropolyzonamine
from the millipede *Polyzonium rosalbum* (Diplopoda)). Alkaloids have even
been found in mammals, mostly in the form of hydroxylated tyrosine or tryptamine derivatives. Thus, they occur in all types of living organisms.

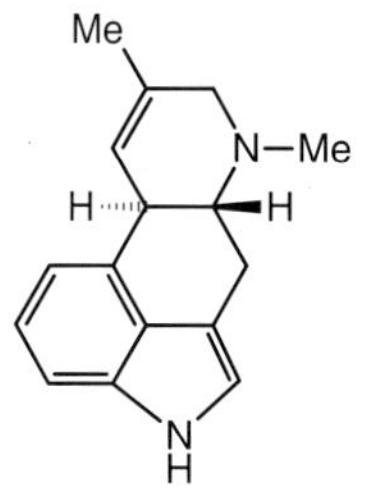

A central property of the substance class is basicity. Even this, when considered more closely, however, is problematic; it is necessary to take account of
the fact that, in nature, there are found not only primary, secondary, and tertiary nitrogen bases (such as tenuilobine from *Oncinotis tenuiloba* (Apocynaceae), phellodendrine iodide from *Phellodendron amurense* (Rutaceae), or
ricinine from *Ricinus communis* (Euphorbiaceae)), but that there are also
internal salts such as stachydrine from *Stachys sieboldi* (Labiatae), amine ox-
ides (*N*-oxides) such as kopsinoline from *Pleiocarpa mutica* (Apocynaceae)), hydroxylamine derivatives such as nareline from *Alstonia scholaris*
(Apocynaceae)), nitro compounds such as aristolochic acid I from various
Aristolochia species (appearing as natural degradation products of the corresponding aporphine alkaloids), and amino acetals such as samandarine.
Whether a meaningful comparison of basicities can and should be made under such circumstances is an open question.

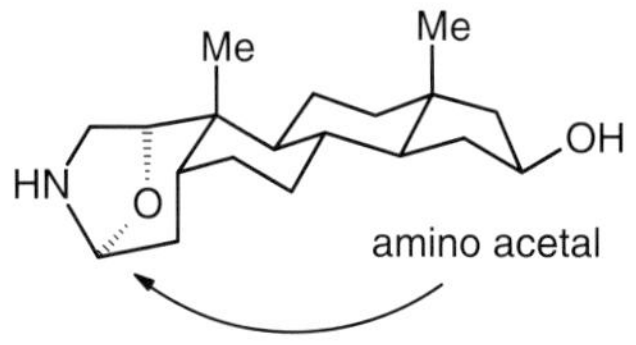

In view of the fact that pure amino acids, peptides, nucleic acids, and synthetic organic nitrogen bases such as aniline are not counted among the alkaloids, the end result may be the following, general definition:

***Alkaloids are nitrogen-containing organic substances of natural origin
with a greater or lesser degree of basic character.***

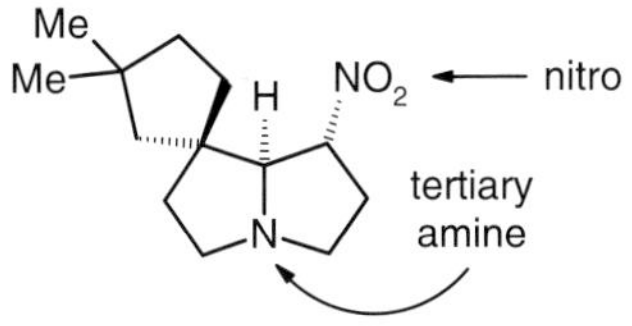

1.2. On Alkaloid Terminology and Nomenclature

As with other natural product classes (flavonoids, terpenoids, *etc.*), to date
no unified system for deriving alkaloid trivial names has become generally
accepted. In many cases, the alkaloid name is partly based on the systematic plant name. Hence, an alkaloid obtained from a *Papaver* species receives
the name *papaverine*. Terms such as *atropine* (from *Atropa belladonna*, and
so from a generic name), *cocaine* (from *Erythroxylum coca*, a species name),
and *spegazzinine* (from *Aspidosperma chakensis* SPEGAZZINI[10], the 'author'

(–)-Agroclavine   (abs. config.)

Samandarine   (abs. config.)

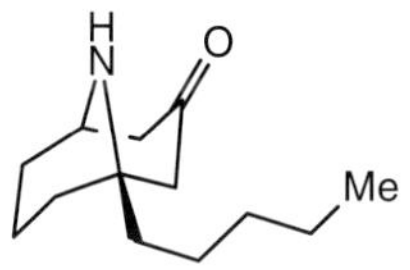

(–)-Adaline   (abs. config.)

Nitropolyzonamine

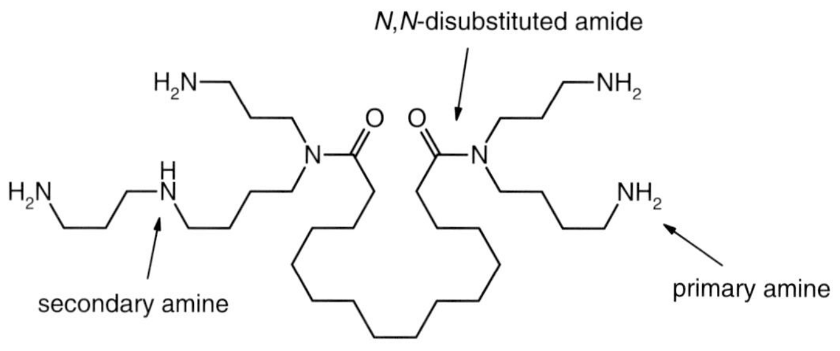

_N_,_N_-disubstituted amide

H₂N

secondary amine

primary amine

Tenuilobine

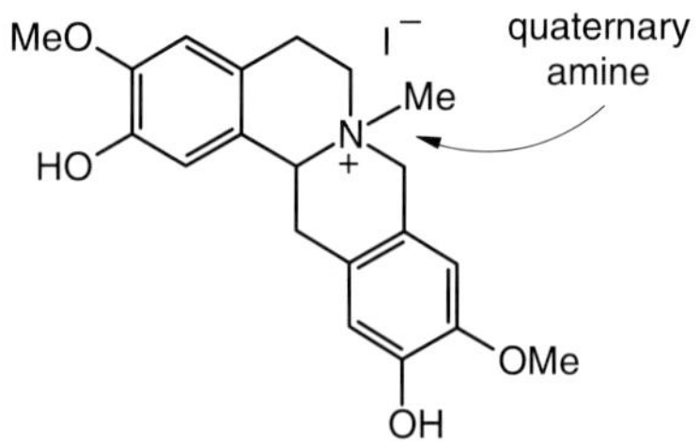

quaternary amine

(±)-Phellodendrine Iodide

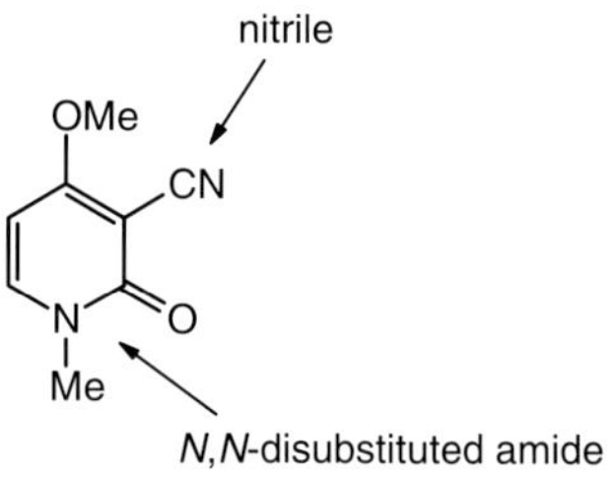

nitrile

_N_,_N_-disubstituted amide

Ricinine

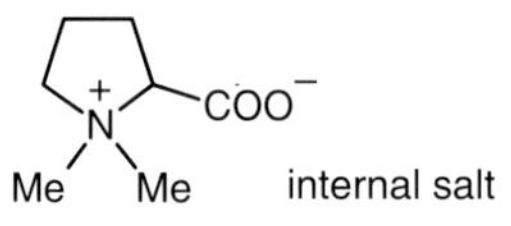

(±)-Stachydrine   internal salt

amine oxide

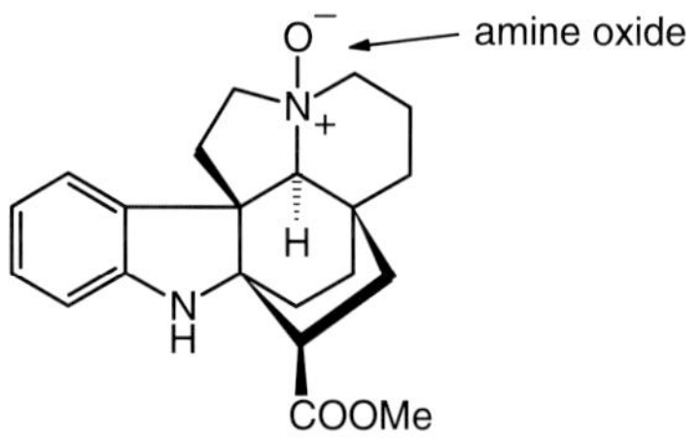

Kopsinoline   (abs. config.)

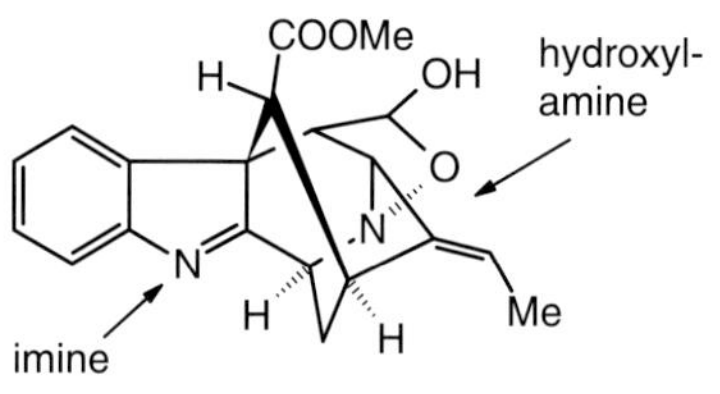

hydroxyl-amine

imine

Nareline

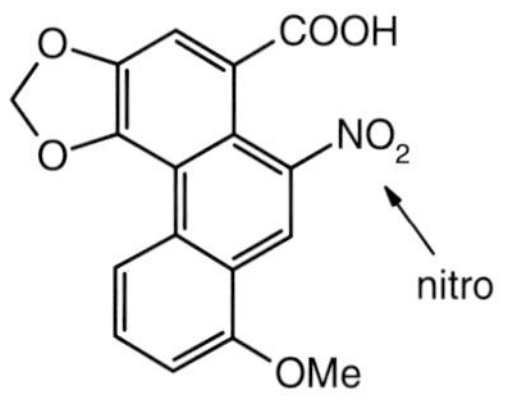

nitro

Aristolochic Acid I

name of the plant discoverer) appeared in analogous fashion. In English and French, the '-ine' suffix is commonly used; this becomes '-in' in German.

Often, several alkaloids can be isolated from a single plant. Depending on the chemist's imagination and way with words, their naming is then carried out with a greater or lesser degree of judiciousness. To distinguish between the various isolated alkaloids, the selected root word is generally modified either with suffixes (such as '-ine', '-idine', '-anine', '-aline', '-inine', *etc.*), or with combinations of letters and/or numbers. The case of *Vinca* (Apocynaceae) provides a cautionary example of how this system of naming on the basis of part of the plant name can be carried *ad absurdum*. A number of species, particularly *V. major*, *V. minor*, *V. difformis*, *V. erecta*, *V. herbacea*, and naturally *V. rosea* (= *Catharanthus roseus*), were studied by several research groups at much the same time during the 1960s. The result was the announcement, over time, of 86 alkaloid names containing the root syllable '*Vin*' (41 of them alone '*Vinca*'). That this *simply had to* give rise to both phonetic confusion and the use of identical names for different substances (given the lack of sufficient information exchange between the researchers) is self-evident.

The naming procedure for alkaloids from animals or fungi is essentially the same as that for plant alkaloids; there are no differences in the principles involved. For compounds isolated from fungi, however, along with general names such as alkaloid A or B, there is also a 'hybrid nomenclature' in use; this means that the names may also include terms derived from certain pharmacological properties (as in the drug name penitrem A–F for the *trem*organic indole alkaloids from *Penicillium* species). When expedient, even geographical references relating in some way to the plant, animal, or fungus have found application in alkaloid nomenclature. When more compounds were isolated from the Tasmanian *Aristotelia penducularis* (Elaeocarpaceae) than were namable just with the aid of the systematic plant name, the researchers turned to local geography and created names such as *tasmanine*, *hobartine* (after *Hobart*, the state capital), and *sorreline* (after *Lake Sorrel*). On occasion, following the example of the barbiturates (derived from *Barbara*), a few women's names have also been used for naming alkaloids (one example is *Salhiha → macrosalhine*, isolated from *Alstonia macrophylla*).

As a rule, alkaloid trivial names contain very little information. From the name 'strychnine', for example, it is possible, given a fair wind, to see that the compound originates from *Strychnos*, but to infer the structure of the compound from the name is utterly impossible; either you know it or you don't. Even less is to be garnered from a geographically derived name, although still more than from a term based on the pleasant sounding name of a pretty woman.

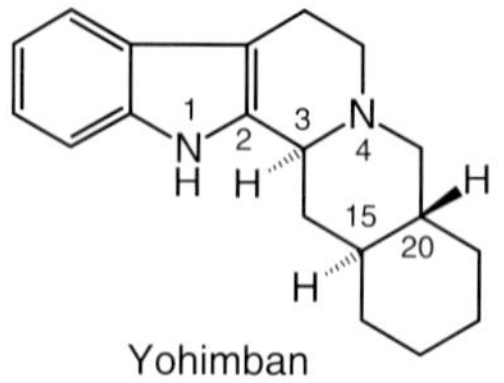

Yohimban

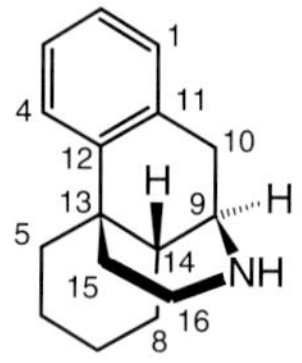

Morphinan

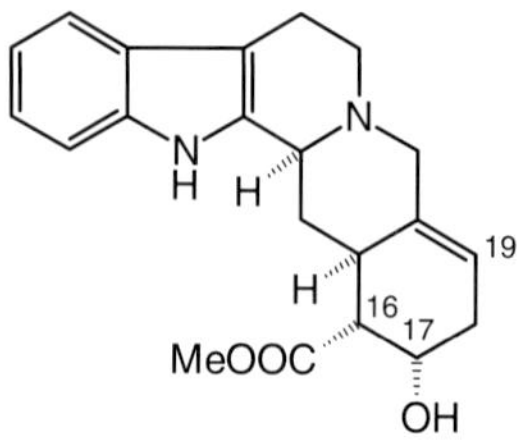

19-Dehydroyohimbine

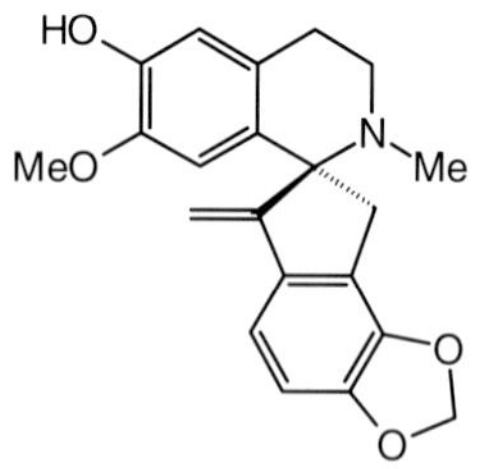

MeOOC

OH

(+)-Ochotensine

The first alkaloids, isolated at the beginning of the 19th century, had to be given trivial names, since their structures were still unknown, and so no systematic nomenclature could be established. That apart, there was not yet any perceived need for such a thing. Today, though, there are more than enough grounds for requiring systematic naming, and so the use of the *IUPAC* (*International Union of Pure and Applied Chemistry*) nomenclature should be emphasized here.

Monocyclic alkaloids can usually be named systemically without problems. In addition, the names produced are generally short and clear enough also to be used in the laboratory. One example is the hemlock poison (–)-coniine, the systematic name of which is (–)-(*S*)-2-propylpiperidine. Most ant alkaloids (often pyrazine derivatives) are also named according to the *IUPAC* rules, without being assigned interim trivial names for the phase between isolation and characterization. Systematic naming of polycyclic systems, however, can become appreciably more complicated. To make the naming of these compounds easier, many of the more commonly occurring key alkaloid ring systems have been assigned names of their own, implicitly conveying clearly defined information about the constitution and configuration of the molecule, together with a prescribed atom numbering system. A ring system defined in such a way is also known as a *stereoparent*. Examples are yohimban and morphinan (provisional *IUPAC* recommendation, 1976).

If this model is applied to the naturally occurring, traditional 19-dehydroyohimbine, then the systematic *IUPAC* name is methyl 19,20-didehydro-17-hydroxyyohimban-16-carboxylate. The situation is much more complicated for (+)-ochotensine, isolated from various *Corydalis* and *Dicentra* species, for which there is no defined ring system to fall back on; adherence to *IUPAC* nomenclature rules in this case provides the systematic name (+)-3′,4′,6,8-tetrahydro-7′-methoxy-2′-methyl-6-methylidenespiro[7*H*-indeno[4,5-*d*]-[1,3]dioxol-7,1′(2′*H*)-isoquinolin]-6′-ol. To derive the structure of (+)-ochotensine from this name would doubtless be asking much more of a nomenclatorially less trained chemist than in the first example, for which a stereoparent is available.

Certainly, the two systems will have to co-exist also in the future, although the decision of whether to use a fully systematic name or a semisystematic one is essentially dependent on the existence of an *IUPAC*-approved, stereochemically defined ring system. In dubious cases, however, the following rule always applies:

If general organic nomenclature is applicable, then it should be used!

References and Notes

1. W. Meissner, '*Über ein neues Pflanzenalkali (Alkaloid)*', *J. Chem. Phys. (Schweiger)*, **1819**, *25*, 379.
2. C. Löwig, '*Chemie der organischen Verbindungen*', 3 Vols., F. Schulthess-Verlag, Zürich, 1839–1841.
3. J. J. Berzelius, '*Lehrbuch der Chemie*', German translation by F. Wöhler, 10 Vols., Arnoldische Buchhandlung, 3rd edn., Dresden, 1835–1841.
4. a) '*Handwörterbuch der Chemie*', Ed. A. Ladenburg, 13 Vols., Verlag E. Trewendt, Breslau, 1882–1885. b) *ibid.*, Vol. 1, p. 213.
5. I. Guareschi, '*Einführung in das Studium der Alkaloide mit besonderer Berücksichtigung der vegetabilischen Alkaloide und der Ptomaine*', Gaertners Verlagsbuchhandlung, Berlin, 1896.
6. E. Winterstein, G. Trier, '*Die Alkaloide*', Verlag Bornträger, Berlin, 1910.
7. A. Stoll, E. Jucker, '*Alkaloide*', in *Ullmanns Encyklopädie der technischen Chemie*, 3rd edn., Vol. 3, *177*, 1953.
8. P. Karrer, *Lehrbuch der Organischen Chemie*, 13th edn., Thieme Verlag, Stuttgart, 1959.
9. S. W. Pelletier in *Alkaloids, Chemical and Biological Perspectives*, Ed. S.W. Pelletier, John Wiley & Sons, New York, 1983.
10. If the Latin name of a plant or animal is merely mentioned, the author name is generally omitted.

2. Classification of Alkaloids

2.1. General

The total number of alkaloids so far isolated from (or detected in) plant and animal organisms, fungi, or natural folk medicines (calabash curare, tubocurare, opium *etc.*) is enormous. According to rough estimates, taking all naturally occurring structural types into account, a good 10,000 different alkaloids exist. To cope with so many compounds, it is necessary to perform some form of sub-classification. This is difficult, though, since there is probably no other class of organic natural products that displays such a huge breadth of structural variation. In comparison, steroids, flavonoids, and saccharides are confined to only a few basic skeleton types.

The criteria currently used for alkaloid classification are: biogenesis, structural relationship, biological origin, and spectroscopic/spectrometric properties (chromophores in UV spectroscopy, ring systems in mass spectrometry). No unified, general taxonomic principle that would permit consistent classification exists; all of the classification guidelines so far developed have to accept compromises in certain borderline situations.

Those compounds that may be subdivided according to chromophore or basic skeleton type include – among others – indole alkaloids, quinoline alkaloids, and isoquinoline alkaloids. At first glance, it seems paradoxical that, for example, not all representatives of the isoquinoline alkaloids (such as rhoeadine and its derivatives) are required to contain an isoquinoline chromophore to be qualified as part of this class. Resolution of this paradox, though, is only to be achieved by revision of the criteria and enlargement of the class to include '*alkaloids with indole, quinoline, or isoquinoline (etc.) chromophores plus biogenetically related bases*'. A completely different means of classification of these compounds is by reference to their biological origin (plants, animals, and fungi). The tobacco and Amaryllidaceae alkaloids are examples of this. In general, members of these groups are found not only in plants of the tobacco or Amaryllidaceae families, but also in plants bearing no relationship to those species. This type of subdivision also suffers from the drawback that individual members within a class may sometimes have very different skeletons (Amaryllidaceae alkaloids are one example). Hence, we can see that neither subdivision by 'basic skeleton' nor by 'origin' provides any clear means of classification of all alkaloids.

In the case of the so-called peptide alkaloids, which consist of a (poly)cyclic peptide part and an attached amine component, the predominant bond type serves as a criterion. If the amine component is lysergic acid, then the name 'ergot alkaloids' can be used. With these compounds, interestingly, the tendency is to view them structurally more as indole alkaloids than as peptide alkaloids; although we are dealing here clearly with one of the borderline cases mentioned above. The so-called biogenic amines, representatives of which are structurally extremely heterogeneous, are similarly hard to classify.

The fact that scarcely a treatise on alkaloids appears without a chapter entitled '*Miscellaneous*' or '*Unclassified Alkaloids*' reflects the lack of any generally valid taxonomic principle on one hand, and the difficulty of defining the term '*alkaloid*' itself on the other. It is sometimes extremely problematic to define precise structural boundaries between true alkaloids, peptides, amino acid derivatives, antibiotics, steroids, and terpenes, respectively. In this book, the immediate environment of the N-atom in the individual alkaloid is put forward as the guiding principle, and so alkaloids are classified on the basis of their N-containing structural elements. If a number of N-atoms are present in a compound, then, as a rule, characteristic structural elements are assigned priority. Exceptions to this are made only with steroid, terpene, spermidine, spermine, and peptide alkaloids. Essentially, we can draw distinctions between the following five alkaloid classes, according to the position of the N-atom in the main structural element:

- Heterocyclic alkaloids
- Alkaloids with N-atoms in exocyclic position including aliphatic amines
- Putrescine, spermidine, and spermine alkaloids
- Peptide alkaloids
- Terpene and steroid alkaloids

If the complete set of alkaloids is divided up along these lines, then the resulting picture is very uneven: the bulk of compounds belong to the heterocycles, while the smallest classes are constituted by the putrescine, spermidine, and spermine bases.

2.2. Heterocyclic Alkaloids

In representatives of this class, the N-atom constitutes part of a heterocycle.

2.2.1. Pyrrolidine Alkaloids[1–3]

Besides simple *N*-acylated derivatives of pyrrolidine (such as 3-methoxycinnamic acid pyrrolidide from *Piper* (Piperaceae), a few alkylated compounds also belong to this group, for example hygrine (*cf. Fig. 2.1*) from *Erythroxylum* (Erythroxylaceae), *Atropa*, *Datura*, *Hyoscyamus*, and *Nicandra* (all Solanaceae), *Dendrobium* (Orchidaceae), and *Cochlearia* (Cruciferaceae).

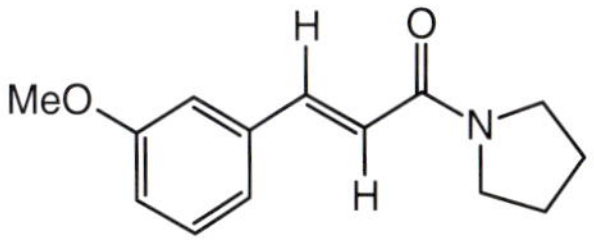

Biosynthetically, these alkaloids all include some component derived from the amino acid ornithine. Hence, for example, stachydrine (from plants of the genera Capparidaceae, Compositae, Labiatae, Leguminosae, Liliaceae, and Rutaceae) is synthesized *via* proline as shown in *Scheme 2.1*. In addition, other biosynthetic pathways to these compounds have also been found; one example is acetate incorporation during the formation of shihunine (from *Dendrobium pierardii* (Orchidaceae)). Extensive investigations have unequivocally shown that the N-atom has to originate from a methylamine equivalent (*Scheme 2.2*).

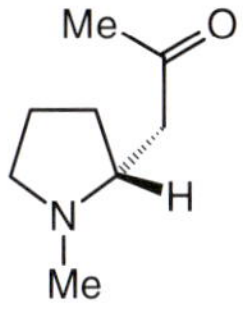

3-Methoxycinnamic Acid Pyrrolidide

(+)-(*R*)-Hygrine

Scheme 2.1

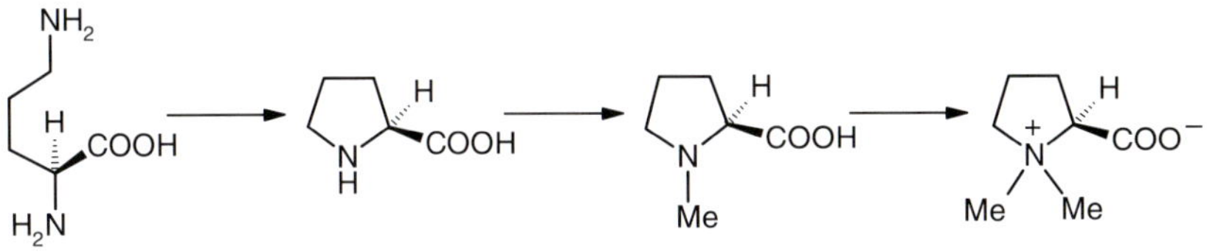

Ornithine   Proline   (–)-Stachydrine (abs. config.)

Scheme 2.2

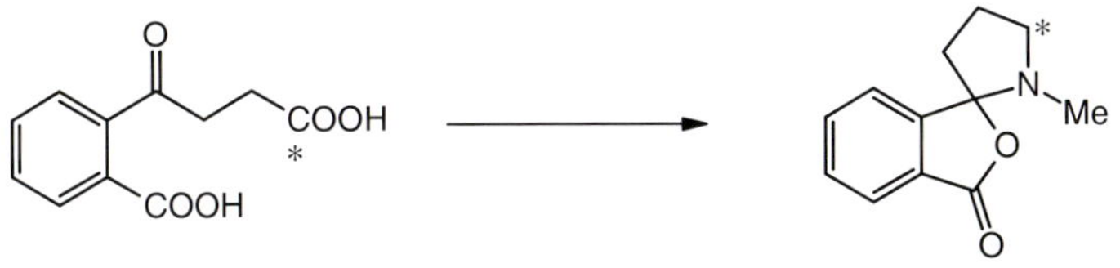

Shihunine

Fig. 2.1. *The pyrrolidine alkaloid hygrine and several tropane derivatives* (e.g., 8-methyl-8-azabicyclo[3.2.1]oct-3-yl 2-methylbut-2-enoate) *occur* inter alia *in the roots of the popular dried flower Physalis alkekengi* L. (Solanaceae), *also called Chinese or Japanese Lantern* (ground cherry) (photo *M. Hesse*)

Mesembrenol

(–)-Dendrobine

In mesembrenol from *Sceletium* (Aizoaceae), and dendrobine from *Dendrobium*, the N-atom is similarly present as part of a five-membered ring. Dendrobine and its derivatives may also, in view of their carbon skeletons, be regarded as sesquiterpene alkaloids.

2,5-Dialkylpyrrolidines (*e.g.*, 2-butyl-5-pentylpyrrolidine) have been isolated from the poison glands of ants of both the genera *Solenopsis* and *Monomorium* (Myrmicinae). Methyl 4-methylpyrrole-2-carboxylate is found in the two species *Atta* and *Acromyrmex* (Myrmicinae)[4], in which it serves as a trail pheromone.

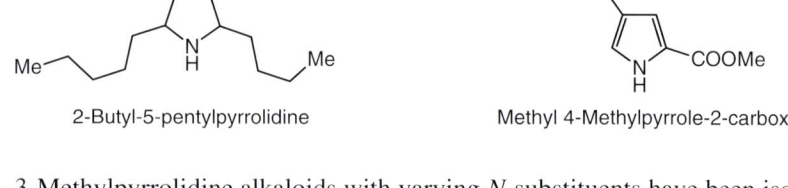

2-Butyl-5-pentylpyrrolidine

Methyl 4-Methylpyrrole-2-carboxylate

Fig. 2.2. *Slavemaking ants Harpagoxenus sublaevis* (Leptothoracini). One female raises her tail and secretes a drop from the poison gland. The basic active substances in the secretion are 3-methylpyrrolidine alkaloids, which attract males ready for mating (photo *H. J. Veith*).

3-Methylpyrrolidine alkaloids with varying *N*-substituents have been isolated from the poison gland secretions of the so-called slavemaking ants *Harpagoxenus sublaevis* (Leptothoracini, Myrmicinae, see *Fig. 2.2*). The compounds serve the creatures as intraspecies-specific signaling substances and sex attractants[5].

(abs. config.)

2.2.2. Indole Alkaloids[6]

The total number of known indole alkaloids is *ca*.1,500. This figure includes both those compounds that incorporate the actual indole chromophore and those containing its derivatives: namely indoline (dihydroindole), indolenine, α-methylideneindoline, pseudoindoxyl, and oxindole. Also members of this group are alkaloids in which the nucleus incorporates an additional benzene or pyridine ring, for instance, carbazole or β- and γ-carbolines and their derivatives.

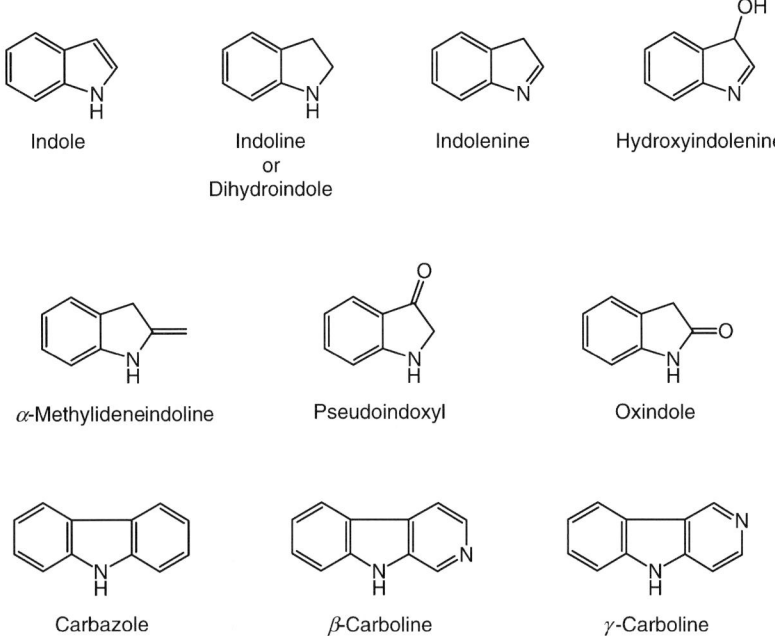

Indole   Indoline or Dihydroindole   Indolenine   Hydroxyindolenine

α-Methylideneindoline   Pseudoindoxyl   Oxindole

Carbazole   β-Carboline   γ-Carboline

For further subclassification of indole alkaloids, structural and biogenetic criteria are applied.

Simple Indole Bases[7]

The simplest natural representatives incorporating the indole basic skeleton are the biogenetic amines tryptamine and serotonin. Gramine – with a side chain shorter by one CH_2 group – also belongs to this class. Psilocybine, a derivative of serotonin, is the psychotropic agent of the Mexican narcotic mushroom teonánacatl (*Psilocybe mexicana*). Hydrolysis converts psilocybine into the dephosphorylated psilocine, which also occurs in the mushroom. The two compounds are not very toxic, but they both induce states of euphoria similar to those brought on by LSD, yet some 100 times more weakly, when taken orally or intramuscularly. Representatives of other simple indole bases are apparicine (from a variety of genera of Apocynaceae) and vallesamine from *Vallesia* (Apocynaceae).

Some ergot fungi, growing on meadow grasses in Australia, New Zealand, and the south-western USA, produce alkaloids of the paspaline type (such as paspalicine and paspalinine from *Claviceps paspali, Penicillium paxilli*). These and similar alkaloids are toxic to grazing animals. They act as tremorgens and are responsible for a number of diseases.

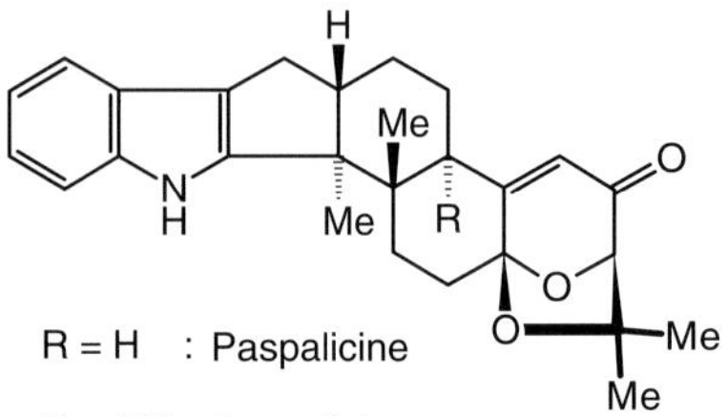

R = H : Paspalicine

R = OH : Paspalinine

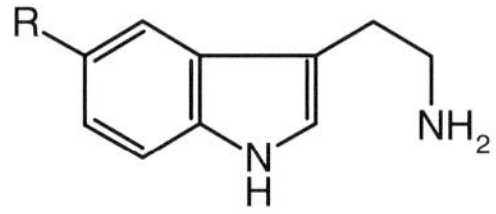

R = H : Tryptamine

R = OH : Serotonin

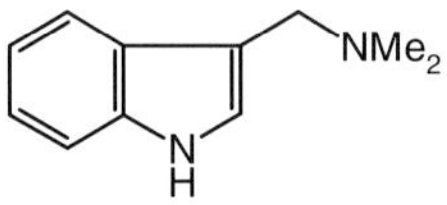

Gramine

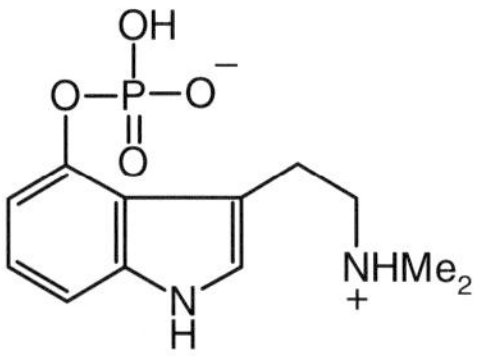

Psilocybine

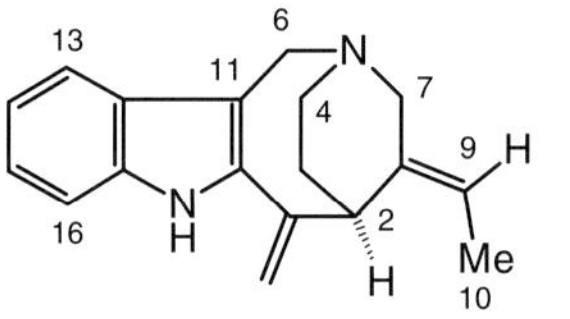

Apparicine

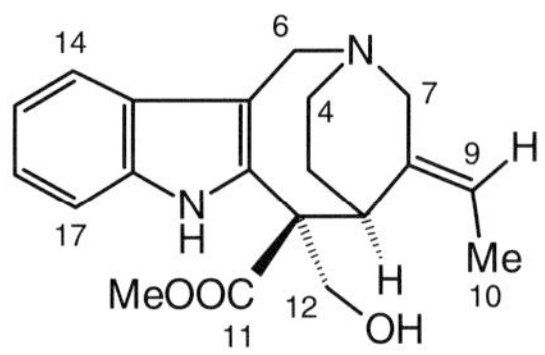

Vallesamine (abs. config.)

Alkaloids with Carbazole Skeletons

The distinction drawn below between alkaloids with carbazole skeletons and those with β-carboline skeletons tells nothing about the biogenesis of these compounds. The majority of alkaloids in both groups are derived biogenetically from tryptamine.

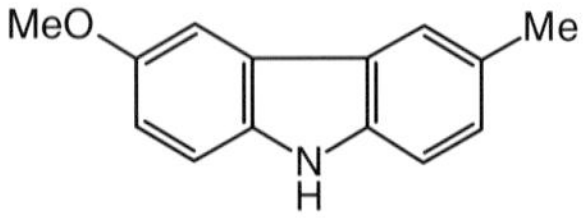

Carbazole Alkaloids[8]

A simple representative of this alkaloid type is glycozoline (from *Glycosmis* (Rutaceae)). The *Murraya* genus, also part of the Rutaceae family, is the

Glycozoline

source of various carbazole derivatives, among them mahanimbine. Another group of compounds derives from carbazole through annelation of a pyridine ring.

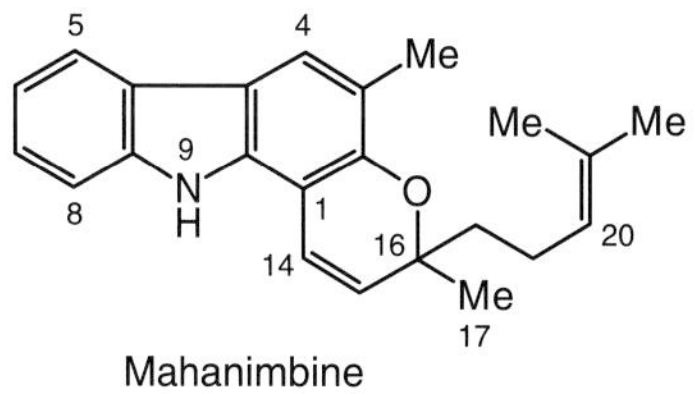

This group also includes olivacine (from various species of *Aspidosperma* (Apocynaceae)) and ellipticine (from *Ochrosia* species (Apocynaceae)), which, due to their antitumor properties, are of great interest both to medicinal chemists (preparation of analogous substances) and to medical researchers[9].

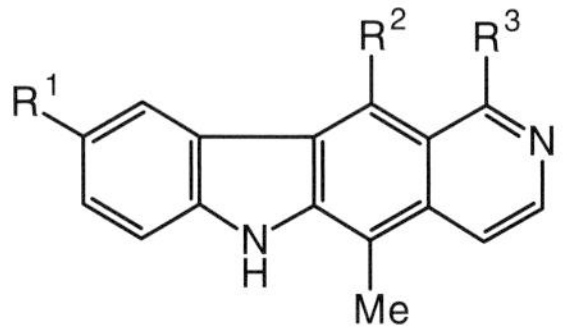

Aspidospermidine and Biogenetically Related Bases

To date, *ca.* 80 alkaloids that may be viewed as part of this subgroup have been isolated. They occur exclusively in the Apocynaceae plant family (*cf.* also *Chapt. 7*). The skeleton types are represented by the following structures: 15,16,17,20-tetrahydrosecodine (from *Rhazya*), quebrachamine (from various species of the genera *Amsonia*, *Aspidosperma*, *Gonioma*, *Melodinus*, *Pleiocarpa*, *Rhazya*, *Stemmadenia*, and *Vinca*), aspidospermidine (from various species of the genera *Amsonia*, *Aspidosperma*, *Gonioma*, and *Rhazya*), vincadifformine (from various species of the genera *Amsonia*, *Melodinus*, *Rhazya*, *Tabernaemontana*, *Vallesia*, and *Vinca*; *cf. Sect. 5.4*), aspidoalbine (from *Aspidosperma* species), pleiocarpine (from *Hunteria* and *Pleiocarpa* species), tuboxenine (from *Pleiocarpa*), kopsine (from *Kopsia*), fruticosine (from *Kopsia*), and neblinine (from *Aspidosperma*).

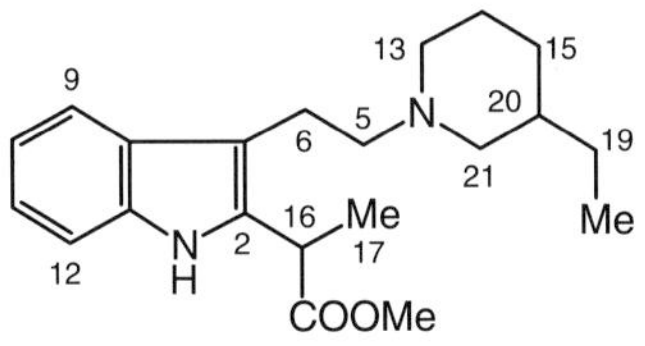

Mahanimbine

R[1] = R[2] = H, R[3] = Me : Olivacine

R[1] = MeO, R[2] = Me, R[3] = H : Ellipticine

15,16,17,20-Tetrahydrosecodine

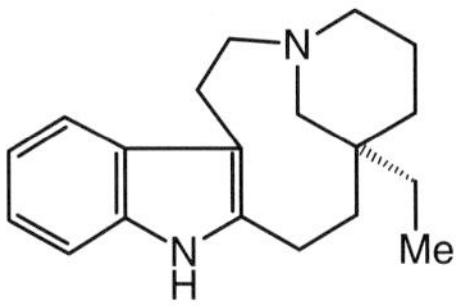

(+)-Quebrachamine

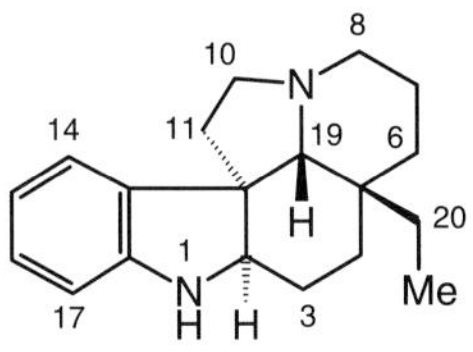

(+)-Aspidospermidine

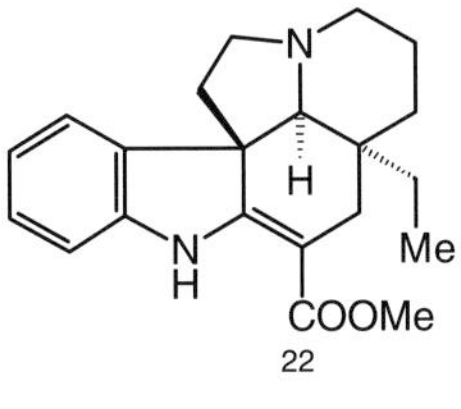

(−)-Vincadifformine

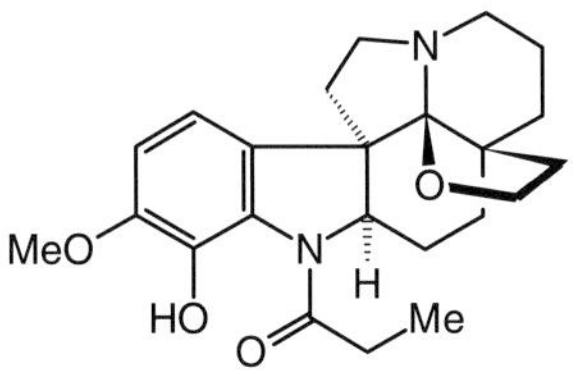

(+)-Aspidoalbine

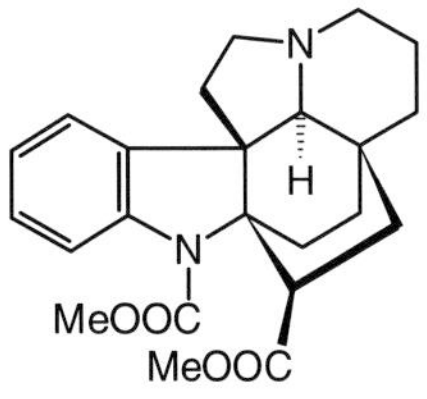

(−)-Pleiocarpine

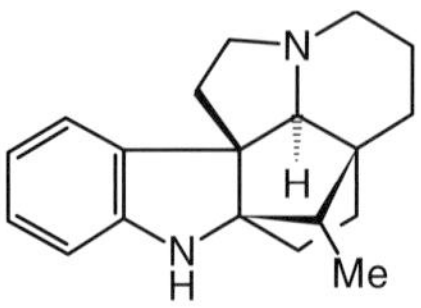

Tuboxenine

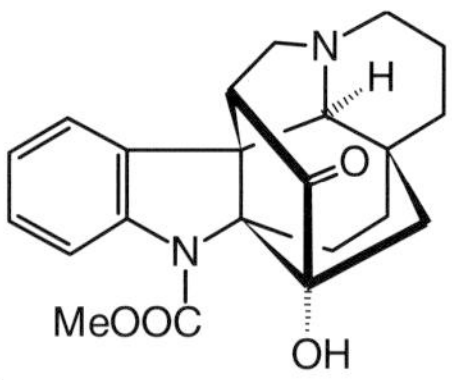

(–)-Kopsine

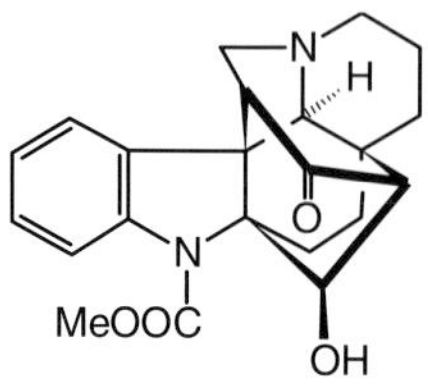

(–)-Fruticosine

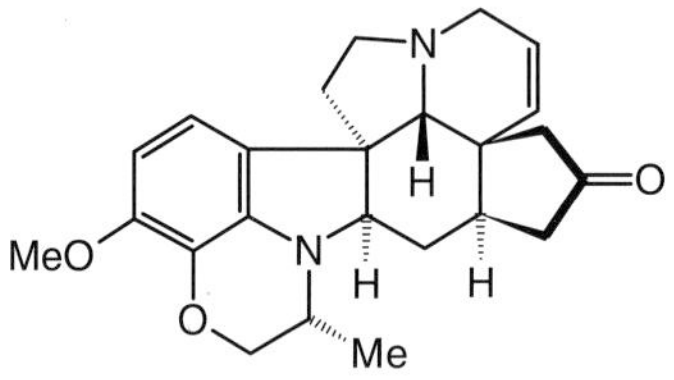

(all abs. config.)

(–)-Neblinine

There are several different derivatives of these basic types which, a few not yet mentioned exceptions aside, do not exhibit any variation in the C,N-skeleton. Mostly, representatives with additional double bonds, oxygen functions, *N*-substituents, and ether rings (*e.g.*, beninine from *Hedranthera*) have been isolated.

The alkaloid vincatine (from *Vinca*) is derived from alkaloids of the vinca-difformine type by cleavage of the C(2)–C(3) bond.

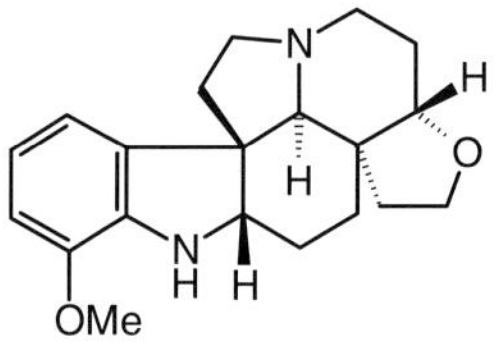

Beninine (abs. config.)

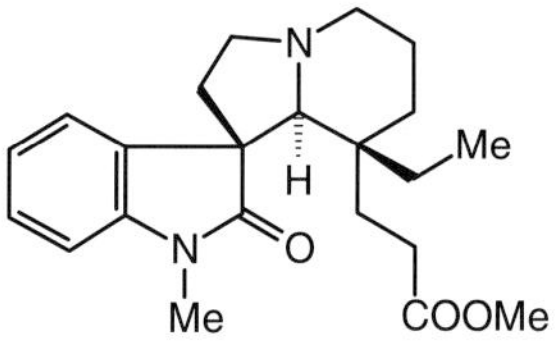

Vincatine

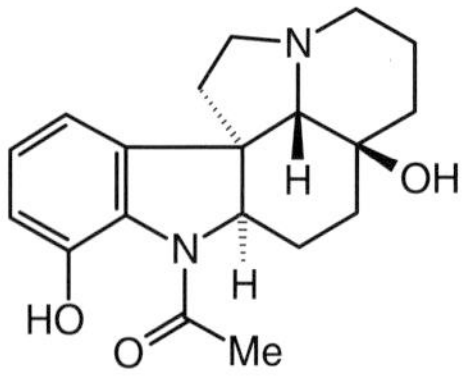

(+)-Aspidodispermine (abs. config.)

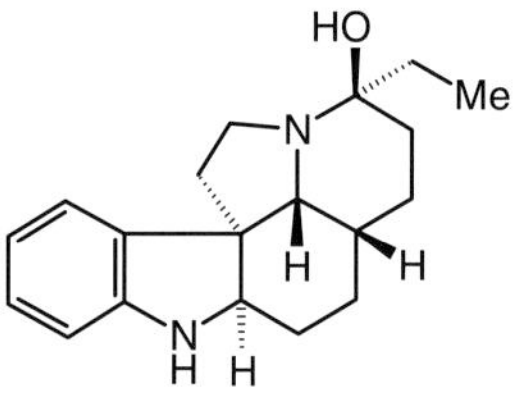

(+)-Pandoline

Unlike the aspidospermidine alkaloids so far listed, aspidodispermine (from *Aspidosperma*) contains no alkyl side chain, while pandoline (from *Pandaca* (Apocynaceae)) bears its side chain at a different location.

Aspidospermine is also a component in quite a large number of bisindole alkaloids, in which it may be present either singly or, occasionally, in a double variant.

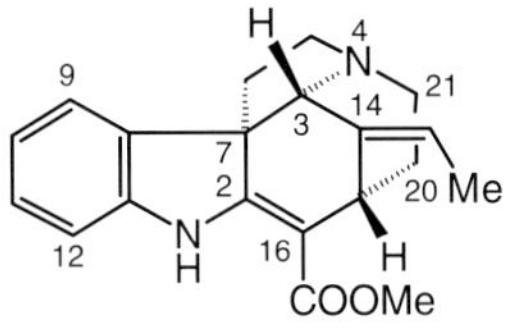

Alkaloids of the Condylocarpine Type

The most important representative of this alkaloid subgroup is condylocarpine (from *Diplorrhynchus* (Apocynaceae)). Unlike condylocarpine, most other members of this group possess no C-substituent at C(16). Precondylocarpine, however, has an additional CH$_2$OH group in this position; and dichotine (from *Vallesia* (Apocynaceae)) contains two additional rings.

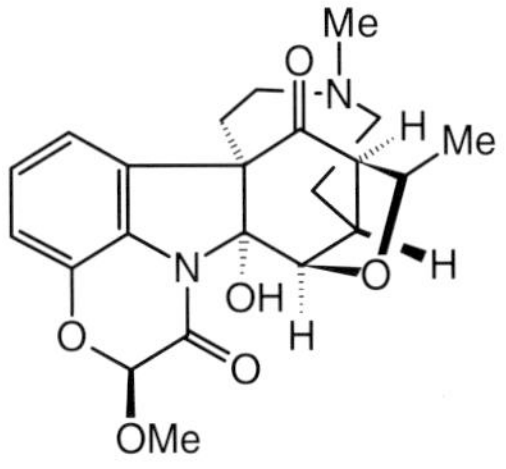

Strychnos *Alkaloids and Related Bases*[10]

Again, numerous bases also belong to this subgroup of 'indole' alkaloids. The most important skeletons are represented by the following compounds: stemmadenine (from *Diplorrhynchus, Melodinus, Stemmadenia,* and *Vallesia* species (Apocynaceae)), tubifolidine (from *Pleiocarpa* (Apocynaceae)), akuammicine (from *Picralima* and *Vinca* species (Apocynaceae)), *Wieland–Gumlich* aldehyde (from *Strychnos* (Loganiaceae)), and also two compounds extended by one C$_2$-moiety: strychnine (a powerful neurotoxin from *Strychnos* species) and vomicine (also from *Strychnos* species). Tsilanine (from *Strychnos* (Loganiaceae)) contains an additional ring, compared with akuammicine.

The calabash curare alkaloids (*cf. Sect. 2.7* and *Chapt. 6.7*), among other compounds, are derived from the *Wieland–Gumlich* aldehyde, a degradation product of strychnine. The natural derivatives of these bases possess no additional C-atoms in the C,N-skeleton, but constitute reduced, oxidized, or *O*- or *N*-alkylated/acylated products.

An extract from the 1869 *Handbuch der gerichtlichen Chemie*[11] (*Handbook of Forensic Chemistry*) by *F. L. Sonnenschein* takes up the theme:

'*Strychnine N^{2}C^{42}H^{22}O^4*

(+)-Condylocarpine (abs. config.)

(+)-Dichotine (abs. config.)

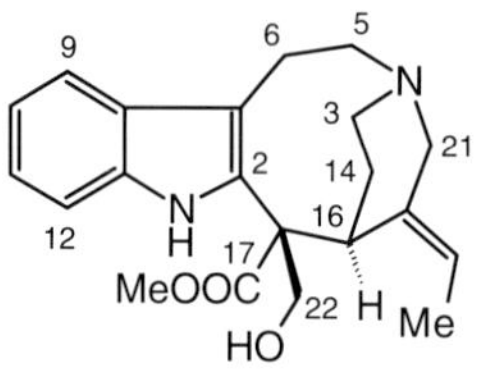

(+)-Stemmadenine
(abs. config.)

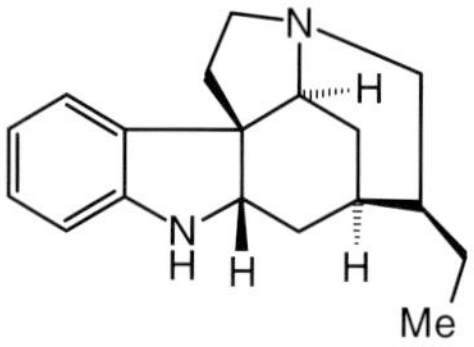

(–)-Tubifolidine
(abs. config.)

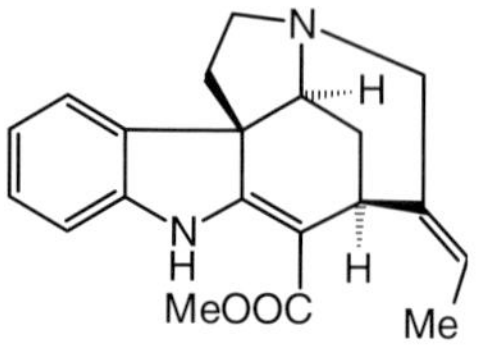

Akuammicine
(abs. config.)

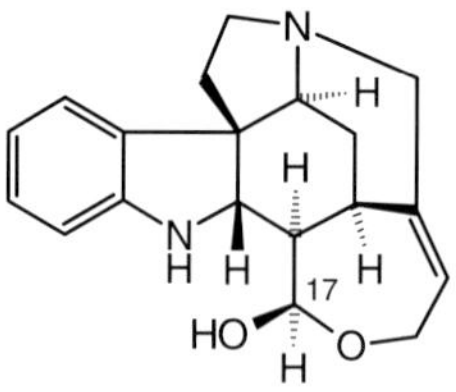

(–)-*Wieland–Gumlich* aldehyde
(abs. config.)

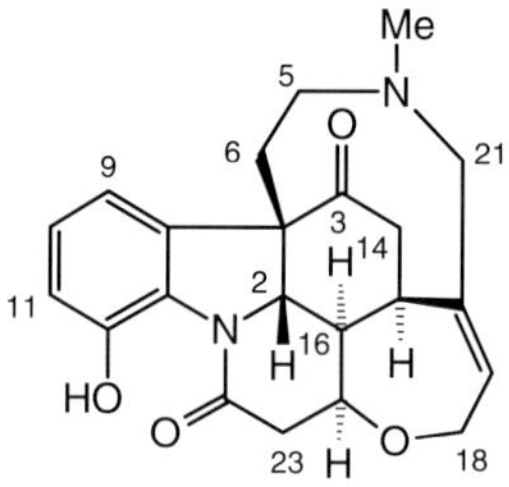

(+)-Vomicine
(abs. config.)

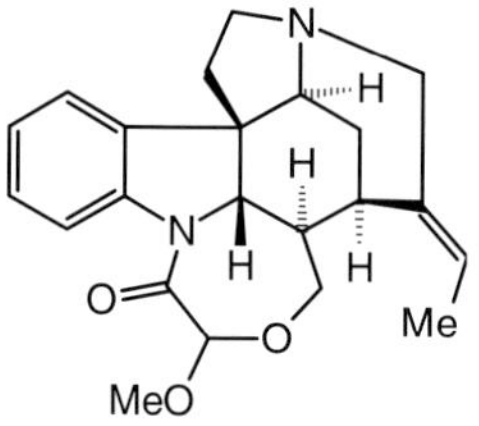

(+)-Tsilanine

This alkaloid, occurring in various Strychnos species, is one of the most terrible poisons. There are many known cases in which poisoning by this agent has been observed. Even at the onset of symptoms, poisoning by the same can be recognized with reasonable certainty. Usually, the poisoned subject is suddenly, over a period of a few minutes to an hour, afflicted by difficulty in breathing and suffocation, sometimes entirely without warning symptoms. Convulsions, trembling, etc. set in. As well as this, tetanic symptoms appear, and the body adopts a bow-shaped form, as the back curves and the body comes to be resting on the head and heels'.

And a short crime story:

In a small town, a young child sickened with very violent symptoms. This same child had been sent some cakes by some unknown person. As well as this, the family had dined on Eierkuchen [a German pancake], *after which they all became ill. A cat, fed some of the cake, died immediately after consumption of the same. The interesting thing about the situation was as follows. The husband, the father of the sick child, had previously had an intimate relationship with another person, who, in an act of vengeance, had sent the child toys and cakes under a false name and ad-*

dress. Of these, the latter, as intimated, were poisoned. Once it had been ascertained that the young lady's brother, who had supposedly sent the articles, knew nothing of the entire affair, precautionary steps were taken. Later on, a young maid was sent out to fetch sugar. As she came past the house of the accused, she was asked by the accused to put down her basket, with the bag of sugar in it, and to pick up fruit in the orchard. The girl assented to this request, leaving the basket and its contents unattended. She later took the sugar to her mistress, who sprinkled some of it on the baked Eierkuchen, after which the above-mentioned events were to take place. After chemical analysis, 3.576 Grans of strychnine were found in the cake, and 29.649 Grans in one pound [500 g] of sugar. The declared purpose of the strychnine in this case had been for poisoning foxes.

(1 Gran (old apothecary's measure) varied regionally between 0.061 g (Saxony) and 0.073 g (Austria)).

Uleine Alkaloids

The small subgroup of uleine alkaloids is derived from the condylocarpine type. In these compounds, the ethylene group between C(7) and N(4) is missing. Uleine (from *Aspidosperma* (Apocynaceae)) is given as an example.

Alkaloids with the β-Carboline Skeleton[12]

β-Carboline derivatives of low molecular weight are often found in nature. Their widespread occurrence presumably has to do with their simple biogenesis from tryptamine (or tryptophan) and carbonyl compounds. Harman, for example, is found in six different plant families: Rubiaceae (*Arariba*, *Ophiorrhiza*, and *Sickingia*), Polygonaceae (*Calligonum*), Passifloraceae (*Passiflora*; *Fig. 2.3*), Symplocaceae (*Symplocos*), Zygophyllaceae (*Zygophyllum*), and Elaeagnaceae (*Elaeagnus*). Not a few representatives have additional C-substituents on the ring skeleton, for instance harman-3-carboxylic acid (from *Aspidosperma* (Apocynaceae)). Also, brevicolline (from *Carex* (Cyperaceae)) contains a third N-atom.

In the course of chronic alcoholism, it is quite remarkable that both the oxidation products acetaldehyde (MeCHO) and formaldehyde (HCHO, from methanol (MeOH) present in the ethanol) react with endogenous amines to yield tetrahydroisoquinolines or tetrahydro-β-carbolines, which are detectable at elevated levels in the serum, in the urine, and in the blood[13]. Ingest-

Uleine

Harman

(−)-Brevicolline
(abs. config.)

21

ed alcohol is not the only source of the aldehydes, though. The formation of tetrahydro–β–carbolines proceeds, as shown in *Scheme 2.3*, through a *Pictet–Spengler* cyclization[14].

Fig. 2.3. *Passiflora caerulea* L. (Passifloraceae), *a variety sold as a house plant, is distinguished by its harman derivative content* (photo F. Strauss)

Scheme 2.3

R = H or Me
1,2,3,4-Tetrahydroisoquinoline derivative

R = H or Me
1,2,3,4-Tetrahydro-β-carbolines

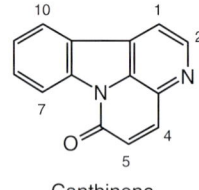

Canthinone

Alkaloids of the Canthine Type

A typical representative of this small group of bases is canthinone (from *Pentaceras* and *Xanthoxylum* (Rutaceae) as well as from *Picrasma* (Simaroubaceae)).

Eburnamine Alkaloids[15]

The alkaloids that provide the basic frameworks for the structural variations found in this subgroup comprise eburnamine (from *Amsonia, Gonioma, Haplophyton, Hunteria, Pleiocarpa, Rhazya*, and *Vinca* species (Apocynaceae)), vincamine[16] (from *Vinca* species; *cf. Chapt. 6.6*), and schizozygine (from *Schizozygia* (Apocynaceae)). About 80 more natural derivatives of these three alkaloids are known.

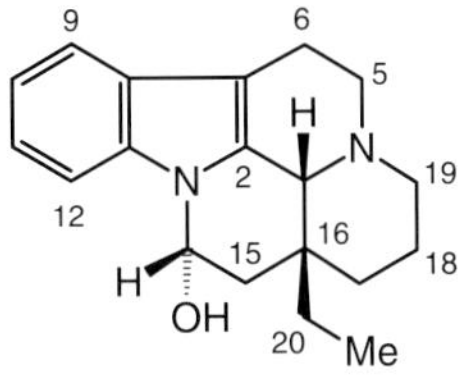

(−)-Eburnamine
(abs. config.)

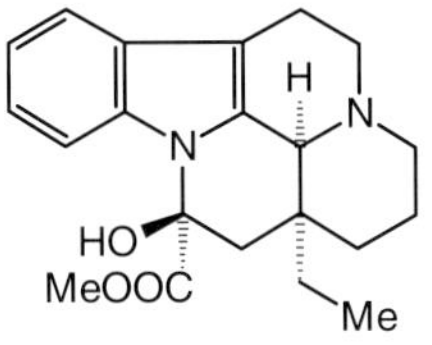

(+)-Vincamine
(abs. config.)

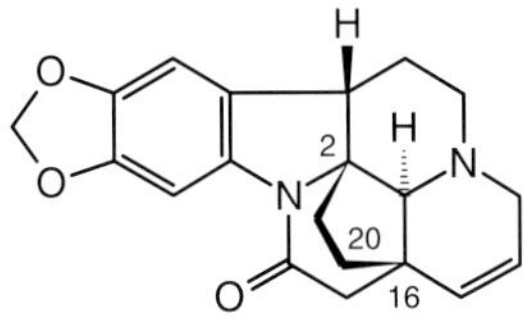

(+)-Schizozygine

Eburnamine alkaloids are also occasionally encountered as components of bisindole alkaloids.

Heteroyohimban and Biogenetically Related Bases[17]

This subgroup of alkaloids can be subdivided further. Particularly clear is the biological relationship of the C,N-skeletons of the following bases: corynantheidine (from *Mitragyna* (Rubiaceae) and *Pseudocinchona* (Rubiaceae)), ajmalicine (as an example of a heteroyohimban alkaloid; from various plants of the genera *Catharanthus*, *Rauwolfia*, *Stemmadenia*, *Vinca*, and *Tonduzia* (Apocynaceae), *Corynanthe* and *Pausinystalia* (Rubiaceae), and *Mitragyna* (Rubiaceae)), pleiocarpamine (from *Gonioma*, *Hunteria*, *Pleiocarpa*, and *Vallesia* species (Apocynaceae)), sarpagine (from both *Rauwolfia* and *Vinca* species (Apocynaceae)[18]), vobasine (from *Gabunia*, *Hazunta*, *Peschiera*, and *Voacanga* species (Apocynaceae)), ajmaline[19] (from various plants of the genera *Aspidosperma*, *Rauwolfia*, and *Tonduzia* (Apocynaceae)[18]), alstophylline (from *Alstonia* species (Apocynaceae)), and akuammiline (from *Conopharyngia* and *Picralima* species (Apocynaceae)).

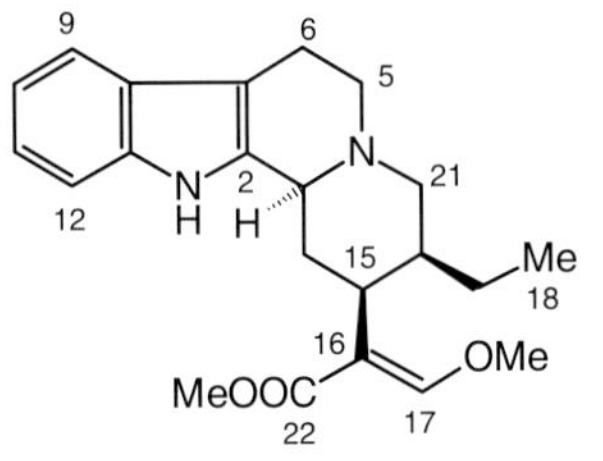

(−)-Corynantheidine (abs. config.)

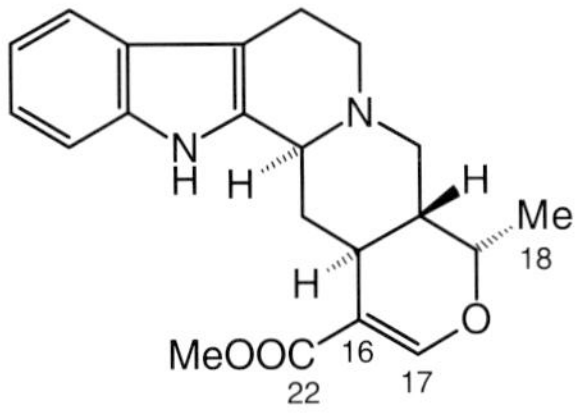

(−)-Ajmalicine (abs. config.)

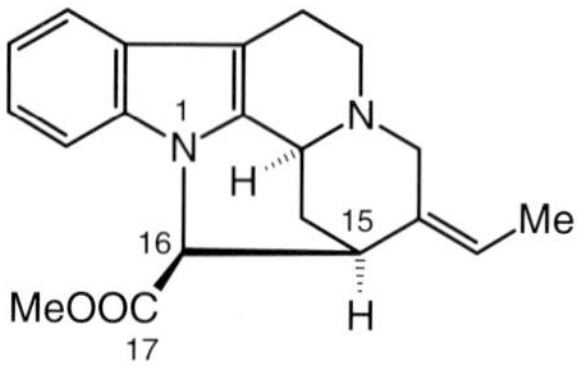

(+)-Pleiocarpamine (abs. config.)

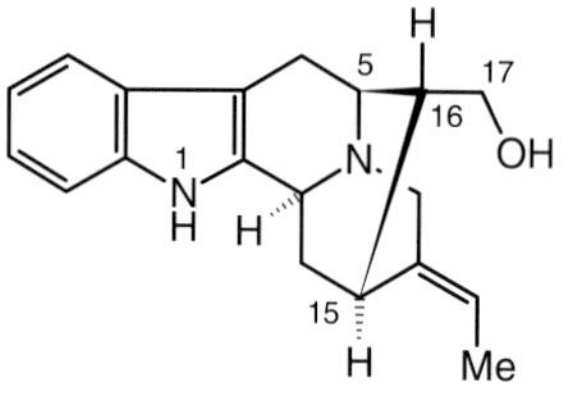

(+)-Sarpagine (abs. config.)

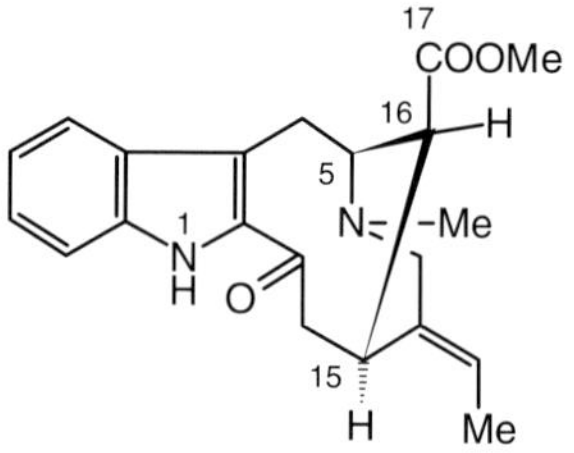

(−)-Vobasine (abs. config.)

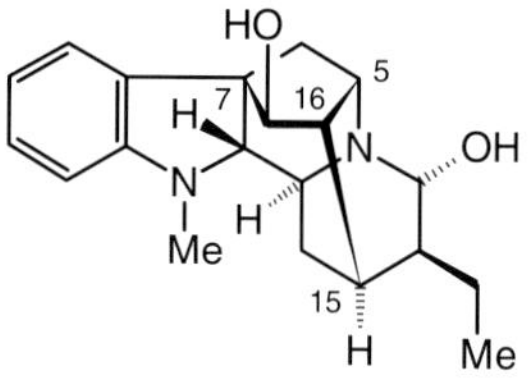

(+)-Ajmaline (abs. config.)

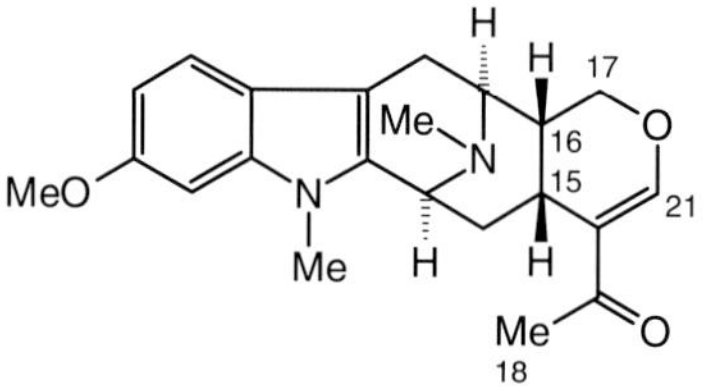

(−)-Alstophylline (abs. config.)

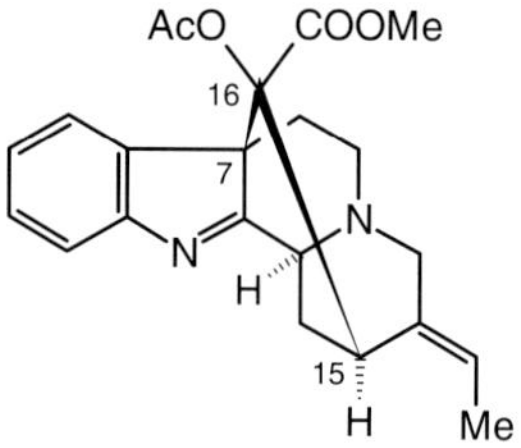

(+)-Akuammiline (abs. config.)

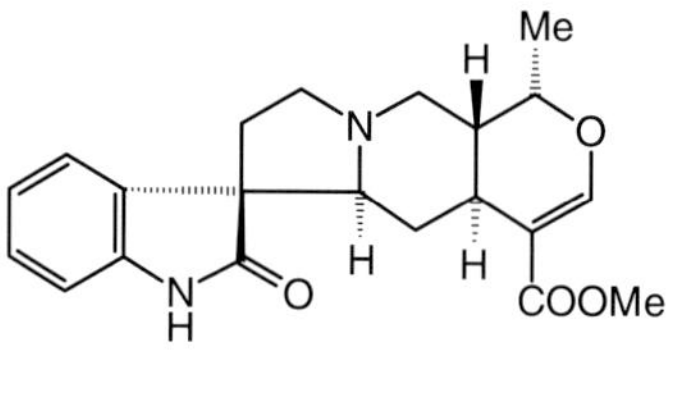

(−)-Mitraphylline (abs. config.)

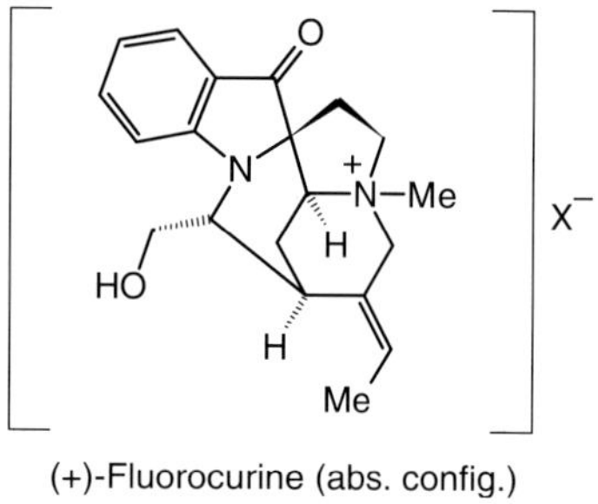

(+)-Fluorocurine (abs. config.)

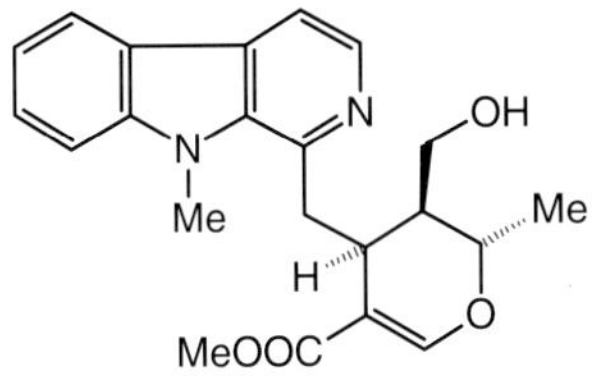

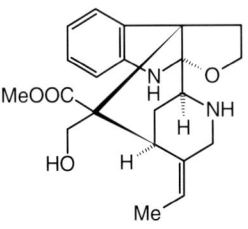

(−)-Alstonidine

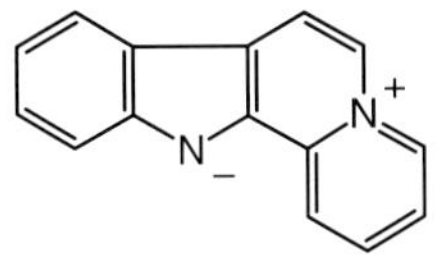

Indolopyridicoline

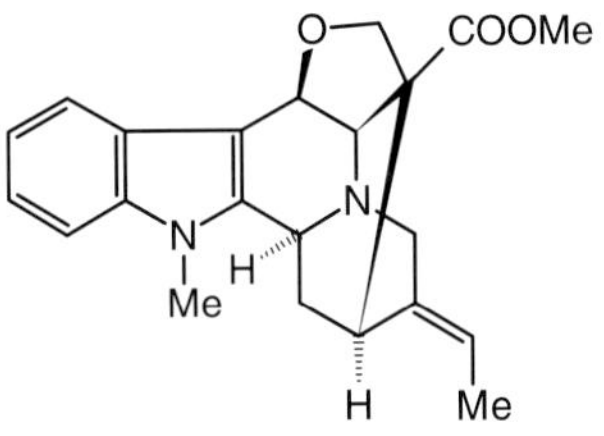

(+)-Dehydrovoachalotine

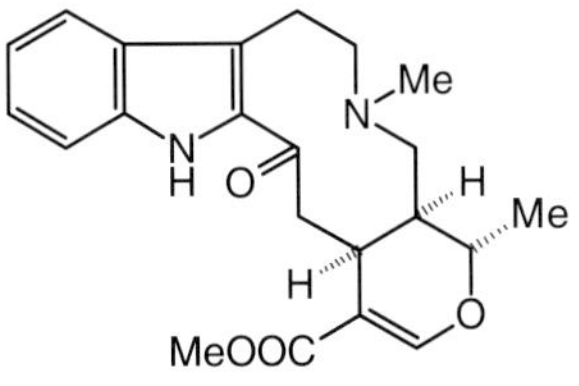

(−)-Picraphylline (abs. config.)

(−)-Aspidodasycarpine
(abs. config.)

There are also a few subtypes of heteroyohimban bases, in which C(22) may or may not be present. Naturally occurring rearrangement products have also been characterized.

Mitraphylline (an ajmalicine derivative from *Mitragyna* and *Vinca* species) can be cited as an example of an oxindole derivative (*cf. Chapt. 9.7*). Other such derivatives, though, are also known among alkaloids of the corynantheidine and vobasine types. The one known naturally occurring indoxyl alkaloid, fluorocurine (from calabash curare and *Strychnos* species (Loganiaceae)), however, is of the pleiocarpamine type.

As some illustrative representatives of other structural variants, one can also cite alstonidine (from *Alstonia* (Apocynaceae)), indolopyridocoline (from *Gonioma* (Apocynaceae)), dehydrovoachalotine (from *Voacanga* (Apocynaceae)), picraphylline (from *Picralima* (Apocynaceae)), and aspidodasycarpine (from *Aspidosperma* (Apocynaceae)).

There are also some additional alkaloid groups closely related to the heteroyohimban subcategory, for which vallesiachotamine (from *Vallesia* (Apocynaceae)), talbotine (from *Pleiocarpa* (Apocynaceae)), adifoline (from *Adina* (Rubiaceae)), and cinchonamine (from *Remijia* (Rubiaceae)[20]) are given as examples.

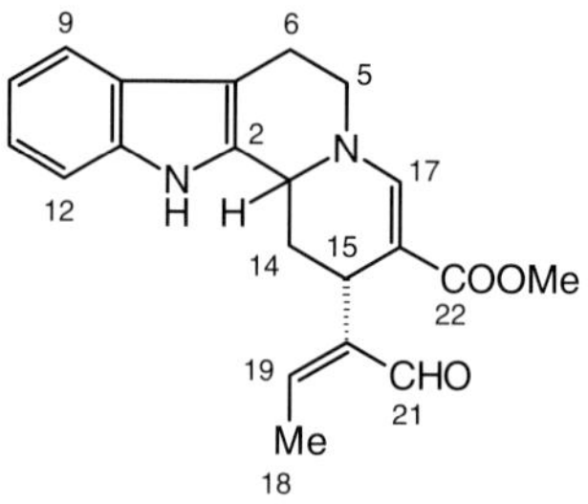

Vallesiachotamine

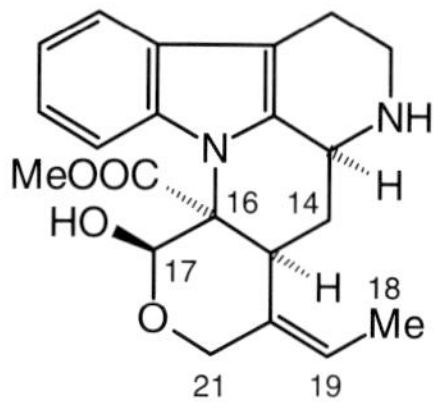

(–)-Talbotine

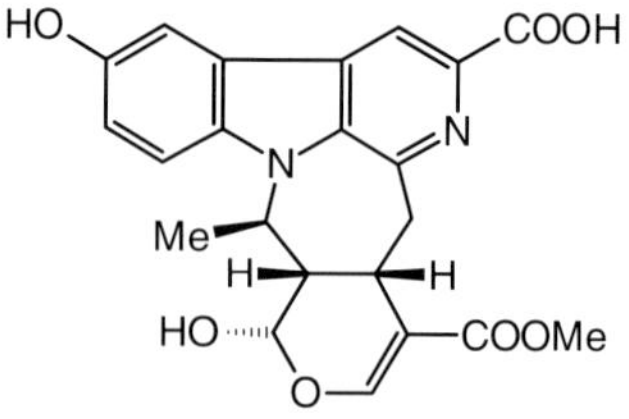

Adifoline

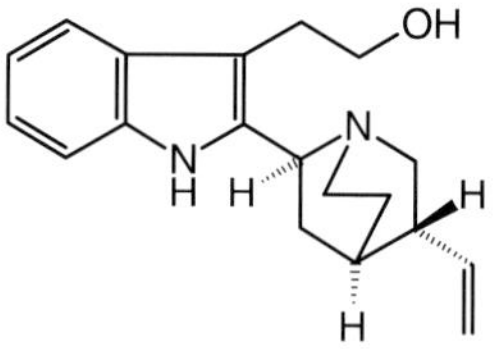

(+)-Cinchonamine (abs. config.)

Heteroyohimban alkaloids and related bases are also found as components of bisindole alkaloids.

Yohimban Alkaloids[21]

This subgroup is characterized by two main alkaloids: yohimbine[22] (from plants of the genera *Alchornea* (Euphorbiaceae), *Aspidosperma, Catharanthus, Diplorrhynchus, Rauwolfia*, and *Vinca* (Apocynaceae), *Corynanthe* and *Pausinystalia* (Rubiaceae)), as well as reserpine[23] (from various species of *Alstonia, Excavatia, Ochrosia, Rauwolfia, Tonduzia, Vallesia*, and *Vinca*

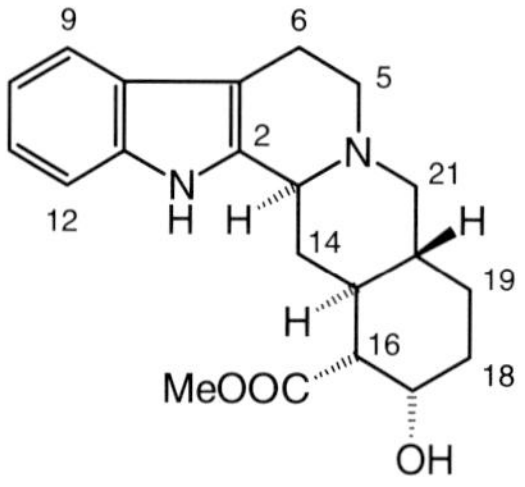

(+)-Yohimbine (abs. config.)

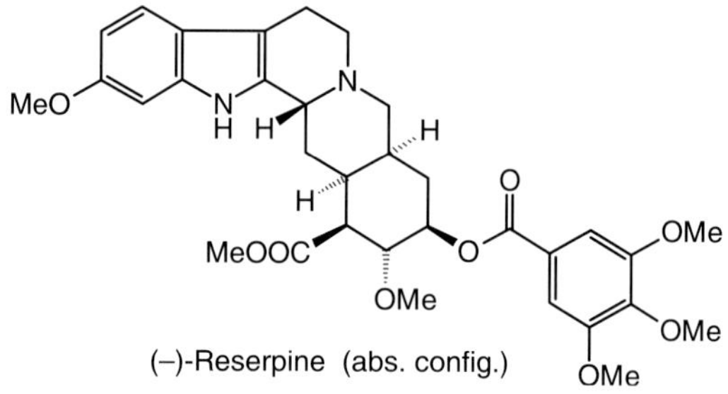

(–)-Reserpine (abs. config.)

(Apocynaceae)). For both yohimbine and reserpine various stereoisomers, regioisomers, and numerous other derivatives exist. Alstonilidine, isolated from *Alstonia* (Apocynaceae), can be viewed as a derivative of this substance class.

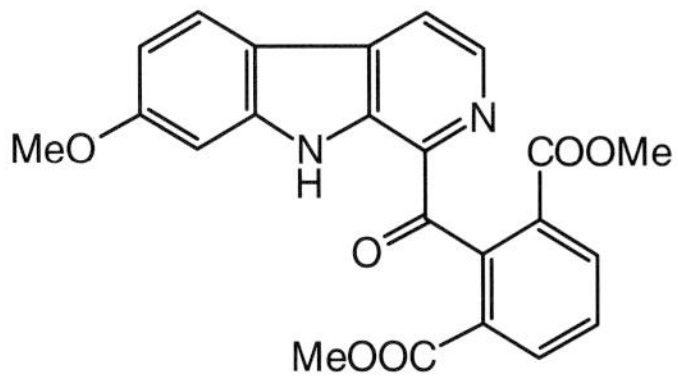

Alkaloids with the γ-Carboline Skeleton[7]

The only alkaloid belonging to this group of compounds known to date is cryptolepine (from *Cryptolepis* (Periplocaceae)).

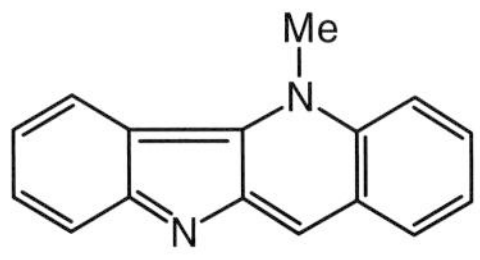

Iboga *Alkaloids*[6]

In this alkaloid group, the indole nucleus is formally fused with a N-containing, seven-membered ring. Characteristic of the class are ibogamine (from species of *Gabunia, Hazunta, Stemmadenia, Tabernaemontana,* and *Tabernanthe* (Apocynaceae)) and voacangine (from species of *Callichilia, Conopharyngia, Gabunia, Hedranthera, Peschiera, Rejoua, Stemmadenia, Tabernaemontana,* and *Voacanga* (Apocynaceae)). Oxidation products are also encountered in this alkaloid class, as with the heteroyohimban alkaloids. Hydroxyindolenine-voacangine (from *Voacangine* (Apocynaceae)), voaluteine (from *Rejoua* (Apocynaceae)), and crassanine (from *Tabernaemontana* (Apocynaceae)) are listed here as representatives.

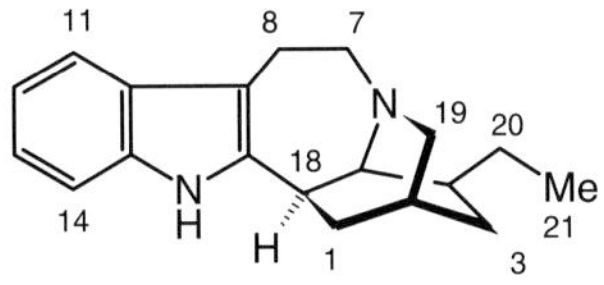

Alstonilidine

Cryptolepine

(–)-Ibogamine (abs. config.)

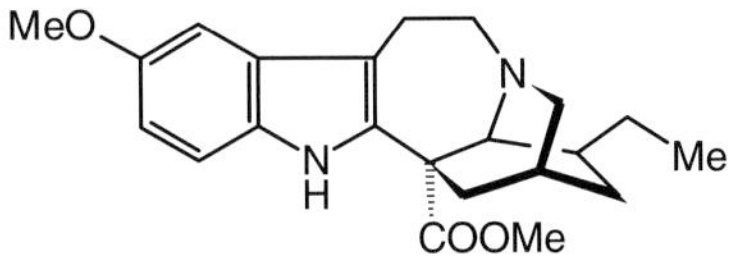

(–)-Voacangine (abs. config.)

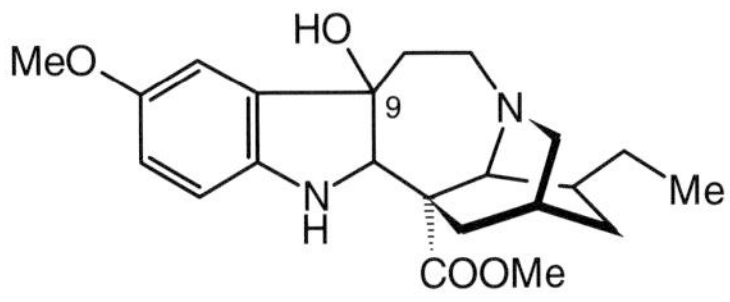

(–)-Hydroxyindolenine-voacangine

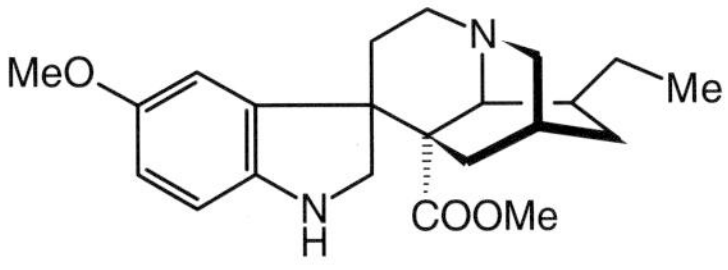

(–)-Voaluteine (abs. config.)

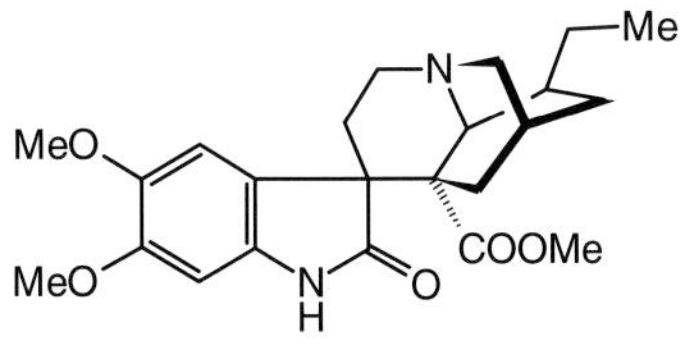

(+)-Crassanine

Besides these compound types, a number of other alkaloids with different substitution patterns have also been found. In addition, bisindole alkaloids, in which one component base is an *Iboga* alkaloid, have been isolated from Apocynaceae (*cf. Chapt. 9.8*).

Indole Alkaloids with the Pyrrolo-Indole Skeleton

Eserine Type[24]

Only two alkaloid skeletal variants of eserine are known to date: physostigmine (= eserine) (from *Dioclea* and *Physostigma* (Leguminosae) and from *Hippomane* (Euphorbiaceae)), and geneserine (from *Physostigma*).

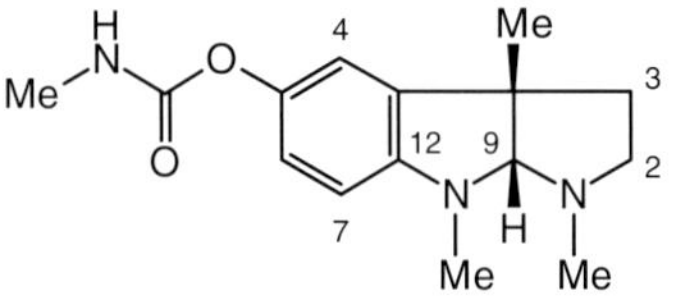

(–)-Physostigmine (= Eserine)
(abs. config.)

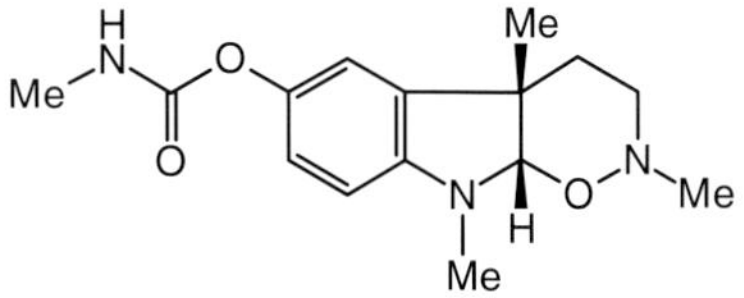

(–)-Geneserine
(abs. config.)

This subclass of plant bases also includes compounds from *Calycanthus* species (Calycanthaceae)[25], which contain indole skeletons. However, monomeric members of this group are unknown (to date only dimeric, trimeric, and tetrameric compounds have been isolated), and so they will be examined more closely in *Sect. 2.7* (dimeric alkaloids).

Echitamine-Erinine Type

The biogenetic relationship of the alkaloids echitamine (from *Alstonia* species (Apocynaceae)), corymine (from *Hunteria* species (Apocynaceae)), and

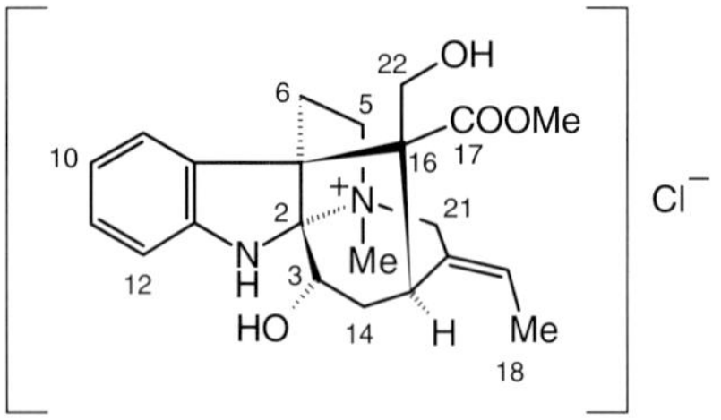

(–)-Echitamine Chloride   (abs. config.)

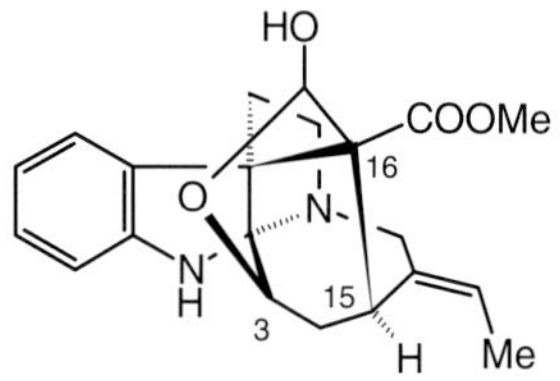

(+)-Corymine (abs. config.)

erinine (from *Hunteria* (Apocynaceae)) to the heteroyohimban alkaloids is obvious. Because of our classification criteria, however, they are listed at this point.

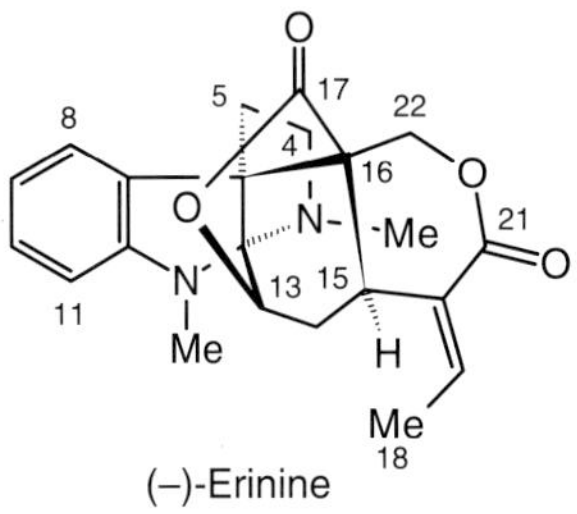

(–)-Erinine

Ergot Alkaloids[26]

The fundamental building block of these alkaloids, isolated from *Claviceps purpurea*, is always lysergic acid or isolysergic acid (= 8-*epi*-lysergic acid)[27]. Two more basic types – agroclavine and chanoclavine – are derived from the former compound. Some peptide alkaloids, *e.g.*, ergotamine and ergocornine, also belong to this group (*cf. Chapt. 11.4*), together with simpler amides of lysergic acid (such as ergobasine or LSD).

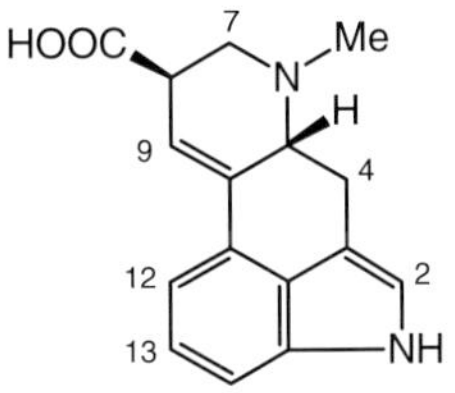

(+)-Lysergic Acid
(abs. config.)

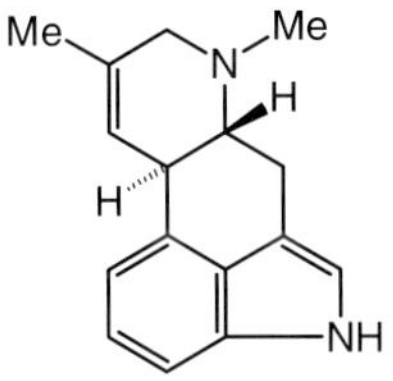

(–)-Agroclavine
(abs. config.)

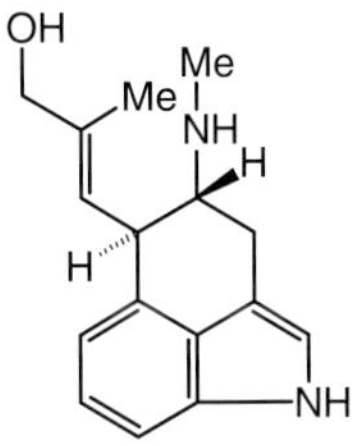

(–)-Chanoclavine
(abs. config.)

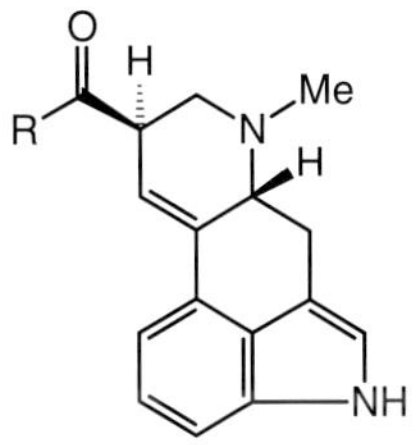

R = HO–CH$_2$–CH(Me)–NH: Ergobasine

R = Et$_2$N : LSD

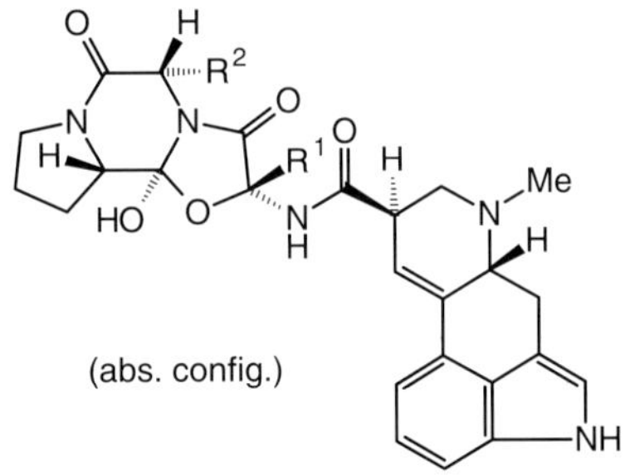

(abs. config.)

R^1= Me R^2 = PhCH$_2$: (–)-Ergotamine
R^1= Me R^2 = i-Bu   : (–)-Ergosine
R^1= Me R^2 = s-Bu   : (–)-β-Ergosine
R^1= Me R^2 = i-Pr   : (–)-Ergovaline
R^1= Et R^2 = PhCH$_2$: (–)-Ergostine
R^1= Et R^2 = i-Bu   : (–)-Ergoptine
R^1= Et R^2 = s-Bu   : (–)-β-Ergoptine
R^1= Et R^2 = i-Pr   : (–)-Ergonine
R^1= i-Pr R^2 = PhCH$_2$: (–)-Ergocristine
R^1= i-Pr R^2 = i-Bu   : (–)-α-Ergocryptine
R^1= i-Pr R^2 = s-Bu   : (–)-β-Ergocryptine
R^1= i-Pr R^2 = i-Pr   : (–)-Ergocornine

Alkaloids of the Evodiamine Type[28]

Evodiamine (from *Evodia* (Rutaceae)) is a prototype of this relatively small alkaloid group, which, in structural terms, could also be assigned without problem to the quinazoline alkaloids.

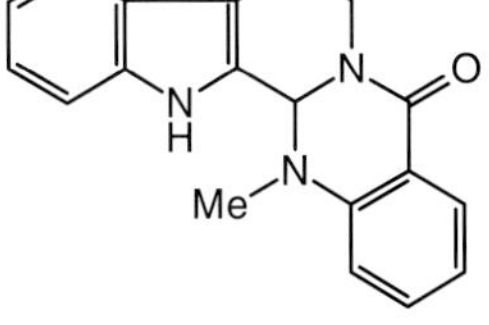

Evodiamine

Aristotelia *Alkaloids*

Plants (bushes or small trees) of the genus *Aristotelia* (Elaeocarpaceae), of which five species are known today, are found in New Zealand, Australia, and Chile. They contain about 30 alkaloids, incorporating tryptamine and a C_{10}-monoterpene moiety. The monoterpene component here is clearly different from secologanin, the corresponding fundamental unit in the huge group of alkaloids isolated from Apocynaceae, Rubiaceae, and Loganiaceae.

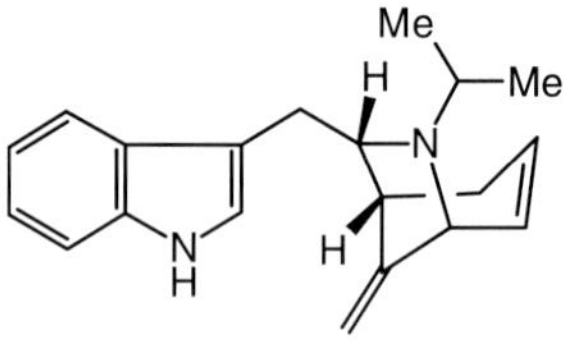

The individual alkaloids from *Aristotelia* species differ from one another primarily by structural variation in the terpene part. The following four examples attest to the wealth of forms in this substance class, in which the primary alkaloid peduncularine[29] (from *A. peduncularis*) is particularly worth mentioning: the terpene segment is divided by the N-atom into three (*N*-isopropyl) plus seven C-atoms.

Fruticosonine (from *A. fruticosa*) contains a fragment only slightly altered from the hypothetical biogenetic precursor, while serratoline and aristoteline (both from *A. serrata*) are pentacyclic derivatives already (*Scheme 2.4*)[30].

(−)-Peduncularine

Scheme 2.4

Hypothetical Biogenetic Precursor

Serratoline

(+)-Fruticosonine

(+)-Aristoteline (abs. config.)

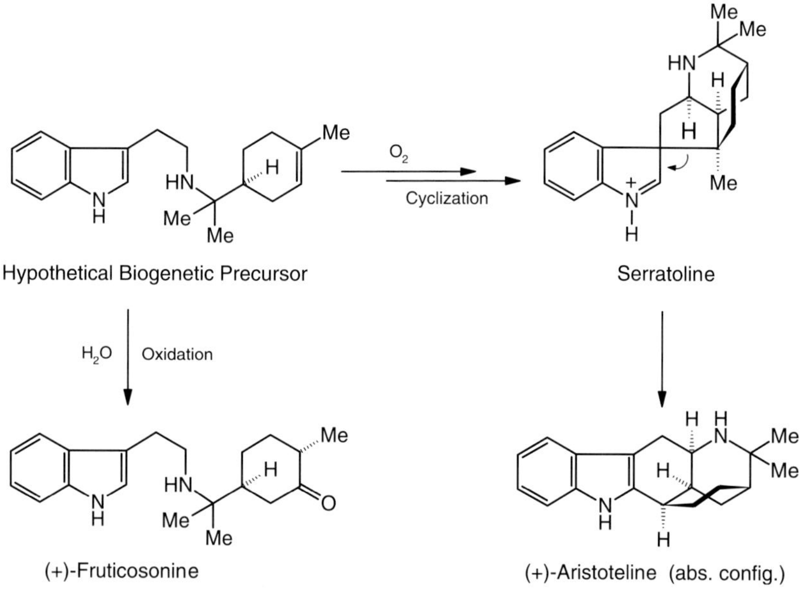

2.2.3. Piperidine Alkaloids[1,31]

In the next group of compounds to be discussed, the N-atom forms part of a six-membered ring. The simplest piperidine derivatives are comparable with their corresponding pyrrolidine counterparts. Thus, the structure of piperine (from *Piper* (Piperaceae)) is very similar to that of 3-methoxycinnamic acid pyrrolidide. Until 1997, 145 alkaloids had been isolated from plants of the *Piper* genus. As an example of an α-alkylpiperidine derivative, coniine (from *Aethusa* and *Conium* species) is offered (*cf.* also *Chapts. 3.3, 6.6,* and *11.2*), while anaferine (from *Withania* (Solanaceae)), on the other hand, provides an example of a compound with two piperidine nuclei. The skytanthines (α-skytanthine from *Skytanthus* (Apocynaceae)) are made up out of a monoterpene moiety and methylamine. Lobeline, for its part, was isolated from *Lobelia* (Campanulaceae)).

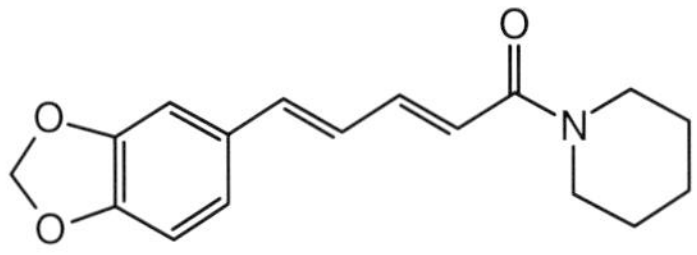

Piperine

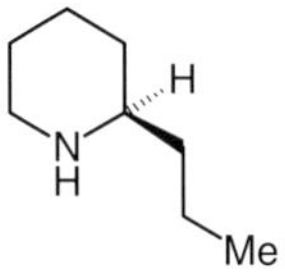

(+)-Coniine
(abs. config.)

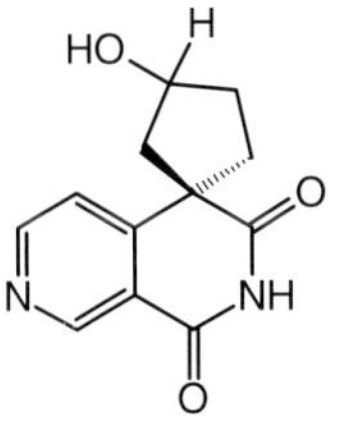

(+)-Sesbanine

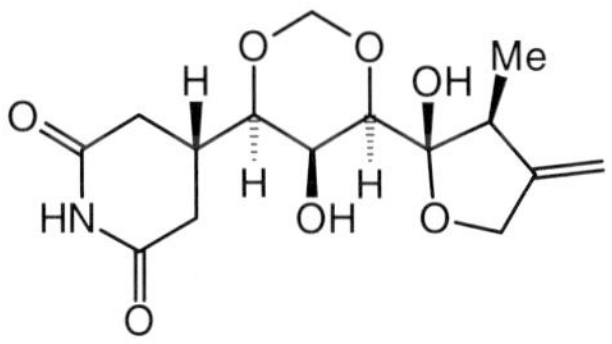

(–)-Sesbanimide

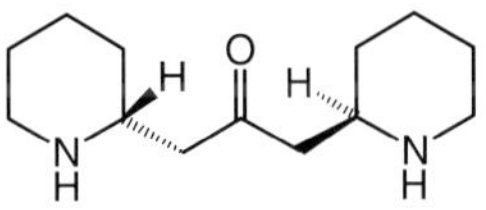

(–)-Anaferine (abs. config.)

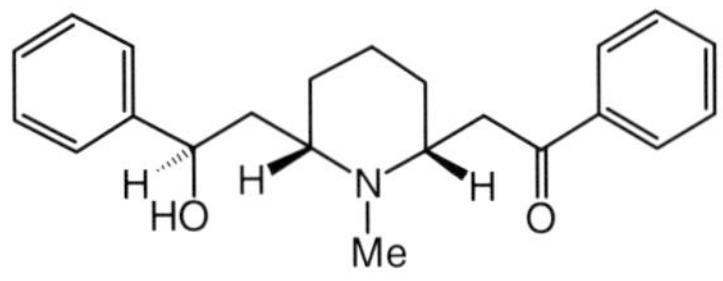

(–)-Lobeline (abs. config.)

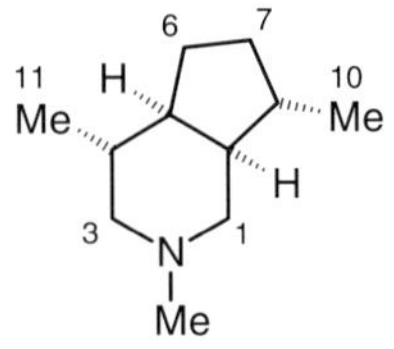

(+)-α-Skytanthine (abs. config.)

The *Sesbania* alkaloids contain an additional pyridine (or hydrogenated pyridine) nucleus and, therefore, mark the transition to the next group. They may also be classified as cyclic urea derivatives. Sesbanine and sesbanimide, isolated from the seeds of *Sesbania drummondii*, a leguminous plant growing in the southern United States, are poisonous to ruminant grazing animals[32].

(−)-Cantleyine

Trigonelline

Pyridine Alkaloids[1,33]

The pyridine skeleton is contained, *inter alia*, in cantleyine (from *Cantleya* (Icacinaceae) and *Jasminum* (Oleaceae)), which we can immediately identify structurally as a dehydrogenated relative of skytanthine. Both these compounds are occasionally also counted among the terpene alkaloids, as their skeletons originate from a monoterpene. Of the structurally simple pyridine alkaloids, the widely distributed ricinine (*cf. Chapt. 1.1*) and trigonelline (from, *e.g., Trigonella* (Leguminosae)) should also be mentioned. Other nicotinic acid derivatives are discussed in *Sect. 2.3*.

An intermediate position between the pyrrolidine, piperidine, and pyridine bases is occupied by a group of compounds generally referred to as tobacco alkaloids (*cf. Chapt. 11.9*). They are represented here by nicotine[34] (from genera of various families: *Asclepias* (Asclepiadaceae); *Atropa, Duboisia, Lycopersicon, Nicotiana, Withania, Cestrum* (all Solanaceae); *Eclipta, Zinnia* (Compositae); *Lycopodium* (Lycopodiaceae); *Macuna* (Leguminosae); *Sedum* and *Sempervivum* (Crassulaceae)), and by anabasine (especially common in *Nicotiana* species, but also found in *Sclerobunus robustus* (Opiliones (Arachnida)). Technical production of nicotine is mostly from waste tobacco.

Bases of the halfordinol type (from *Halfordia* (Rutaceae)) also show a certain kinship with the tobacco alkaloids.

Piperidine, pyridine, pyrrolidine, pyrrole, and indolizidine derivatives, displaying great structural diversity, also occur in insects. *Hymenoptera* (ants and wasps) are known for their physiologically active monocyclic and bicyclic alkaloids. The poisons of the Fire Ant (*Solenopsis*), for example, can cause grain harvest failures, and can also harm livestock. In humans, the results can be edemas, necrosis, and rashes. As well as this, the poisons often exhibit phytotoxic, insecticidal, antibacterial, and fungicidal properties, primarily due to biogenic 2,6-dialkylpiperidines.

Also not to be neglected are 1,2-dehydroanabasine (from *Aphaenogaster* species, Myrmicinae ants), actinidine (from *Actinidia* (Actinidiaceae), *Tecoma* (Bignoniaceae), and *Valeriana* (Valerianaceae) species), and a mevalonic acid derived repellant substance from *Conomyrma* species (Myrmicinae), from *Philanthus* species, and from other species of Staphylinidae.

Alkaloids from Ants

n = 8, 10, 12 or 14
cis- and *trans*-configuration
of ring substituents

Actinidine

(–)-Nicotine

(–)-Anabasine (abs. config.)

Halfordinol

2.2.4. Tropanes and Related Bases[35]

Bicyclic systems of both the tropinone (from *Nicandra* (Solanaceae)) and the Ψ-pelletierine (from the pomegranate tree *Punica* (Punicaceae)) types also feature a piperidine ring. The tropinone ring system occurs in an extremely large number of pharmacologically significant alkaloids.

Atropine (from *Atropa belladonna* (*cf. Chapt. 11.7*), from *Datura stramonium*, *Hyoscyamus niger* (both Solanaceae), and from other species; *Fig. 2.4*), for example, causes pupil dilation, while also having been used for treatment of asthma (asthma cigarettes!). Cocaine (known only in leaves of the South American coca plant *Erythroxylum coca* (Erythroxylaceae)) once served as an important local anesthetic (*cf. Chapt. 11.6*). Both the widely distributed hyoscyamine (*e.g.*, from *Hyoscyamus* (Solanaceae), *Fig. 2.5*) and scopolamine (*e.g.*, from *Scopolia* (Solanaceae)) generally work in a similar manner to atropine, acting to calm the central nervous system. Taken in high doses, alkaloids of this type are powerful poisons.

Fig. 2.4. *Hyoscyamus niger* L. (Solanaceae) *contains hyoscyamine, formerly used as a narcotic* (photo *T. Gözler*)

33

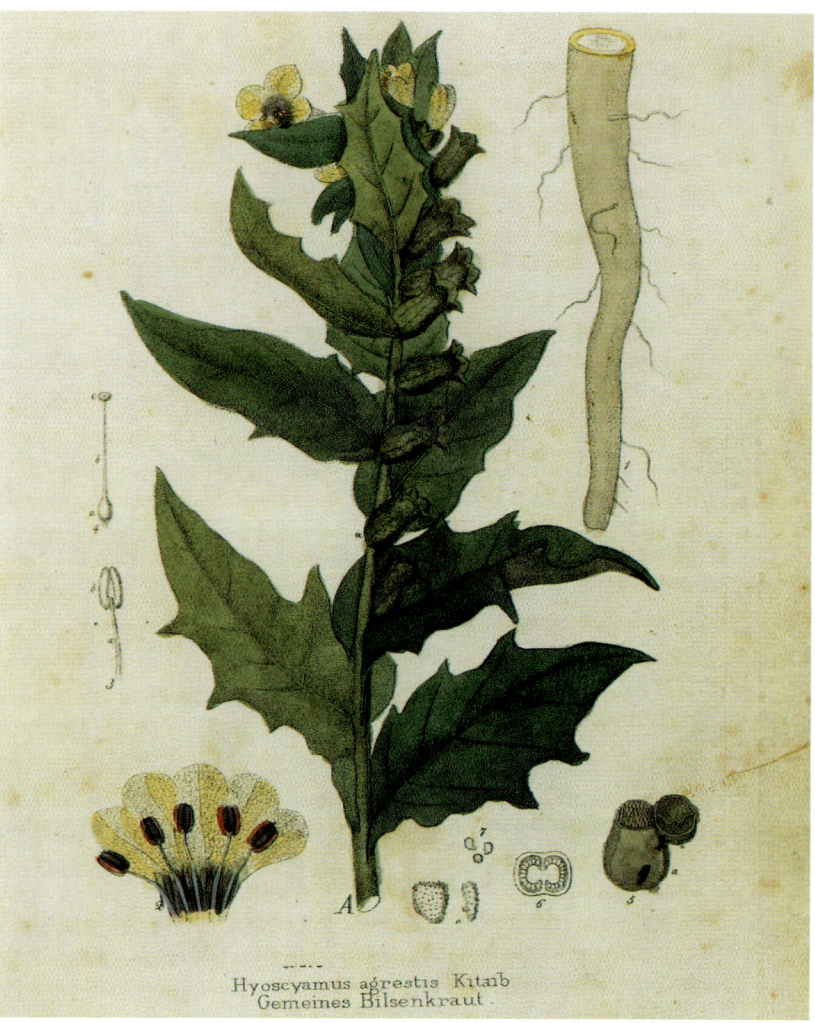

Fig. 2.5. *Hyoscyamus agrestis* KITAIB, *Common Henbane* (from *J. Pecirka, Giftgewächse*, Verlag K. André, Prague, 1859). It is probably also *H. niger* (private collection).

(±)-Atropine

(−)-Hyoscyamine (abs. config.)

(−)-Cocaine (abs. config.)

Fig. 2.6. *Excerpt from L. Oken, Naturgeschichte, Abbildungsband* (Natural History, illustrated volume), *Hoffmannsche Verlagsbuchhandlung, Stuttgart, 1843, Blumen-Pflanzen (Classe XII), Tab. XVIII* (hand-colored). Among the plants pictured, there are many that are known as alkaloid producers (private collection).

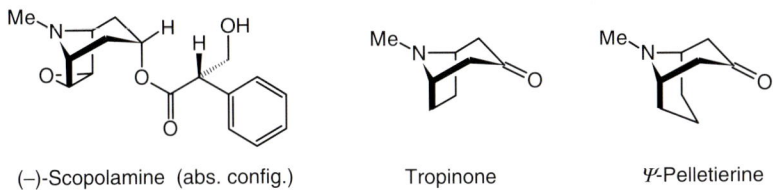

(−)-Scopolamine (abs. config.)   Tropinone   Ψ-Pelletierine

2.2.5. Histamine, Imidazole, and Guanidine Alkaloids[36]

This relatively small group of bases commonly features a five- or six-membered ring containing two N-atoms. Glochidine, isolated from *Glochidion* (Euphorbiaceae), is an example of a histamine derivative, while alchornine (from *Alchornea* (Euphorbiaceae)) may be viewed as a pyrimidine derivative.

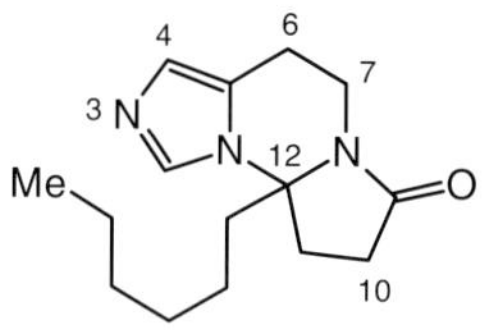

(±)-Glochidine 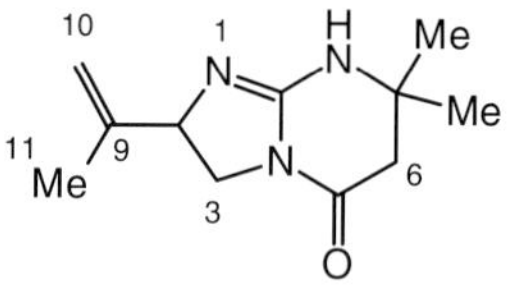   Alchornine 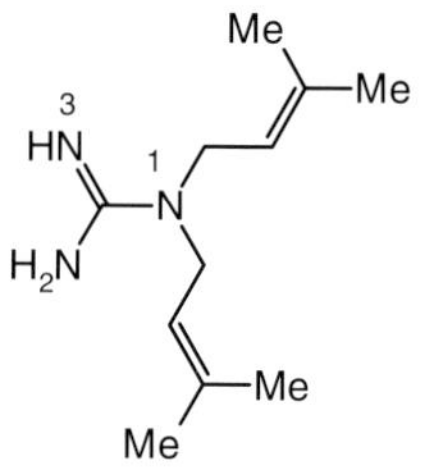 N^1,N^1-Diisopentenylguanidine

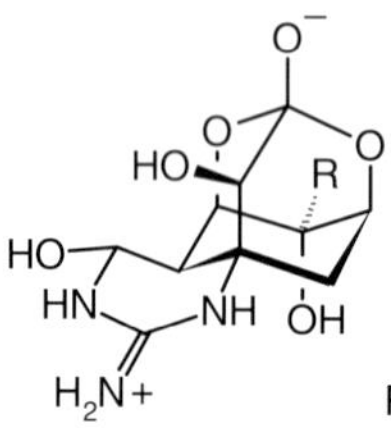

Truly small, though, is the number of natural alkaloids incorporating guanidine. Examples are N^1,N^1-diisopentenylguanidine (from *Alchornea* (Euphorbiaceae)) and the extremely toxic tetrodotoxin (from the puffer fish *Fugu rubripes rubripes*).

(–)-Tetrodotoxin   R = CH_2OH

2.2.6. Isoquinoline Alkaloids[37]

The total number of isoquinoline alkaloids known today amounts to *ca.* 1,200, comparable with the number of indole alkaloids. As already mentioned, some plant bases that no longer contain an isoquinoline chromophore as such – for example, the rhoeadine, papaverrubine, and the protopine alkaloids – are also attributed to this class for biogenetic reasons.

Simple Isoquinoline Alkaloids

These alkaloids derive from tetrahydroisoquinoline, and usually possess a short carbon side-chain, often a C_1-moiety, at C(1). Typical examples are hydrohydrastinine (from *Corydalis* (Fumariaceae)), salsoline (from *Salsola* (Chenopodiaceae)), and lophocerine (from *Lophocereus* (Cactaceae)). Lactam alkaloids are also observed in this group. The tricyclic peyoglutam found in peyote (= *Lophophora williamsii* (Cactaceae)) is counted among the simple isoquinoline alkaloids.

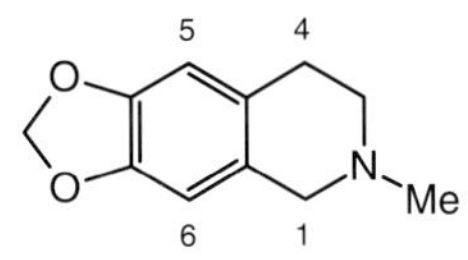

Hydrohydrastinine

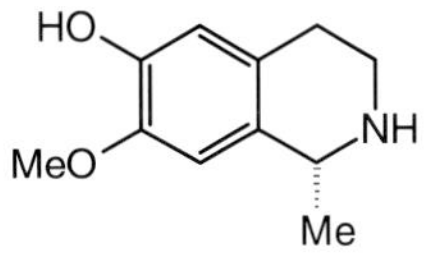

(+)-Salsoline

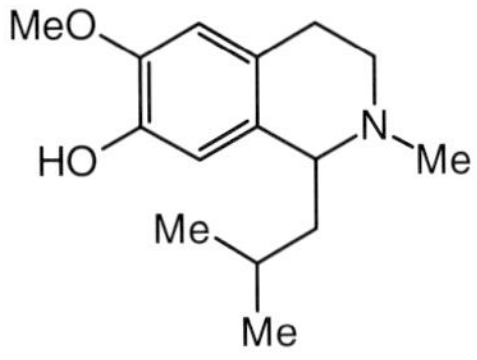

Lophocerine

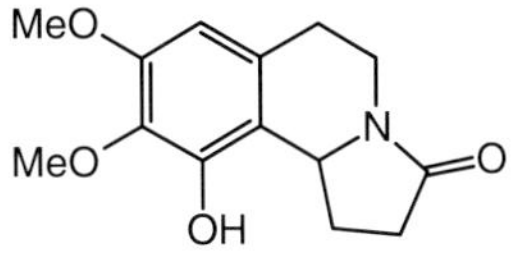

Peyoglutam

Naphthylisoquinoline Alkaloids[38]

Although the *ca.* 25 representatives of this group do contain isoquinoline chromophores, they differ significantly from other members of the main group, thanks to their biogenesis. *Scheme 2.5* illustrates a possible pathway in which six acetate units and a free or bound 'NH$_3$' participate (through a polyketide intermediate) in the biogenesis of ancistrocladine, isolated from *Ancistrocladus* species (Ancistrocladaceae)).

As well as ancistrocladine, *N*-methyltriphyophylline (from *Triphyophyllum, Dionchophyllum* (both Dionchophyllaceae)), with a different type of connection between the two ring systems (naphthalene and isoquinoline), also belongs in this category. What makes these bases especially fascinating is the fact that free rotation about the isoquinoline–naphthalene bond ((C(5)–C(1′)) is sterically hindered, allowing separation of the correspond-

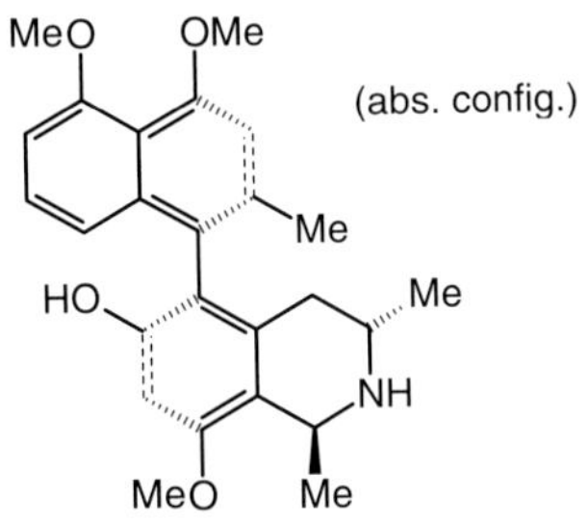

(−)-Ancistrocladine
$[\alpha]_D = -21$ (CHCl$_3$)

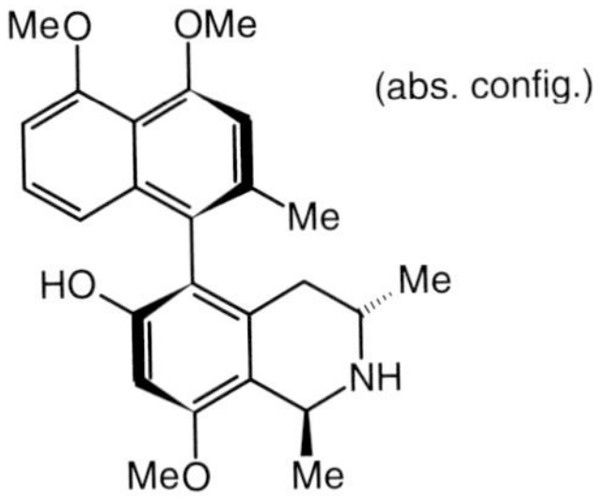

(+)-Ancistrocladine or (+)-Hamatine
$[\alpha]_D = +66$ (CHCl$_3$)

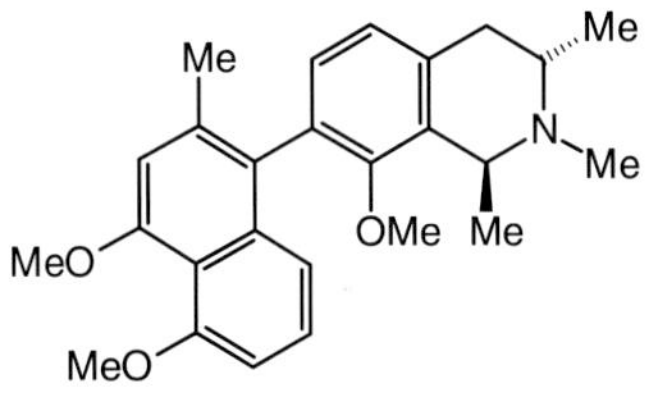

N-Methyltriphyophylline

Scheme 2.5

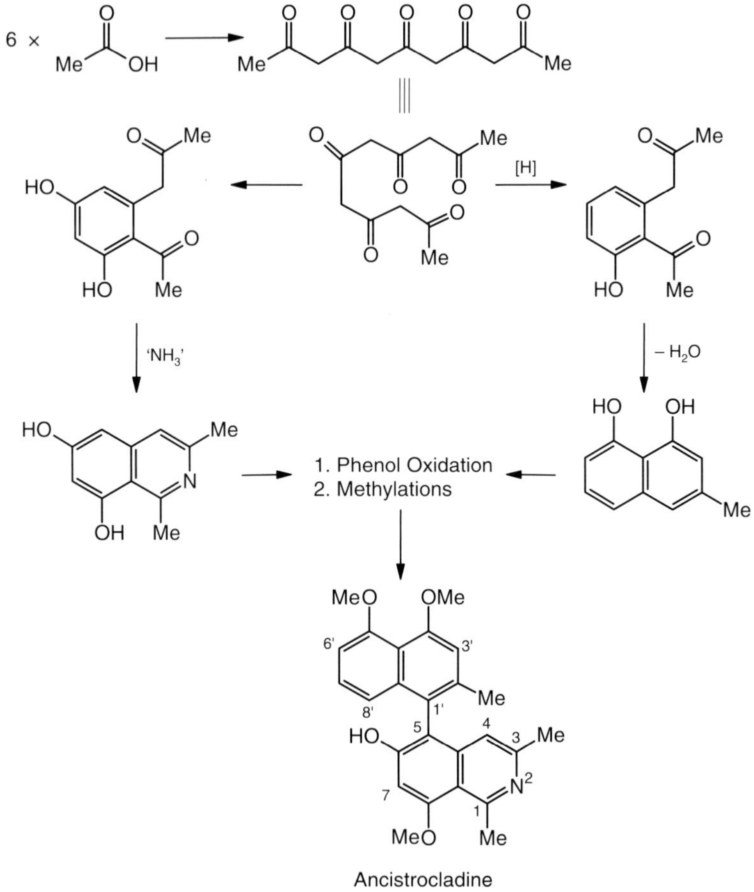

Ancistrocladine

ing (diastereoisomeric) atropisomers. For instance, (+)-ancistrocladine (or (+)-hamatine), isolated from *Ancistrocladus* species (Ancistrocladaceae) and from *Celosia argentia* (Amaranthaceae), as well as (−)-ancistrocladine from *Ancistrocladus* species (Ancistrocladaceae) have been isolated. The biogenesis of the main group of isoquinoline alkaloids is discussed in *Chapt. 8.2.*

Benzylisoquinoline Alkaloids[6, 39]

Only a few alkaloids of this class contain an aromatic B-ring (for example papaverine[40] from *Papaver* (Papaveraceae)). Most representatives are derivatives of benzyltetrahydroisoquinoline, such as laudanosine (from *Papaver*).

An example of a 'real' *N*-benzylisoquinoline alkaloid is sendaverine (from *Corydalis* (Fumariaceae)). Homologues of the benzylisoquinoline alkaloids, such as autumnaline (from *Colchicum* (Liliaceae)) are also known, however.

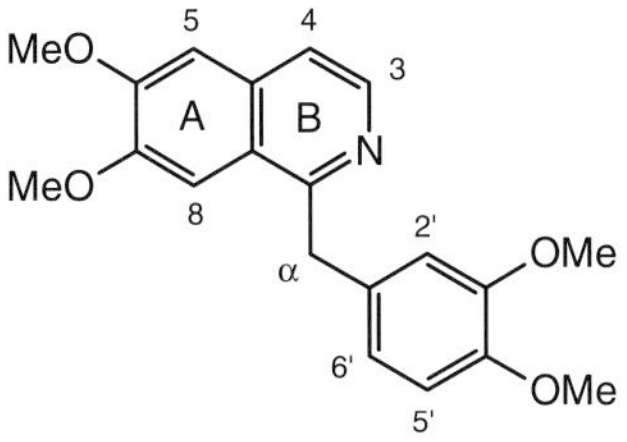

Papaverine

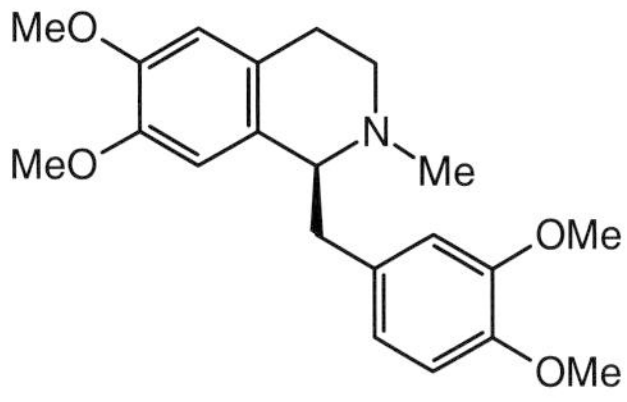

(+)-Laudanosine (abs. config.)

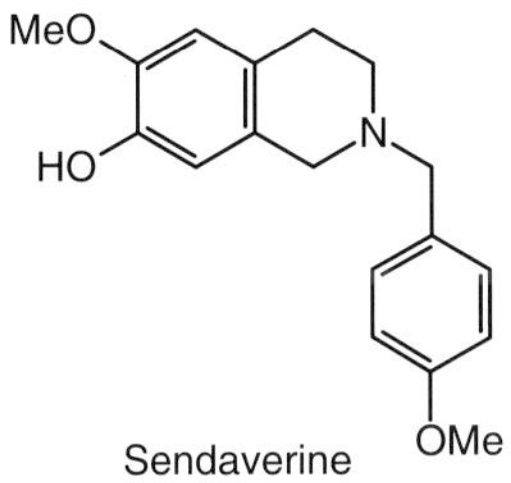

Sendaverine

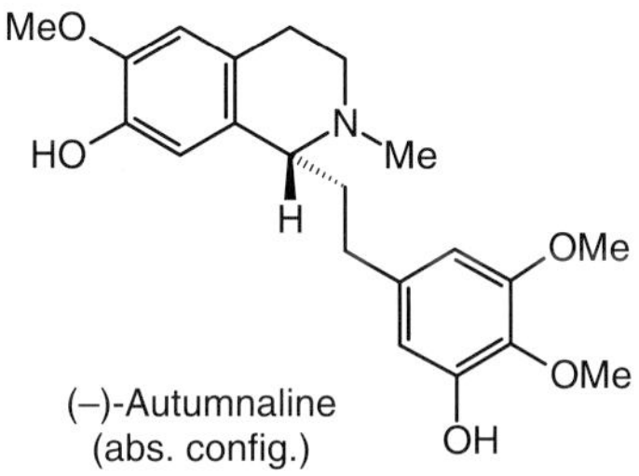

(–)-Autumnaline (abs. config.)

As with other isoquinoline alkaloid subgroups, a few quaternary nitrogen compounds are also encountered here. Found almost even more commonly in nature than the monomeric bases, though, are the so-called bis(benzylisoquinoline) alkaloids (*cf. Sect. 2.7*).

Phenyltetrahydroisoquinoline Alkaloids

As the name implies, this group is similar to the benzylisoquinoline alkaloids, but it has a CH_2 unit missing between the isoquinoline component and the additional benzene ring. A typical example is cryptostyline I (from *Cryptostylis* (Orchidaceae)).

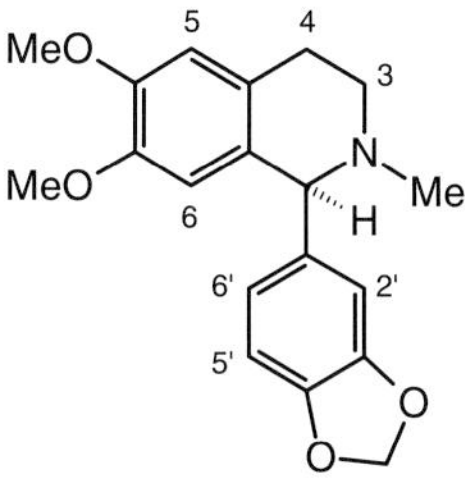

Phthalideisoquinoline Alkaloids[41]

This group, numbering about 50 representatives, has both the phthalideisoquinolines themselves and the *seco*-phthalideisoquinolines assigned to it.

(+)-Cryptostyline I (abs. config.)

Members of the first group possess two adjacent stereogenic centers, and so both *erythro*- and *threo*-forms can exist. Both occur in nature, as demonstrated in the case of hydrastine: the *threo*-form corresponds to (+)-(1S,1'S)-α-hydrastine (isolated from *Fumaria* species (Fumariaceae)) and the *erythro*-form to (+)-(1S,1'R)-β-hydrastine (from *Corydalis* species (Fumariaceae) and from *Hydrastis canadensis* (Hydrastidaceae)). Racemic α- and β-hydrastine have both been isolated from *Fumaria schleicherii*.

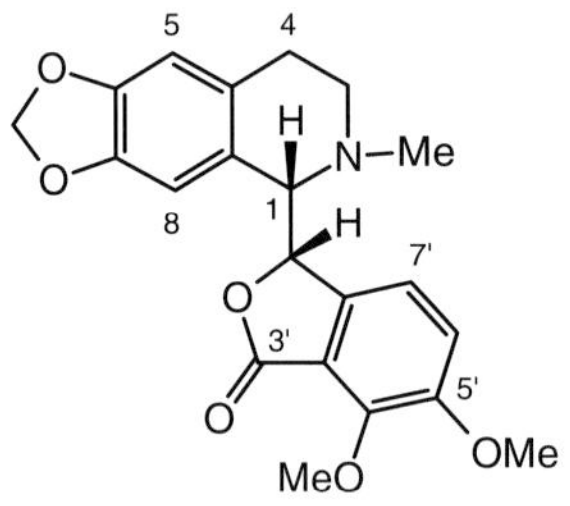

A

(+)-(1S,1'S)-α-Hydrastine
(abs. config.)

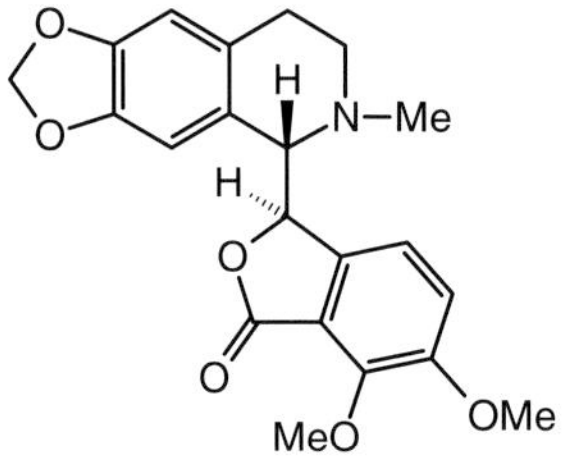

B

(+)-(1S,1'R)-β-Hydrastine
(abs. config.)

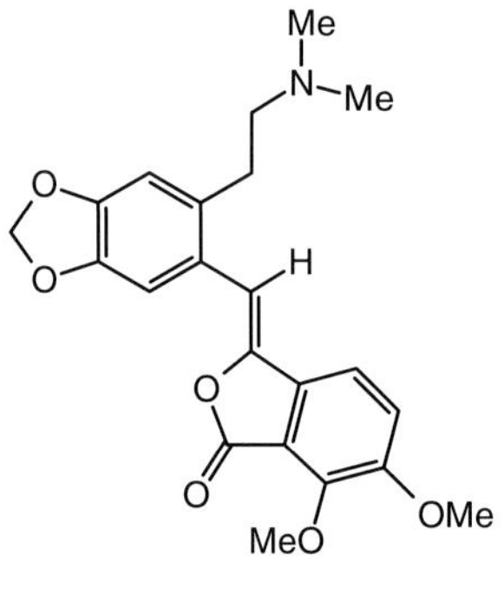

C

(Z)-N-Methylhydrastine

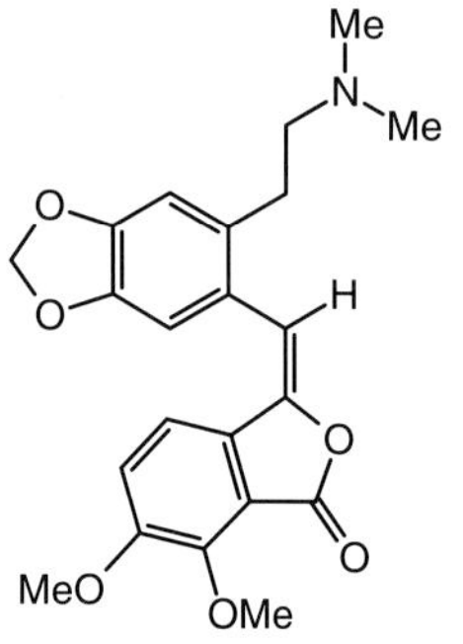

D

(E)-N-Methylhydrastine

The alkaloids α- and β-narcotine (= 8-methoxyhydrastine, from *Corydalis* and *Papaver* species (Papaveraceae), and also from the dried seed cases of *P. somniferum*) both display antitussive (anti-cough) properties.

seco-Phthalideisoquinolines are formally derived from N-methylphthalideisoquinolinium compounds, from which, as can be verified in the laboratory, they arise in nature through *Hofmann* degradation[42]. Hence, (Z)-N-methylhydrastine (**C**) (from *Corydalis* and *Fumaria* species) is preferential-ly produced from the *erythro*-isomer **B**[43], while the (*E*)-isomer **D** stems from α-hydrastine (**A**).

Rhoeadine and Papaverrubine Alkaloids

This group of compounds (*ca.* 35 members) is counted biogenetically among the isoquinoline alkaloids, although its members possess benzazepine rather than isoquinoline skeletons. Representative examples of this compound class are rhoeadine (with the (1*R*,2*R*,14*S*)-configuration, from *Bocconia, Meco-nopsis, Papaver* (all Papaveraceae), first isolated in 1865 from *Papaver rhoeas*[44], and featuring a *cis*-junction between rings B and D) and papaver-rubine A (from *Meconopsis* or *Papaver*, with a *trans*-junction between rings B and D)[45].

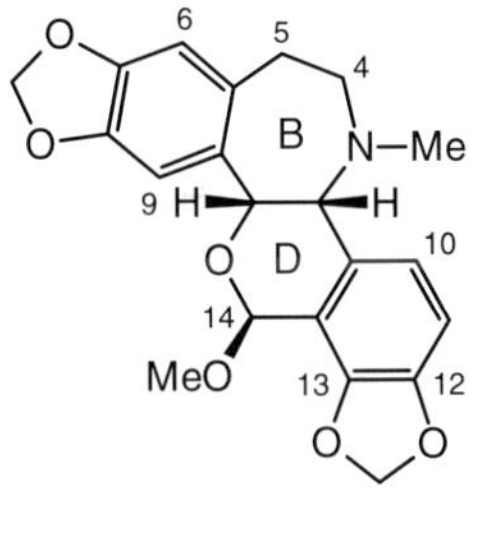

(+)-Rhoeadine (abs. config.) 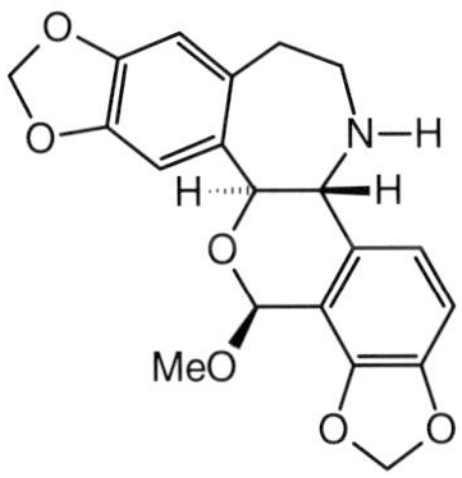   (+)-Papaverrubine A (abs. config.)

Ipecacuanha Alkaloids[46]

As well as compounds of the protoemetine type (from *Psychotria* and *Ura-goga* species (Rubiaceae))[47] and from the ipecoside type (from *Psychotria*) types, the term 'ipecacuanha alkaloids' also includes compounds possessing two N-atoms, as, for instance, emetine (from *Alangium* (Alangiaceae) and from *Borreria, Bothriospora, Capirona, Ferdinandusa, Hillia, Psychotria*,

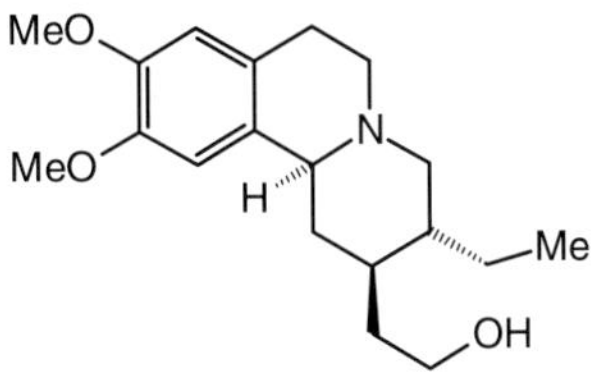

(–)-Protoemetine 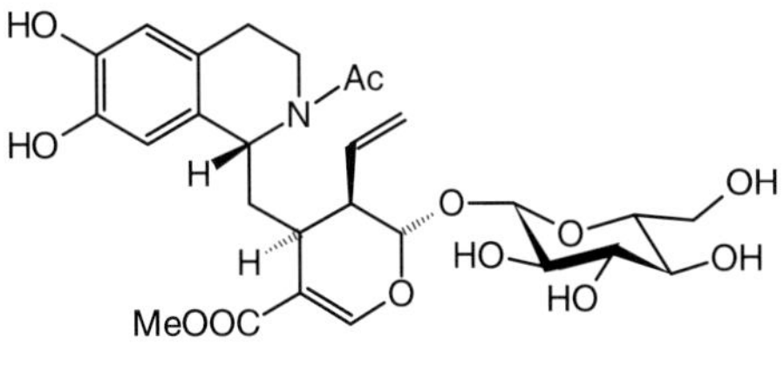   (–)-Ipecoside (abs. config.) 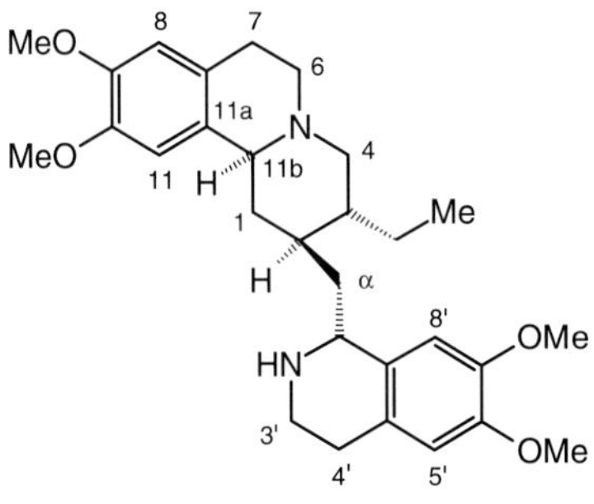   (–)-Emetine (abs. config.)

Remijia, and *Tocoyena* species (Rubiaceae)). The structural relationship between the biogenetic precursor ipecoside and the indole alkaloid vincoside (*cf. Sect. 7.2.4*) should also be noted. Other bases, possessing a close relationship to the ipecacuanha alkaloids, are dealt with later in the 'bis-alkaloids' section.

(–)-Cryptaustoline (abs. config.)

Cryptaustoline Type

As a typical representative of this very small group, cryptaustoline itself (from *Cryptocarya* (Lauraceae)) is shown.

Aporphine and Homoaporphine Alkaloids[48]

Examples of aporphine alkaloids (*ca.* 250 members) include glaucine (from *Corydalis* and *Dicentra* (Fumariaceae) (*Fig. 2.7*), and also from *Glaucium*

Fig. 2.7. *Often found in front gardens, the evergreen Mahonia (Mahonia aquifolium (Pursh) Nutt., oregon grape), has small, yellow blooms in the spring and blue, non-poisonous berries in August.* The plant contains tetrahydroisoquinoline alkaloids, berberine in particular, in its roots (photo *M. Hesse*).

species (Papaveraceae)) and corydine (from species of Annonaceae, Berberidaceae, Fumariaceae, Lauraceae, Menispermaceae, Papaveraceae, Ranunculaceae, and Rutaceae). Additionally, however, oxidized aporphine alkaloids have also been characterized: among others liriodenine (from several species of Annonaceae, Araceae, Atherospermataceae, Eupomatiaceae, Lauraceae, Magnoliaceae, Menispermaceae, and Rutaceae; *cf. Fig. 2.8*). Thalphenine, isolated from *Thalictrum* (Ranunculaceae), also belongs to this class.

Fig. 2.8. *The Common Fumitory (Fumaria officinalis L. (Fumariaceae)) is used as a component of infusions for liver and gall bladder ailments, due in part to its content of fumaricine derivatives* (from F. Bianchini, F. Corbetta, M. Pistoia, *Der grosse BLV Heilpflanzenatlas*, Munich, 1983; drawing *Marilena Pistoia*)

Homoaporphine alkaloids such as multifloramine (from *Kreysigia* (Liliaceae)) contain a seven-membered C-ring.

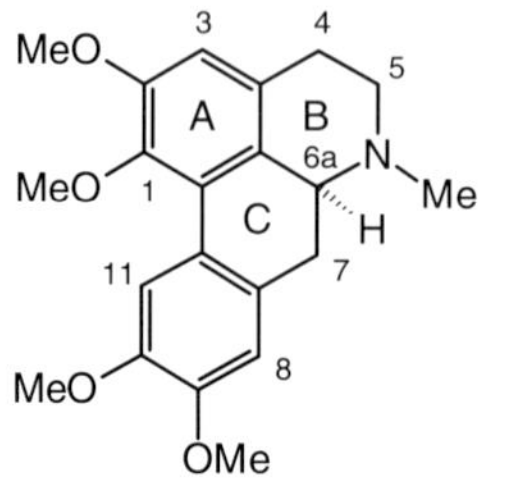

(+)-Glaucine (abs. config.) 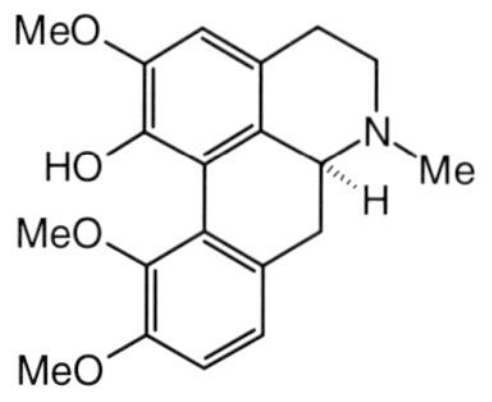　　　(+)-Corydine (abs. config.) 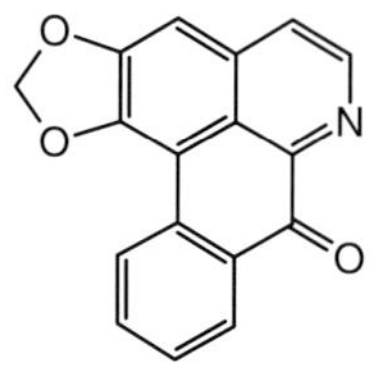　　　Liriodenine

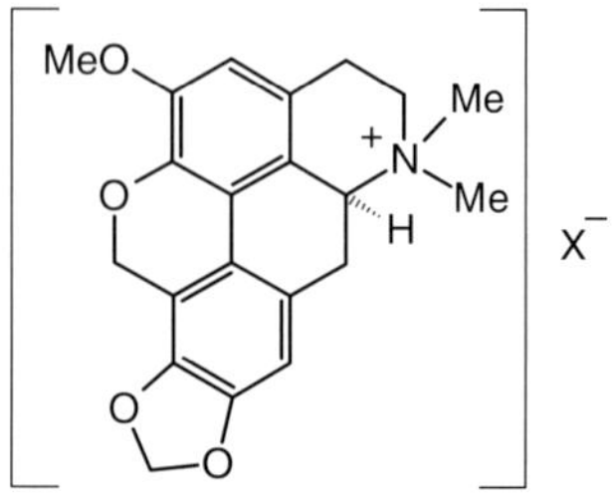

(+)-Thalphenine (abs. config.) 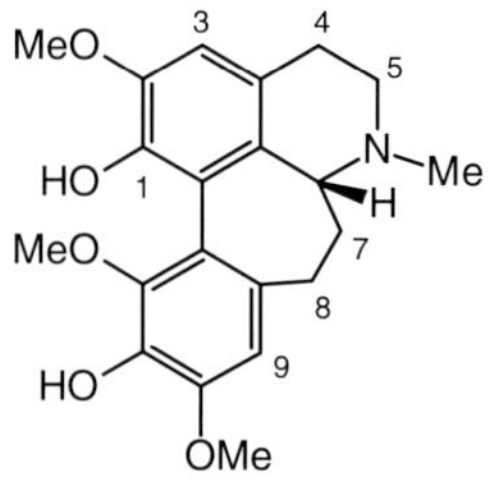　　　(−)-Multifloramine (abs. config.)

Other members possess Me substituents at C(7), a C(7)=C(6a) bond, or additional oxygen functionalities at C(4), C(5), or even at the N-atom. One notable aporphine base is arosinine (from *Glaucium flavum* var. *vestitum* (Papaveraceae)), which forms dark green needles upon crystallization. It is also green colored in neutral or alkaline solution. In acidic media, however, it turns red, which can be explained by the formation of mesomeric quinone-iminium forms and aromatic betaines, corresponding to the formulas **A** and **B**, or **C** and **D**, respectively (*Scheme 2.6*).

Existing in a certain biogenetic relationship with the aporphine alkaloids, one finds compounds such as eupolauramine (from *Eupomatia laurina* (Eupomatiaceae)). The distribution of the Australian native *Eupomatia laurina* is locally very limited; the plant has no near relatives, which partially explains why the structures of its component substances are encountered only very rarely. It can be assumed that the putative biogenetic precursor **F** could be formed through *1*) oxidative opening of ring A, *2*) transformation

Scheme 2.6

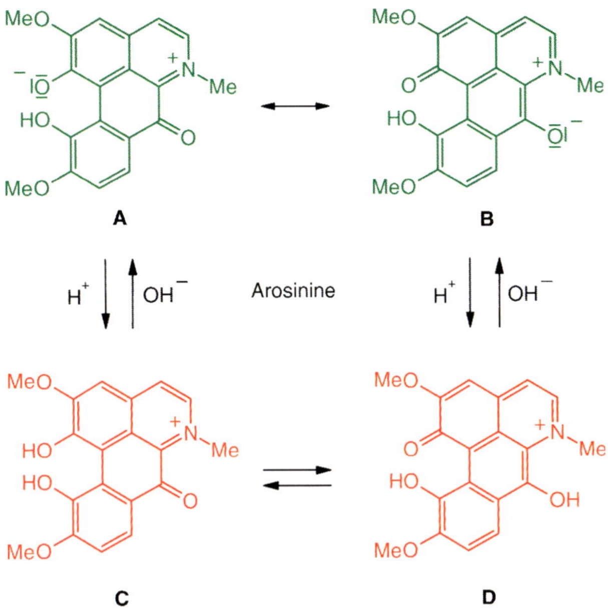

to the corresponding amino acid, and *3*) benzil/benzilic acid rearrangement of an intermediate 1,2-dicarbonyl compound in ring B (*Scheme 2.7*)[49].

Scheme 2.7

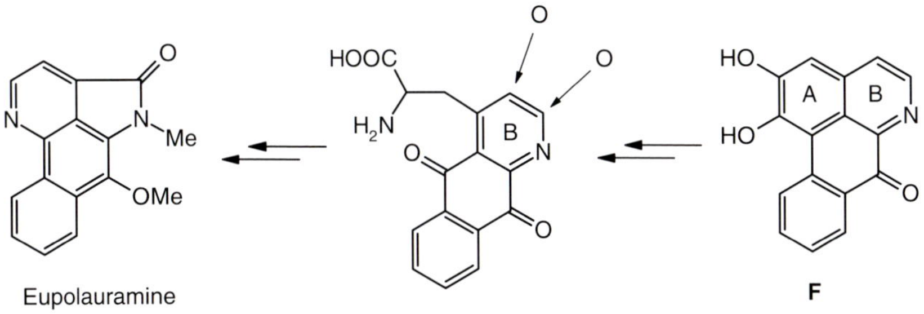

Eupolauramine

F

Proaporphine and Homoproaporphine Alkaloids[50]

A typical representative of this substance class is pronuciferine (from species of *Croton* (Euphorbiaceae), *Nelumbo* (Nelumbonaceae), *Papaver* (Papaveraceae), and *Stephania* (Menispermaceae)). Furthermore, alkaloids with different *O*-substitution patterns are also found, while natural derivatives with (partially) hydrogenated D-rings are known as well. In homoproaporphine alkaloids, the C-ring is enlarged by one CH_2 unit (*cf.* bulbocodine from *Bulbocodium* (Liliaceae)).

(+)-Pronuciferine (abs. config.)   (+)-Bulbocodine

Cularine Alkaloids

The occurrence of this group of isoquinoline alkaloids, consisting of *ca.* 30 representatives, is restricted to a few plant genera: *Corydalis* (*Fig. 2.9*), *Dicentra* (*Fig. 2.10*), and *Sarcocapnos* (all Fumariaceae), as well as *Guatteria* (Annonaceae) and *Berberis* (Berberidaceae). A typical representative is cularine (from *Corydalis* and *Dicentra* species). In addition, some interesting structural variants are also found, such as the red-colored yagonine (from *Sarcocapnos enneaphylla*), the isoindole alkaloid aristoyagonine (from *S.enneaphylla*), and the oxidation product noyaine (from *Corydalis claviculata*), the latter, for obvious reasons, being structurally related to cularine.

Fig. 2.9. *The plant Corydalis solida is known for its content of isoquinoline alkaloids (photo T. Gözler)*

(+)-Cularine (abs. config.)   Yagonine

Aristoyagonine

Noyaine

Fig. 2.10. *Dicentra spectabilis* L. Lᴇᴍ. (Fumariaceae), *Bleeding Heart, prized in the garden as an early blooming summer flower, contains a variety of 1,2,3,4-tetrahydroisoquinoline alkaloids* (photo *M. Hesse*)

Interestingly, basic treatment of yagonine with Ba(OH)$_2$ yields aristoyago-nine and 1,α-dihydroyagonine (arising from benzil/benzilic acid rearrange-ment followed by a disproportionation)[51].

Formally arising from certain isoquinoline alkaloids by *Hofmann* degrada-tion, the so-called phenanthrene alkaloids (*ca.* 160 known structures) may pass as derivatives of the aporphine alkaloids (such as atherosperminine from species of the genera *Annona* (Annonaceae), *Atherosperma* (Athero-spermataceae), *Cryptocarya* (Lauraceae), *Duguetia, Enantia, Fissistigma, Guatteria*, and *Monodora* (all Annonaceae)) or as derivatives of the mor-phine alkaloids. Presumably, though, a proportion of these compounds are artifacts (*Hofmann* elimination) arising in the course of the (basic) extraction of the relevant plants[52].

Protoberberine Alkaloids

Representatives of the berberines are widely distributed in the plant families Alangiaceae, Annonaceae, Berberidaceae, Fumariaceae, Hydrastidaceae, Lauraceae, Leguminosae, Menispermaceae, Papaveraceae, Ranunculaceae, and Rutaceae. The most conspicuous is undoubtedly berberine[53], which has been isolated in the form of yellow crystals from a great many genera of the plant families mentioned above.

Besides alkaloids of the canadine type (from species of *Corydalis* (Fumaria-ceae), *Hydrastis* (Hydrastidaceae), and *Xanthoxylum* (Rutaceae)), there also occur oxidized derivatives (*inter alia*, ophiocarpine (from *Corydalis* (Fuma-riaceae)), as well as C$_1$-substituted derivatives such as mecambridine (from *Papaver* (Papaveraceae) and corydaline (from *Corydalis*).

Atherosperminine

Berberine

(–)-Canadine (abs. config.)

(–)-Ophiocarpine

(–)-Mecambridine (abs. config.)

(+)-Corydaline (abs. config.)

(–)-Lienkonine (8*S*,13a*S*)
(abs. config.)

The formation of (–)-lienkonine in *Corydalis* involves the incorporation of acetaldehyde into the 1-benzyltetrahydroisoquinoline structure in place of formaldehyde that would give rise to the berberine skeleton.

In solidaline (from *Corydalis solida*), we once more have an alkaloid that can exist either in an iminium or in a carbinolamine form, depending on pH (*Scheme 2.8*). This transition between the two states can be followed by UV spectroscopy[54].

Scheme 2.8

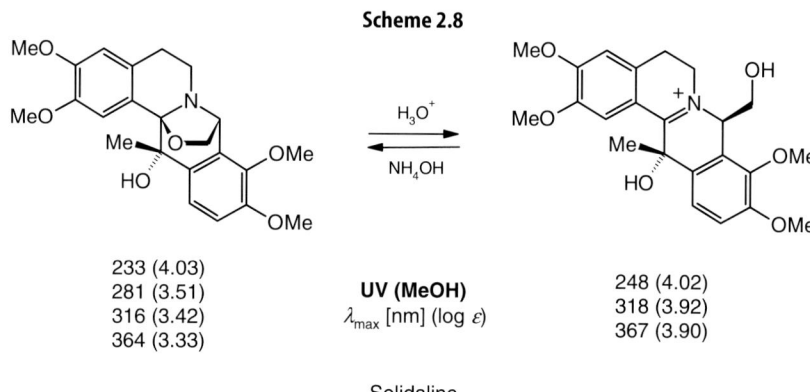

233 (4.03)
281 (3.51)
316 (3.42)
364 (3.33)

UV (MeOH)
λ_{max} [nm] (log ε)

248 (4.02)
318 (3.92)
367 (3.90)

Solidaline

Protopine Alkaloids[55]

The principal alkaloid of this group is protopine (widely distributed in the Fumariaceae, Hypecoaceae, Nandinaceae, Papaveraceae, and Pteridophyllaceae families). Alkaloids of the oxomuramine type (from *Papaver* (Papaveraceae)) and of the corycavidine type (from *Corydalis* (Fumariaceae)) are also known. All three alkaloids display similar structures compared to analogous substances from the berberine class, from which they can be prepared synthetically.

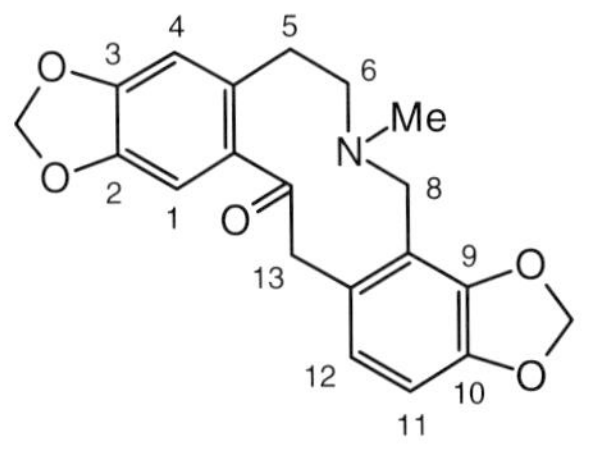

Protopine 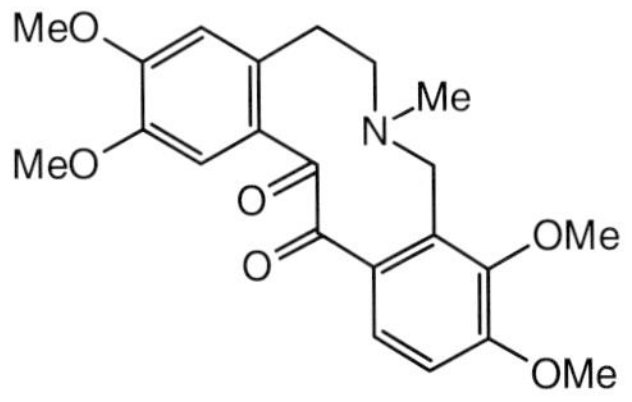   Oxomuramine

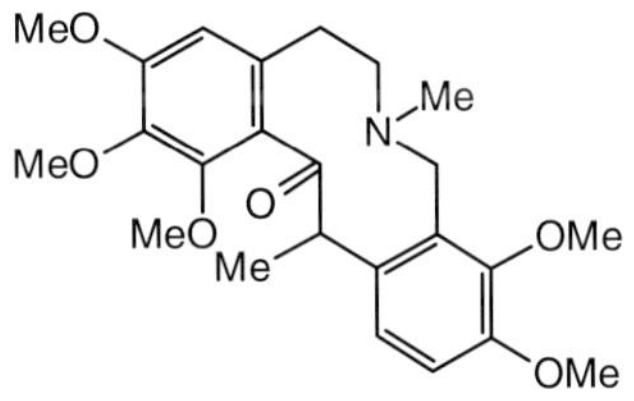

Corycavidine

Benzophenanthridine Alkaloids[56]

Typical here are sanguinarine (widely distributed in the Fumariaceae, Hypecoaceae, Papaveraceae, and Rutaceae families), oxynitidine (from *Xanthoxylum* (Rutaceae)), corynolamine (from *Corydalis* (Fumariaceae)), and corynoloxine (from *Corydalis*). Benzophenanthridine derivatives may also be found incorporated into bis-alkaloids.

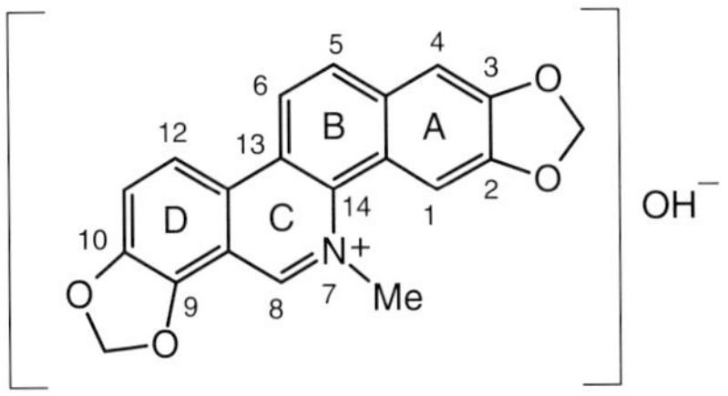

Sanguinarine 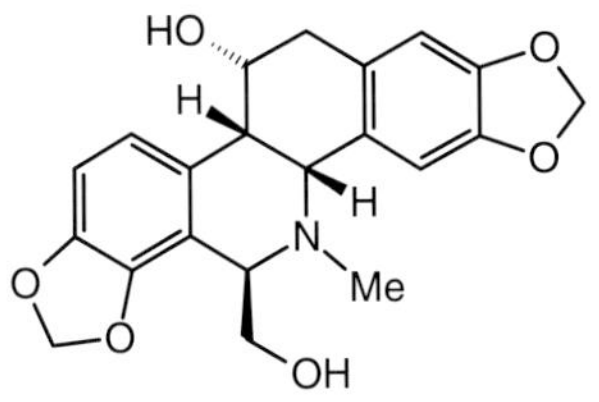   Corynolamine

Fig. 2.11. *Papaver argemone* L. Field poppy, corn poppy (*Argemone capitulo longiore* (Papaveraceae)). Watercolor by *Johann Heinrich Timroth*, 1756, Museum Heidecksburg, Rudolstadt, Thüringen (Thuringia).

Oxynitidine

Corynoloxine

Spirobenzylisoquinoline Alkaloids[57]

Two characteristic representatives of this alkaloid group are fumaricine (from *Fumaria* (Fumariaceae)) and ochotensine (from *Corydalis* and *Dicentra* species (Fumariaceae)).

(–)-Fumaricine

(+)-Ochotensine (abs. config.)

Pavine and Isopavine Alkaloids[58]

While argemonine (from *Argemone* (Papaveraceae); *Fig. 2.11*) serves as an example of a pavine alkaloid, amurensine (from *Papaver* (Papaveraceae)) is a typical isopavine alkaloid (*Fig. 2.12*).

(–)-Argemonine (abs. config.)

(–)-Amurensine (abs. config.)

Fig. 2.12. *Leontice leontopetalum* L. (Leonticaceae) *is a Mediterranean plant producing tetrahydroisoquinoline alkaloids* (photo *T. Gözler*)

Morphine Alkaloids[33,59]

This famous and notorious class of isoquinoline alkaloids is broken down into a few subgroups. While resembling the tetracyclic sinomenine[60] (from *Menispermum* and *Sinomenium* species (Menispermaceae)), morphine[61] (from *Papaver somniferum* (Papaveraceae); *Fig. 2.13*) features an additional ether ring (*cf. Chapts. 5.6* and *11.5*). Also structurally related to the true morphine alkaloids are the hasubanone alkaloids (such as hasubanonine from *Stephania japonica* (Menispermaceae)), the ring-enlarged homomorphinanes (such as androcymbine from *Androcymbium* (Liliaceae) and kreysiginine from *Kreysigia* (Liliaceae)), and also alkaloids of the acutumine type (*cf.* acutumine itself from *Menispermum* (Menispermaceae)).

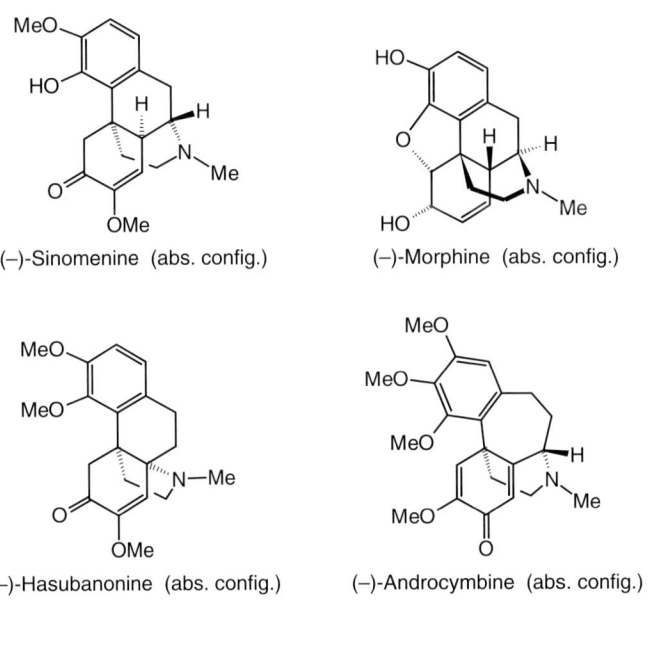

(–)-Sinomenine (abs. config.)   (–)-Morphine (abs. config.)

(–)-Hasubanonine (abs. config.)   (–)-Androcymbine (abs. config.)

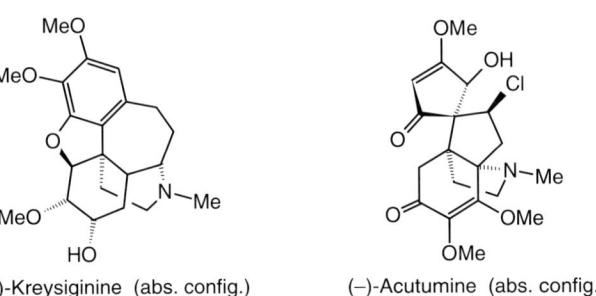

(+)-Kreysiginine (abs. config.)   (–)-Acutumine (abs. config.)

Fig. 2.13. *Papaver somniferum* L. (from *W. v. Hamm, T. Schwartze, H. Wagner, J. Zöllner,* 'Die Chemie des täglichen Lebens' ('Chemistry in Everyday Life'), in *Das neue Buch der Erfindungen, Gewerbe und Industrien* (The New Book of Discoveries, Trade, and Industry), 6th edn., Vol. 5, O. Spamer, Leipzig, 1873)

Amaryllidaceae Alkaloids[62]

The ordering principle behind this subclass of the isoquinoline alkaloids is their common occurrence in plants of the Amaryllidaceae families (*Figs. 2.14* and *2.15*). Distinctions are drawn between six different skeletal types, all of

Fig. 2.14. *Galanthus nivalis* L. (Amaryllidaceae), *snowdrop*. The plant contains a number of so-called Amaryllidaceae alkaloids of the galanthamine, crinine, and lycorenine types (photo *M. Hesse*).

Fig. 2.15. *Sternbergia lutea* (L.) Ker-Gawl. ex Schult. F. (*Amaryllidaceae*), *native to western Anatolia, contains alkaloid types also found in other Amaryllidaceae species* (photo *T. Gözler*)

them derived from one basic structure: namely lycorine. While the number of further skeletal variants is small, many individual alkaloids are found to incorporate the following basic types: lycorine (lycorine type, widely distributed, *cf. Chapt. 5.5*), ambelline (ambelline type, from *Amaryllis, Ammocharis, Brunsvigia,* and *Buphane* species), hippeastrine (hippeastrine type, from *Amaryllis, Brunsvigia, Clivia,* and *Crinum* species), tazettine (tazettine type, from various genera), galanthamine (galanthamine type, from *Amaryllis, Cooperanthes,* and *Crinum* species), and montanine (montanine type, from *Haemanthus* species).

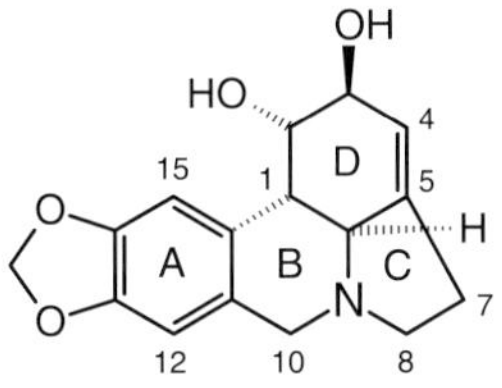

(−)-Lycorine (abs. config.)

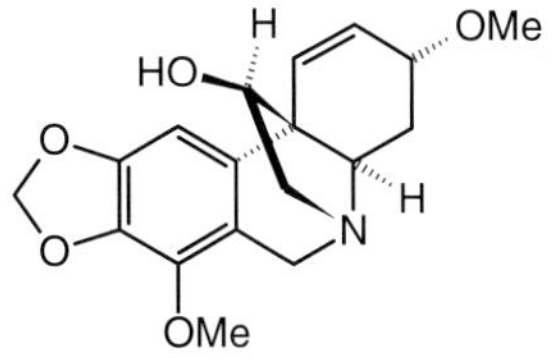

(+)-Ambelline

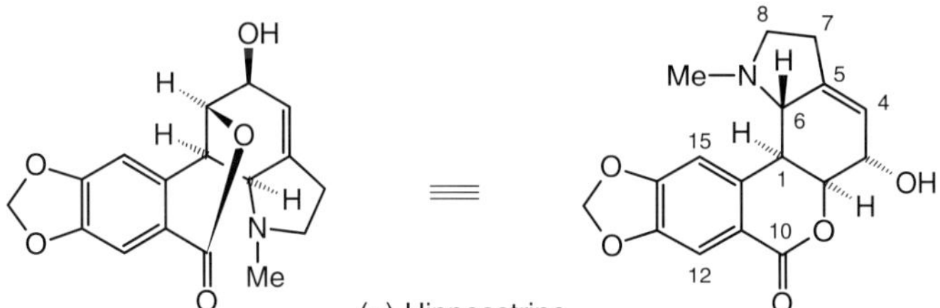

(+)-Hippeastrine
(abs. config.)

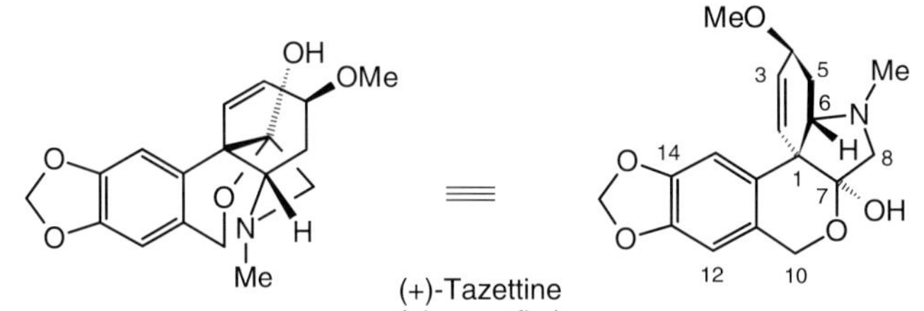

(+)-Tazettine
(abs. config.)

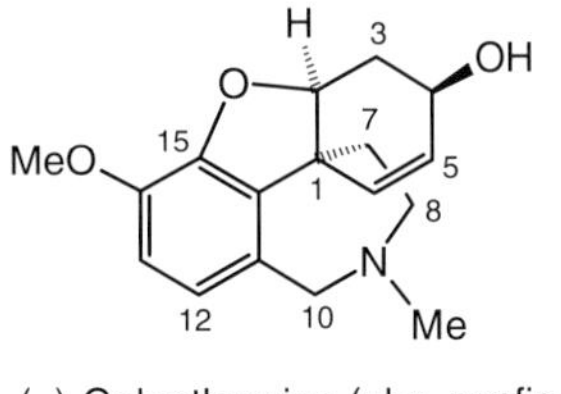

(−)-Galanthamine (abs. config.)

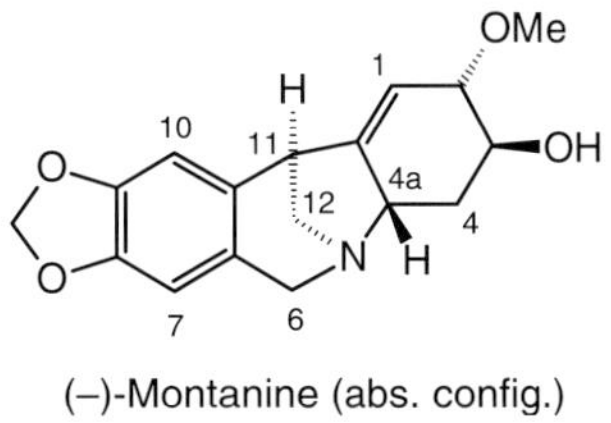

(−)-Montanine (abs. config.)

OH

MeO

HO

N—Me

(−)-Cherylline (abs. config.)

Fig. 2.16. *Erythrina crista-galli* L. (Coral Tree (Leguminosae)), *indigenous to Brazil, is a popular ornamental plant in Central Europe*. The most diverse *Erythrina* alkaloids have been isolated from the plant (photo *H. Reinhard*).

Cherylline Type

The principal representative of this group is cherylline (from *Crinum* species (Amaryllidaceae)), the relationship of which to the Amaryllidaceae bases is immediately apparent.

Erythrina *Alkaloids*[63]

In principle, the *Erythrina* alkaloids could (according to our classification criteria) be listed either under the isoquinoline alkaloids or under the izidine alkaloids. Again, biogenetic perspectives are the decisive factor behind their appearance at this point. Two group-specific skeleton systems are represented in erysodine and β-erythroidine (both from *Erythrina* species (Leguminosae); *Fig. 2.16*).

The erysodine skeleton appears in many *Erythrina* alkaloids, although frequently undergoing modifications in degree of saturation and *O*-substitution pattern. Higher homologues (such as schelhammerine from *Schelhammera* (Liliaceae)) and lactone derivatives of these homoerythrinan alkaloids have also been isolated. Isomeric with the *Erythrina* alkaloids in terms of the ring skeleton are the *Cephalotaxus* alkaloids (*ca.* 20 representatives), to which deoxyharringtonine (from *Cephalotaxus* (Cephalotaxaceae)) belongs[64].

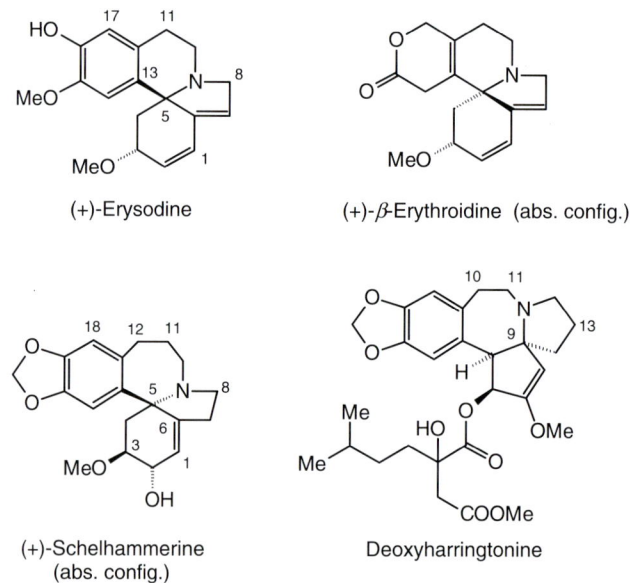

(+)-Erysodine

(+)-β-Erythroidine (abs. config.)

(+)-Schelhammerine
(abs. config.)

Deoxyharringtonine

2.2.7. Quinoline Alkaloids[65]

The total number of alkaloids of this class amounts to *ca.* 150. In case of the natural quinoline alkaloids, distinction can be made between the following chromophores:

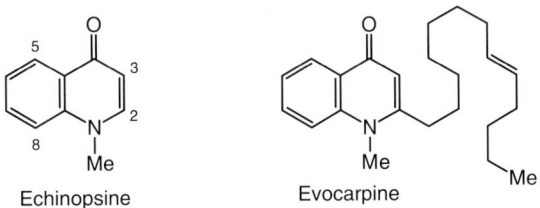

Quinoline   2-Quinolone   4-Quinolone

1,2,3,4-Tetrahydroquinoline   Furoquinoline   Furoquinolone

2-Quinolone and 4-Quinolone Derivatives

Together with simple compounds such as echinopsine (from *Echinops* species (Compositae)) and *N*-methyl-2-quinolone (from *Galipea* (Rutaceae)), both *O*- and *C*-alkylated derivatives are also found. The nature of the alkyl substituents varies wildly: in evocarpine (from *Evodia* (Rutaceae)) an unbranched C_{13} chain is attached at C(2), in galipine (from *Galipea*) a 2-phenylethyl residue, and in lunamarine (from *Lunasia* (Rutaceae)) a phenyl residue, while in ptelefolidine (from *Ptelea* (Rutaceae)) an isoprenyl residue is found at C(3). Numerous structural variants of this alkaloid type have been found in nature (*Fig. 2.17*).

Fig. 2.17. *Haplophyllum myrtifolium* Boiss. (Rutaceae; southern Turkey). This species contains structurally simple quinoline alkaloids (photo *T. Gözler*).

Echinopsine   Evocarpine

55

Galipine

Lunamarine

Ptelefolidine

Furoquinoline Alkaloids[66]

This subgroup also encompasses both simple representatives such as dictamnine (from various genera of Flindersiaceae and Rutaceae) or skimmianine (from *Skimmia* (Rutaceae); *Fig. 2.18*), as well as members that still contain

Fig. 2.18. *Skimmia japonica* Thunb. (Rutaceae), *cultivated in Europe as an ornamental plant, contains various quinoline alkaloids, skimmianine among them* (photo *M. Hesse*)

the complete isoprene moiety of ptelefolidine (*e.g.*, lunacrine from *Lunasia* (Rutaceae)). Acrophylline (from *Acronychia* (Rutaceae)) bears an isoprene residue on the N-atom. Quaternary quinolidinium salts have also been isolated from plants.

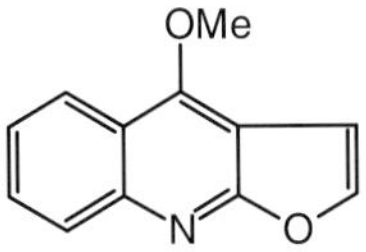

Dictamnine

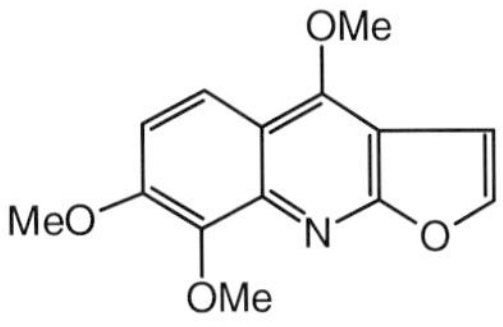

Skimmianine

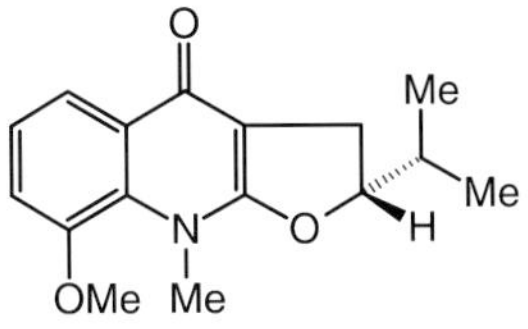

(–)-Lunacrine
(abs. config.)

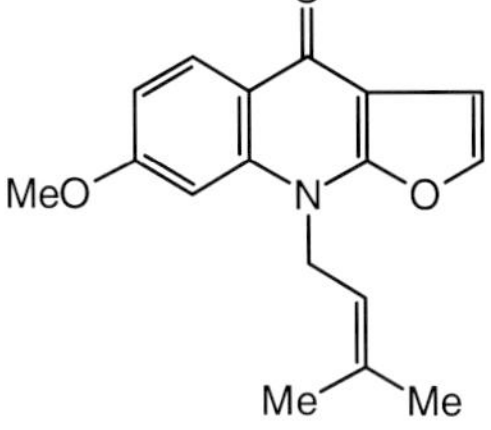

Acrophylline

Quinine Alkaloids (*Cinchona* Alkaloids)[20,67]

Cinchona alkaloids have been known from very early times. They are characterized by two skeleton types: quinine[68] (from *Cinchona* and *Remijia* species (Rubiaceae); *cf. Chapt. 11.11*) and quinotoxine (from *Cinchona*), in which the quinuclidine ring is opened.

There is a close biogenetic relationship between these compounds and *Cinchona* alkaloids incorporating indole moieties.

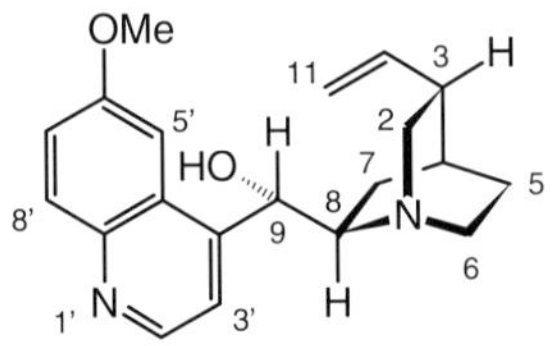

(–)-Quinine (abs. config.)

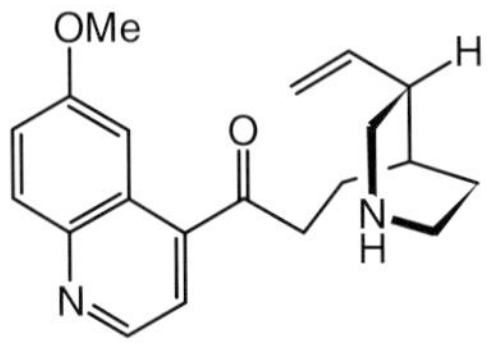

(+)-Quinotoxin (abs. config.)

Melodinus *Alkaloids Incorporating Quinoline Moieties*

Although biogenetically belonging to the aspidospermidine type of indole alkaloids and endowed with a 2-quinolone chromophore, this small group of alkaloids is listed separately at this point with meloscine (from *Melodinus* (Apocynaceae)) shown as an example.

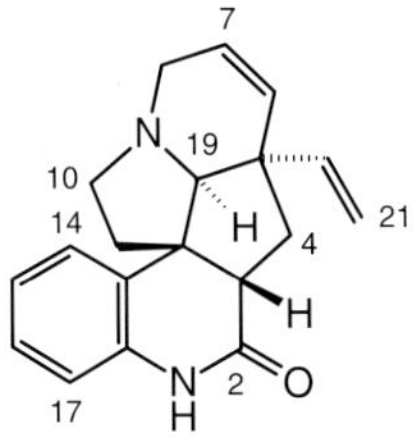

(+)-Meloscine
(abs. config.)

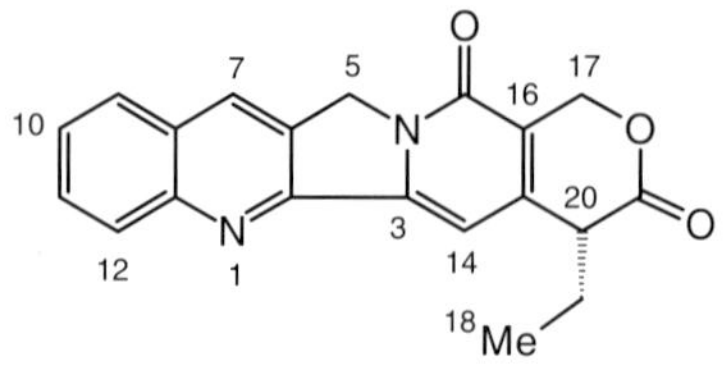

(+)-Camptothecin
(abs. config.)

Camptotheca Bases[69]

Like meloscine, camptothecin (from *Camptotheca* (Nyssaceae) and *Mappia* (Olacaceae) species), the primary representative of this structural type, is biogenetically related to the indole alkaloids.

Alkaloids with the camptothecin skeleton differ from the indole alkaloids containing C_9/C_{10} monoterpene units in so far that the rings B and C in the indole alkaloids are five- and six-membered, respectively, while they are six- and five-membered in the former compounds. Synthetically, indole alkaloids can be transformed in four steps into the corresponding camptothecin structures by *1*) photooxidation, *2*) treatment with mild base ($NaHCO_3$), *3*) addition of $SOCl_2$, and *4*) reduction to the quinoline nucleus (*Scheme 2.9*).

Scheme 2.9

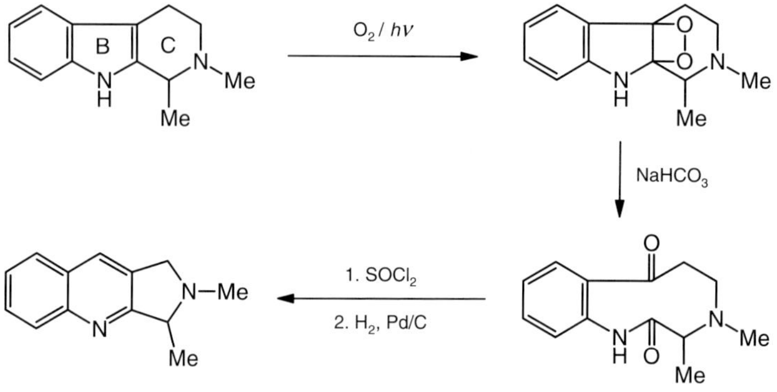

Bis-alkaloids incorporating tetrahydroquinoline chromophores have also been isolated from *Calycanthus* species.

Acridine Alkaloids[70]

This group comprises *ca.* 80 bases. Together with simpler derivatives of acridine (*e.g.*, evoxanthine from *Evodia* and *Teclea* species (Rutaceae)), isoprene-substituted compounds such as atalaphylline (from *Atalantia* (Rutaceae)) are also known. From the latter, both tetra- and pentacyclic structures are derived through ether ring-closure.

Acronycine (from *Acronychia baueri* (Rutaceae)) displays antitumor activity. On treatment with methanolic hydrogen chloride, the compound astonishingly produces – together with the expected *O*-demethylacronycine – dimers, trimers, tetramers, and even pentamers (as shown).

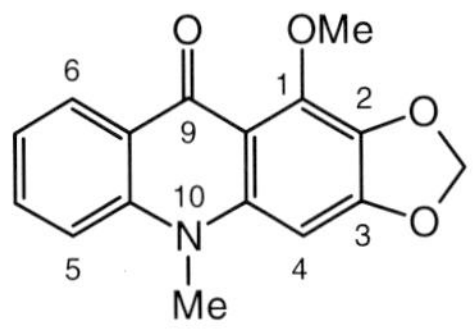

Evoxanthine

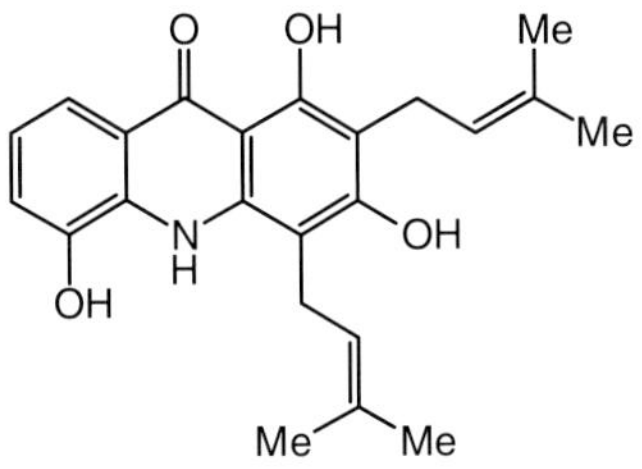

Atalaphylline

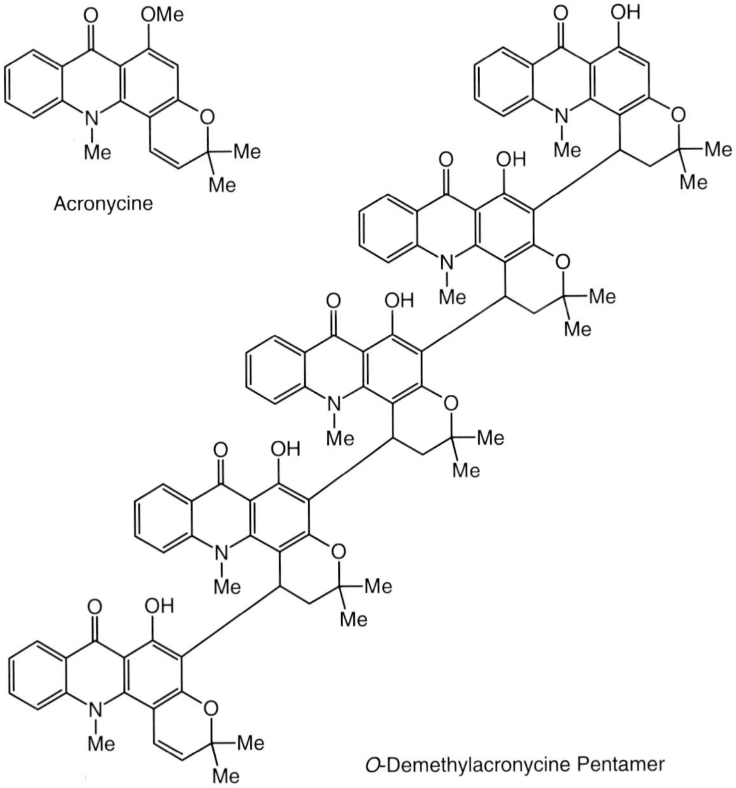

Acronycine

O-Demethylacronycine Pentamer

2.2.8. Quinazoline Alkaloids[71]

To date, *ca.* 50 alkaloids incorporating the quinazoline chromophore have been isolated. Together with simpler bases such as glomerine (a component of the defensive secretion of the millipede *Glomeris marginata*) or glycos-micine (from *Glycosmis arborea* (Rutaceae)), there are also known derivatives with one or two additional rings. Examples of these include vasicolinone (from *Adhatoda vasica*) and vasicine (= peganine, isolated from species of *Adhatoda* (Acanthaceae), *Galega* (Leguminosae), *Linaria* (Scrophulariaceae), *Lythrum* (Lythraceae), *Peganum* (Zygophyllaceae; *Fig. 2.19*), and *Sida* (Malvaceae); (*cf. Scheme 2.10*)). Febrifugine (from *Dichroa* and *Hydrangea* (Hydrangeaceae)) represents another structural variant. Attention should also be paid to alkaloids of the evodiamine type; these have already been introduced under the indole alkaloids, but they also incorporate quinazoline structural elements.

Glomerine

Glycosmicine

Vasicine

Vasicolinone

Candidine

(+)-Febrifugine (abs. config.)

Quinazoline

Fig. 2.19. *Peganum harmala* L. (Zygophyllaceae; central Anatolia) *is notable for containing many simply constituted indole alkaloids of the harman type.* A second group of alkaloids contained in the plant are counted as belonging to the peganine/vasicinone type (photo *T. Gözler*).

The intensely colored candidine, from culture solutions of *Candida lipolytica* (UV/VIS: λ_{max} 244 nm (log ε 4.34), 250 (4.36), 282 (4.21), 538 (3.95), 573 (4.11)) is a diindoxyl condensation product.

Biogenetically, anthranilic acid is an important precursor for this group of compounds. As can be seen from *Scheme 2.10*, the second N-atom of vasicine originates from ornithine or aspartic acid.

Scheme 2.10

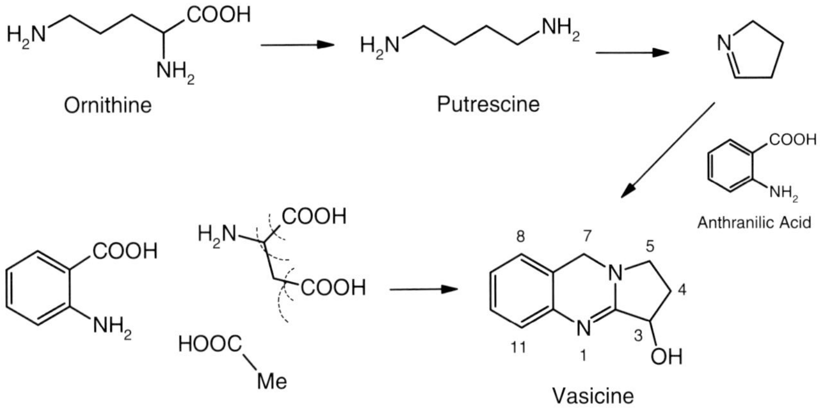

Ornithine

Putrescine

Anthranilic Acid

Vasicine

2.2.9. Benzoxazines and Benzoxazoles

Representatives of these two compound types seem to occur ubiquitously in the plant kingdom. The reasons for this, and their significance, are addressed elsewhere. Besides 3,4-dihydro-2,4-dihydroxy-1,4-benzoxazine-3(2*H*)-one (DIBOA) and its 4-deoxy and 7-methoxy derivatives, the corresponding 2-glucosides have also been isolated (*cf. Chapt. 9.4*).

Anthranilic acid and ribose-5-phosphate react together in plants (under formation of a *Schiff* base and after *Amadori* rearrangement) to yield the intermediate **A** (*Scheme 2.11*), which is converted into the phenol **B** by oxidative decarboxylation. The intermediate **C** has been observed in the cyclization of **B** to DIBOA; as far as exact knowledge of this biosynthetic pathway is concerned, however, there still seem to be certain holes present. In *Zea mays* (Graminaceae), on the other hand, indoles have recently been identified as DIBOA precursors.

During the course of hydrolysis, formic acid (HCOOH) can split off from the hydroxamic acid part of DIBOA, whereupon the naturally occurring 2(3*H*)-benzoxazolines (BOA) are produced. Amine oxidation in nature proceeds with the aid of the enzyme *N*-monooxygenase.

Scheme 2.11

1
Anthranilic Acid

2
Ribose-5-phosphate

A

B

C

DIBOA

+ H₂O
− HCOOH

BOA

HBOA

3-Hydroxyoxindole

Oxindole

Indole

Izidines

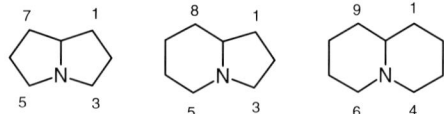

The term 'izidines' stands for all those alkaloids that possess a ring skeleton in which a single N-atom constitutes part of two or even three rings. Encompassing the pyrrolizidine, indolizidine, and quinolizidine alkaloids, this category of bases numbers at least 500 members.

2.2.10. Pyrrolizidine Alkaloids (*Senecio* Alkaloids)[35, 72]

In principle, it is possible to distinguish between four different alkaloid types within this group: *a*) Bases containing the necine skeleton as their essential structural component (*e.g.*, retronecanol, from *Crotolaria* (Leguminosae)), a relatively small subgroup; *b*) monoesters (*e.g.*, heliotrine from *Heliotropium* (Boraginaceae)); *c*) diesters with two independent acid components (*e.g.*, lasiocarpine from species of *Heliotropium*, *Lappula*, and *Symphytum* (Boraginaceae)); and *d*) cyclic diesters, consisting of a necine part and a dicarboxylic acid (*e.g.*, monocrotaline from *Crotolaria*).

Structural variants such as ammonium compounds, *N*-oxides, and derivatives with a carboxy group at C(9) are all found, as well as representatives with opened pyrrolizidine rings (*inter alia*, clivorine from *Ligularia* (Compositae)).

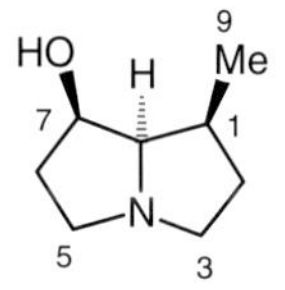

(–)-Retronecanol 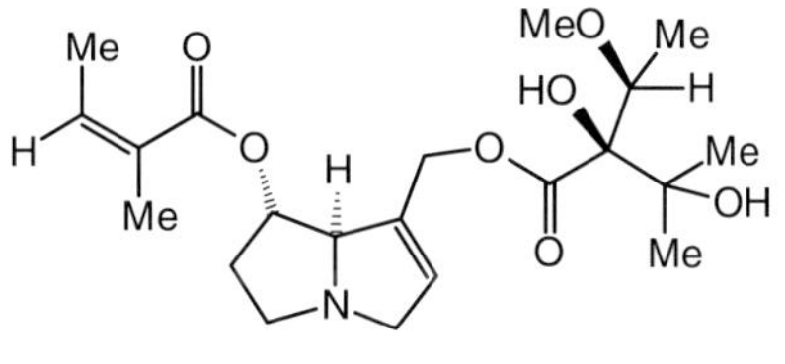   Lasiocarpine

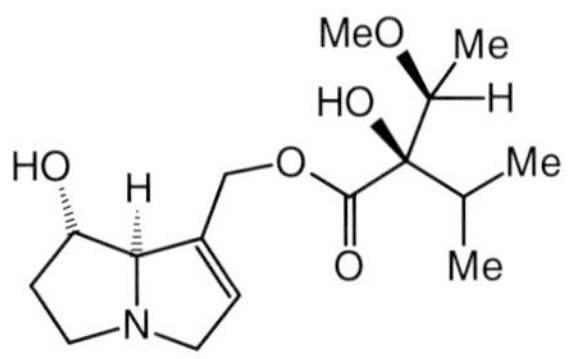

(+)-Heliotrine   (–)-Monocrotaline   (+)-Clivorine
(abs. config.)   (abs. config.)   (abs. config.)

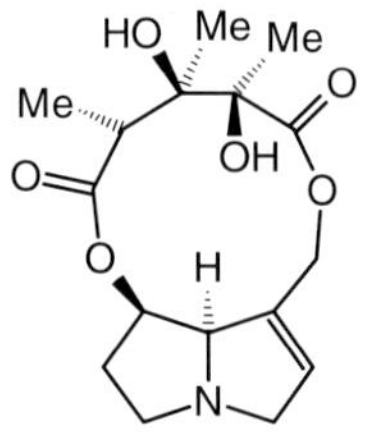

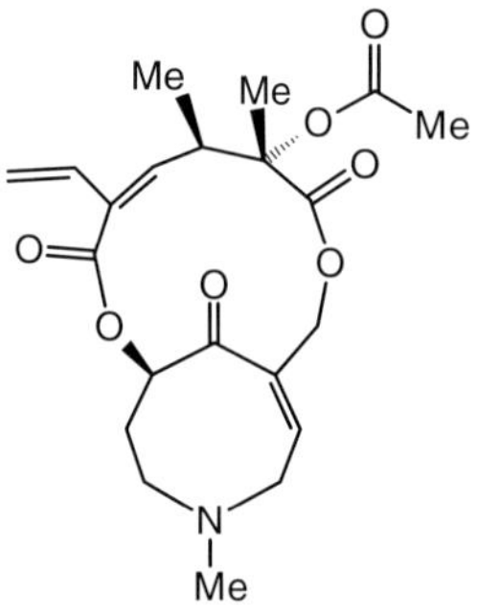

Today, there are *ca.* 390 alkaloids known from this group. It should be borne in mind, though, that the astonishing structural diversity actually merely reflects the breadth of variation in the carboxylic acid components.

The pyrrolizidine alkaloid danaidone (from *Lycorea ceres ceres* and *Danaus gilippus berenice* (Danainae)) acts as a sex attractant (pheromone) for butterflies of various *Danaus* species and is stored by the males in glands. Danaidone is biologically synthesized from lycopsamine, which the insects take up from dead or damaged pyrrolizidine-containing plants such as *Helio-*

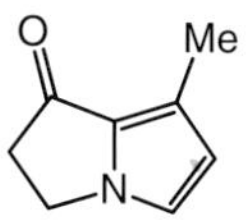

Danaidone 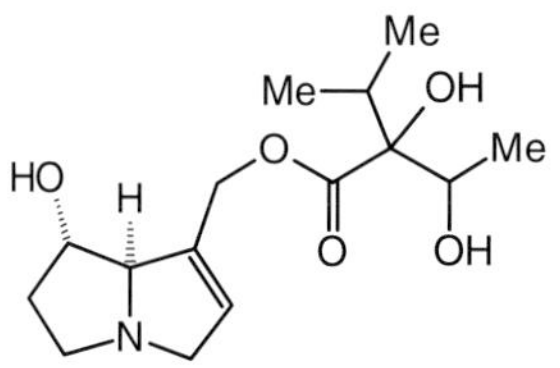   Lycopsamine

(+)-Indicine *N*-Oxide

tropium (Boraginaceae). At the same time, danaidone serves the butterflies as a self-defense deterrent.

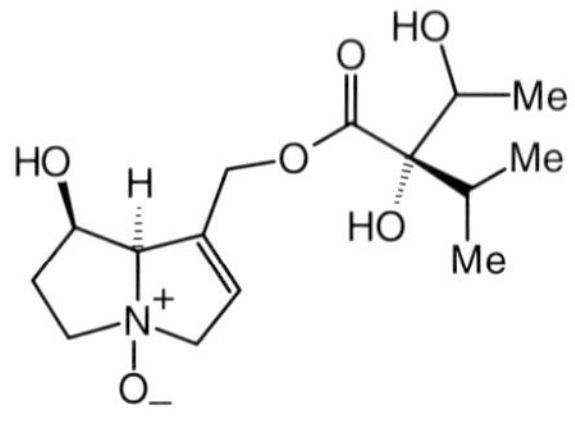

Pyrrolizidine alkaloids are poisonous (often hepatotoxic) to humans, other mammals, and birds. The plant *Senecio vulgaris* (Compositae) was known as early as 400 B. C. as a treatment for 'tumors on the foot and sinew' and was in use for treatment of cancer around 1000 A.D. In fact, several pyrrolizidine derivatives are active against various tumors. Particularly noteworthy in this respect is indicine *N*-oxide (from *Heliotropium indicum* (Boraginaceae)).

In Central Africa (Congo Basin), pyrrolizidine alkaloid extracts from *Crotolaria* species were put to a completely different use, namely as arrow poisons.

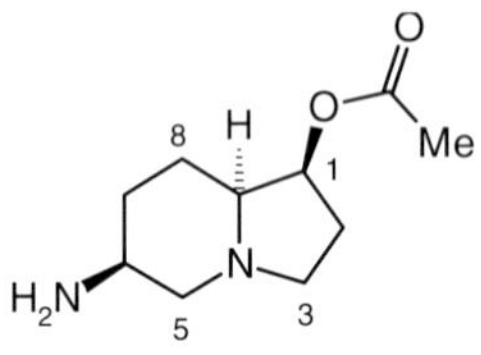

(−)-Slaframine
(abs. config.)

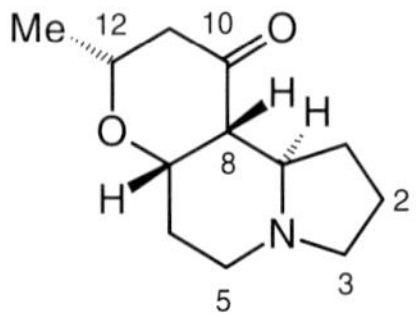

(+)-Elaeokanine E

2.2.11. Indolizidine Alkaloids

This class comprises alkaloids of the following types: slaframine (from *Rhizoctonia*, a fungus), elaeokanine E (from *Elaeocarpus* (Elaeocarpaceae)), elaeocarpine (from *Elaeocarpus*)[73], crepidamine (from *Dendrobium* (Orchidaceae)), serratinine (from *Lycopodium* (Lycopodiaceae), and securitinine (from Securinega (Euphorbiaceae)).

Norsecurinine, in structural terms actually a pyrrolizidine alkaloid, belongs, together with securitinine, to the *Securinega* alkaloids and so is mentioned at this point. Crepidamine may be an artifact. Finally, mention should be made of other alkaloids that, while resembling the isoquinoline alkaloids biogenetically, possess indolizidine or quinolizidine structural elements; among these are tylophorine (from *Cynanchum, Tylophora, Vincetoxicum* (Asclepiadaceae), and *Ficus* (Moraceae)), cryptopleurine (from *Cryptocarya* (Lauraceae)), and protostephanine (from *Stephania* (Menispermaceae))[74]. Indolizidine alkaloids have also been isolated from ants of both the genera *Monomorium* and *Solenopsis* (*cf. Chapt. 9*).

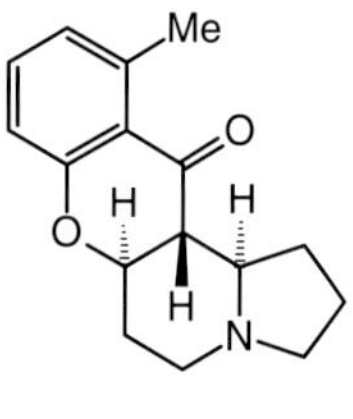

(+)-Elaeocarpine (abs. config.)

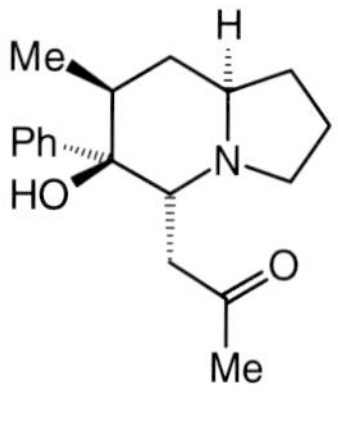

(±)-Crepidamine

(–)-Serratinine

Securitinine

(+)-Norsecurinine
(abs. config.)

(–)-Tylophorine
(abs. config.)

(–)-Cryptopleurine
(abs. config.)

Protostephanine

2.2.12. Quinolizidine Alkaloids

Lupine Alkaloids[75]

Into the category of the so-called lupine alkaloids fall the bicyclic lupinine (from both *Anabasis* (Chenopodiaceae) and *Lupinus* (Leguminosae) species), the tricyclic cytisine (a strong poison, widely distributed in the Leguminosae family), tinctorine (from *Genista* (Leguminosae)), and the tetracyclic sparteine (widely distributed in Leguminosae, also from *Chelidonium* (Papaveraceae) and Leonticaceae). (–)-Sparteine (= lupinidine) has been isolated from *Adenocarpus*, *Anagyris*, *Chamaecytisus*, *Piptanthus*, and from *Sarothamnus* species, as well as from *Lupinus luteus*; its mirror image (+)-sparteine (= pachycarpine) from species of *Cytisus*, *Genista*, *Sophora*, and from *Lupinus pusillus* (all Leguminosae) (*Figs. 2.20*, *2.21*, and *2.22*). The racemic form has also been obtained from Leguminosae species. Furthermore, matrine (from *Sophora* (Leguminosae)) and related substances, and also 4β-*O*-[(pyrrol-2-yl)carbonyl]epilupinine (from *Virgilia* (Leguminosae)), lamprolobine (from both *Sophora* and *Lamprolobium* (Leguminosae)), presumably an oxidation product of sparteine, and camoensidine (from *Maackia* (Leguminosae)), formally a ring-contraction product of sparteine, are counted in this group. As far as the absolute configuration of the

Fig. 2.20. *Bongardia chrysogonum* (L.) Griseb. (Leonticaceae). Alkaloids with quinolizidine skeletons (such as leonticine) have been found in this eastern Anatolian plant (photo *T. Gözler*).

Fig. 2.22. *A rare tunnel of Golden Chain (Laburnum anagyroides* MEDIK. *(Leguminosae)), in Bodnant Garden, Gwynnedd, Wales, UK.* The plants, especially their seeds though, contain poisonous quinolizidine alkaloids, cytisine among them. In an adult, ingestion of as little as three seeds leads to vomiting of blood, diarrhea, and disruption of motor functions. The lethal dose is *ca.* 15 – 20 seeds (photo *A. Williams*).

Fig. 2.21. *Spartium junceum* (Leguminosae; western Mediterranean). The plant contains quinolizidine alkaloids such as *N*-methylcytisine (photo *T. Gözler*).

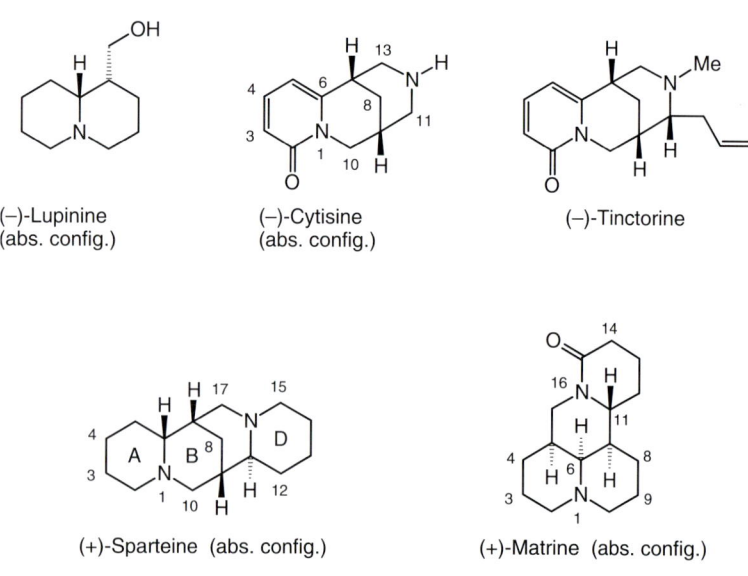

(−)-Lupinine
(abs. config.)

(−)-Cytisine
(abs. config.)

(−)-Tinctorine

(+)-Sparteine (abs. config.)

(+)-Matrine (abs. config.)

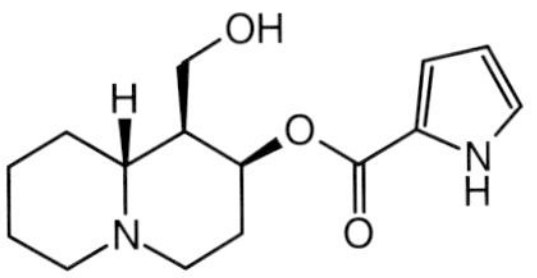

4β-O-[(Pyrrol-2-yl)carbonyl]epilupinine

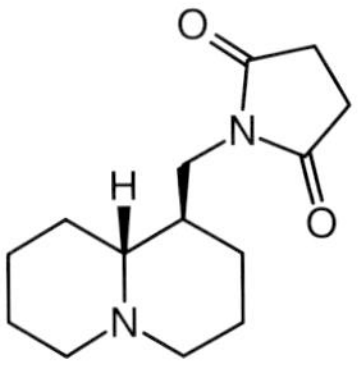

(+)-Lamprolobine
(abs. config.)

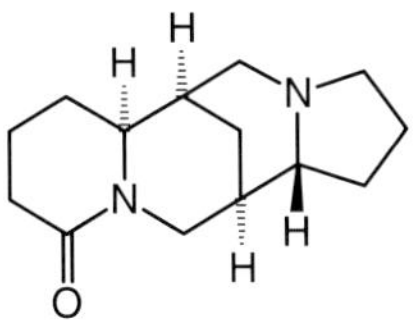

(−)-Camoensidine

basic skeleton is concerned, camoensidine (6S,7R,9R,11R) and sparteine (6R,7S,9S,11S) are mirror images of one another. Numerous natural derivatives of all these listed types exist.

Biogenetically, the lupine alkaloids are derived from the α-amino acid L-lysine, three equivalents of which, decarboxylated to cadaverine, are involved in the biosynthesis of sparteine. The biogenetic pathway has been established by using radiolabeled cadaverine and lysine in feeding experiments[76].

Nuphar Alkaloids[77]

Most *Nuphar* alkaloids incorporate a quinolizidine skeleton. Deoxynupharidine (from waterlilies of the *Nuphar* genus (Nymphaeaceae)) may be regarded as the parent compound. A number of derivatives are known, some of them also containing sulfur (such as neothiobinupharidine oxide (from *Nuphar luteum*)).

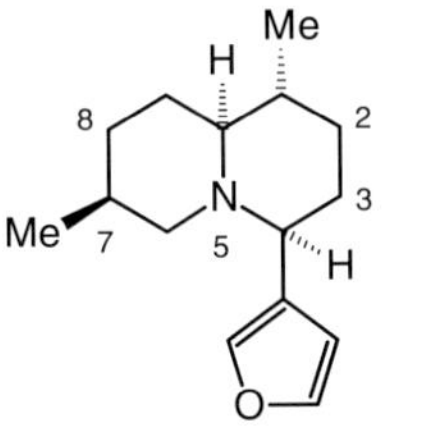

(−)-Deoxynupharidine
(abs. config.)

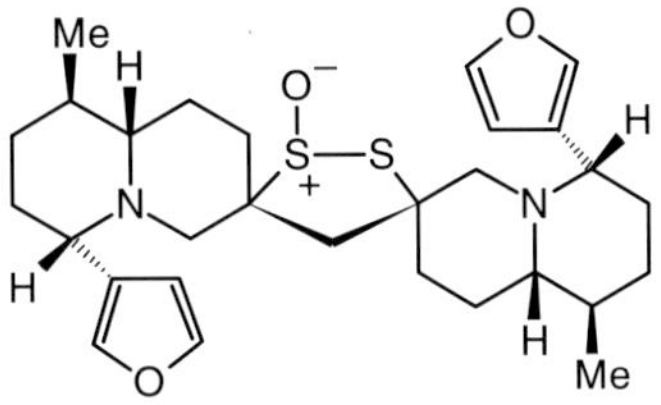

Neothiobinupharidine Oxide

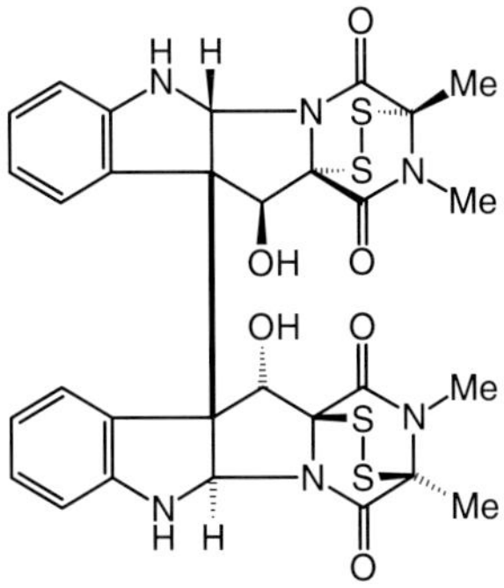

(+)-Verticillin A

Many naturally occurring bases apart from the *Nuphar* alkaloids also contain sulfur[78]. Examples are verticillin A (from *Verticillium* species, displaying antimicrobial activity against Gram-positive bacteria and mycobacteria) or zapotidine (from the seeds of *Casimiroa edulis* (Rutaceae)).

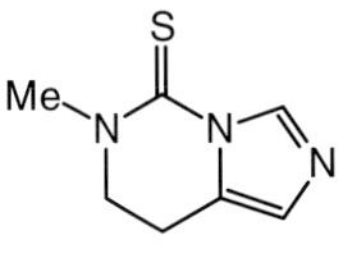

Zapotidine

Lythraceae Alkaloids

Lythrine (from *Decodon* and *Heimia* (Lythraceae)) and lythrancine III (from *Lythrum* (Lythraceae)) are depicted as the basic types of this alkaloid group.

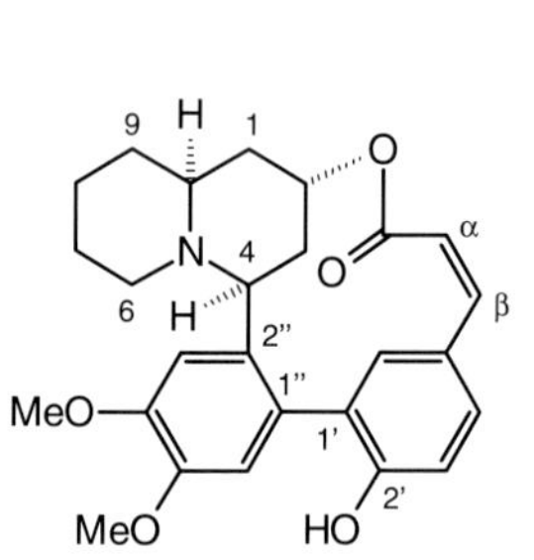

(+)-Lythrine 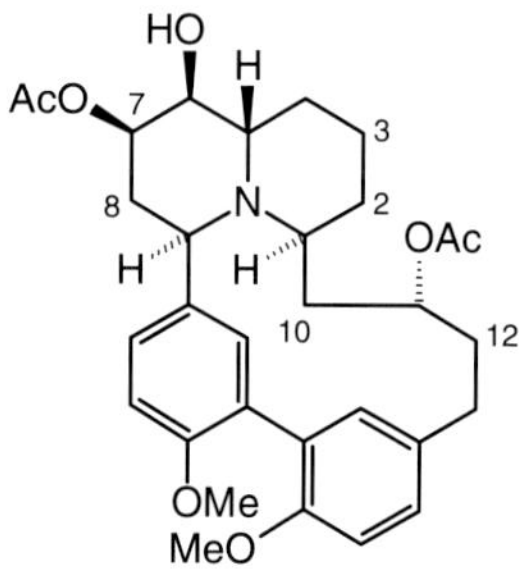   (+)-Lythrancine III

Lycopodium *Alkaloids*[79]

The alkaloids isolated from *Lycopodium* species (Lycopodiaceae) contain a broad range of different skeleton types but frequently include a quinolizidine nucleus. One representative of each of the different groups is depicted below.

Tricyclic compounds incorporating a central N-atom, which may also be described as bridged quinolizidine alkaloids, have been isolated both from insects (such as *Coccinella* species, *e.g.*, ladybug) and from plants (such as *Poranthera corymbosa, cf. Chapt. 6*).

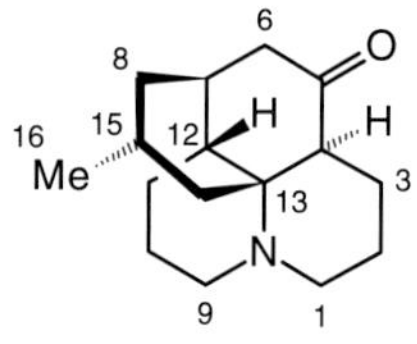

(−)-Lycopodine   (+)-Lyconnotine

(Lycopodine Type)   (Lyconnotine Type)

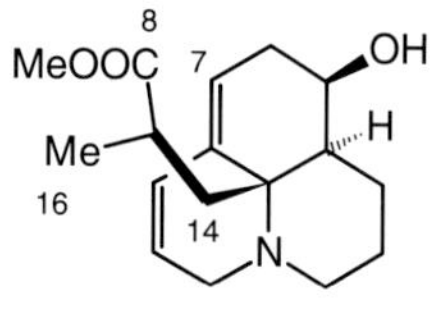

(−)-Annotine (abs. config.)

(Annotine type)

( )-Annotinine

(Annotinine type)

(−)-Lycodine (abs. config.)

(Lycodine type)

(+)-Luciduline

(Luciduline type)

Alolycopine

(Lycopine type)

(−)-Selagine

(Selagine type)

Carolinianine

(Cernuine type)

(−)-Alopecurine

(Inundatine type)

Betalains, a Group of Chromo Alkaloids[80]

Betalains certainly represent the most colorful group of alkaloids. As the main pigment of red beet (*Beta vulgaris* (Chenopodiaceae); *Fig. 2.23*), of the *Bougainvillea* species (Nyctaginaceae; *Fig. 2.24*) and of the pokeweed (pokeberry, poke, inkberry, pigeonberry; *Phytolacca americana* (Phytolaccaceae)), betanine, the aglycon of which is betanidine, enjoys particular prominence. Betanidine constitutes the structural basis of all betacyanines. A second group of natural dies is represented by the betaxanthines. These are yellow-orange colored compounds, for which indicaxanthine (from the cactus fruit *Opuntia ficus-indica* (Cactaceae)) serves as a typical representative. The essential component of these iminium salts is betalamic acid. Its ring-enlargement product, muscaflavine, is one of the pigments of Fly Agaric (*Amanita muscaria* (Amanitaceae)). A large number of iminium compounds are also derived from betalamic acid, arising from condensation of the aldehyde functionality with different α-amino acids (proteinogenic and non-proteinogenic) or with primary amines. Condensation of betalamic acid with an amine results in a significant shift of the UV absorption maximum to longer wavelengths ('bathochromic shift') in the product (such as betanine); (*Fig. 2.25*).

Fig. 2.23. *The red color of the Red Beet (beetroot, Beta vulgaris* L. var. *conditiva (Chenopodiaceae)) can be traced to betanine (photo M. Stoll in Der Brückenbauer)*

Fig. 2.24. *The violet pigment of the leafless climbing plant 'Bougainvillea' (Nyctaginaceae) belongs to the betanines (photo A. Kögel)*

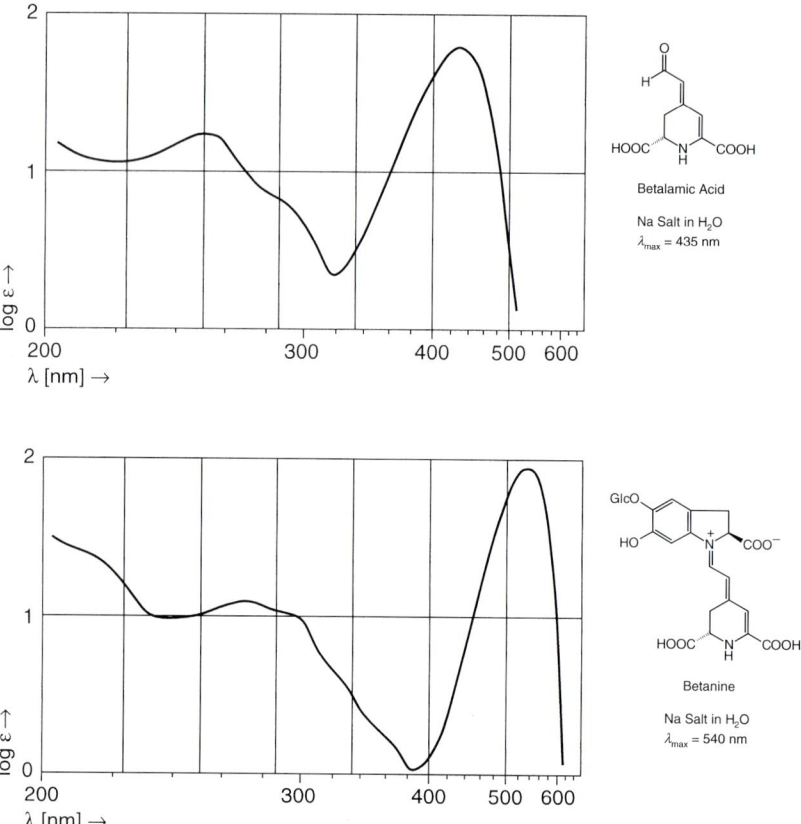

Fig. 2.25. *UV/VIS Spectra of the sodium salts of betalamic acid* (top) *and betanine* (below) *in water*

Betalains are very widely distributed in flowers, fruits, and in other colored plant organisms. Because of this, the pigments have been used as chemotaxonomic markers in the Caryophyllales, which are accordingly subclassified into the betalain-producing Chenopodiineae (Aizoacea, Amaranthaceae, Basellaceae, Cactaceae, Chenopodiaceae, Didiereaceae, Nyctaginaceae, Phytolaccaceae, Portulacaceae) and into the anthocyanine-producing Caryophyllineae (Caryophyllaceae, Molluginaceae). Biogenetically, betalamic acid derives from L-DOPA (L-3,4-dihydroxyphenylalanine (*Scheme 2.12*).

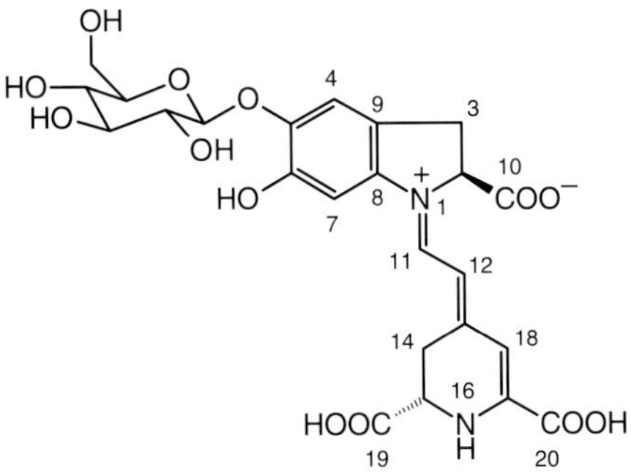

Betanine (a Betacyanine)
(abs. config.)

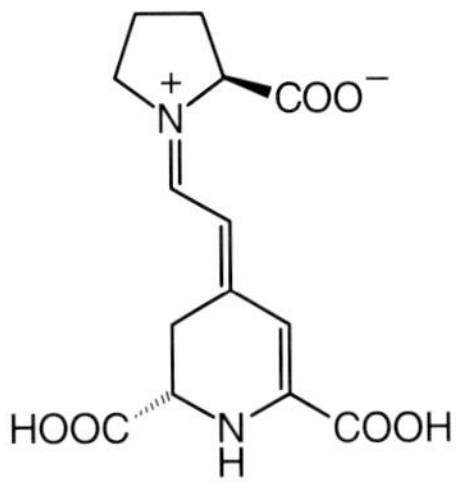

Indicaxanthine (a Betaxanthine)
(abs. config.)

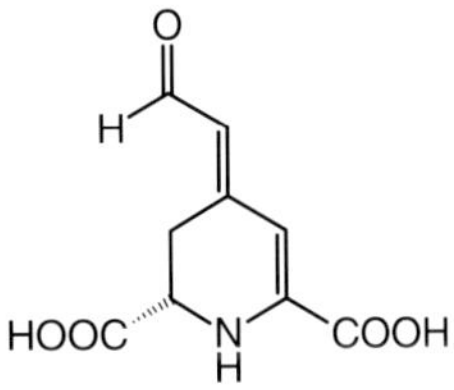

Betalamic acid

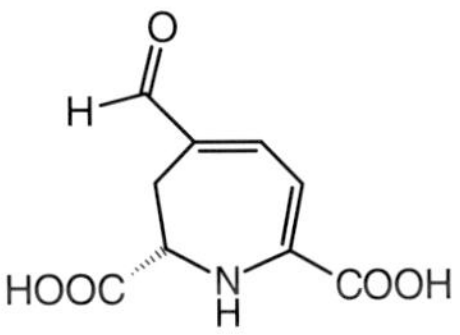

Muscaflavine

UV (H_2O) λ_{max} = 420 nm (ε = 58.4)

Scheme 2.12

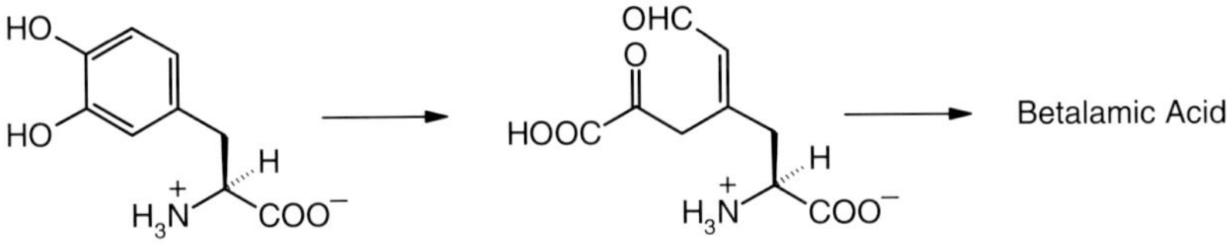

L-DOPA

Betalamic Acid

2.2.13. Pyrazine Alkaloids

As shown in *Scheme 2.13*, it is generally accepted that amides of α-amino acids (such as **A**) condense on biogenesis with 1,2-diones of type **B** to yield (after reduction) the corresponding pyrazines **C**.

Scheme 2.13

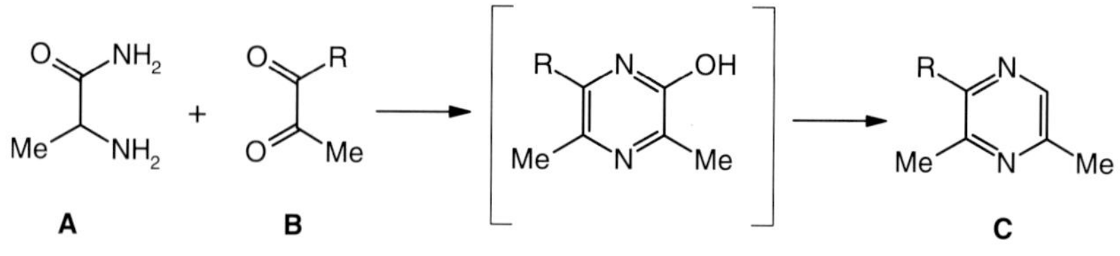

R = Alkyl

Pyrazine alkaloids have been isolated from numerous insect species, often serving as attractant pheromones. Examples are 3-ethyl-2,5-dimethylpyrazine (from *Acromyrmex*, *Atta*, *Myrmica*, *Tetramorium* (Myrmicinae), *Iridomyrmex* (Dolichoderinae) → ants) or 3-isopentyl-2,5-dimethylpyrazine from various genera of *Eumenidae* and *Sphecidae* (both wasps). 2-Methoxy-3-methylpyrazine has been isolated from *Metriorrhynchus rhipidius* (beetle).

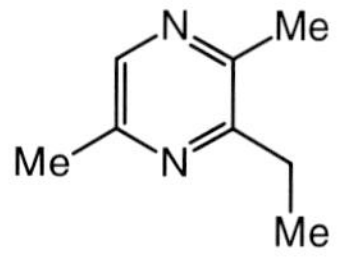

3-Ethyl-2,5-dimethyl-
pyrazine

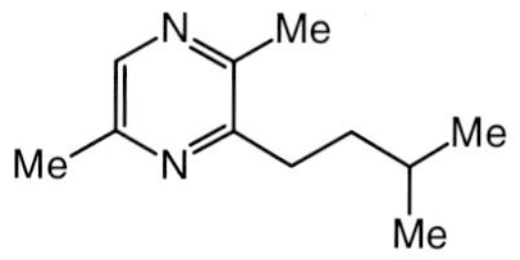

3-Isopentyl-2,5-dimethyl-
pyrazine

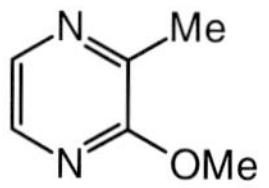

2-Methoxy-3-methyl-
pyrazine

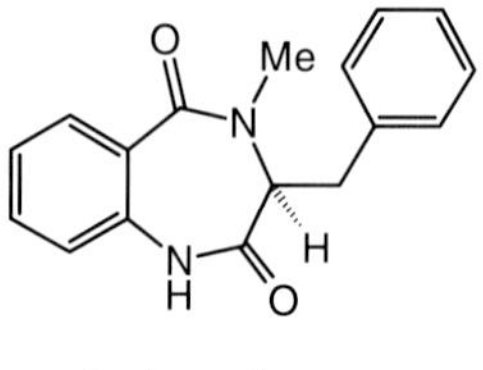

A small, related group of *ca.* 20 compounds contains the benzodiazepine skeleton (*e.g.*, cyclopeptine from *Penicillium* species)[81]. Such compounds have, up to now, exclusively been isolated from three microorganism genera (*Penicillium*, *Streptomyces*, and *Aspergillus*). Biogenetically, they arise from the reaction between anthranilic acid and an α-amino acid (such as L-phenylalanine, L-tyrosine, L-tryptophan, or L-glutamine).

Cyclopeptine

2.2.14. Purine Bases[82]

Purines undoubtedly belong among the borderline cases of alkaloid-like natural substances. Together with the bases adenine and guanine, which occur freely in nature, and which partly constitute the mononucleotides (which are not included in the scope of this work), a number of compounds of unambiguously alkaloidal character occur in both the plant and animal kingdoms. To this group belong the hydroxylated and methylated derivatives of purine itself: xanthine (contained in animal secretions (urine) and in blood, and also as a plant product accompanying caffeine), uric acid (the end product and principal excretiatory product of purine degradation in humans, other primates, birds, and insects), and hypoxanthine (similarly widely distributed in the animal and plant kingdoms).

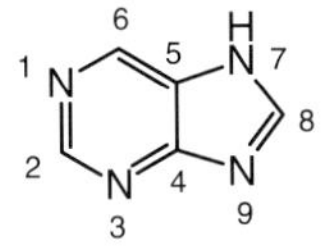

Purine 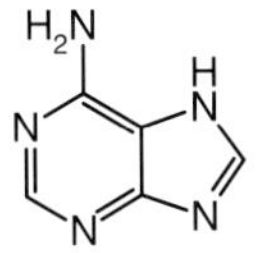   Adenine 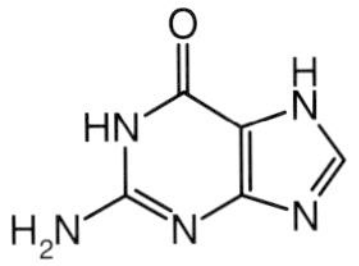   Guanine

Theophylline (= 1,3-dimethylxanthine) is present in small quantities in the leaves of tea (*inter alia, Camellia sinensis (= Thea viridis)* (Theaceae (= Camelliaceae))). Theobromine is the most important alkaloid of the cocoa bean (from *Theobroma cacao* and other *Theobroma* species (Sterculiaceae)), while caffeine (also known as theine)[83] can be isolated from coffee beans (*Coffea arabica* (Rubiaceae), up to 1.9%), leaf tea (up to 5%), cocoa beans, South American maté (*Ilex paraguariensis* (Aquifoliaceae)), the cola nut (*Cola nitida, C. accuminata* (Sterculiaceae)), and from a great variety of other plants (*cf. Chapts. 11.2 – 11.4*).

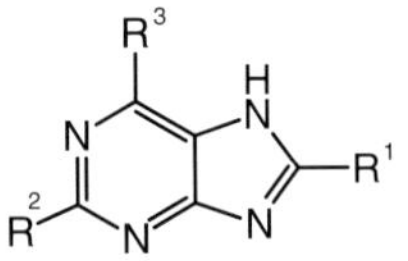

$R^1 = R^2 = H, R^3 = OH$: Hypoxanthine

$R^1 = H, R^2 = R^3 = OH$: Xanthine

$R^1 = R^2 = R^3 = OH$: Uric Acid

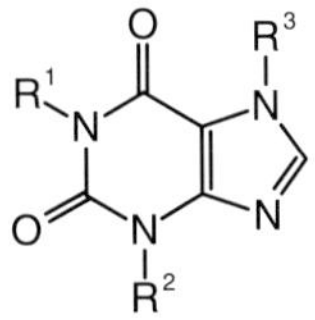

$R^1 = R^2 = Me, R^3 = H$: Theophylline

$R^1 = H, R^2 = R^3 = Me$: Theobromine

$R^1 = R^3 = Me, R^2 = H$: Paraxanthine

$R^1 = R^2 = R^3 = Me$: Caffeine

2.2.15. Pteridines

Derivatives with a pteridine basic structure are found in great numbers in the animal kingdom, particularly in the wing pigments of insects, especially butterflies (*Lepidoptera*). In this context, the following orders are worth mentioning: *Hymenoptera* (common among Formicinae (ants), *Vespidae* (wasps), and *Apidae* (bees)), *Lepidoptera* (butterflies, especially their occurrence in the *Pieridae, Bombycidae* (silkworm moths), *Pyralidae* (snout and grass moths), *Diptera* (two-winged insects, especially *Drosophilidae*, *Trypetidae*, *Calliphoridae*), *Orthoptera* (the cockroach order of insects, especially *Acrididae* (*Locustii*)), *Hemiptera*, and *Coleoptera* (beetles). Important color pigments are: leucopterin (white)[84], xanthopterin (yellow), chrysopterin (yellow), erythropterin (orange-red), pterorhodine (red), ekapterin (colorless), lepidopterin (colorless), sepiapterin, and biopterin[4, 82] (*Figs. 2.26 – 2.31*).

Fig. 2.26. *Araschnia levana* L. (Map Butterfly), *spring generation* (the summer generation is black/brown → seasonal dimorphism) *is found in damp regions with stinging nettle colonies* (photo *A. Krebs*)

Fig. 2.27. *Zerynthia polyxena* D. & S., *Southern Festoon, on its food plant Aristolochia clematitis* L. (Aristolochiaceae), *which contains 1-benzyl-1,2,3,4-tetrahydroisoquinoline alkaloids.* Occurrence particularly in northern Italy, especially the Po plain. The wing pigments contain pteridines (photo *J.-P. Zuber*).

Fig. 2.28. *Cupido osiris* MEIGEN, *Osiris Blue; male, upper side.* A Mediterranean species, occurring in Valais/Switzerland at altitudes up to 2,000 m. The wing pigments contain pteridines (photo *D. Jutzeler*).

Fig. 2.29. *Plebejides pylaon trappi* VRTY., *Spanish Blue.* Mating on their sole food plant *Astragalus exscapus* L. (Leguminosae); left male, right female; in Switzerland found only in Valais. The wing pigments contain pteridines (photo *D. Jutzeler*).

Fig. 2.30. *Pyronia tithonus* L., *Gatekeeper; male, upper side.* Found in southern and western Switzerland up to altitudes of 1,000 m. The wing pigments contain pteridines (photo *D. Jutzeler*).

Fig. 2.31. *Tyria jacobaeae* L., *cinnabar moth, on the alkaloid-producing ragwort* (*Senecio jacobaea* L. (Compositae)). The wing pigments contain pteridines (photo *A. Krebs*).

Biogenetically, the pteridines are derived from purines and are essential components of several coenzymes (such as folic acid derivatives). 'Pterin' itself refers to 2-amino-3,4-dihydro-4-oxopteridine. Derivatives that are hydroxylated in the pterin skeleton, particularly leucopterin, co-exist in the corresponding tautomeric keto forms.

Pteridine Leucopterin Xanthopterin Pterorhodine

Chrysopterin Erythropterin Pterin

75

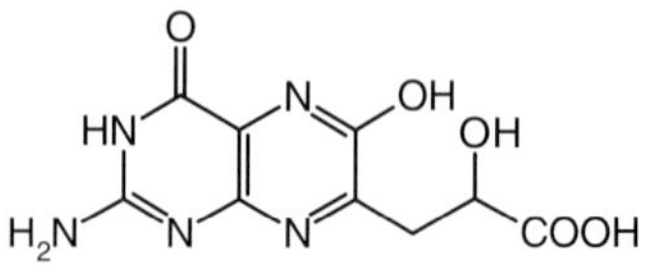

Ekapterin

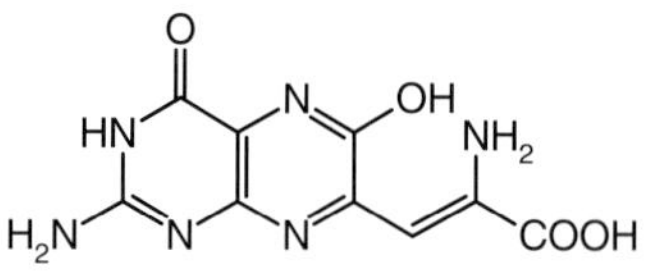

Lepidopterin

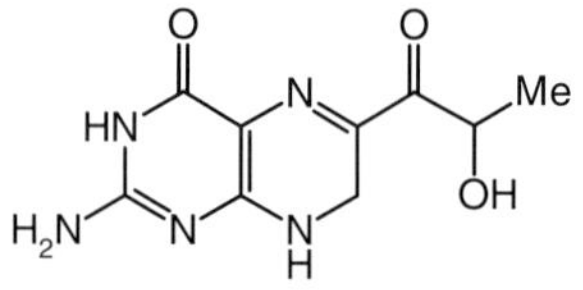

Sepiapterin

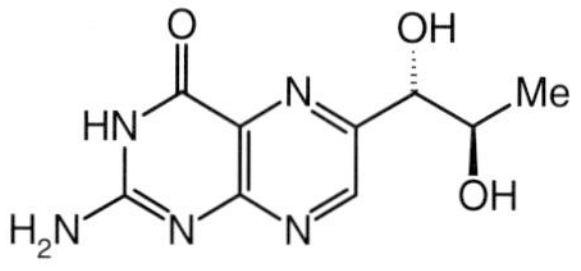

(–)-Biopterin (abs. config.)

The significance of pterins (especially that of 5,6-dihydro- and 5,6,7,8-tetrahydro derivatives) as components of Fe complexes in biological redox processes is great, although not yet fully understood in every detail[85].

2.3. Alkaloids with Exocyclic N-Atoms and Aliphatic Amines

Alkaloids possessing exocyclic N-atoms are differentiated into six categories: *Erythrophleum* alkaloids, phenylalkylamine derivatives, benzylamine-type alkaloids, colchicines, khat alkaloids, and muscarines.

2.3.1. *Erythrophleum* Alkaloids[86]

The *Erythrophleum* alkaloids, with *ca.* 15 members (*e.g.*, cassaine from *Erythrophleum* (Leguminosae)), are derivatives of aminoethanol. These substances display *Digitalis*-like activity on the heart and are also powerful local anesthetics.

2.3.2. Phenylalkylamine Derivatives[35, 87]

The phenylalkylamine alkaloids (*inter alia* ephedrine from plants of the genus *Ephedra* (Ephedraceae)) include a number of compounds, the origin of which may be explained by *Hofmann* degradation of the corresponding isoquinoline alkaloids (*e.g.*, uvariopsamine from *Uvariopsis* (Annonaceae)). *Ephedra* drugs have been an ingredient of Chinese folk medicine for centuries. (–)-Ephedrine itself is an effective oral sympathomimetic.

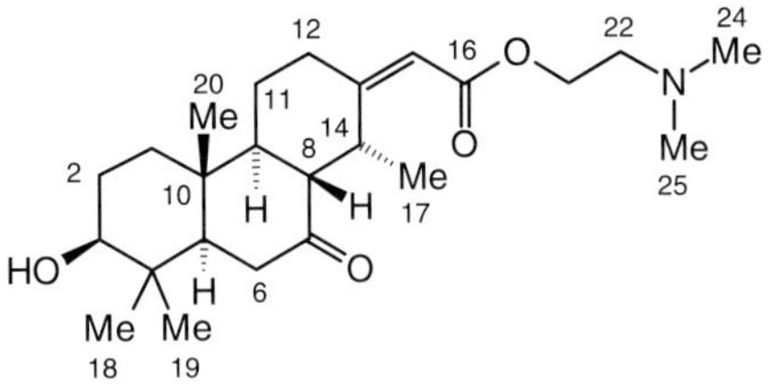

(–)-Cassaine (abs. config.)

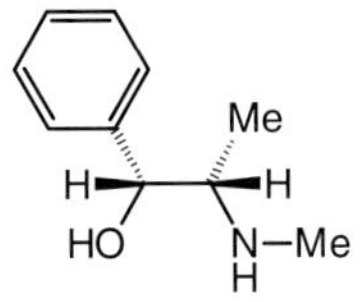

(–)-Ephedrine (abs. config.)

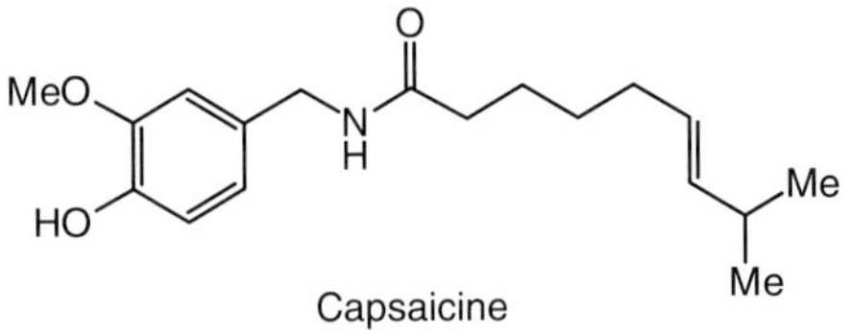

Capsaicine

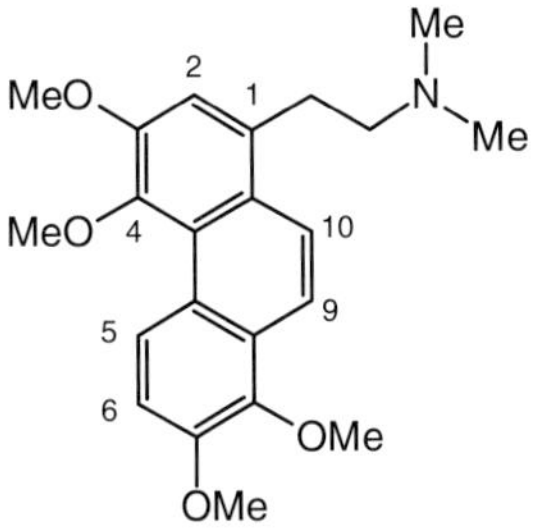

Uvariopsamine

2.3.3. Benzylamine-Type Alkaloids

Capsaicine (from both *Piper* (Piperaceae) and *Capsicum* (Solanaceae) species) serves as an example of a benzylamine-type alkaloid[88].

2.3.4. Colchicines[89]

The colchicines constitute a further group possessing exocyclic N-atoms. Although listed at this point, colchicines are biogenetically very closely related to the isoquinoline alkaloids.

The very poisonous colchicine, from the meadow saffron *Colchicum autumnale* (Liliaceae) (*Fig. 2.32*) and from other *Colchicum* species, possesses both one stereogenic center at C(7) and a chirality axis, since the two rings A and C are not positioned in coplanar fashion (→ atropisomerism). In naturally occurring (–)-(aR,7S)-colchicine (**A**), the two rings are oriented in a clockwise manner, while in the less energetically favored conformer **B** ((+)-(aS,7S)-colchicine) they are oriented anticlockwise (*Scheme 2.14*).

Fig. 2.32. *Colchicum autumnale* L., *meadow saffron* (Liliaceae). *a)* Flower, September 1983, *b)* Fruit, June 1984. The plant contains the poisonous colchicine (photo *M. Hesse*).

Scheme 2.14

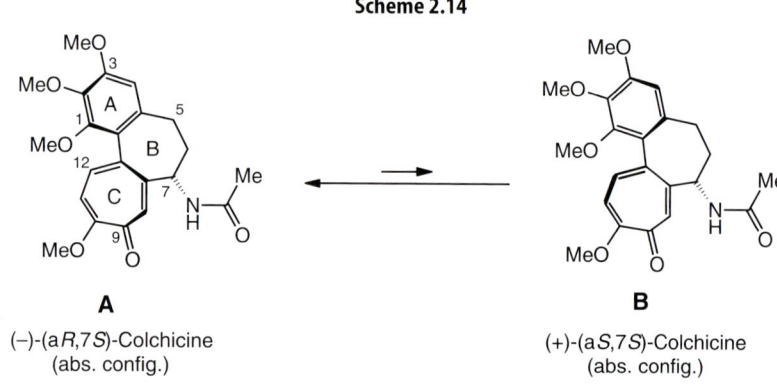

A

(−)-(a*R*,7*S*)-Colchicine
(abs. config.)

B

(+)-(a*S*,7*S*)-Colchicine
(abs. config.)

The (aR)/(aS) rearrangement rate, measured for analogues of colchicine in CDCl$_3$ and C$_6$D$_6$ at 22° C, lies in the range of 10^{-4} to 10^{-5} [s^{-1}], which corresponds to an activation energy of *ca.* 92 – 100 kJ/mol. According to ^{1}H-NMR studies, the equatorial position of the acetamido group is energetically more favored by only 4 – 12 kJ/mol compared to the axial position; nevertheless, this is the decisive factor in controlling the equilibrium between the (aR)- and the (aS)-conformations. The isomerization rate in general depends essentially on the magnitude of the steric interaction between the substituents at C(1) and H–C(12)[90]. Other members of this small group of alkaloids are allocolchicine (from *C. cornigerum*) and androbiphenyline (from *C. ritchii*).

2.3.5. Khat

Khat or Kath (also known variously as al-Qat, quat, chat, jat, tschatt, miraa, and many other names) is the shrub *Catha edulis* (Celastraceae), which grows in the highland valleys of Somalia and Yemen. Important areas of cultivation also exist in Ethiopia, between Diredana and Harar, and in Kenya, on the eastern slopes of Mount Kenya. Its freshly cut leaves are chewed, giving rise to euphoria and gentle arousal (mild intoxication); tiredness disappears, physical effort and talking become easier, and hunger is suppressed. However, routine use – the rule in Yemen – attacks the stomach lining and leads to general loss of appetite, impotence, and constipation – '*a euphoria that cripples an entire people*' (*C. E. Buchalla*)[91] (*Fig. 2.33*). Khat is first mentioned in *ca.* 1300. Its principle active ingredient is (–)-(*S*)-cathinone[92], which, pharmacologically, exerts a stimulatory effect (like that of (+)-amphetamine) on the central nervous system. Accompanying alkaloids are merucathine ((3*R*,4*S*)-4-amino-1-phenylpent-1-en-3-ol) and its other isomers and oxidation products. Together, these compound types are termed 'khatamines'[93]. Finally, *ca.* 20 other compounds have been isolated from khat (so-called 'cathedulines'), which are structurally related to the pyridine ester alkaloids evonine (from the *Evonymus* species Cardinal's Hat or Spindle Tree (Celastraceae); *Fig. 2.34*) and maytoline (from *Maytenus ovatus* (Celastraceae)). Catheduline E3 constitutes a typical catheduline alkaloid, as the name implies. The structural similarity of evonine, maytoline, and catheduline – composed of nicotinic acid and a bicyclic sesquiterpene moiety – is interesting in terms of chemotaxonomy[94], since all the substances have been isolated from plants of the Celastraceae family.

Allocolchicine (abs. config.)
R^1 = MeOOC, R^2 = H

Androbiphenyline (abs. config.)
R^1 = MeO, R^2 = OH

Fig. 2.33. *Khat.* Once having reached adulthood, Yemeni men, like this nobleman, traditionally chew khat (*Catha edulis* (Vahl) Forsk. *ex* Endl. (Celastraceae)) from about late afternoon. The rest of the day is spent floating in a mild state of euphoria. The essential active substance is cathinone (*W. M. Weiss, K. M. Westermann, Der Basar* (The Bazar), Brandstätter, Vienna, 1994; edn. in English, Thames and Hudson Ltd., London, 1998).

Fig. 2.34. *Fruit of the Cardinal's Hat, or Spindle Tree* (*Evonymus europaeus* L. (Celastraceae)). The plant contains the alkaloid evonine, among others (photo *C. Werner*).

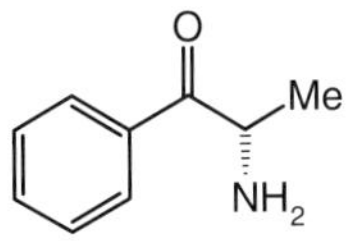

(−)-(S)-Cathinone (abs. config.)

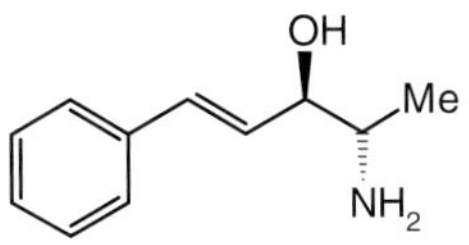

Merucathine

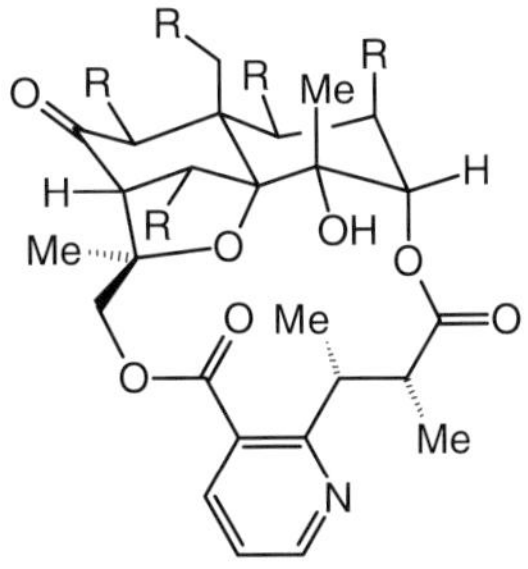

(+)-Evonine

R = AcO

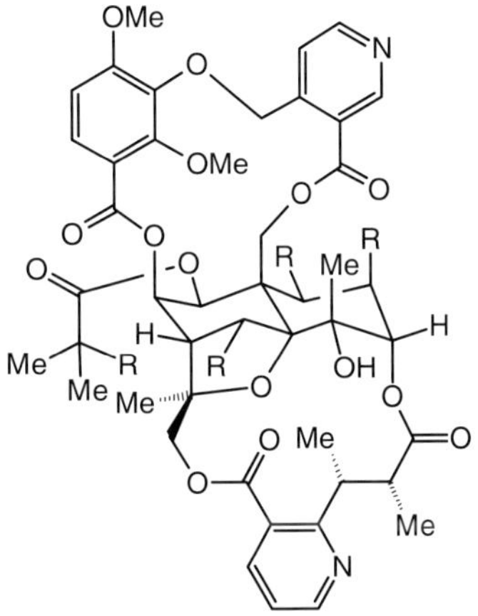

(−)-Catheduline E3

R = AcO

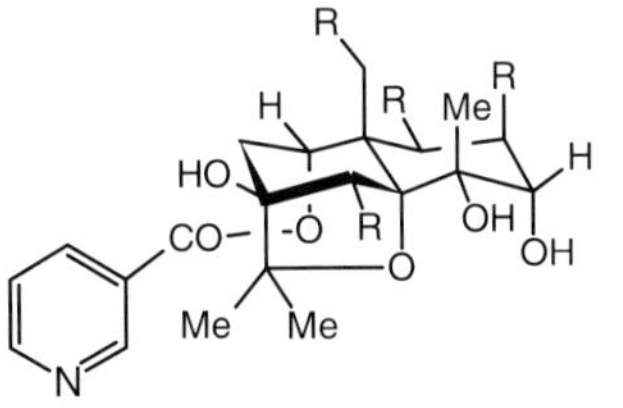

(0)-Maytoline

R = AcO

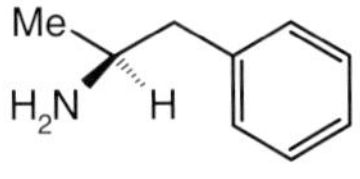

(+)-(S)-Amphetamine

2.3.6. Muscarines

The N-atom is also in an exocyclic position in (+)-(2S,3R,5S)-muscarine and in its diastereoisomers (−)-(2S,3R,5R)-allomuscarine, (+)-(2S,3S,5S)-epi-muscarine, and (+)-(2S,3S,5R)-epiallomuscarine (all isolated from *Amanita muscaria, A. pantherina* (Amanitaceae), *Inocybe* (Cortinariaceae), or from *Clitocybe* (Tricholomataceae) species)[95] (*cf. Chapt. 11.3*).

(+)-Muscarine
(abs. config.)

(−)-Allomuscarine
(abs. config.)

(+)-Epimuscarine
(abs. config.)

(+)-Epiallomuscarine
(abs. config.)

2.4. Putrescine, Spermidine, and Spermine Alkaloids[96]

As mentioned previously, the three aliphatic bases putrescine, spermidine, and spermine count among the ubiquitously occurring biogenic amines: their derivatives (mostly stemming from fatty acids and cinnamic acids) are also widely distributed, but their occurrence becomes rarer with increasing structural complexity (additional rings *etc.*). Typical examples of such complex structures are the putrescine derivative paucine (from *Pentaclethra* (Leguminosae)), the spermidine derivative inandenin-12-one (from *Oncinotis* (Apocynaceae); *Fig. 2.35*), lunarine (from *Lunaria biennis* (Cruciferaceae); *Fig. 2.36*), codonocarpine (from *Codonocarpus* (Gyrostemonaceae)), and the spermine derivatives chaenorhine (from *Chaenorhinum* (Scrophulariaceae); *Fig. 2.37*), verbascenine (from *Verbascum* species (Scrophulariaceae); *Fig. 2.38*), and aphelandrine (from *Aphelandra* species (Acanthaceae); *Fig. 2.39*). Other polyamine derivatives of natural origin are described in *Chapts. 3.5, 6.5,* and *9.6*.

Fig. 2.35. *Oncinotis tenuiloba* Stapf, *belonging to the Apocynaceae family, is a creeper found in Kenya. Its spermine and spermidine alkaloid contents (inandenines, for example) are relatively high in all parts of the plant (photo G. M. Mungai).*

Paucine

Inandenin-12-one

(+)-Lunarine (abs. config.)

Codonocarpine

(+)-Chaenorhin
(abs. config.)

(−)-Verbascenine
(abs. config.)

(+)-Aphelandrine
(abs. config.)

Fig. 2.36. *The biennial Lunaria biennis* MOENCH *(Cruciferae) (honesty, money plant), ca. 1 m high, with violet (occasionally white) flowers in spring, contains the macrocyclic spermidine alkaloids lunarine and lunaridine. The seedcases are popular in dried flower arrangements* (photo *M. Hesse*).

Fig. 2.37. *The Dwarf Snapdragon (USA, Canada), or Small Toadflax (UK) (Chaenorhinum minus L. (Scrophulariacae)), an inconspicuous garden weed, synthesizes macrocyclic spermine alkaloids such as chaenorhine* (photo *M. Hesse*)

Fig. 2.38. *Common Mullein (Verbascum (Scrophulariaceae)), a biennial plant containing spermine/cinnamic acid derivatives in the roots and throughout the body. Pictured is a flowering V. thapsus L. in a rural Swiss garden* (photo *M. Hesse*).

Fig. 2.39. *Aphelandra fuscopunctata* MARKGRAF (Acanthaceae), *indigenous to northern South America and, like some other Aphelandra species, grown as a houseplant in Europe, contains macrocyclic spermine alkaloids of the aphelandrine and chaenorhine types in its roots (photo C. Werner)*

2.5. Peptide Alkaloids[97]

One peptide alkaloid subgroup – the ergot alkaloids – has already been described under the indole bases. Here, we should also introduce the Rhamnaceae (buckthorn) alkaloids, of which two main structural types should in turn be pointed out: first, integerrine (from *Ceanothus* (Rhamnaceae)), made up of the amino acids tryptophan, *N,N*-dimethylvaline, and phenylserine, plus 4-hydroxystyrylamine as an amine component; second, mucronine A (from *Zizyphus* (Rhamnaceae)), comprising the amino acids phenylalanine, isoleucine, and a condensation product of *N,N*-dimethylalanine and 4-methoxystyrylamine.

Integerrine (abs. config.)

Mucronine A

2.6. Terpene and Steroid Alkaloids

2.6.1. Diterpene Alkaloids[98]

A few monoterpene and sesquiterpene alkaloids have already been introduced elsewhere (*cf.* pyrrolidine, piperidine, and *Nuphar* alkaloids). The diterpene alkaloids encompass two large groups, containing either a C_{20} or a C_{19} skeleton. The four basic skeletal types comprise the ring systems of *1*) *veatchine* (such as veatchine itself, isolated from *Garrya* (Garryaceae)), *2*) *atisine* (*e.g.*, atisine from *Aconitum* (Ranunculaceae), among others), *3*) *lycoctonine* (as in aconitine[99] from *Aconitum napellus* (Ranunculaceae); *Figs. 2.40 – 2.42*), and *4*) heteratisine (such as heteratisine itself from *Aconitum*). The truly numerous naturally occurring diterpene alkaloids (*ca.* 200 representatives) almost all derive from substituent variations of these basic types.

Fig. 2.40. *Aconitum napellus* ssp. *compactum* (Rchb.) Gayer, *aconite, or monkshood*. One of its toxic components is the alkaloid aconitine (photo *M. Hesse*).

Fig. 2.41. *Aconitum napellus* L. (Ranunculaceae). Monkshood is one of the most poisonous plants native to western Europe, even in virgin shoots (© Karl Blossfeldt Archiv – Ann and Jürgen Wilde, Köln/Prolitteris Zürich, 1997).

Fig. 2.42. *Aconitum vulparia* Rchb., *or yellow monkshood* (flowers and fruit). The plant is just as poisonous as its blue cousin (photo *M. Hesse*).

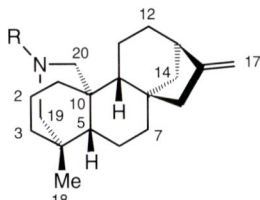

Veatchine Skeleton

(−)-Veatchine (abs. config.)

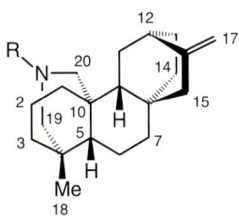

Atisine Skeleton

(−)-Atisine (abs. config.)

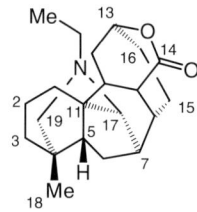

Lycoctonine Skeleton

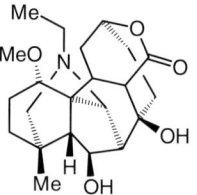

(+)-Aconitine (abs. config.)

Heteratisine Skeleton

(+)-Heteratisine (abs. config.)

2.6.2. *Daphniphyllum* Alkaloids

In Japan, *Daphniphyllum humile* (Daphniphyllaceae), after ingestion in cattle-feed, proved itself to be a cause of potentially fatal cattle sickness (jaundice, colic, photophobia). About 40 alkaloids have been isolated from *Daphniphyllum* species; they belong to six different ring system types. One example, daphniphylline, is shown together with its biogenesis from squalene (*Scheme 2.15*). The C_{30} skeleton is preserved in this ring system, whereas some other representatives of this substance class possess significantly altered ring systems with smaller numbers of C-atoms[100].

2.6.3. *Taxus* Alkaloids[101]

The evergreen yew tree (*Taxus baccata* (Taxaceae); *Fig. 2.43*), widely grown in European gardens and parks, contains a poison deadly to humans, dogs, and horses, but not, however, to ruminants. Roughly 50 alkaloids have been isolated from *T. baccata* and from other *Taxus* species, as well as from *Austrotaxus spiccata*. Structurally, these compounds can be viewed as mono- or oligoesters of diterpene alkaloids. As shown in *Scheme 2.16*, *Taxus* alkaloids are derived from geranylgeranyl pyrophosphate through the intermediate **A**, which is subsequently converted into the basic ring systems **B** – **D**.

Scheme 2.15

Squalene
$C_{30}H_{50}$

(+)-Daphniphylline
$C_{30}H_{46}NO_4 \cdot COMe$

Fig. 2.43. *Berry of Taxus baccata* L. (Taxaceae), *yew.* All parts of this European and South-west Asian ever-green tree are poisonous, except for the red fruit mantle (Arillus). The plant's toxicity is mainly due to the *Taxus* alkaloids (photo *C. Werner*).

Scheme 2.16

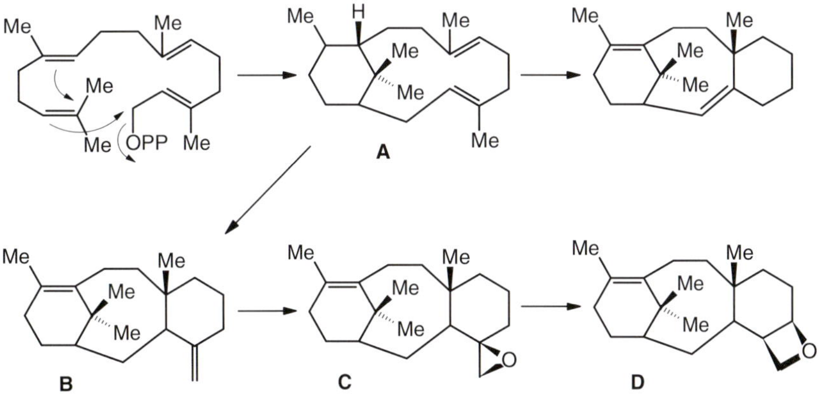

All *Taxus* alkaloids known to date are variants on the above fundamental structures. Shown are the well-known ( )-taxol[102] as well as taxine A to-gether with the two direct terpene derivatives baccatin and taxusin.

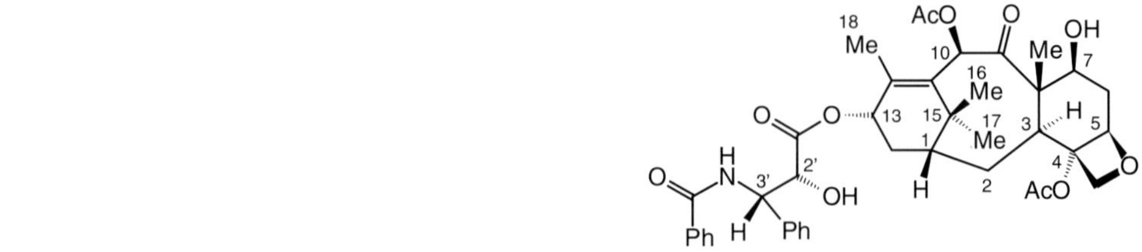

( )-Taxol (abs. config.)

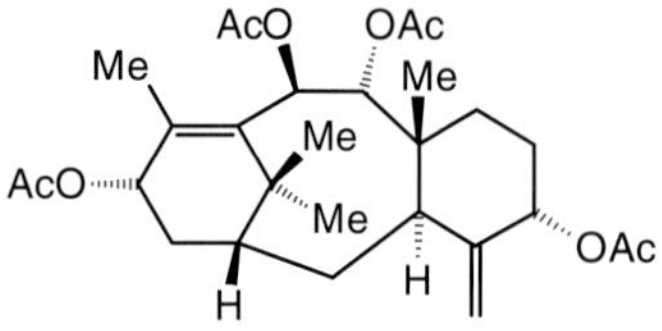

Taxusin (abs. config.)

Taxine A

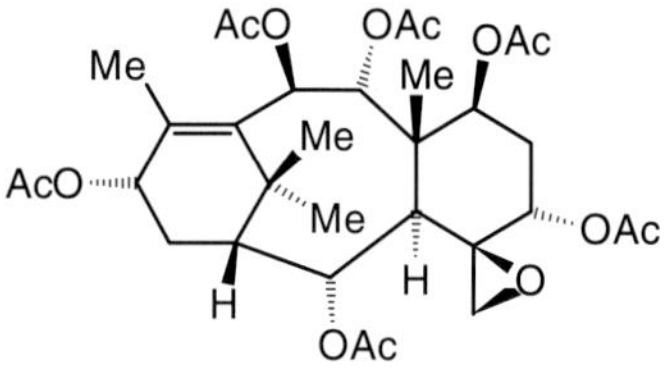

Baccatin

2.6.4. Steroid Alkaloids[103]

Steroid alkaloids constitute a class of compounds distinguished by their (intact or modified) steroid skeletons in conjunction with one or several amine functionalities. The N-atoms may in principle be present either as parts of rings, or attached to the main skeleton as side chains. Since the steroid ring system is common to all these bases, they are grouped together into a single class irrespective of the nature and number of incorporated N-atoms.

Steroid alkaloids occur both in the animal and in the plant kingdoms. A variety of skeleton types can be identified, depending both on the nature of the steroid part and the means of nitrogen incorporation. One representative example of each type is given here, namely funtumine (from *Funtumia* (Apocynaceae)), paravallarine (from *Paravallaris* (Apocynaceae)), terminaline (from *Pachysandra* (Buxaceae)), holarrhimine (from *Holarrhena* (Apocynaceae)), conessine (from *Holarrhena*), veralkamine (from *Veratrum* (Liliaceae)), solasodine (from *Solanum* (Solanaceae); *Fig. 2.44*), rubijervine (from *Veratrum*), and solanocapsine (from *Solanum* species).

Jervine (from *Veratrum* and *Zigadenus* (Liliaceae)) and verticine (from *Fritillaria* (Liliaceae)) both possess 'C-nor-D-homo' steroid ring systems. The *Buxus* alkaloid buxenine G contains a B ring enlarged to include C(19), while cycloxobuxidine F (from *Buxus*) is special due to its extra cyclopropane ring.

Fig. 2.44. *Solanum nigrum* L. (Solanaceae), *Black Nightshade, grows both in Europe and Asia on wasteground and waysides.* The plant contains a number of poisonous steroid alkaloids (photo *S. Johne*).

Funtumine (abs. config.)

(−)-Paravallarine (abs. config.)

Terminaline (abs. config.)

Holarrhimine (abs. config.)

89

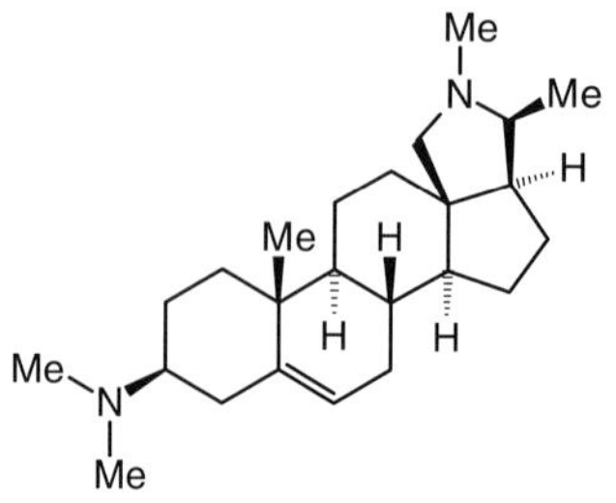

(+)-Conessine (abs. config.)

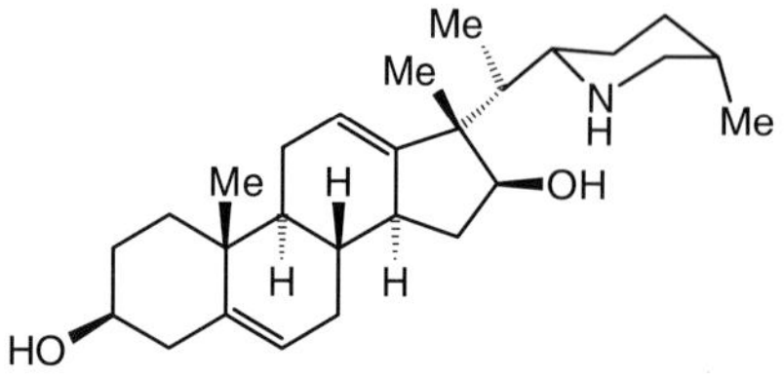

(−)-Veralkamine (abs. config.)

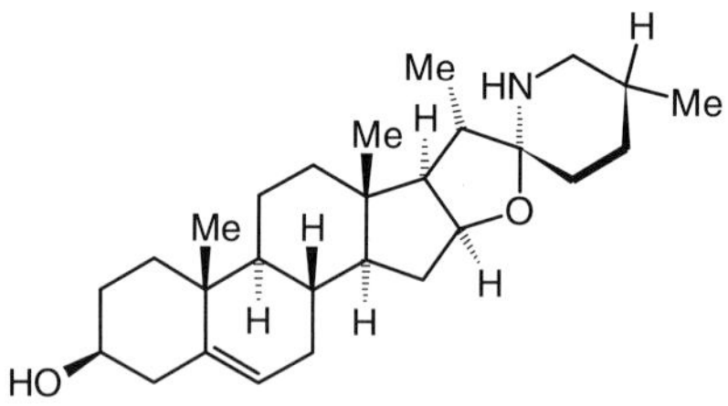

(−)-Solasodine (abs. config.)

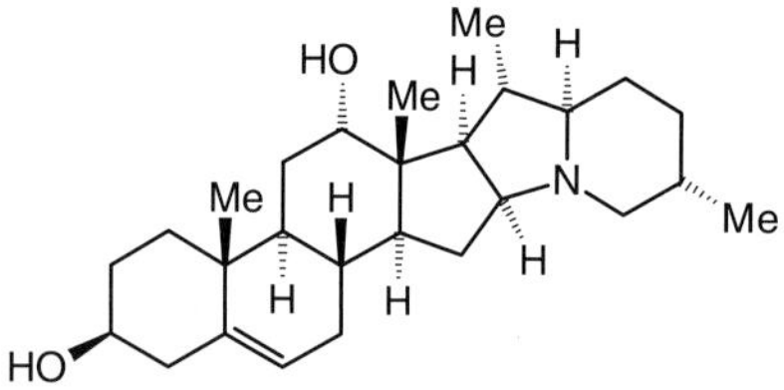

(+)-Rubijervine (abs. config.)

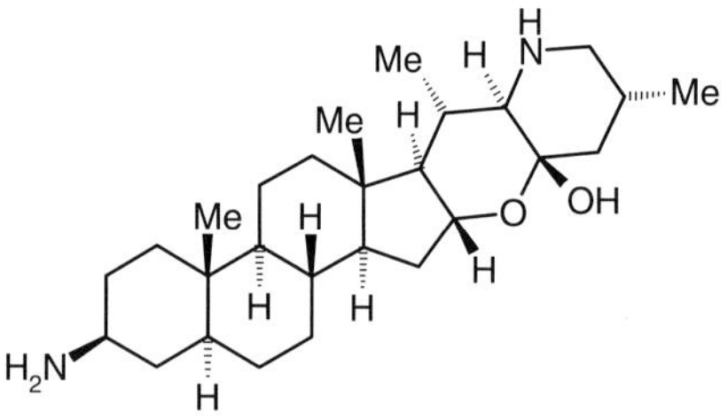

(+)-Solanocapsine (abs. config.)

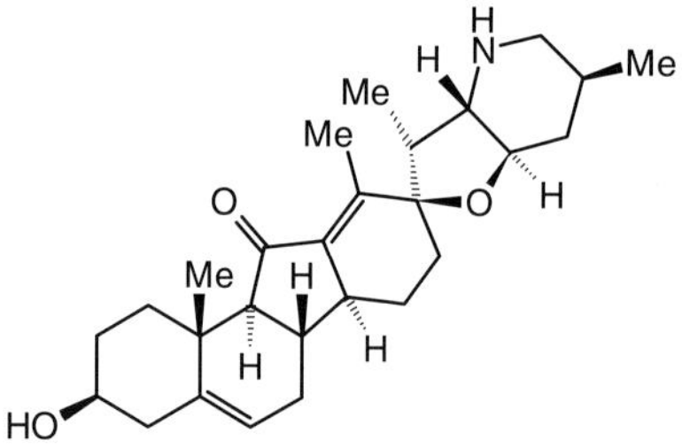

(−)-Jervine (abs. config.)

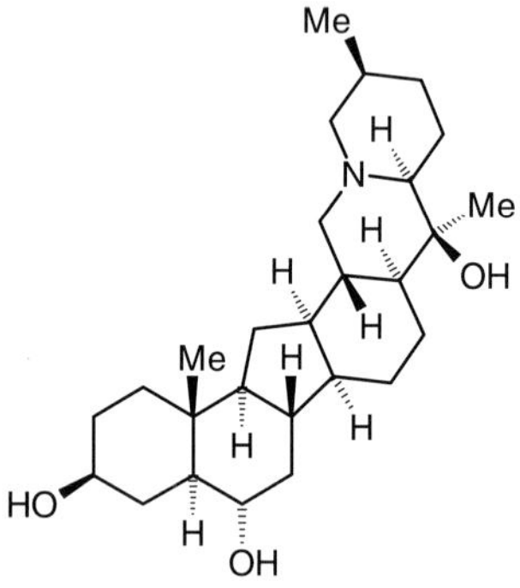

(−)-Verticine (abs. config.)

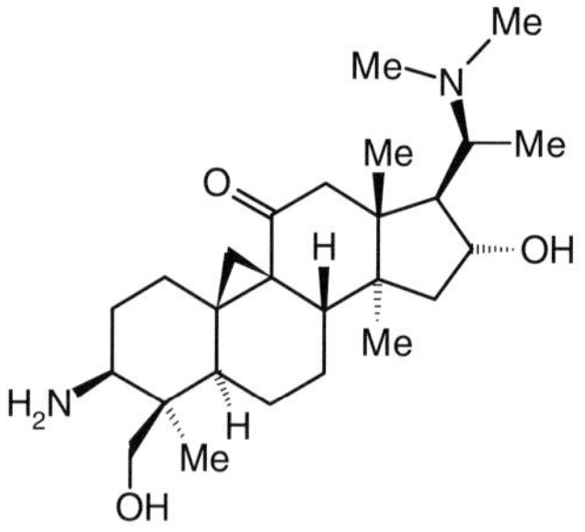

Cycloxobuxidine F (abs. config.)

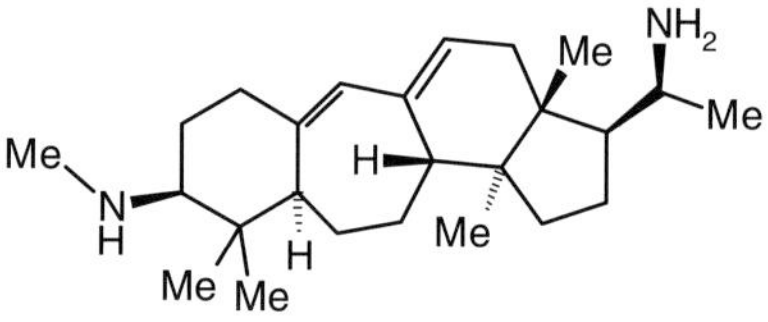

( )-Buxenine G (abs. config.)

The *Veratrum* alkaloid group[104] is made up of *ca.* 400 members, based almost exclusively on the cevane ring system. *Veratrum album* (previously *Helleborus albus*, white hellebore; *Fig. 2.45*) was mentioned by *Pliny* in a comment that '*the Gauls make use of arrows dipped in hellebore when hunting*'. This poison has also been used in other parts of Europe to improve the success in hunting[105].

A classic example of an animal-kingdom steroid alkaloid is samandarine. Samandarine itself (*cf. Chapt. 1*) and related compounds are found in skin gland secretions of the Fire Salamander *Salamandra maculosa*, acting on the central nervous system as a paralytic poison. Their broad activity against both bacteria and fungi is similarly noticeable.

One remark to conclude this section: although an impressive number of important alkaloid skeletons has now been mentioned, it should at this point once more expressly be emphasized that this survey and classification of alkaloids only represents a *selection* of important structural types and so in no way can or wishes to make any claim to completeness.

2.7. Dimeric Alkaloids ('Bis-Alkaloids')

The preceding sections have offered a perspective onto the great structural diversity of the alkaloids. That, however, is by no means the end of the matter; in addition to the 'monomeric' structures discussed, 'dimers', more rarely 'trimers', and in a very few cases even 'tetramers' are also found. Inspiring though this may be in a chemical sense, it causes major headaches for the linguists among the researchers. The term 'dimeric alkaloid', and also the equally common 'bis-alkaloid' are simply inadequate to take account of all aspects of the diversity of the overall compound class. Since the problem cannot be solved without major effort, though, it is necessary, for lack of any better expression, to use these terms.

Some 500 known structures are attributed to the bis-alkaloid class, with the bisindole and bisisoquinoline alkaloids alone each contributing *ca.* 200 bases. The remainder are distributed over all the other structure types. It would seem then that there is a notable preference for the indole and isoquinoline alkaloids; this, however, might not be the case, because a particularly large number of plants containing the corresponding monomers have been examined in the past. As mentioned above, there are a few (*ca.* ten) known alkaloids in which three units of the same structural type are combined in a single molecule, together with, lastly, only two alkaloids known to date arising from the conjunction of four identical units. It is to be assumed that the imag-

Fig. 2.45. *Veratrum album* L. (Liliaceae), *White Hellebore.* The plant grows in alpine meadows, in high shrubland, and in fens. Poisonous *Veratrum* alkaloids are found in all parts of the plant, especially in the roots (*E. Wendelberger, Alpenblumen,* BLV Naturführer, 6th edn., Munich, 1990; photo *E. Forisch*).

Cevane Skeleton

inativeness of nature is greater than as has been established up to now, and so there are probably still significantly more 'polymeric' representatives. Their relatively high molecular weights, though, *in tandem* with numerous polar functional groups, makes the chromatographic separation and purification of these substances difficult. There has, therefore, always been a bias towards isolating the much more conspicuous 'monomeric' alkaloids, while the remaining plant extracts, as a rule, have been discarded.

Apart from a few exceptions still to be discussed, the bis-alkaloids are all composed of two alkaloids of the same type. Accordingly, there exist bisindole[106], bisisoquinoline, bisisoquinolizidine alkaloids, and so on. The observed preference for combinations of equivalent ring system types may be at least partly explained by the composition of the 'monomer' populations contained in each plant. Thus, *Catharanthus roseus* (Apocynaceae), probably the most intensively studied plant, contains indole alkaloids exclusively, and the bis-alkaloids, so far isolated from this plant, are also bisindole alkaloids. Analogous situations apply for a number of other alkaloid-producing plants. Therefore, it seems reasonable to conclude that the synthesis of bis-alkaloids in plants proceeds in such a manner that the 'monomeric' representative is synthesized initially, and that its 'dimerization' or combination into a bis-alkaloid follows afterwards. It is often possible to observe this second reaction step under laboratory conditions, under which the syntheses of some substances of this type may be performed in good yield. Contrarily, reversal of this synthetic pathway allows the bis-alkaloid to be cleaved into its two components (*cf. Chapt. 3*).

The biosynthesis of bis-alkaloids in plants is not compelled to take place on the completed monomeric alkaloid, though. Different synthetic pathways are also conceivable. To the best of our knowledge, no direct evidence for the preference of one out of the two possibilities has been gained so far.

What kind of reactions might now explain the origin of these 'dimers', however? In the following sections, some possible biosynthetic principles are listed and illustrated with relevant examples.

Mannich *Reaction*

A large number of bis-alkaloids appear to owe their origin to a *Mannich* reaction, *e.g.*, the bisindole alkaloid pleiomutine (from *Pleiocarpa mutica* (Apocynaceae)), as shown in *Scheme 2.17*.

Scheme 2.17

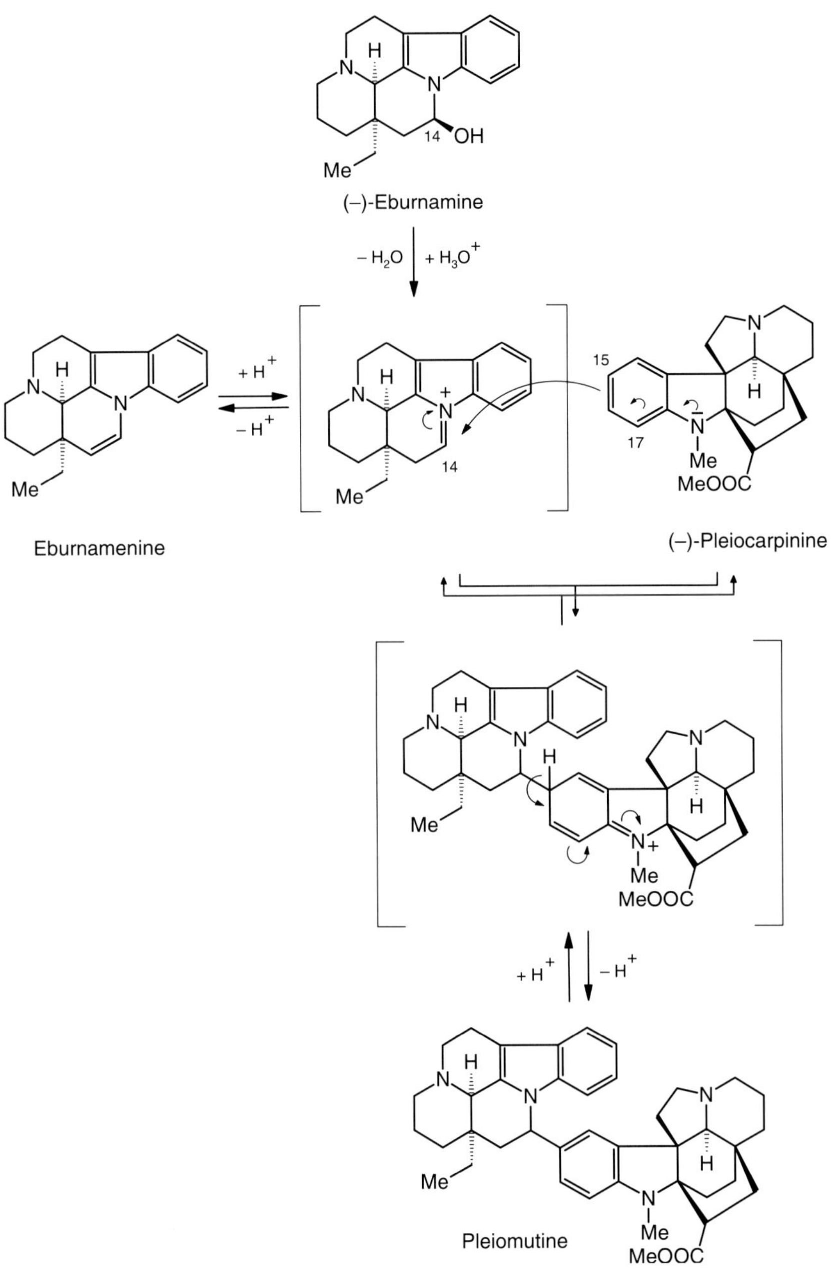

(−)-Eburnamine

− H₂O │ + H₃O⁺

Eburnamenine

(−)-Pleiocarpinine

Pleiomutine

Of the two constituent parts, one possesses an eburnamine ring system, the other a pleiocarpinine skeleton. It is possible to cleave pleiomutine back into eburnamenine and pleiocarpinine, using dilute hydrochloric acid. The two compounds are also found individually in *P. mutica*. The synthesis of pleiomutine from pleiocarpinine and eburnamenine can also successfully be performed in a hydrochloric acid environment. This involves initial formation of an eburnamenium ion, in which the C(14) atom has electrophilic character. The atoms C(15) and C(17) of the aromatic ring possess partial negative charges, since they are either in *para*- or in *ortho*-position relative to the indole N-atom. Both for steric reasons and due to its higher nucleophilicity, however, only C(15) reacts successfully. The loss of a proton then results in the re-aromatization of the benzene ring in the pleiocarpinine component, leading to the formation of pleiomutine.

By analogous mechanisms, the origins of a large number of other bisindole alkaloids, among them voacamine (from various genera of Apocynaceae) can be explained. This natural product is accessible synthetically from the *Iboga* alkaloids voacangine and vobasinol by treatment with dilute hydrochloric acid. The C(3) atom in vobasinol becomes electrophilic during the course of the reaction. The OH group at C(3) thus first becomes protonated and is then eliminated (–H$_2$O) with the assistance of the indole N-atom (allylic position). On the other hand, were the amine N-atom also to take part in the elimination, then a quaternary ammonium salt would be produced, which (under the reaction conditions) would be stable and not react any further.

The alkaloid sanguidimerine (from *Sanguinaria* (Papaveraceae)) consists of two equivalents of the benzophenanthridine alkaloid sanguinarine chloride plus one equivalent of acetone. Obviously, sanguidimerine is formed through a double *Mannich* reaction, and, self-evidently, two molecules of HCl split off in the course of this reaction. Whether this compound is really a natural product or – as is more likely – rather an artifact from sanguinarine chloride and acetone, is not going to be discussed here. The case of the alkaloid dendrocrepine (from *Dendrobium* (Orchidaceae)) is, by the way, similar.

Vobasinol

Voacangine

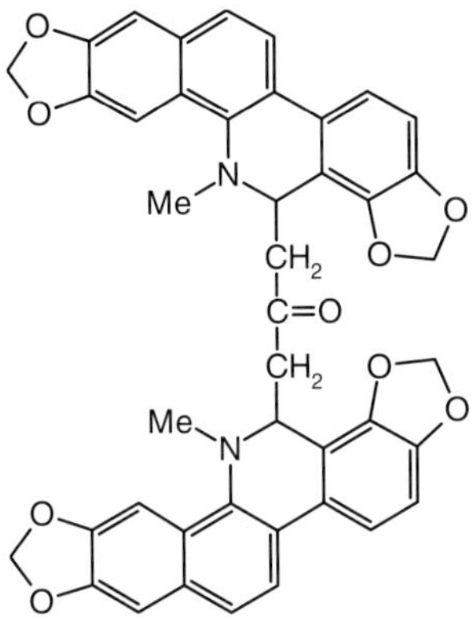

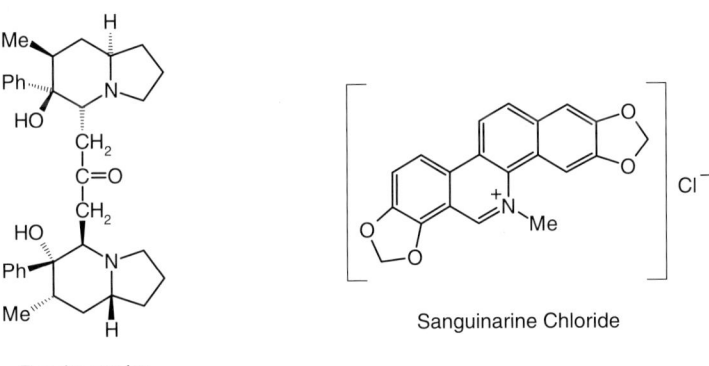

Sanguidimerine

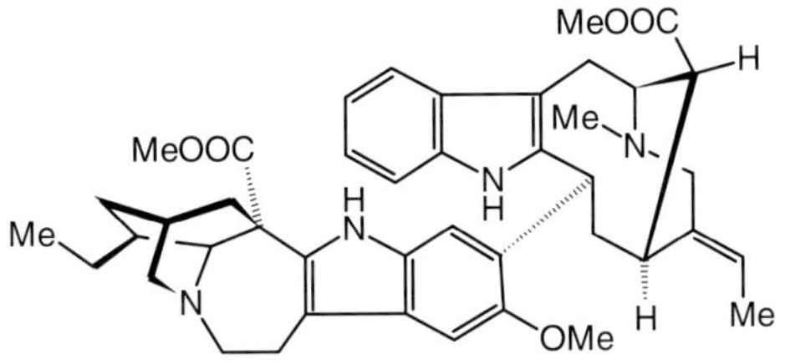

Voacamine

Dendrocrepine

Sanguinarine Chloride

Michael *Reaction*

The *Michael* reaction also plays a role in the origination of 'dimeric' alkaloids. One example is villalstonine, discussed in detail elsewhere (*cf. Sect. 3.4*). The actual coupling step is shown in *Scheme 2.18*.

It is similarly possible to postulate a *Michael* reaction in the biosynthesis of other macroline-containing *Alstonia* bis-alkaloids. These include, for example, alstomacroline and macralstonine (both from *Alstonia macrophylla* (Apocynaceae)).

Scheme 2.18

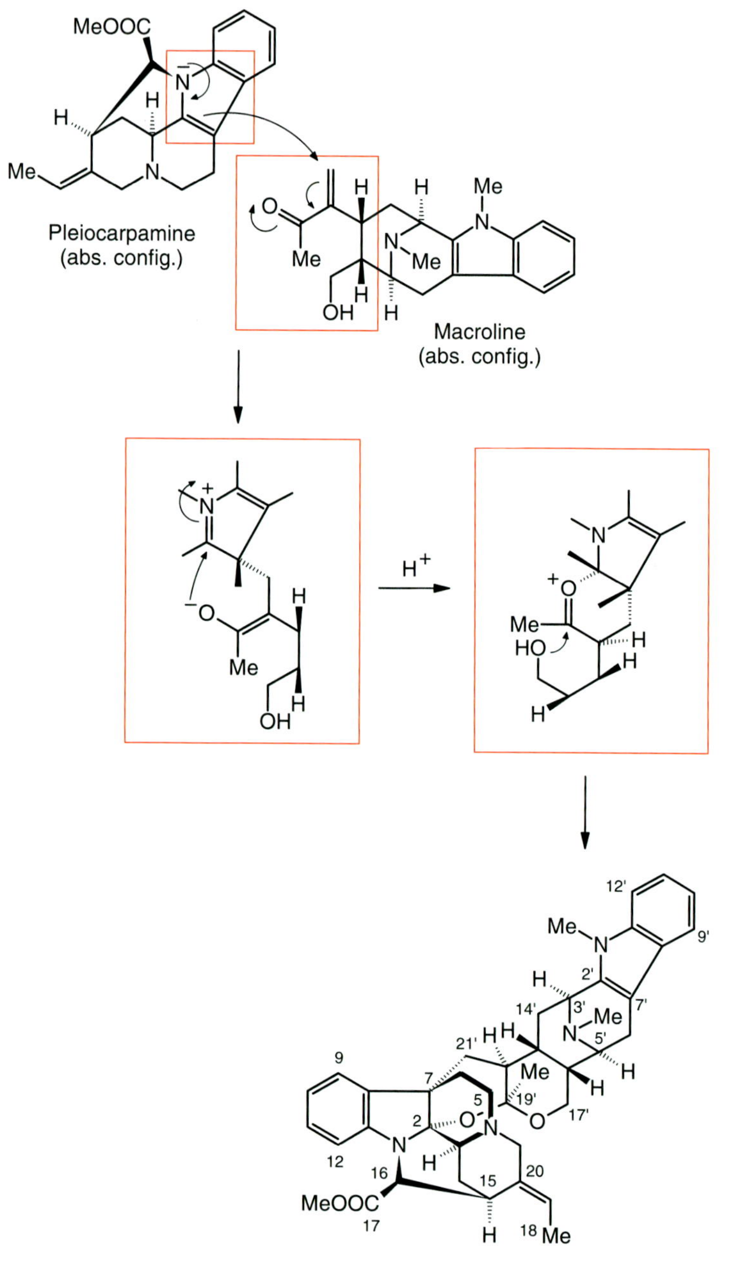

Pleiocarpamine
(abs. config.)

Macroline
(abs. config.)

Villalstonine (abs. config.)

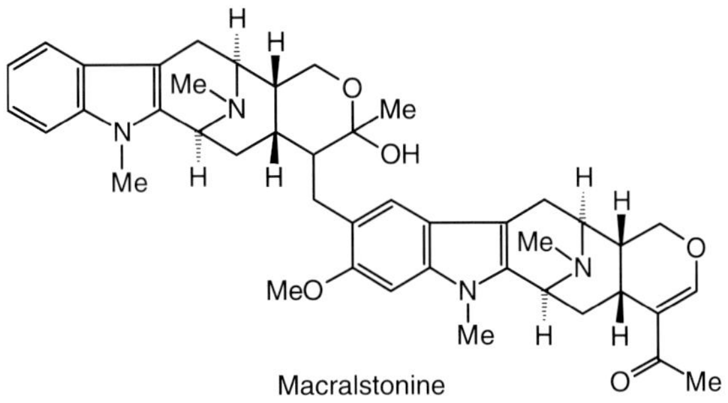

Macralstonine

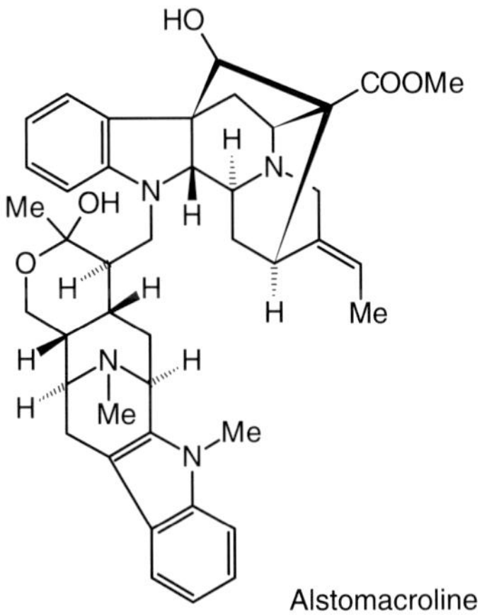

Alstomacroline

Aldehyde–Amine Condensation

A large group of 'dimeric' alkaloids results from the condensation of a primary or secondary amine of one component base with an aldehyde group in the other. The curare alkaloid dihydrotoxiferine (from calabash curare and South American *Strychnos* species (Loganiaceae)) is also the result of such a process[10]. The amino-aldehyde hemidihydrotoxiferine most probably serves as the monomeric precursor; it is very easily transformed into dihydrotoxiferine upon warming in dilute acetic acid. The acid-catalyzed reverse process, leading back to hemidihydrotoxiferine, is synthetically also feasible (*cf. Sect. 6.7*).

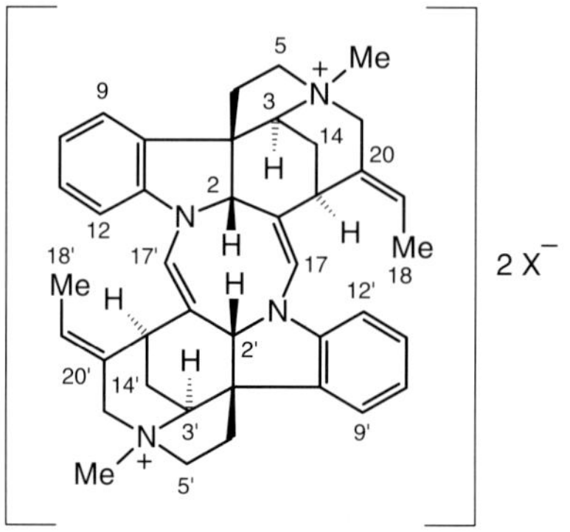

Dihydrotoxiferine (abs. config.)

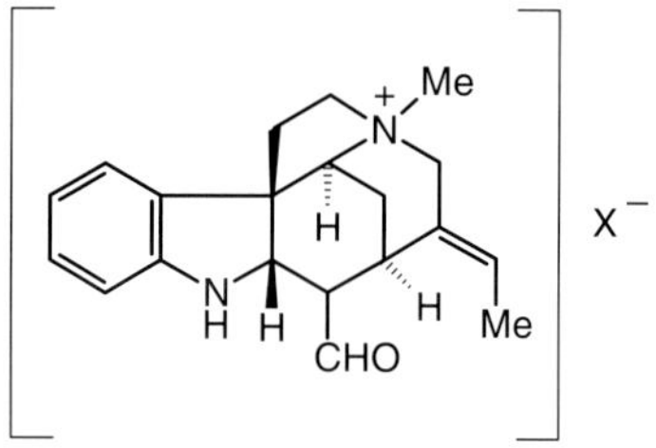

Hemidihydrotoxiferine (abs. config.)

The chemical resistance of dimeric alkaloids towards cleavage is improved, if the momomeric condensation products are chemically altered at the point of connection. A naturally occurring oxidation product of dihydrotoxiferine, *e.g.*, is calebassine (occurrence as for dihydrotoxiferine), which is no longer acid-sensitive to breakdown into its component bases (*cf. Chapt. 4.7*). Calebassine, produced in good yield by oxidation of dihydrotoxiferine, contains – compared to its precursor – two additional OH groups at C(2) and C(2′) plus a new bond between C(17) and C(17′), crucial for its altered behavior in mineral acid. In other words: cleavage of the 'dimer' without prior opening of the pertinent C,C bond is virtually impossible.

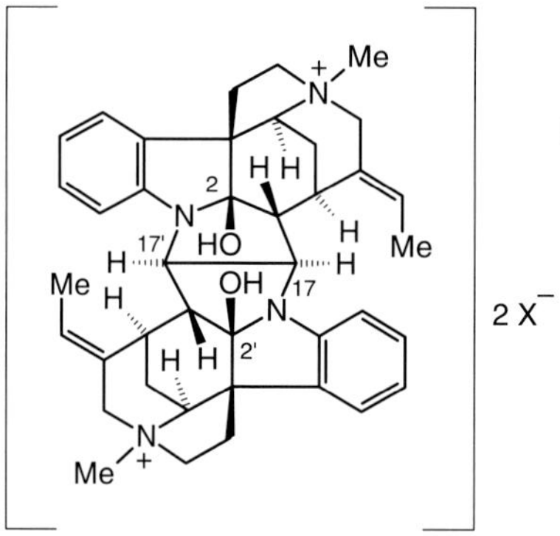

Calebassine (abs. config.)

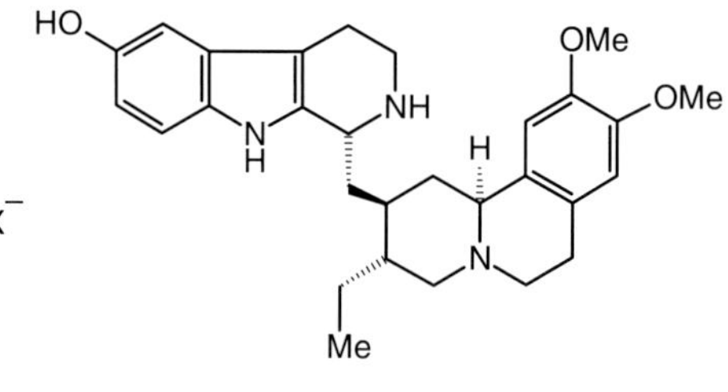

Tubulosine

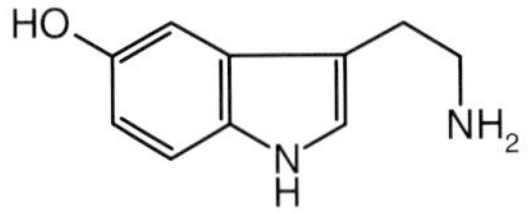

5-Hydroxytryptamine 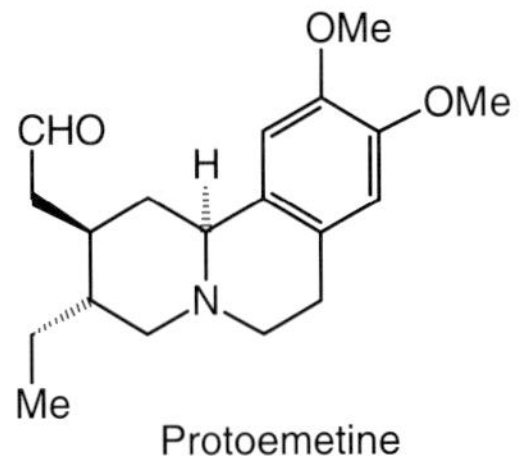   Protoemetine

Unlike the dimeric curare alkaloids, made from two complete monomers, alkaloids of the tubulosine type (from both *Alangium lamarckii* (Alangiaceae) and *Pogonopus tubulosus* (Rubiaceae)) occupy a position intermediate between dimeric and monomeric bases. Tubulosine consists of the isoquinoline alkaloid protoemetine (from *Psychotria* (Rubiaceae)) and 5-hydroxytryptamine. Important for the corresponding condensation reaction are both the aldehyde group at C(16) in protoemetine and the primary amine of 5-hydroxytryptamine. In tubulosine, only one aliphatic residue (a C_9 derivative of secologanin) is present, together with two aromatic constituents (tryptamine and tyramine derivatives).

As far as their formation is concerned, alkaloids of the tubulosine type resemble the 'monomeric' *Ipecacuanha* alkaloids, which contain an additional tyramine residue in place of the tryptamine moiety. In addition, a number of alkaloids are known, which display the same or a similar construction as tubulosine, among them roxburghine D (from *Uncaria* (Rubiaceae)).

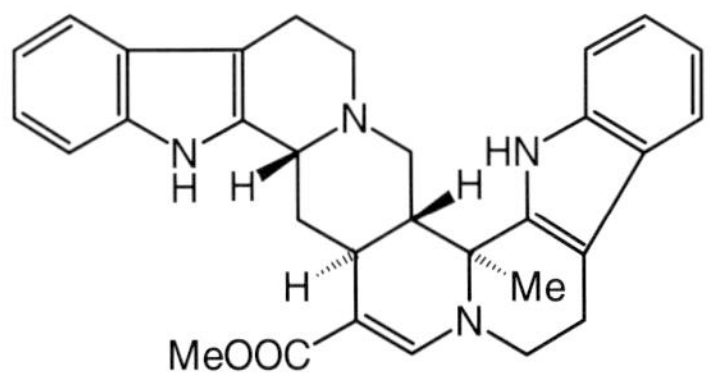

Roxburghine D

Coupling by Phenol Oxidation[107]

The origin of almost all bis(benzylisoquinoline) alkaloids[108] (almost 200 members) may be explained by one or more oxidative phenol coupling reactions. 1-Benzyltetrahydroisoquinoline alkaloids with phenolic OH groups can be regarded as suitable starting materials. The large number of alkaloids of this type can be explained, on one hand, by the numerous possible connecting points between the two bases, and, on the other hand, by variations in the substituent patterns. The following basic bis(benzylisoquinoline) alkaloid types are categorized by structure.

The simplest representatives involve those alkaloids with their constituent bases joined by a *tale-to-tale* bond[109]. One example of this type is dauricine (from *Menispermum* species (Menispermaceae)).

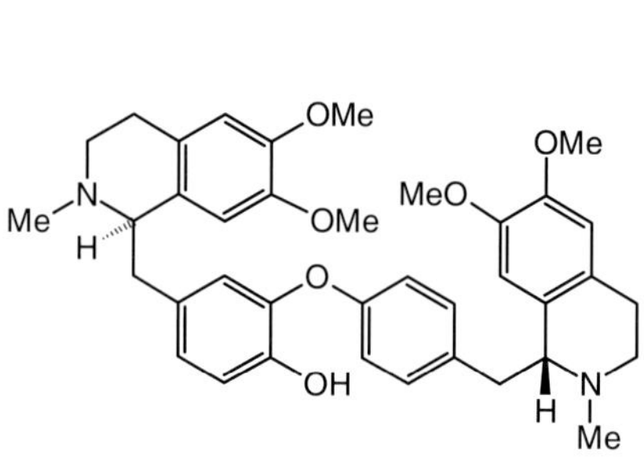

Dauricine

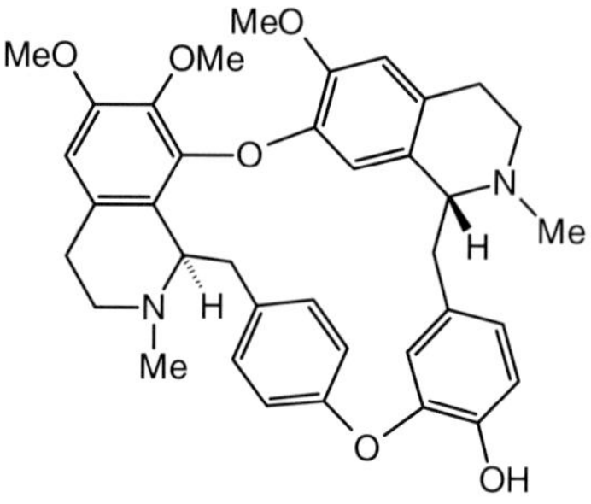

Oxyacanthine

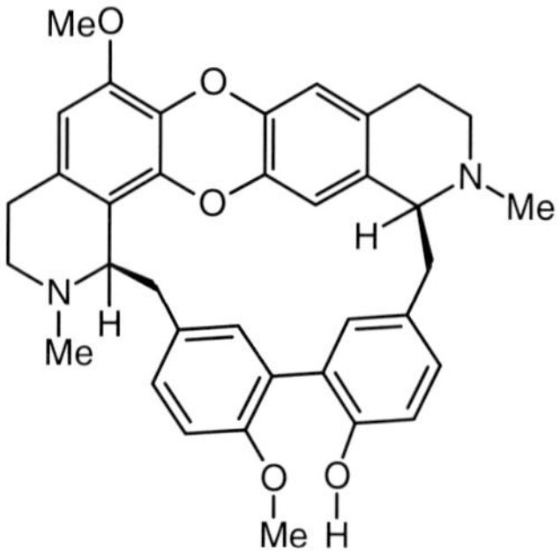

(+)-Tiliacorine

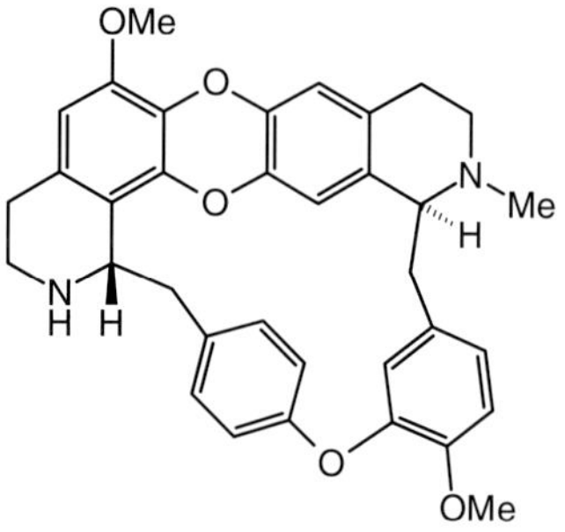

Trilobine

Two bonds (*head-to-head* and *tail-to-tail* fashion) connect the constituent bases in oxyacanthine, isolated from species of the genera *Berberis* and *Mahonia* (Berberidaceae), *Magnolia* and *Michelia* (Magnoliaceae), and from *Cocculus* (Menispermaceae).

In alkaloids of the oxyacanthine type, there may also be two bonds present between the two 'heads' of the component bases, as found, for example, in trilobine (from *Cocculus* species (Menispermaceae)).

As well as diphenyl ether bonds between the component bases, there are also structures containing additional C,C bonds between the two 'monomer' bases. Tiliacorine (from *Tiliacora acuminata* (Menispermaceae)) belongs to this group.

A third group of bis(benzylisoquinoline) alkaloids consists of *head-to-tail*- and *tail-to-head*-linked component bases. Tubocurarine[110] (from both *Chondodendron* (Menispermaceae) and tubocurare arrow poisons) is given as a typical example. This type, by the way, also includes numerous subtypes containing one, two, or even three ether bonds between the two components.

As far as synthesis is concerned, some bis(benzylisoquinoline) bases have been prepared under basic conditions by treatment of the corresponding monomeric alkaloids with potassium hexacyanoferrate(III) ($K_3[Fe(CN)_6]$); yields have not been very high, however.

Since the actual coupling mechanism is an oxidation, it is not possible to cleave bis(benzylisoquinoline) alkaloids into their constituent bases under acidic reaction conditions. Here, it is necessary to operate under reductive conditions, for which Na/NH_3 (liq.) has proven to be an effective reagent.

Pilocerine (from *Lophocereus*, *Pachycereus*, and *Pilocereus* species (Cactaceae)) is an isoquinoline alkaloid, in which three component bases are joined together by oxidative phenol coupling. Unlike the other alkaloids listed in this section, its bases are not of the 1-benzylisoquinoline type, but are 1-isobutyl-1,2,3,4-tetrahydroisoquinoline alkaloids.

Oxidative coupling reactions are not necessarily confined to phenols. Two N^b-methyltryptamine moieties may unite in straightforward fashion by oxidative coupling to give the bisindolenine shown in *Scheme 2.19*. This dimerization product is equivalent to the hypothetical tetraaminodialdehyde that would arise from the 'hydrolysis' of bisindolenine. Reaction of the nucleophilic and electrophilic centers N^a and N^b with C(2'), and of $N^{a'}$ and $N^{b'}$ with C(2), finally gives rise to the bisquinoline derivative. If, though, centers N^a and N^b join with C(2), and $N^{a'}$ and $N^{b'}$ with C(2') (a reaction, starting from bisindolenine, which is easier to follow), then the bisindoline derivative is obtained. Both compounds – calycanthine (from *Calycanthus* and *Chimonanthus* (Calycanthaceae; *Fig. 2.46*) and from *Bhesa* (Celastraceae)) and chimonanthine (from *Chimonanthus*) – have been isolated from plants belonging to the Calycanthaceae family.

(+)-Tubocurarine

Pilocerine

Quadrigemine A

Scheme 2.19

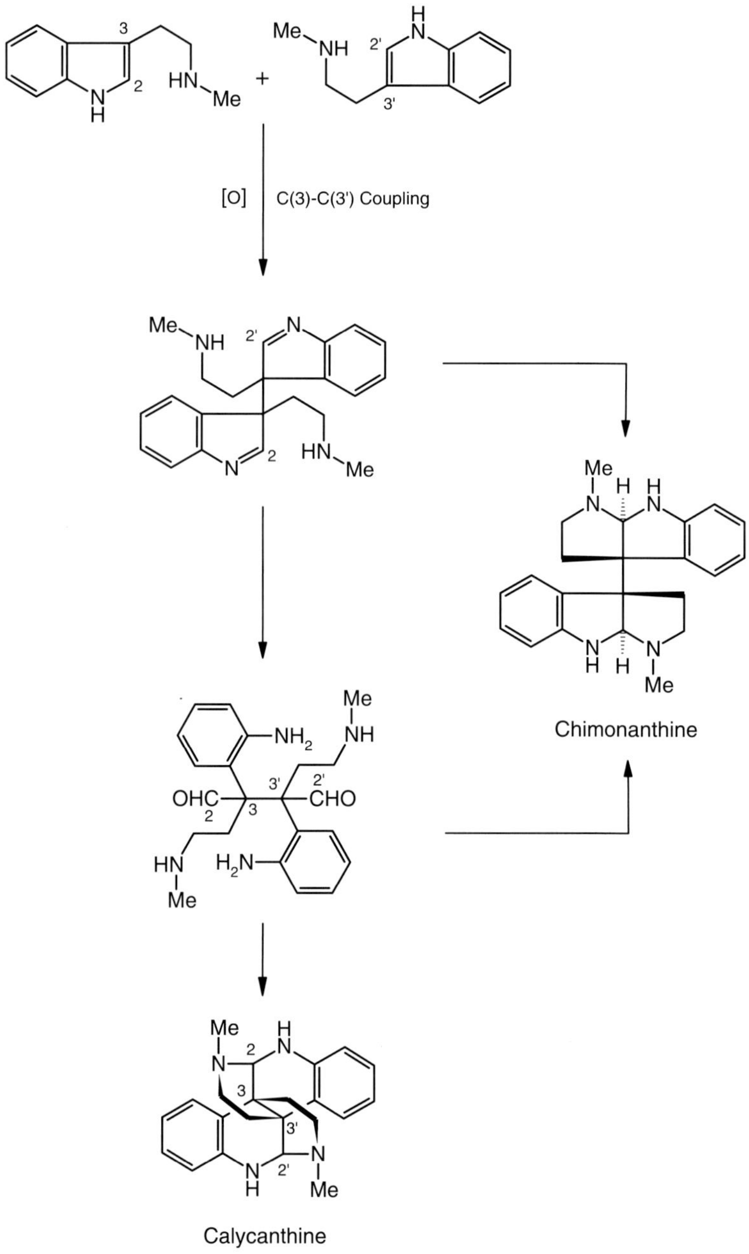

Chimonanthine

Calycanthine

Fig. 2.46. *Chimonanthus praecox* (L.) Link. (Calycanthaceae), *Japan Allspice.* A shrub, 2 – 3 m high, with yellow-brown, highly perfumed flowers, in Switzerland blooming during January/February (watercolor *R. Gfeller*, 1998).

As well as these, still other naturally occurring calycanthine derivatives are known. Trimeric and tetrameric representatives have also been produced by reactions of N^b-methyltryptamine, for instance, quadrigemine A (from *Hodgkinsonia frutescens* (Rubiaceae)).

A radical coupling mechanism might lie behind the formation of dithermamine (from *Thermopsis* (Leguminosae)), a dimeric alkaloid of the sparteine type.

Dithermamine

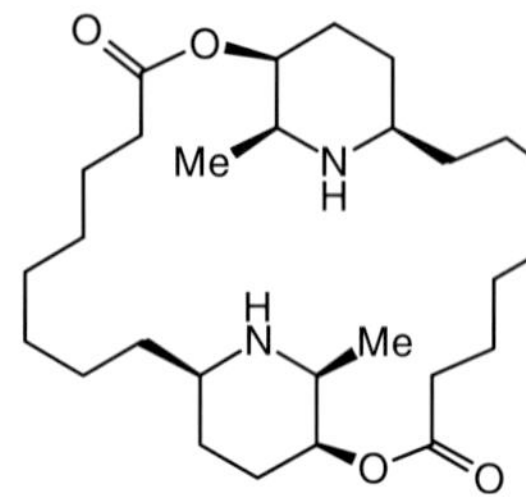

(+)-Carpaine

Lactonization

Dimerization of monomeric alkaloid components (possessing both OH and COOH groups) by means of double esterification is also possible, as seen in carpaine (isolated from *Azima* (Salvadoraceae), *Cerbera* (Apocynaceae), and *Carica* (Caricaceae) species). However, only a few examples of this type are currently known.

Complex Mechanisms

A special mention is warranted for the alkaloid cancentrine (from *Dicentra canadensis* (Fumariaceae))[111]. The compound is composed of one representative each from the morphine and the cularine alkaloids. A plausible hypoth-

Scheme 2.20

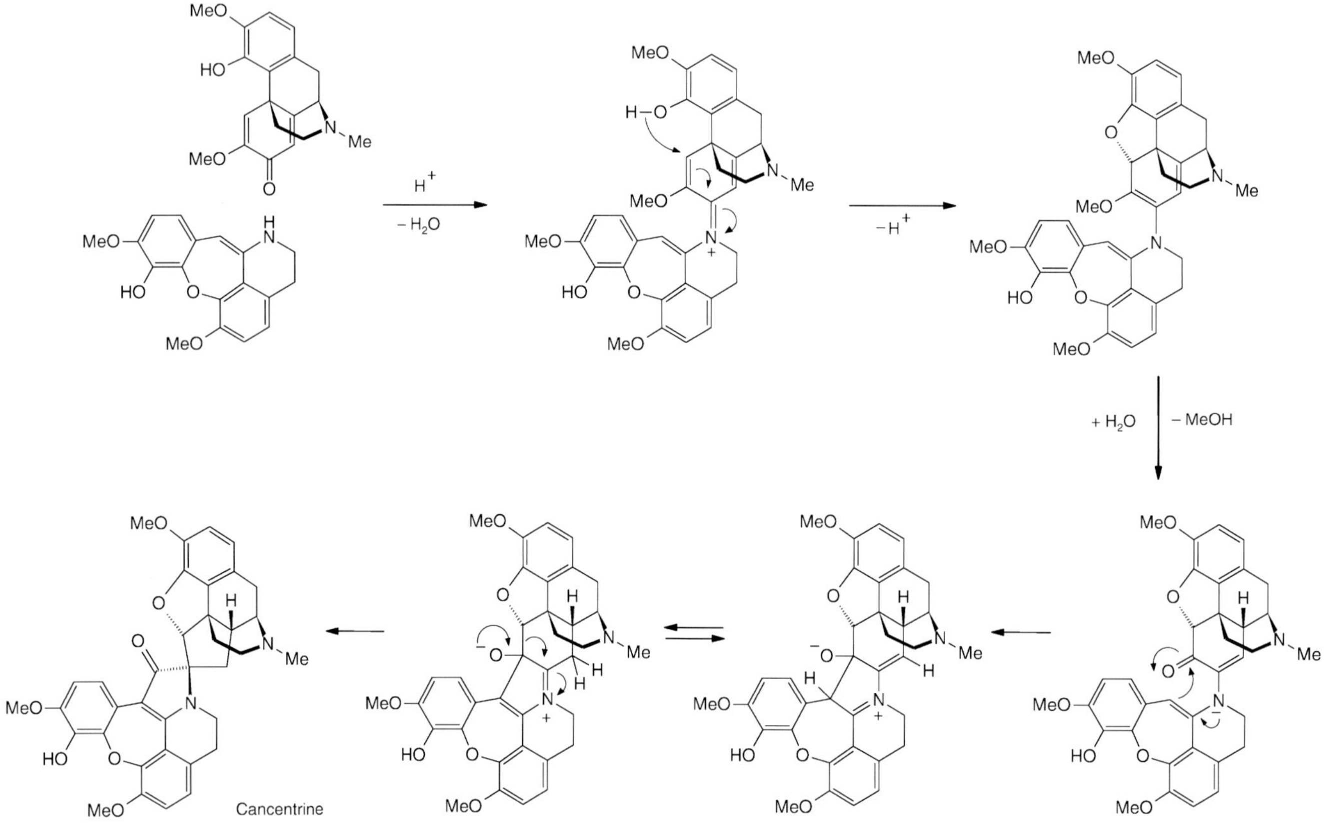

Cancentrine

esis for the biogenesis of cancentrine is shown in *Scheme 2.20*. The postulated mechanism explains the biogenesis of the morphine component, while at the same time offering an explanation for a ring-contraction mechanism leading to the central five-membered ring.

Finally, vinblastine and vincristine, both obtained from *Catharanthus roseus* (Apocynaceae), should be mentioned. These two bisindole alkaloids are used intravenously in treatment of both *Hodgkin*'s disease and childhood leukemia. The production of these compounds is expensive, since only natural sources with a very low drug content are currently available.

It appears that the plant biosynthesis of the above two bases (starting from one *Iboga* and one aspidospermidine alkaloid) involves the cleavage of an assisting functional group (from the *Iboga* moiety) responsible for the initial connection.

R^1 = Me, R^2 = MeOOC : Vinblastine

R^1 = CHO, R^2 = MeOOC : Vincristine

References and Notes

1. R. K. Hill, '*Pyrrolidine, Piperidine, Pyridine and Imidazole Alkaloids*', in S. W. Pelletier (Ed.), *Chemistry of the Alkaloids*, Van Nostrand, New York, 1970, 385.
2. L. Marion, '*The Pyrrolidine Alkaloids*', The Alkaloids **1950**, *1*, 91; ibid. **1960**, *6*, 31.
3. A. Popelak, G. Lettenbauer,'*The Mesembrine Alkaloids*', The Alkaloids **1967**, *9*, 467.
4. A. Numata, T. Ibuka, '*Alkaloids from Ants and Other Insects*', The Alkaloids **1987**, *31*, 194.
5. R. Koob, C. Ruldolph, H. J. Veith , '*The Absolute Configuration of 3-Methylpyrrolidine Alkaloids from Poison Glands of Ants, Leptothoracini (Myrmicinae)*', Helv. Chim. Acta **1997**, *80*, 267.
6. *General overview*: A. R. Battersby, '*Recent Researches on Indole Alkaloids*', Pure Appl. Chem. **1963**, *6*, 471; L. Marion, '*The Indole Alkaloids*', The Alkaloids **1952**, *2*, 369; N. Neuss, '*Indole Alkaloids*', in S. W. Pelletier (Ed.), *Chemistry of the Alkaloids*, Van Nostrand, New York, 1970, 213; J. E. Saxton, '*The Indole Alkaloids*', The Alkaloids **1960**, *7*, 1; J. E. Saxton, '*The Indole Alkaloids excluding Harmine and Strychnine*', Quart. Rev. **1956**, *10*, 108; W. I. Taylor, *Indole Alkaloids, an Introduction to the Enamine Chemistry of Natural Products*, Pergamon Press, Inc., New York, 1966. *Structure and literature overview*: M. Hesse, *Indolalkaloide in Tabellen*, Springer-Verlag, Berlin, 1964, Suppl. Vol. 1968. *Mass spectra*: M. Hesse, *Progress in Mass Spectrometry, Vol. 1, Indolalkaloide*, Verlag Chemie, Weinheim, 1974. *Syntheses*: J. P. Kutney, '*The Total Synthesis of Indol Alkaloids*', in D. H. Hey (Ed.), MTP International Review of Science, Organic Chemistry, Series 1, Vol. 9, 'Alkaloids', Butterworth University Park Press, London, 1973, 27; E. Winterfeldt, 'Stereoselektive Totalsynthese von Indolalkaloiden', *Fortschr. Chem. organ. Naturstoffe* **1974**, *31*, 469; E. Winterfeldt, '*Neuere Aspekte in der Chemie der Indolalkaloide*', Chimia **1971**, *25*, 394. *Biogenesis*: I. Kompis, M. Hesse, H. Schmid, '*An Approach to the Biogenetic Classification of Indole Alkaloids*', Lloydia **1971**, *34*, 269; A. I. Scott, '*Biosynthesis of Indole Alkaloids*', in D. H. Hey (Ed.), MTP International Review of Science, Organic Chemistry, Series 1, Vol. 9, 'Alkaloids', Butterworth

University Park Press, London, 1973, 105. *Specialist Reviews*: *Oxindole Alkaloids*:
J. S. Bindra, 'Oxindole Alkaloids', *The Alkaloids* **1973**, *14*, 84. *Alstonia Alkaloids*:
J. E. Saxton, 'Alkaloids of Alstonia Species', *The Alkaloids* **1970**, *12*, 207; *ibid.*
1965, *8*, 159; J. E. Saxton, 'Alkaloids of Picralima and Alstonia Species', *The Alkaloids* **1973**, *14*, 157. *Apocynaceae Alkaloids*: R. F. Raffauf, M. B. Flagler, 'Alkaloids of the Apocynaceae', *Econ. Botany* **1960**, *14*, 37. *Diplorrhynchus Alkaloids*: B. Gilbert, 'The Alkaloids of Aspidosperma, Diplorrhynchus, Kopsia, Ochrosia, Pleiocarpa, Melodinus and Related Genera', *The Alkaloids* **1965**, *8*, 336; *ibid.* **1968**, *11*, 205.
Geissospermum Alkaloids: R. H. F. Manske, W. A. Harrison 'The Alkaloids of Geissospermum Species', *The Alkaloids* **1965**, *8*, 679. *Gelsemium Alkaloids*: J. E. Saxton, 'Alkaloids of Gelsemium Species', *The Alkaloids* **1965**, *8*, 93. *Haplophyton Alkaloids*: J. E. Saxton, 'Alkaloids of Haplophyton cimicidum', *The Alkaloids* **1965**, *8*, 673. *Iboga Alkaloids*: W. I. Taylor, 'The Iboga and Voacanga Alkaloids', *The Alkaloids* **1965**, *8*, 203; *ibid.* **1968**, *11*, 79. *Kopsia Alkaloids*: B. Gilbert, 'The Alkaloids of Aspidosperma, Diplorrhynchus, Kopsia, Ochrosia, Pleiocarpa, Melodinus and Related Genera', *The Alkaloids* **1965**, *8*, 336; *ibid.* **1968**, *11*, 205. *Melodinus Alkaloids*: B. Gilbert, 'The Alkaloids of Aspidosperma, Diplorrhynchus, Kopsia, Ochrosia, Pleiocarpa, Melodinus and Related Genera', *The Alkaloids* **1965**, *8*, 336; *ibid.* **1968**, *11*, 205. *Mitragyna Alkaloids*: J. E. Saxton, 'Alkaloids of Mitragyna and Ourouparia Species', *The Alkaloids* **1965**, *8*, 59; *ibid.* **1968**, *10*, 521; J. E. Saxton, 'Alkaloids of Mitragyna and Related Genera', *The Alkaloids* **1973**, *14*, 123. *Ochrosia Alkaloids*: B. Gilbert, 'The Alkaloids of Aspidosperma, Diplorrhynchus, Kopsia, Ochrosia, Pleiocarpa, Melodinus and Related Genera', *The Alkaloids*, **1965**, *8*, 336; *ibid.* **1968**, *11*, 205. *Ourouparia Alkaloids*: J. E. Saxton, 'Alkaloids of Mitragyna and Ourouparia Species', *The Alkaloids* **1965**, *8*, 59; *ibid.* **1968**, *10*, 521. *Picralima Alkaloids*: J. E. Saxton, 'Alkaloids of Picralima and Alstonia Species', *The Alkaloids* **1973**, *14*, 157. J. E. Saxton, 'Alkaloids of Picralima Nitida', *The Alkaloids* **1965**, *8*, 119; *ibid.* **1968**, *10*, 501. *Pleiocarpa Alkaloids*: B. Gilbert, 'The Alkaloids of Aspidosperma, Diplorrhynchus, Kopsia, Ochrosia, Pleiocarpa, Melodinus and Related Genera', *The Alkaloids* **1965**, *8*, 336; *ibid.* **1968**, *11*, 205. *Pseudocinchona Alkaloids*: R. H. F. Manske, 'Alkaloids of Pseudocinchona and Yohimbe', *The Alkaloids* **1965**, *8*, 694. *Vinca Alkaloids*: W. I. Taylor, 'The Vinca Alkaloids', *The Alkaloids* **1965**, *8*, 272 ; *ibid.* **1968**, *11*, 99. *Voacanga Alkaloids*: W. I. Taylor, 'The Iboga and Voacanga Alkaloids', *The Alkaloids* **1965**, *8*, 203; *ibid.* **1968**, *11*, 79. *Yohimbe Alkaloids*: R. H. F. Manske, 'Alkaloids of Pseudocinchona and Yohimbe', *The Alkaloids* **1965**, *8*, 694; H. J. Monteiro, 'Yohimbine and Related Alkaloids', *The Alkaloids* **1968**, *11*, 145.

7. J. E. Saxton, 'The Simple Bases', *The Alkaloids* **1965**, *8*, 1; J. E. Saxton, 'The Simple Indole Bases', *The Alkaloids* **1968**, *10*, 491; B. B. Stowe, 'Occurrence and Metabolism of Simple Indoles in Plants', *Fortschr. Chem. organ. Naturstoffe* **1959**, *17*, 248. H.-P. Husson, 'Simple Indole Alkaloids including β-Carbolines and Carbazoles', *The Alkaloids* **1985**, *26*, 1.

8. R. S. Kapil, 'The Carbazole Alkaloids', *The Alkaloids* **1971**, *13*, 273.

9. G. W. Gribble, 'Synthesis and Antitumor Activity of the Ellipticine Alkaloids and Related Compounds', *The Alkaloids* **1990**, *39*, 239; M. Suffness, G. A. Cordell, 'Antitumor Alkaloids', *The Alkaloids* **1985**, *25*, 1.

10. A. R. Battersby, H. F. Hodson, 'Alkaloids of Calabash Curare and Strychnos Species', *The Alkaloids* **1965**, *8*, 515; *ibid.* **1968**, *11*, 189; K. Bernauer, 'Alkaloide aus Calebassencurare und südamerikanischen Strychnosarten', *Fortschr. Chem. organ. Naturstoffe* **1959**, *17*, 183; A. R. Battersby, H. F. Hodson, 'Alkaloids of Calabash Curare and Strychnos Species', *Quart. Rev.* **1960**, *14*, 77; H. L. Holmes, 'The Strychnos Alkaloids', *The Alkaloids* **1950**, *1*, 375; *ibid.* **1952**, *2*, 513; J. Bosch, J.

Bonjoch, M. Amat, 'The Strychnos Alkaloids', *The Alkaloids* **1996**, *48*, 75; G. F. Smith, 'Strychnos Alkaloids', *The Alkaloids* **1965**, *8*, 592.

11. F. L. Sonnenschein, *Handbuch der gerichtlichen Chemie*, Verlag A. Hirschwald, Berlin, 1869, p. 233.

12. R. H. F. Manske, 'The Carboline Alkaloids', *The Alkaloids* **1965**, *8*, 47; H.-P. Husson, 'Simple Indole Alkaloids including β-Carbolines and Carbazoles', *The Alkaloids* **1985**, *26*, 1.

13. F. Musshoff, 'Formaldehyd-Neuroamin-Kondensationsprodukte bei chronischen Alkoholikern', *Toxichem. & Krimtech.* **1995**, *62*, 79.

14. A. Brossi, 'Mammalian Alkaloids', *The Alkaloids* **1993**, *43*, 119.

15. M. Lounasmaa, A. Tolvanen, 'Eburnamine-Vincamine Alkaloids', *The Alkaloids* **1992**, *42*, 1.

16. Pharmacologically, (+)-vincamine, which aids cerebral blood circulation, is of particular significance.

17. A. Chatterjee, 'Rauwolfia Alkaloids', *Fortschr. Chem. organ. Naturstoffe* **1953**, *10*, 390; A. Chatterjee, S. C. Pakrashi 'Recent Developments in the Chemistry and Pharmacology of Rauwolfia Alkaloids', *Fortschr. Chem. organ. Naturstoffe* **1956**, *13*, 346.

18. W. I. Taylor, 'The Ajmaline-Sarpagine Alkaloids', *The Alkaloids* **1965**, *8*, 789; *ibid.* **1968**, *11*, 41.

19. Ajmaline is used in medicine as a cardiac antiarrhythmic drug.

20. R. B. Turner, R. B. Woodward, 'The Chemistry of the Cinchona Alkaloids', *The Alkaloids* **1953**, *3*, 1; W. I. Taylor, 'The Chemistry of the 2,2'-Indolylquinuclidine Alkaloids', *The Alkaloids* **1965**, *8*, 238; *ibid.* **1968**, *11*, 73; M. R. Uskokovic, G. Grethe, 'The Cinchona Alkaloids', *The Alkaloids* **1973**, *14*, 181.

21. C. Szántay, G. Blaskó, K. Honty, G. Dörnyei, 'Corynantheine, Yohimbine, and Related Alkaloids', *The Alkaloids* **1986**, *27*, 131; E. Schlittler, 'Rauwolfia Alkaloids with Special Reference to the Chemistry of Reserpine', *The Alkaloids* **1965**, *8*, 287.

22. Yohimbine is a popular aphrodisiac (especially in veterinary medicine) and a strong vasodilatory cardiac stimulant.

23. Reserpine and its derivatives find application in medicine as antihypertonics.

24. E. Coxworth, 'Alkaloids of the Calabar Bean', *The Alkaloids* **1965**, *8*, 27; B. Robinson, 'Alkaloids of the Calabar Bean', *The Alkaloids*, **1968** *10*, 383; *ibid.* **1971**, *13*, 213.

25. R. H. F. Manske, 'The Alkaloids of Calycanthaceae', *The Alkaloids* **1965**, *8*, 581.

26. D. Gröger, 'Fortschritte der Chemie und Biochemie der Mutterkornalkaloide', *Fortschr. chem. Forschung* **1966**, *6*, 159; A. Stoll, A. Hofmann, 'The Chemistry of the Ergot Alkaloids', in S. W. Pelletier (Ed.), *Chemistry of the Alkaloids*, Van Nostrand, New York, 1970, 267; A. Stoll, A. Hofmann, 'The Ergot Alkaloids', *The Alkaloids* **1965**, *8*, 726; A. Stoll, 'Recent Investigations on Ergot Alkaloids', *Fortschr. Chem. organ. Naturstoffe* **1952**, *9*, 114; P. A. Stadler, P. Stütz, 'The Ergot Alkaloids', *The Alkaloids* **1975**, *15*, 1.

27. Ergot alkaloids are used successfully in many different areas of medicine (such as obstetrics, neurology) (see *Chapt. 9*). The notorious hallucinogen LSD (lysergic acid diethylamide) is actually a synthetic product but can hardly remain unmentioned in the context of the ergot alkaloids.

28. R. H. F. Manske, 'The Quinazolinocarbolines', *The Alkaloids* **1965**, *8*, 55.

29. H.-P. Ros, R. Kyburz, N. W. Preston, R. T. Gallagher, I. R. C. Bick, M. Hesse, 'The Structure of the Alkaloid Peduncularine', *Helv. Chim. Acta* **1979**, *62*, 481.

30. I. R. C. Bick, M. A. Hai, 'Aristotelia Alkaloids', *The Alkaloids* **1985**, *24*, 113; H.-J. Borschberg, 'Aristotelia Alkaloids', *The Alkaloids* **1996**, *48*, 192.

31. D. Gross, '*Vorkommen, Struktur und Biosynthese natürlicher Piperidinverbindungen*', *Fortschr. Chem. organ. Naturstoffe* **1971**, *29*, 1.

32. As well as this, sesbanimide also possesses cytotoxic and antileukemic properties[9].

33. G. M. Strunz, J. A. Findlay '*Pyridine and Piperidine Alkaloids*', *The Alkaloids* **1985**, *26*, 89; W. A. Ayer, T. E. Habgood, '*The Pyridine Alkaloids*', *The Alkaloids* **1968**, *11*, 459; D. Gross, '*Naturstoffe mit Pyridinstruktur und ihre Biosynthese*', *Fortschr. Chem. organ. Naturstoffe* **1970**, *28*, 109.

34. (–)-Nicotine acts in small doses as a central nervous system stimulant, although in higher doses it leads to central nervous system paralysis.

35. G. Fodor, '*Tropane Alkaloids*', in S. W. Pelletier (Ed.), *Chemistry of the Alkaloids*, Van Nostrand, New York, 1970, 431; G. Fodor, '*The Tropane Alkaloids*', *The Alkaloids* **1960**, *6*, 145; *ibid.* **1967**, *9*, 269; *ibid.* **1971**, *13*, 352; G. Fodor, '*New Methods and Recent Developments of the Stereochemistry of Ephedrine, Pyrrolizidine, Granatane and Tropane Alkaloids*', *Recent Developments in the Chemistry of Natural Carbon Compounds* **1965**, *1*, 15; H. L. Holmes, '*The Chemistry of the Tropane Alkaloids*', *The Alkaloids* **1950**, *1*, 271.

36. A. R. Battersby, H. T. Openshaw, '*The Imidazole Alkaloids*', *The Alkaloids* **1953**, *3*, 201.

37. *General overviews*: T. Kametani, '*The Chemistry of the Isoquinoline Alkaloids*', Hirokawa Publishing Co., Inc., Elsevier Publishing Co., Tokyo, 1969; T. Kametani, K. Fukumoto, '*Benzylisoquinoline and Homobenzylisoquinoline Alkaloids*', in D. H. Hey (Ed.), MTP International Review of Science, Organic Chemistry, Series 1, Vol. 9, 'Alkaloids', Butterworth University Park Press, London, 1973, 181; R. H. F. Manske, '*Isoquinoline Alkaloids*', *The Alkaloids* **1960**, *7*, 423. M. Shamma, '*The Isoquinoline Alkaloids*', in S. W. Pelletier (Ed.), *Chemistry of the Alkaloids*, Van Nostrand, New York, 1970, 31; M. Shamma, *The Isoquinoline Alkaloids – Chemistry and Pharmacology*, Academic Press/Verlag Chemie, New York, 1972. *Specialist overviews*: D. Gross, '*Naturstoffe mit Pyridinstruktur und ihre Biosynthese*', *Fortschr. Chem. organ. Naturstoffe* **1970**, *28*, 109; M. Tomita, '*Die Alkaloide der Menispermacea-Pflanzen*', *Fortschr. Chem. organ. Naturstoffe* **1952**, *9*, 175. *Biogenesis*: R. H. F. Manske, '*The Biosynthesis of Isoquinolines*', *The Alkaloids* **1954**, *4*, 1. *Papaveraceae Alkaloids*: R. H. F. Manske, '*Papaveraceae Alkaloids*', *The Alkaloids* **1968**, *10*, 467; V. Preininger, '*The Pharmacology and Toxicology of the Papaveraceae Alkaloids*', *The Alkaloids* **1975**, *15*, 207; F. Santavy, '*Papaveraceae Alkaloids*', *The Alkaloids* **1970**, *12*, 333.

38. G. Bringmann, '*The Naphthyl Isoquinoline Alkaloids*', *The Alkaloids* **1986**, *29*, 141; G. Bringmann, F. Pokorny, '*The Naphthylisoquinoline Alkaloids*', *The Alkaloids* **1995**, *46*, 128.

39. A. Burger, '*The Benzylisoquinoline Alkaloids*', *The Alkaloids* **1954**, *4*, 29; V. Deulofeu, J. Comin, M. J. Vernengo, '*The Benzylisoquinoline Alkaloids*', *The Alkaloids* **1968**, *10*, 402.

40. Papaverine acts as a general relaxant on smooth muscle and so is used medicinally as a general antispasmodic agent.

41. J. Stanek, '*Phthalideisoquinoline Alkaloids*', *The Alkaloids* **1954**, *4*, 433; *ibid.* **1967**, *9*, 117; D. B. MacLean, '*Phthalideisoquinoline Alkaloids and Related Compounds*', *The Alkaloids* **1985**, *24*, 253.

42. It is also possible that a *Hofmann* degradation of quaternary *N*-methylphthalide-isoquinolines might occur in the course of plant extraction. The occurrence of *N*-methylphthalide-isoquinolines in plants is well documented.

43. The alkaloid term for the ring-opened *N*-methylhydrastine is misleading, since the quaternary form of hydrastine might be understood by it.

44. O. Hesse, '*Vorläufige Notiz über Rhoeadin*', *Liebigs Ann. Chem.* **1865**, Suppl. *4*, 50.

45. H. Rönsch, '*Rhoeadine Alkaloids*', *The Alkaloids* **1986**, *28*, 1.

46. A. Brossi, S. Teitel, G. V. Parry, 'The Ipecac Alkaloids', The Alkaloids **1971**, *13*, 189; M.-M. Janot, 'The Ipecac Alkaloids', The Alkaloids **1953**, *3*, 363; R. H. F. Manske, 'The Ipecac Alkaloids', The Alkaloids **1960**, *7*, 419. H. T. Openshaw, 'The Ipecacuanha Alkaloids', in D. H. Hey (Ed.), MTP International Review of Science, Organic Chemistry, Series 1, Vol. 9, 'Alkaloids', Butterworth University Park Press, London, 1973, 85; C. Szántay, 'Structure and Synthesis of Ipecac Alkaloids', Recent Developments in the Chemistry of Natural Carbon Compounds **1967**, *2*, 63.

47. Psychotria ipecacuanha MÜLL. ARG. (= Cephaelis ipecacuanha WILLD. = Uragoga ipecacuanha BAILL.) is the original emetic root (Radix ipecacuanha) used since antiquity by indigenous Brazilians as a treatment for dysentery. The physician Helvetius in Reims was prescribing it in 1686 as a specific remedy for dysentery and later sold his secret for 1,000 Louis d'or (a considerable sum then) to King Louis XIV of France. It was an extremely popular and frequently used medicine used both as an emetic and in treatment of enteritis, dysentery (diarrhea), and internal bleeding.

48. R. H. F. Manske, 'The Aporphine Alkaloids', The Alkaloids **1954**, *4*, 119; T. Kametani, T. Honda, 'Aporphine Alkaloids', The Alkaloids **1985**, *24*, 153; M. Shamma, 'The Aporphine Alkaloids', The Alkaloids **1967**, *9*, 1; M. Shamma, W. A. Slusarchyk, 'The Aporphine Alkaloids', Chem. Rev. **1964**, *64*, 60; M. Shamma, R. L. Castenson, 'The Oxoaporphine Alkaloids', The Alkaloids **1973**, *14*, 226.

49. W. C. Taylor, 'Eupomatia Alkaloids', The Alkaloids **1985**, *24*, 1.

50. K. Bernauer, W. Hofheinz, 'Proaporphin-Alkaloide', Fortschr. Chem. organ. Naturstoffe **1968**, *26*, 245; L. L. Stuart, M. P. Cava 'The Proaporphine Alkaloids', Chem. Rev. **1968**, *68*, 321.

51. L. Castedo, R. Suaun, 'The Cularine Alkaloids', The Alkaloids **1986**, *29*, 287.

52. L. Castedo, G. Tojo, 'Phenanthrene Alkaloids', The Alkaloids, **1990**, *39*, 99.

53. Berberine is used in the treatment of gastric and intestinal illnesses.

54. R. H. F. Manske, R. Rodrigo, H. L. Holland, D. W. Hughes, D. B. MacLean, J. K. Saunders, 'Solidaline. A Modified Protoberberine Alkaloid from Corydalis solida', Can. J. Chem. **1978**, *56*, 383.

55. R. H. F. Manske, 'The Protopine Alkaloids', The Alkaloids **1954**, *4*, 147.

56. R. H. F. Manske, 'α-Naphthaphenanthridine Alkaloids', The Alkaloids **1954**, *4*, 253; ibid. **1968**, *10*, 485.

57. S. McLean, J. Whelan, 'Spirobenzylisoquinoline Alkaloids', in D. H. Hey (Ed.), MTP International Review of Science, Organic Chemistry, Series 1, Vol. 9, 'Alkaloids', Butterworth University Park Press, London, 1973, 161; M. Shamma, 'The Spirobenzylisoquinoline Alkaloids', The Alkaloids **1971**, *13*, 165.

58. B. Gözler, 'Pavine and Isopavine Alkaloids', The Alkaloids **1987**, *31*, 317.

59. K. W. Bentley, 'The Morphine Alkaloids', in S. W. Pelletier (Ed.), Chemistry of the Alkaloids, Van Nostrand, New York, 1970, 117; K. W. Bentley, 'The Chemistry of the Morphine Alkaloids', Oxford University Press, London, 1954; K. W. Bentley, 'The Morphine Alkaloids', The Alkaloids **1971**, *13*, 3; C. Szántay, G. Dörnyei, G. Blaskó, 'The Morphine Alkaloids', The Alkaloids **1994**, *45*, 127; H. L. Holmes, 'The Morphine Alkaloids I', The Alkaloids **1952**, *2*, 1; H. L. Holmes, G. Stork 'The Morphine Alkaloids II', The Alkaloids **1952**, *2*, 161; G. Stork 'The Morphine Alkaloids', The Alkaloids **1960**, *6*, 219; E. S. Stern, 'Synthetic Approaches to the Morphine Structure', Quart. Rev. **1951**, *5*, 405.

60. H. L. Holmes, 'Sinomenine', The Alkaloids **1952**, *2*, 219.

61. Morphine is used as a highly effective means of blocking pain reception in the central nervous system. Because of its massively addictive character, greatest caution is shown in repeated application. The morphine derivative codeine (morphine monomethyl ether) is known as a specific antitussive (anti-cough) agent.

62. J. W. Cook, J. D. Loudon, '*Alkaloids of the Amaryllidaceae*', *The Alkaloids* **1952**, *2*, 331; C. Fuganti, '*The Amaryllidaceae Alkaloids*', *The Alkaloids* **1975**, *15*, 83; P. W. Jeffs, '*The Amaryllidaceae Alkaloids*', in D. H. Hey (Ed.), MTP International Review of Science, Organic Chemistry, Series 1, Vol. 9, 'Alkaloids', Butterworth University Park Press, London, 1973, 273; W. C. Wildman, '*Amaryllidaceae Alkaloids*', in S. W. Pelletier (Ed.), *Chemistry of the Alkaloids*, Van Nostrand, New York, 1970, 151; W. C. Wildman, '*Alkaloids of the Amaryllidaceae*', *The Alkaloids* **1960**, *6*, 289; *ibid.* **1968**, *11*, 307.

63. V. Boekelheide, '*The Erythrina Alkaloids*', *The Alkaloids* **1960**, *7*, 201; R. K. Hill, '*The Erythrina Alkaloids*', *The Alkaloids* **1967**, *9*, 483; A. Mondon, '*Erythrina Alkaloids*', in S. W. Pelletier (Ed.), *Chemistry of the Alkaloids*, Van Nostrand, New York, 1970, 173; Y. Tsuda, T. Sano, '*Erythrina and Related Alkaloids*', *The Alkaloids* **1996**, *48*, 249.

64. *Cephalotaxus* alkaloids possess antileukemic properties.

65. H. T. Openshaw, '*Quinoline Alkaloids, other than those of Cinchona*', *The Alkaloids* **1953**, *3*, 65; *ibid.* **1960**, *7*, 229; *ibid.* **1967**, *9*, 223.

66. P. J. Scheuer, '*The Furoquinoline Alkaloids*', in S. W. Pelletier (Ed.), *Chemistry of the Alkaloids*, Van Nostrand, New York, 1970, 355.

67. W. Solomon, '*The Cinchona Alkaloids*', in S. W. Pelletier (Ed.), *Chemistry of the Alkaloids*, Van Nostrand, New York, 1970, 301; R. Verporte, J. Schripsema, T. v.d. Leer, '*Cinchona Alkaloids*', *The Alkaloids* **1988**, *34*, 332.

68. Quinine has achieved great significance as an antimalarial and antipyretic drug, while also being used for flavoring 'tonic water'.

69. *Camptotheca* alkaloids possess antileukemic properties and thus have been intensively studied.

70. J. R. Price, '*Acridine Alkaloids*', *The Alkaloids* **1952**, *2*, 353.

71. S. Johne, '*Quinazoline Alkaloids*', *The Alkaloids* **1986**, *29*, 99.

72. N. J. Leonard, '*Senecio Alkaloids*', *The Alkaloids* **1950**, *1*, 107; *ibid.* **1960**, *6*, 35; F. L. Warren, '*Senecio Alkaloids*', *The Alkaloids* **1970**, *12*, 246; F. L. Warren, '*The Pyrrolizidine Alkaloids*', *Fortschr. Chem. organ. Naturstoffe* **1955**, *12*, 198; *ibid.* **1966**, *24*, 329.

73. S. R. Johns, J. A. Lamberton, '*Elaeocarpus Alkaloids*', *The Alkaloids* **1973**, *14*, 326.

74. T. R. Govindachari, '*Tylophora Alkaloids*', *The Alkaloids* **1967**, *9*, 517.

75. F. Bohlmann, D. Schumann, '*Lupine Alkaloids*', *The Alkaloids* **1967**, *9*, 176; F. Galinovsky, '*Lupinen-Alkaloide und verwandte Verbindungen*', *Fortschr. Chem. organ. Naturstoffe* **1951**, *8*, 245; N. J. Leonard, '*Lupine Alkaloids*', *The Alkaloids* **1953**, *3*, 119; *ibid.* **1960**, *7*, 253.

76. W. M. Golebiewski, I. D. Spenser, '*Biosynthesis of the Lupine Alkaloids. II. Sparteine and Lupanine* ', *Can. J. Chem.* **1988**, *66*, 1734; S. Ohmiya, K. Saito, I. Murakoshi '*Lupine Alkaloids*', *The Alkaloids* **1995**, *47*, 1.

77. J. T. Wróbel, '*Nuphar Alkaloids*', *The Alkaloids* **1977**, *16*, 181.

78. J. T. Wróbel, K. Wojtasiewicz '*Sulfur-Containing Alkaloids*', *The Alkaloids*, **1985**, *26*, 53; *ibid.* **1992**, *42*, 249.

79. W. A. Ayer, '*The Lycopodium Alkaloids Including Synthesis and Biosynthesis*', in D. H. Hey (Ed.), MTP International Review of Science, Organic Chemistry, Series 1, Vol. 9, 'Alkaloids', Butterworth University Park Press, London, 1973; D. B. MacLean, '*The Lycopodium Alkaloids*', in S. W. Pelletier (Ed.), *Chemistry of the Alkaloids*, Van Nostrand, New York, 1970, 469; *idem*, *The Alkaloids* **1968**, *10*, 306; R. H. F. Manske, '*The Lycopodium Alkaloids*', *The Alkaloids* **1960**, *7*, 505; D. B. MacLean, '*The Lycopodium Alkaloids*', *The Alkaloids* **1973**, *14*, 348; K. Wiesner, '*Structure and Stereochemistry of the Lycopodium Alkaloids*', *Fortschr. Chem. organ. Naturstoffe* **1962**, *20*, 271.

80. W. Steglich, D. Strack, '*Betalains*', *The Alkaloids* **1990**, *39*, 1; H. Döpp, H. Musso, '*Isolierung und Chromophore der Farbstoffe aus Amanita muscaria*', *Chem. Ber.* **1973**, *106*, 3473.

81. W. Roos, '*Benzodiazepine Alkaloids*', *The Alkaloids* **1990**, *39*, 63.

82. A. Rahman, M. I. Choudhary '*Purine Alkaloids*', *The Alkaloids* **1990**, *38*, 226.

83. Caffeine acts to stimulate and invigorate heart activity; it stimulates the cerebral cortex, and the respiratory and circulatory centers. It also acts to dilate blood vessels and promotes diuresis. Therapeutically, it is used as a tonic, pyschoanaleptic, and diuretic, although theophylline and theobromine are significantly stronger diuretics.

84. Leucopterin was isolated in 1926 as the first pteridine from 200,000 (!) specimens of *Pieris brassicae* L. (Pieridae) (large Cabbage White) by *Wieland* and *Schöpf.*

85. A. Schäfer, B. Fischer, H. Paul, R. Bosshard, M. Hesse, M. Viscontini, '*Electrospray-Ionization Mass Spectrometry: Detection of a Radical Cation Present in Solution: New Results on the Chemistry of (Tetrahydropteridinone)-Metal Complexes*', *Helv. Chim. Acta* **1992**, *75*, 1955.

86. G. Dalma, '*The Erythrophleum Alkaloids*', *The Alkaloids* **1954**, *4*, 265; R. B. Morin, '*Erythrophleum Alkaloids*', *The Alkaloids* **1968**, *10*, 287.

87. L. Reti, '*β-Phenethylamines*', *The Alkaloids* **1953**, *3*, 313; ibid. **1953**, *3*, 339; T. Suzuki, K. Iwai, '*Constituents of Red Pepper Species: Chemistry, Biochemistry, Pharmacology, and Food Science of the Pungent Principle of Capsicum Species*', *The Alkaloids* **1984**, *23*, 227.

88. T. Kametani, M. Koizumi, '*Phenethyl Isoquinoline Alkaloids*', *The Alkaloids* **1973**, *14*, 265.

89. J. W. Cook, J. D. Loudon, '*Colchicine*', *The Alkaloids* **1952**, *2*, 261; W. C. Wildman, '*Colchicine*', in S. W. Pelletier (Ed.), *Chemistry of the Alkaloids*, Van Nostrand, New York, 1970, 199; W. C. Wildman, '*Colchicine and Related Compounds*', *The Alkaloids* **1960**, *6*, 247; ibid. **1968**, *11*, 407.

90. O. Boyé, A. Brossi, '*Tropolonic Colchicum Alkaloids and Allo Congeners*', *The Alkaloids* **1992**, *41*, 125; U. Berg, H. Bladh, '*The Absolute Configuration of Colchicine by Correct Application of the CIP Rules*', *Helv. Chim. Acta* **1998**, *81*, 323

91. W. Schmidbauer, J. vom Scheidt, *Handbuch der Rauschdrogen*, Nymphenburger Verlagsbuchhandlung GmbH, Munich, 1981.

92. The loss of activity of leaves of *C. edulis* on storing or drying is probably attributable to the loss of cathinone due to formation of **A** and its oxidation product **B** (3,6-dimethyl-2,5-diphenylpyrazine).

93. J.-P. Wolf, H. Pfander, '*Synthese und Strukturaufklärung von Merucathinon und Synthese von Cathinon. Inhaltsstoffe von Catha edulis FORSK.*', *Helv. Chim. Acta* **1986**, *69*, 1498

94. L. Crombie, W. M. L. Crombie, D. A. Whiting, '*The Alkaloids of Khat (Catha edulis)*', *The Alkaloids* **1990**, *39*, 139.

95. R. Antkowiak, W. Z. Antkowiak, '*Alkaloids from Mushrooms*', *The Alkaloids* **1991**, *40*, 189.

96. M. M. Badawi, K. Bernauer, P. van den Broek, D. Gröger, A. Guggisberg, S. Johne, I. Kompis, F. Schneider, H.-J. Veith, M. Hesse, H. Schmid, *'Macrocyclic Spermidine and Spermine Alkaloids'*, *Pure Appl. Chem.* **1973**, *33*, 81; M. Hesse, H. Schmid, *'Macrocyclic Spermidine and Spermine Alkaloids'*, in D. H. Hey (Ed.), MTP International Review of Science, Organic Chemistry, Series II, Vol. 9, 'Alkaloids', Butterworth University Park Press, London, 1976; 265; A. Guggisberg, M. Hesse, *'Natural Polyamine Derivatives – New Aspects of their Isolation, Structure Elucidation, and Synthesis'*, *The Alkaloids* **1998**, *50*, 219.

97. R. Tschesche, E. U. Kaussmann, *'The Cyclopeptide Alkaloids'*, *The Alkaloids* **1975**, *15*, 165; E. W. Warnhoff, *'Peptide Alkaloids'*, *Fortschr. Chem. organ. Naturstoffe* **1970**, *28*, 162.

98. F.-P. Wang, X.-T. Liang, *'Chemistry of the Diterpenoid Alkaloids'*, *The Alkaloids* **1992**, *42*, 152.

99. Aconitine and related bases are highly toxic compounds, which act on the central nervous system. They were used in Europe by hunters and warriors as arrow poisons.

100. S. Yamamura, *'Daphniphyllum Alkaloids'*, *The Alkaloids* **1986**, *29*, 265.

101. S. Blechert, D. Guenard, *'Taxus Alkaloids'*, *The Alkaloids* **1990**, *39*, 195.

102. Both taxol and certain of its (less effictive) derivatives are of great medical interest thanks to their powerful antitumoral activities.

103. *General:* V. Cerny, F. Sorm, *'Steroid Alkaloids: Alkaloids of Apocynaceae and Buxaceae'*, *The Alkaloids* **1967**, *9*, 305; G. Habermehl, *'Steroid Alkaloids'*, in D. H. Hey (Ed.), MTP International Review of Science, Organic Chemistry, Series I, Vol. 9, 'Alkaloids', Butterworth University Park Press, London, 1973; 235; J. McKenna, *'Steroidal Alkaloids'*, *Quart. Rev.* **1953**, *7*, 231; Y. Sato, *'Steroidal Alkaloids'*, in S. W. Pelletier (Ed.), Chemistry of the Alkaloids, Van Nostrand, New York, 1970, 591. *Anomalous steroid alkaloids*: K. S. Brown, *'Abnormal Steroidal Alkaloids'* in S. W. Pelletier (Ed.), Chemistry of the Alkaloids, Van Nostrand, New York, 1970, 631. *Alkaloids from Apocynaceae*: V. Cerny, F. Sorm, *'Steroid Alkaloids: Alkaloids of Apocynaceae and Buxaceae'*, *The Alkaloids* **1967**, *9*, 305. *Alkaloids from Holarrhena*: O. Jeger, V. Prelog, *'Steroid Alkaloids: The Holarrhena Group'*, *The Alkaloids* **1960**, *7*, 319. *Alkaloids from Buxus*: J. Tomco, Z. Voticky, *'Steroid Alkaloids: The Veratrum and Buxus Groups'*, *The Alkaloids* **1973**, *14*, 1. *Alkaloids from Solanum*: V. Prelog, O. Jeger, *'The Chemistry of Solanum and Veratrum Alkaloids'*, *The Alkaloids* **1953**, *3*, 247; V. Prelog, O. Jeger, *'Steroid Alkaloids: The Solanum Group'*, *The Alkaloids* **1960**, *7*, 343; K. Schreiber, *'Steroid Alkaloids: The Solanum Group'*, *The Alkaloids* **1968**, *10*, 1. *Alkaloids from Veratrum*: O. Jeger, V. Prelog, *'Steroid Alkaloids: The Veratrum Group'*, *The Alkaloids* **1960**, *7*, 363; S. M. Kupchan, A. W. By, *'Steroid Alkaloids: The Veratrum Group'*, *The Alkaloids* **1968**, *10*, 193; K. J. Morgan, J. A. Barltrop, *'Veratrum Alkaloids'*, *Quart. Rev.* **1958**, *12*, 34; C. R. Narayanan, *'Newer Developments in the Field of Veratrum Alkaloids'*, *Fortschr. Chem. organ. Naturstoffe* **1962**, *20*, 298. *Salamander Alkaloids*: G. Habermehl, *'Steroid Alkaloids: The Salamandra Group'*, *The Alkaloids* **1967**, *9*, 427.

104. J. V. Greenhill, P. Grayshan, *'The Cevane Group of Veratrum Alkaloids'*, *The Alkaloids* **1992**, *41*, 177.

105. L. Lewin, *Die Pfeilgifte*, J. A. Barth-Verlag, Leipzig, 1923.

106. A. Chatterjee, G. Ganguli, *'The Chemistry of Bisindole Alkaloids'*, *J. Sci. Ind. Res. (India)* **1964**, *23*, 178; A. A. Gorman, M. Hesse, H. Schmid, *'Bisindole Alkaloids'*, in *The Alkaloids*, Vol. 1, Specialist Periodical Reports, The Chemical Society, London, 1971, 201.

107. W. I. Taylor, A. R. Battersby, *Oxidative Coupling of Phenols*, Marcel Dekker Inc., New York, 1967.

108. M. Curcumelli-Rodostamo, M. Kulka, *'Bisbenzylisoquinoline and Related Alkaloids'*, The Alkaloids **1967**, *9*, 133; M. Curcumelli-Rodostamo, *'Bisbenzylisoquinoline and Related Alkaloids'*, The Alkaloids **1971**, *13*, 304; M. Kulka, *'Bisbenzylisoquinoline Alkaloids'*, The Alkaloids **1960**, *7*, 439; *ibid.* **1954**, *4*, 199.

109. In 1-benzyltetrahydroisoquinoline alkaloids, the isoquinoline component is designated the 'head' and the benzyl moiety the 'tail'.

110. Tubocurarine, which displays no activity when taken orally, is a depolarization inhibiting muscle relaxant used in surgical operations to suppress involuntary muscle spasm.

111. R. Rodrigo, *'The Cancentrine Alkaloids'*, The Alkaloids **1973**, *14*, 407.

3. Structure Elucidation of Alkaloids

3.1. Introduction

Towards the end of the 18th century, once the 'soporific principle' (morphine) had been isolated from opium, researchers, on a broad front, began to search for pharmacologically interesting substances from other plants as well. At the same time, more and more information was collected about the compounds so far isolated – crystals or oils. In these early years, at the beginning of the 19th century, organic chemistry for the purpose of structure elucidation was first to be developed.

Until late in the 20th century, the only options available to organic chemists for the structure elucidation of unknown compounds were derivatization, controlled chemical degradation, and combustion analysis. Success was nevertheless achieved in determining the simpler compounds through a combination of the highest experimental artistry, dedication, and sharp intellect. With the development of UV, IR, and, especially, both NMR spectroscopy and mass spectrometry between *ca.* 1960 and 1980, this picture was to alter fundamentally: as the structures (to be analyzed) became ever more complex, the time required to interpret them became ever shorter. The truly enormous progress during that period, however, was taking place in another technique: X-ray crystal-structure determination. The evaluation of crystallographic data to obtain relevant information, formerly a true labor of *Sisyphus*, can today, with the aid of computers, be performed in a matter of seconds after measurement. In fact, at the current state of technology, growing suitable single crystals takes significantly longer than structure analysis itself. The great majority of known alkaloid structures have been elucidated or confirmed by this technique.

Building on a broad base of knowledge, it is nowadays possible to both *separate* and structurally *characterize* the smallest quantities of alkaloids in natural mixtures in one step using HPLC-UV-MS techniques, supported by spectral comparison and analogy.

From the beginnings of alkaloid chemistry to the present day, almost 200 years have gone by. Hence, to show the kind of dramatic developments that have taken place over this time span, we will discuss some selected examples in detail.

3.2. Isolation of Alkaloids from Plants

As shown in *Chapt. 1*, alkaloids do not constitute a structurally unified substance class. As well as strongly basic compounds, neutral substances and even a few weakly acidic materials are counted among them. Furthermore – and this is particularly important – the solubilities of the various substances strongly depend on the choice of the solvent. In other words, there exists no generally valid recipe for how to isolate alkaloids from natural sources. Processing of each type of plant (or fungus, microorganism, animal secretion) must, therefore, proceed in a manner dictated by the particular properties of the individual alkaloid class.

In plants, alkaloids often exist in the form of hydro salts (salts formed with organic acids). Therefore, alkaloid isolation from plants usually involves extraction of well-dried, ground-up materials with aqueous methanol (MeOH), acidified with acetic acid (AcOH). This offers the advantage that the anions of the naturally occurring hydro salts (as well as any free bases) are after the extraction uniformly present as acetates. In exceptional cases, it may also be necessary to extract fresh plant material; this can be performed under the same conditions as for dried plants, though. The plant material may be cut up with simple kitchen equipment intended for making vegetable pulp.

Ammonium acetates (which includes most alkaloid hydro salts) are known to be very soluble in aqueous MeOH. The extraction procedure is repeated until the filtrate no longer displays any alkaloid-specific coloring in analytical tests (reagents for the detection of alkaloids are described in *Sect. 3.4*). The so-called 'rolling flask' has proven itself as a suitable instrument for extraction. The dried material to be extracted is put in a glass flask, the solvent is poured onto it, so that the flask is a good 2/3 full, and the vessel is securely sealed. It is then positioned horizontally and rolled (mechanically) for several hours.

After this procedure, all the MeOH- and H_2O-soluble substances from the plant are contained in the extract, which (if necessary) is concentrated *in vacuo* at a maximum temperature of 45° C. A rotary evaporator or, in the case of thermally labile compounds, a freeze-drier is used for this purpose. For subsequent procedures, it is important that the extracts are free of MeOH.

To ensure that all the alkaloids are present in the form of their H_2O-soluble salts, dilute (maximum 1N) aqueous hydrochloric acid (HCl) is added and exhaustive extraction with ether (Et_2O) is performed; in this way, all the Et_2O-soluble ingredients, especially fats, hydrocarbons, and other lipophilic, neutral substances are removed. Then, the aqueous phase is treated with sodium hydroxide (NaOH) solution (1N) until alkaline. Use of ammonia (aqueous NH_3) is inadvisable, because of possible artifact formation (*cf. Chapt. 4*). After that, exhaustive extraction with chloroform ($CHCl_3$) is carried out, and the phases are concentrated, dried over anhydrous sodium sulfate (Na_2SO_4) or magnesium sulfate ($MgSO_4$), and, finally, evaporated to dryness ($\rightarrow$ crude extract). If necessary, it is possible to perform some fractionation even at this stage, by extracting the aqueous phase at different pH values.

The majority of bases now find themselves in the $CHCl_3$ extract, while the aqueous phase contains almost only H_2O-soluble alkaloids, mainly quaternary salts. Since concentration of the aqueous phase is highly laborious and time-consuming, it is better to precipitate out quaternary alkaloids after removal of residual organic solvents. Picrates and 'Reineckates' ($[(NH_3)_2Cr(NCS)_4]^-$) are effective precipitation reagents for such compounds.

Picrate   Reineckate

A plant, as a rule, always contains several alkaloids at once, sometimes in very different quantities. Because of this, chromatographic separation is usually carried out on the crude extract, producing uniform, often crystallizable, fractions.

However, the quaternary alkaloid precipitate from the aqueous phase can also consist of many individual components mixed together (*ca.* 60 were found in the case of the curare alkaloids!), which can only be chromatographically separated after conversion into more suitable salts (such as chlorides) by ion exchange.

For successful alkaloid isolation, it is not enough, as a rule, just to follow through the steps described above. There are several points to which particular attention must be given:

1) Before the main proportion of the plant material is processed, it is necessary to carry out preliminary experiments to establish the best conditions for obtaining the highest yields. In this, and for monitoring of the main experiment, testing of the individual steps by thin-layer chromatography is absolutely essential.

2) Depending on the material to be extracted (*e.g.*, with seeds or pollen), preliminary treatment with petroleum ether may be required, for removal of lipid materials. This treatment must be carried out before the aqueous extraction procedure.

117

3) It must always be borne in mind that air (*i.e.*, oxygen) is present during the individual steps of the isolation procedure. Thus, if the substances to be isolated are air-sensitive, special precautions must be taken. It is entirely possible to perform every step under inert gas.

4) Finally, a sufficient quantity of the original material should always be held in reserve, so that it will be possible later to carry out as many control experiments as necessary.

3.3. Structure Elucidation of Coniine

In 1827, *Gisecke*[1] described for the first time the isolation of the active component of hemlock (*Conium maculatum* L.; French *cigué*, German *Schierling*):

> '*By distilling the plant with calcium hydrate* (*Ca(OH)$_2$*) *and water, he obtained an unpleasant-smelling liquid, from which, after saturation with acids and evaporation, it was possible to obtain an alcohol extract that, to a great extent, possessed the poisonous qualities of the hemlock*'.

In 1829, the same author named this substance 'coniine' (French *conicine*). The largest quantities of this alkaloid were obtained by extraction of fruits just before reaching full ripeness.

In 1837, in the journal *Annalen der Pharmazie*, there first appeared the announcement[2]:

> '*Chemists and pharmacists will be pleased to learn that they may acquire coniine, exceptionally pure and crystal clear, at Fl. 24 per ounce, from Herrn Medicinalrath* [an old German academic title] *Merck in Darmstadt*'[3].

Around this time, there also began a period of intensive investigations concerning the compound's true empirical formula. A great many different salts were prepared and analyzed, with different results, as shown below.

Geiger and Liebig (1831)[4]:	$C_{12}H_{14}NO$
Berzelius (1837)[5]:	$C_{13}H_{22}NO$
Ortigosa (1842)[6]:	$C_8H_{16}N$
Gerhardt (1848)[7]:	$C_8H_{17}N$
von Planta and *Kekulé* (1854)[8]:	$C_{16}H_{15}N$
Gerhardt (1849)[9]:	$C_8H_{15}N$
Hofmann (1881)[9]:	$C_8H_{17}N$

At that time, determination of the empirical formula was obviously based on combustion analysis (*Fig. 3.1*). The true composition of coniine, according to modern opinion, is $C_8H_{17}N$, although for more than 40 years two H-atoms fewer were assumed. In justification of this mistake, it may be offered that the plant extracts contained additional compounds besides coniine: namely *N*-methylconiine and, especially, γ-coniceine. The structures of these compounds were, naturally, still unknown, and for a long period it was not even known that they were present in the plant juice. It was only known that coniine stubbornly held on to water.

(+)-Coniine
(abs. config.)

N-Methylconiine
(abs. config.)

'*For purification,* [the alkaloid] *is neutralized with acids and evaporated to dryness, the salt is extracted with alcohol, split up with alkali, and taken up with ether*'[10].

Together with these endeavors to establish the empirical formula, investigations into coniine's structure also naturally made progress.

A. von Planta and *A. Kekulé*[8] summarized the results in 1854, at the end of their highly comprehensive work '*Beiträge zur Kenntnis einiger flüchtigen Basen: Coniin*' [contributions to knowledge of certain volatile bases: coniine] as follows:

'*From the facts communicated above, which, since no more material was for the moment to be found, could not be as complete as we might have liked, we draw the conclusions that:*

1) *Commercial coniine is (mostly) a mixture of two (or more) homologous bases.*

2) *The compound $C_{16}H_{15}N$ designated as coniine, belongs to the second series of volatile organic bases. It contains 1 equiv. of hydrogen, representable as so-called radicals, while the remaining carbon and hydrogen content $C_{16}H_{14}$ plays the role of 2 equiv. H.*

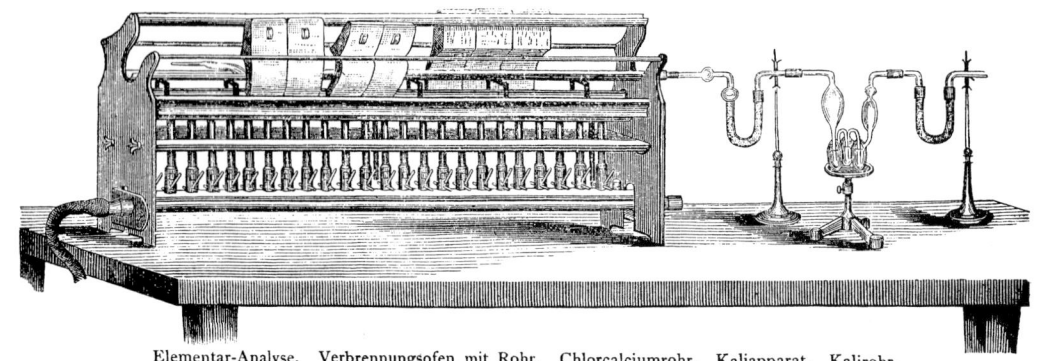

Elementar-Analyse. Verbrennungsofen mit Rohr. Chlorcalciumrohr. Kaliapparat. Kalirohr.

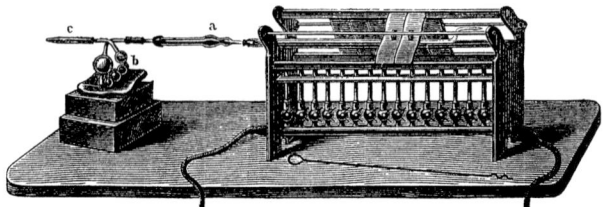

Element.-Anal. Gerade Absorptionsröhren. LIEBIG'scher Kaliapparat.

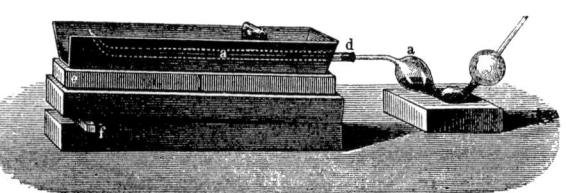

Stickstoffbestimmung nach VARRENTRAPP und WILL.

Exsiccator.

Trockenschrank mit Gasregulator.

Fig. 3.1. *Analytical equipment* (from *A. Ladenburg, Handwörterbuch der Chemie*, Vol. 1, Verlag E. Trewendt, Breslau, 1882 – 1895)

3) *The second base, contained in crude coniine (methylconiine =*
$C_{18}H_{17}N$), is a nitrile base; no more hydrogen in it can be replaced
by so-called radicals; on further substitution it converts into the non-
volatile ammonium base'.

With hindsight, the commercial coniine was indeed a mixture of homologous
secondary N-compounds. The researchers were able to draw this conclusion
after treating coniine with an excess of ethyl iodide (C_2H_5I = EtI). This
gave rise to a 'viscous substance' that did not crystallize. When treated with
KOH/H_2O, it produced an oil ('ethylconiine'), that did, however, possess a
consistent boiling point. Coniine from different sources, on treatment with
EtI, afforded 'ethyl-methyl-coniine iodide = $C_{22}H_{22}NJ$', which could be re-
crystallized. When these crystals were dissolved and heated, decomposition
ensued[8]:

'As already mentioned above, it is possible to evaporate the aqueous so-
lution of the base by boiling without decomposing it; under vigorous
seething nothing more than water escapes. Only when the solution is so
concentrated that it solidifies to something like a paste, on cooling does
decomposition begin. Then, a new, oily base distills over together with
the water, with lively foaming, initially, it is colorless and becomes pale
yellow towards the end of the distillation. At the same time, a permanent
gas, burning with a bright flame, is evolved; this can be identified as eth-
ylene by its behavior with bromine. (It forms a colorless oil, generating
heat in the process, which is heavier than water, has an ether-like smell,
and solidifies in the cooling mixture.) A trace of carbonized material re-
mains in the retort'.

The oil described above is methylconiine ('$C_{18}H_{17}N$'). In a footnote[8] it is explained: *'We once more comment here that the name methylconiine implies*
no more than that the compound contains an additional C_2H_2 relative to co-
niine ($C_{16}H_{15}N$)'.

From a modern perspective, these rather confusing results can be easily
explained as follows. Naturally occurring coniine (a mixture of coniine and
N-methylconiine) reacted with EtI under formation of N-ethylconiine,
N-ethylconiinium hydroiodide, a little of N,N-diethylconiinium iodide, and
N-ethyl-N-methylconiinium iodide, whereupon it was possible to crystallize
the last compound out of the mixture. The noncrystalline components could
be removed with KOH/H_2O, and the crystals were purified. Thermal decomposition then largely produced N-methylconiine. Experience shows
that both MeI and EtI are eliminated in the course of this transformation
(*Scheme 3.1*). Hence, the subsidiary alkaloid was obtained, and the main

alkaloid removed and discarded! In the publication, by the way, no yields were given.

Scheme 3.1

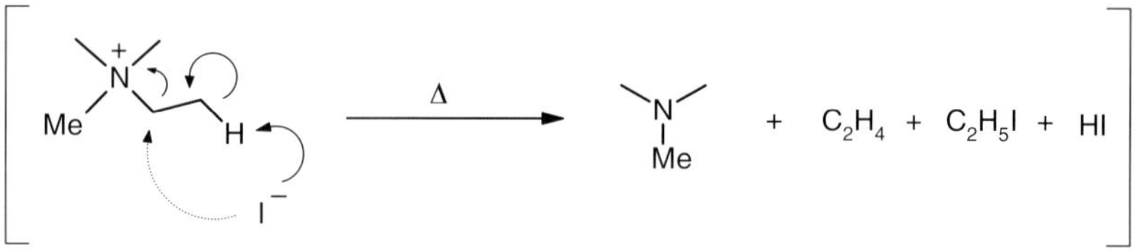

A first more thorough degradation reaction was described by *J. Blyth*[1]. Oxidation experiments, such as treatment of coniine with concentrated nitric acid (HNO_3) or with '*acidic chromium-acidic potassium and sulfuric acid*', or with aqueous bromine (HOBr), gave butyric acid (*Scheme 3.2*), which was '*adequately characterized*' as follows:

– 'aroma' of butyric acid,
– acid reaction,
– cauliflower-like crystals, very deliquescent mass of potassium butyrate,
– the barium butyrate crystallized from alcohol in long prisms, which exhibited '*characteristic rotating motions*' on water,
– '*upon heating with alcohol and concentrated sulfuric acid, butyric acid ether, with the characteristic aroma of pineapple, was immediately given off*'.

Scheme 3.2

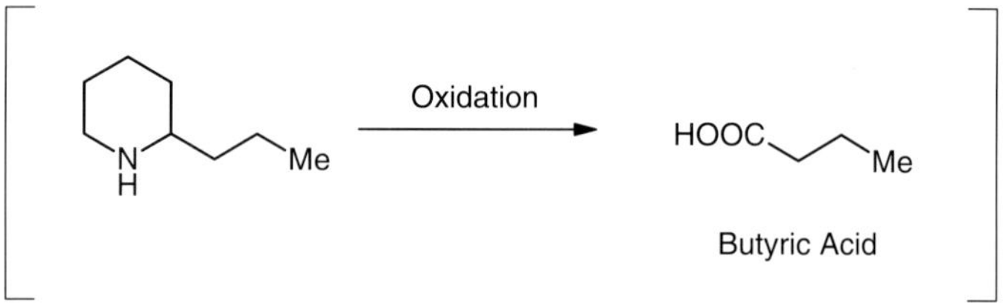

Butyric Acid

A. W. Hofmann (*Fig. 3.2*) found that, on prolonged heating (!) of coniine with hydriodic acid, '*normal octane*' was produced almost quantitatively (*Scheme 3.3*), as presumably identified by its boiling point[11].

Scheme 3.3

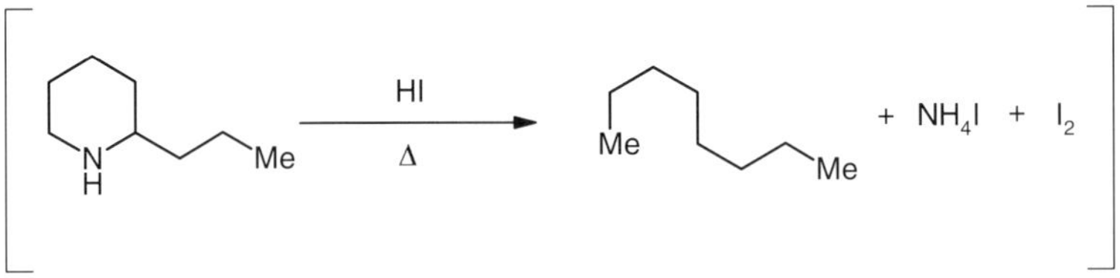

Hofmann had thus observed a reductive variant of the elimination reaction named after him. Pyridine, the structure of which was already known, behaved similarly: '*Upon heating in the presence of concentrated hydriodic acid at temperatures above 300°, there forms the colorless, transparent hydrocarbon quintane (b.p. 35°), which floats on the aqueous liquid in the digestion tube*'. Ammonia and undecomposed pyridine were also detected[12].

The effect of bromine in alkaline solution on coniine was similarly investigated[12]. The '*coniine derivative* $C_8H_{16}NBr$' is produced. With acetic anhydride (Ac_2O), this bromoconiine produces the '*well-characterized base* $C_8H_{14}NH$'. When reducing techniques (not given) were applied, coniine was formed once more.

> '*If reduction is permitted to continue for a protracted period, the coniine is also reduced further, and eventually octylamine and octane are obtained*'[12].

On subjecting coniine to MeI, *Hofmann* obtained '*dimethylconylammonium iodide, which* [displays] *the known properties of completely substituted ammonium iodides*'.

> '*The hydroxide obtained on deiodination with silver oxide, gives on distillation – in analogy with the behavior of the corresponding piperidine compound – neither methyl alcohol nor a hydrocarbon, but a volatile, characteristic base, hardly smelling any more of coniine, which, dehydrated, constantly boils at 182° and which, on analysis, proved to be dimethylconiine, as was not to be expected otherwise. Its composition was established by analysis of a slightly soluble platinum salt, crystallizing in attractive needle form, which melted in the water bath to a deep orange-red liquid, without suffering any decomposition in the process, however. The formula is* $2[(C_8H_{15}CH_3)^{ii}CH_3N \cdot HCl]PtCl_4$'[12].

For successfully deriving the structure of coniine, knowledge of two other related bases, conyrine and conydrine, was crucially important.

Fig. 3.2. *August Wilhelm von Hofmann* (April 8, 1818 – May 5, 1892). Ph.D. in 1841 with Prof. *J. von Liebig*, subject: the chemistry of organic bases of hard coal tar; lecturer in Bonn; 1845–1865 professor at the Royal College of Chemistry, London, then professor at the University of Berlin until his death.

Conydrine also is a hemlock alkaloid, which undergoes reactions analogous to those of coniine. So, for example, on treatment with HI/P at 300° C, octane is likewise formed. If, on the other hand, the temperature does not exceed 150° C, then the 'iodohydrate of an iodinated coniine' is produced[13]. *Hofmann* formulated the following equation for this:

$$'C_8H_{17}NO + 2\,HJ = C_8H_{16}JN \cdot HJ + H_2O'$$

If this salt is then heated with tin and HCl, then coniine is smoothly produced as a reaction product, 'which was identified straightaway on examination of the characteristic, air-stable chlorohydrate'[13].

When 'iodoconiine' was heated with an excess of NaOH while steam was passed through, both α- and β-coniceine were formed[13]. Similar results were obtained when conydrine[14] was treated with concentrated HCl or with phosphoric acid anhydride[15]. The non-naturally occurring compounds α- and β-coniceine can be viewed as isomers of the naturally occurring γ-coniceine. The latter is optically inactive, which is to be expected for a 2,3-dehydroconiine[16].

Finally, *Hofmann* achieved an important breakthrough with the aid of a zinc-powder distillation[17]:

> 'On distilling 10 g of dried hydrochloric coniine with an excess of zinc powder (15 g), there were obtained 1,800 cc [cm³] – in a second experiment with the same amounts 1,780 cc – of a gas that proved on combustion to be hydrogen. […] The distillate, therefore, had to include a base that contains less hydrogen than coniine does, and since the **characteristic smell of the distillate was reminiscent of the pyridine bases** … a base, that is similarly related to coniine as pyridine is to piperidine. […] to which I would like to ascribe the name conyrine ($C_8H_{11}N = C_8H_{17}N - 6H$)'[17].

Oxidation of conyrine with permanganate solution afforded a pyridinecarboxylic acid, 'but not nicotinic acid (which had been obtained by oxidation of a nicotine derivative by *Cahours* and *Etard* in 1881[18], and from nicotine by *Laiblin*[19]), but the isomeric picolinic acid, which Weidel obtained in 1879[20] by oxidation of picoline'[17].

These chemical transformations were confirmed by combustion analysis of several different derivatives (salts). Conyrine could indeed be reductively converted back into coniine[17].

γ-Coniceine

(+)-Conydrine

β-Coniceine

α-Coniceine

In 1885, *Hofmann* was finally able to draw the last conclusions from all his experiments and to graphically illustrate the compositions of the two bases (as shown in *Fig. 3.3*).

Structure determination of the other compounds mentioned, however, could not yet be accomplished definitively. That was achieved later, by *Henry*[21]. For α-coniceine, the least comprehensively characterized dehydroconiine, a bicyclic structure was proposed. The structures *Hofmann* had proposed were confirmed one year later by the synthetic work of *Ladenburg* (*cf. Chapt. 6.2*).

In this context, it is important to note that the structure elucidation of coniine was closely bound up with that of other alkaloids (such as nicotine, among others). Derivation of the basic piperidine structure of coniine was in fact only possible after the pyridinecarboxylic acids had been structurally characterized (around 1880). Then as now, researchers would continually make extensive efforts to keep abreast of new discoveries in the progress of research, so as to be able to solve outstanding problems more efficiently. Over the course of time, this 'mode of operation' was not merely to smooth the path for natural product chemists but was finally going to lead to the development of modern organic chemistry itself.

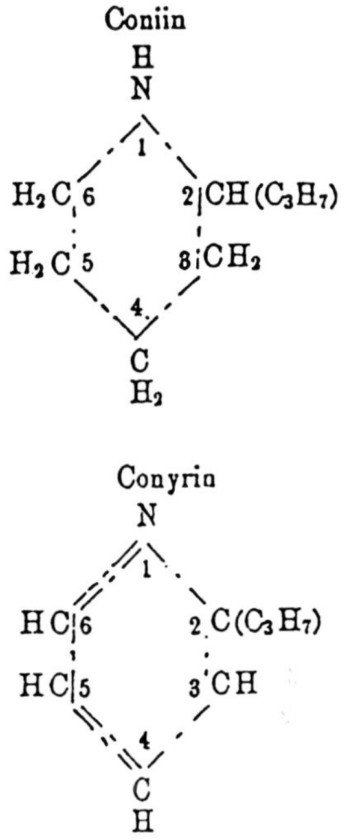

Fig. 3.3. *Graphical formulae of coniine and conyrine (1885)*[16]

3.4. Applications of Chemical and Spectroscopic Methods in Structure Elucidation: Villalstonine

In this section, villalstonine (**1**) will be used for a detailed discussion and explanation of the general methodology involved in the structure elucidation of unknown alkaloids. The highly complicated chemical structure of villalstonine is especially well-suited for demonstrating the kind of logical approaches and techniques that play the key roles in the investigation of unknown compounds. As will be demonstrated, the methodology is quite independent of whether these substances are natural products, synthetic compounds, or environmentally significant chemicals.

3.4.1. Occurrence

The occurrence of villalstonine is restricted to *Alstonia* genus, a member of the Apocynaceae family. The alkaloid was originally found in 1934 in the stem bark of *A. macrophylla* WALL., *A. somersetensis* F. M. BAILEY, and *A. villosa* BLUME (from which species the alkaloid takes its name)[22]. Later, villalstonine was also detected in *A. muelleriana* DOMIN[23], *A. glabriflora*

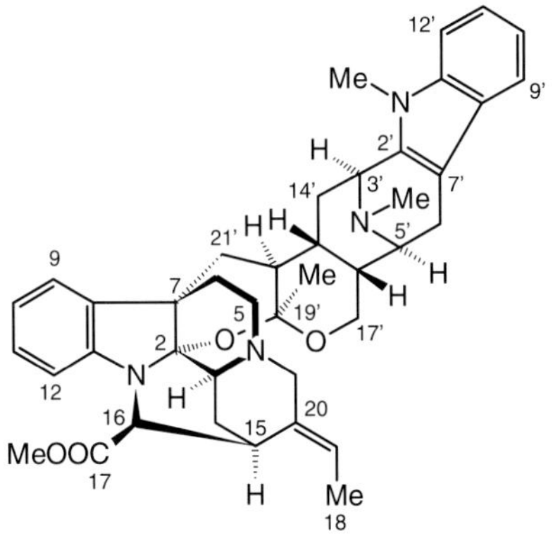

Villalstonine (**1**)

MEF., and *A. spectabilis* R. BR.[24]. The compound does not occur isolated in these plants but is always accompanied by other alkaloids; these can be easily separated by chromatographic techniques, however.

The alkaloid content is commonly not the same in different parts (roots, bark, leaves *etc.*) of a given plant; in many cases, certain compounds are even confined to only one particular part. The alkaloid content may also depends on the season during which the plant was collected (*e.g.*, during or after flowering). Finally, soil quality may also be decisive. Therefore, it is always helpful to give as detailed a description as possible of the drug[25] under examination. The villalstonine used here was obtained from the stem bark of *Alstonia macrophylla*.

3.4.2. Physical Properties

Before a compound is examined chemically, it needs to be physically characterized. That is, its specific properties (such as melting point, optical rotation, spectral behavior *etc.*) must be determined. A *sine qua non* here, however, is that the substance must be pure (uniform). Normally, purity is determined chromatographically (*e.g.*, by thin-layer chromatography (TLC)) in combination with repeated recrystallization and melting point (m.p.) determination. As well as this, it is necessary to ensure that the C,H combustion-analysis values are constant.

Villalstonine (**1**) displays a melting point of $235-270°$ C (decomposition). Its specific optical rotation amounts to $[\alpha]_D^{26} = +79 \pm 5$ ($c = 1.6$, $CHCl_3$). Mass spectrometric analysis[26] of the 'isotopically pure' compound gave for $M^{+\bullet}$ a value of *m/z* 660.3676. Elemental analysis finally provided the following formula: $C_{41}H_{48}N_4O_4$ (M_r 660.864).

Villalstonine shows two pK_a^* values (measured in 89% methylcellosolve): 5.39 and 6.98. From this, it may be concluded that two of the four N-atoms are basic, and that they are located in chemically distinct environments. The remaining two N-atoms appear to possess neither strongly acidic nor basic properties.

The doubly basic nature of the alkaloid is also reflected in its salt formation behavior. Villalstonine bishydrochloride, which precipitates upon mixing an ethereal solution of villalstonine with an anhydrous ethereal HCl solution, is a crystalline compound. *N,N'*-Dimethylvillalstonine diiodide (also called 'villalstonine dimethoiodide') cannot be obtained by dissolving the alkaloid in benzene (containing a little MeOH), adding MeI at room temperature, and leaving the solution to stand for a few hours. However, quaternization proceeds easily, if sodium carbonate (for neutralizing the HI produced from MeI in the protic solvent) is added to the reaction mixture, and the solution is refluxed for several hours. Finally, *Zeisel* determination[27] indicated the presence of 1.1 (*O*)Me and 2.1 (*N*)Me groups.

As determined by the SRB test, Villalstonine displays notable levels of activity against two forms of human lung cancer: MOR-P (adenocarcinoma) and COR-L23 (large cell carcinoma)[28]. Yet, it is a thousand times less effective than, *e.g.*, vinblastine sulfate in this regard[29].

3.4.3. Spectroscopic Analysis and Functional Groups

The infrared (IR) spectrum of villalstonine in $CHCl_3$ shows two intensive bands (*Fig. 3.4*) in the ester carbonyl region at 1754 and 1730 cm^{-1}. At 1610 cm^{-1} there is another strong absorption band. These three bands are discussed in greater detail later. No OH and NH absorptions are observed.

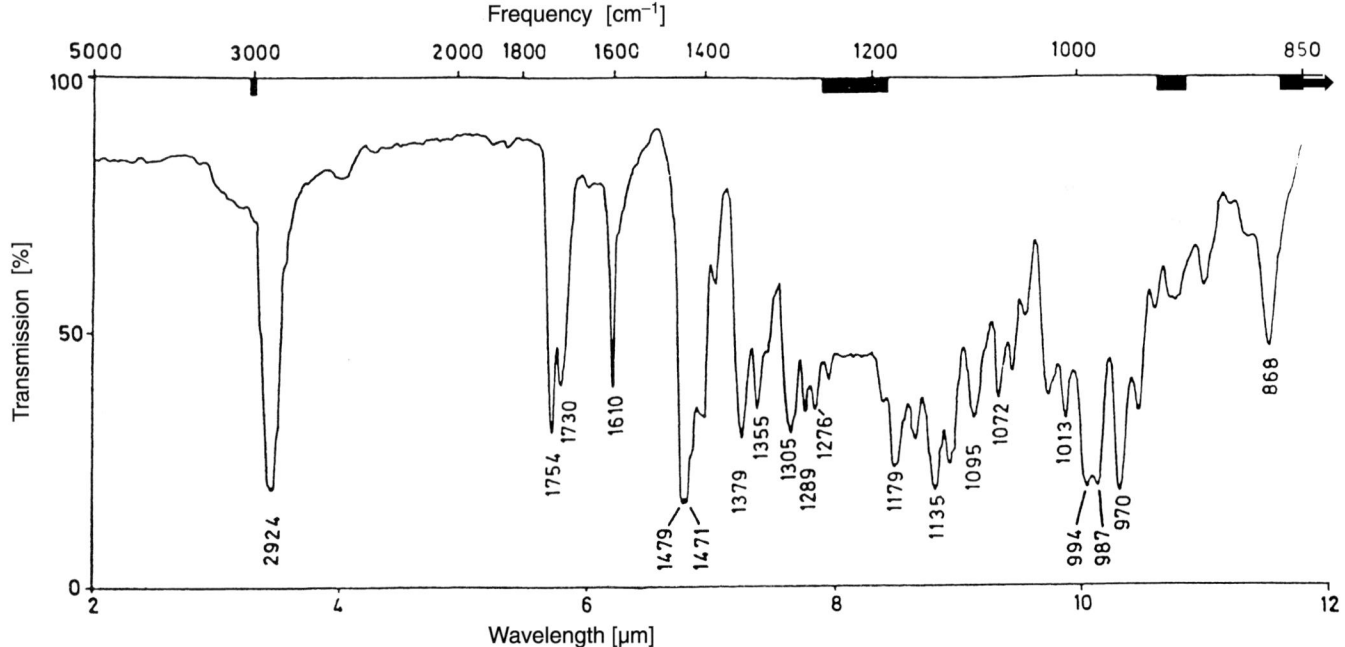

Fig. 3.4. *IR Spectrum of villalstonine in CHCl₃.* Microcell, 0.2 mm; regions indicated in black: solvent absorptions.

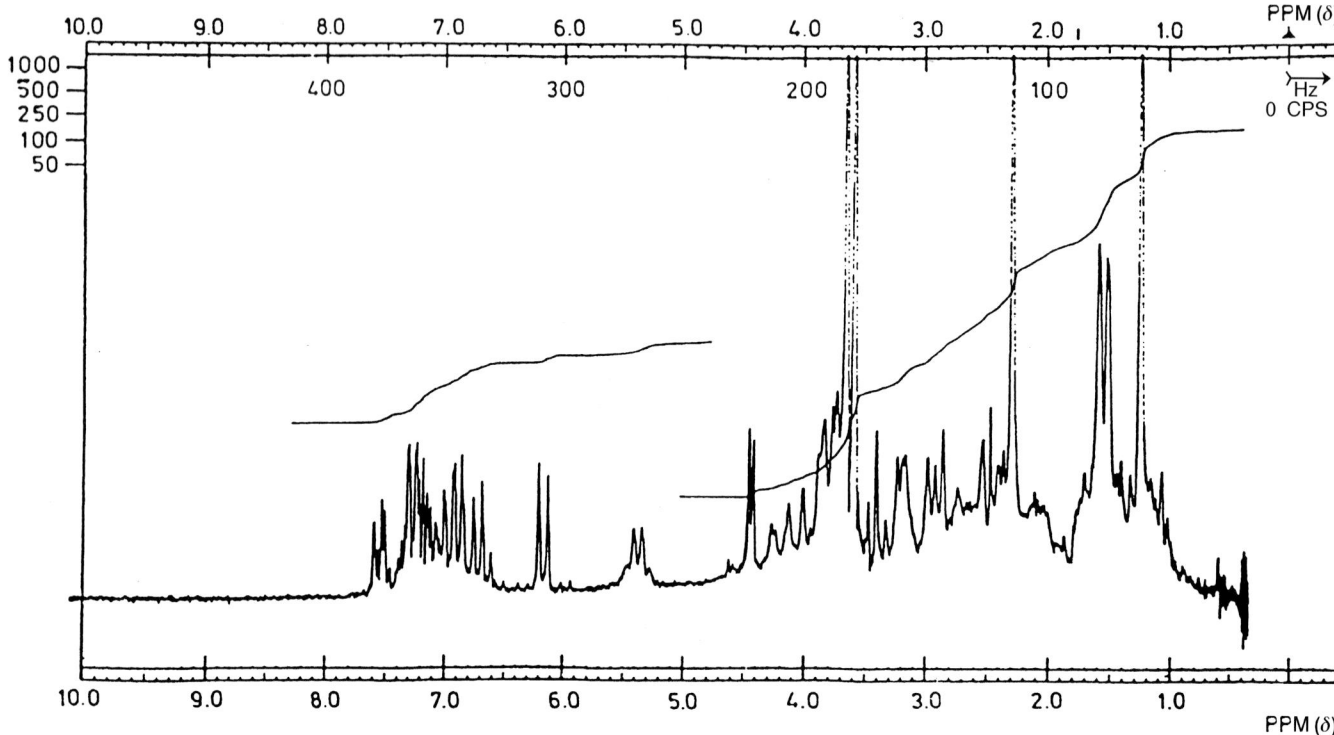

Fig. 3.5. *¹H-NMR Spectrum of villalstonine in CDCl₃ at 100 MHz*

These observations square with the experimental finding that the compound cannot be acetylated under conventional conditions (pyridine/Ac₂O 1:1, 45° C). Unambiguous analysis of the remaining spectral regions does not appear possible.

As to be expected from the large complement of H-atoms, the ¹H-NMR spectrum[30] of villalstonine is truly complicated (*Fig. 3.5*). In the region from 6.5 to 7.7 ppm, there is a *multiplet* corresponding to seven aromatic protons. An extra aromatic proton (H–C(12)) appears as a *doublet* (*J* = 8 Hz) at 6.16 ppm. A poorly resolved *doublet × doublet*, originating from a vinyl proton, is situated at 5.37 ppm. A further signal (H–C(16)) is discernable as another *doublet* (*J* = 4 Hz) at 4.43 ppm. *Singlets* corresponding to Me groups are located at 3.58 and 3.64 ppm (COOMe and indolyl NMe, respectively), at 2.28 ppm (aliphatic NMe), and at 1.23 ppm (Me–C(19′)). In

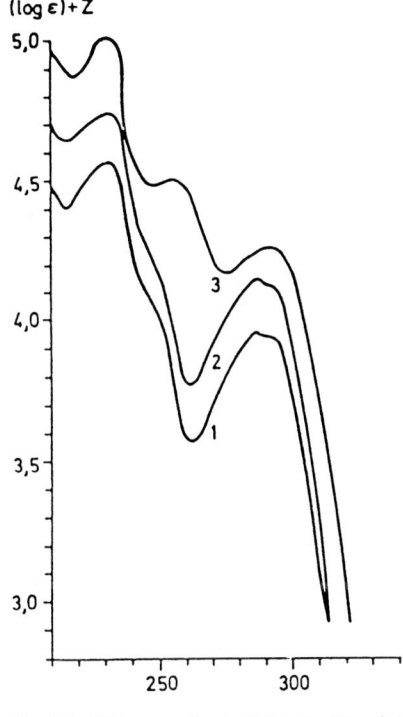

Curve 1:
Villalstonine (**1**; *M* 660)
$c = 4.77 \times 10^{-5}$, $Z = 0.0$

Maxima: 231 (4.57), 286 (3.96)
Minima: 216 (4.40), 261 (3.57)
Shoulder: 250 (4.00), 294 (3.93)

Curve 2:
Villamine (**7**; *M* 660)
$c = 4.62 \times 10^{-5}$, $Z = 0.2$

Maxima: 231 (4.56), 286 (3.97)
Minima: 216 (4.46), 260 (3.60)
Shoulder: 250 (3.96), 294 (3.93)

Curve 3:
Addition Spectrum:
Macroline (**9**) + 2,7-Dihydro-
pleiocarpamine (**6**), $Z = 0.2$

Maxima: 230 (4.61), 252 (4.11)
          292 (3.85)
Minima: 218 (4.47), 247 (4.08)
          274 (3.77)

Fig. 3.6. *UV Spectra of both villalstonine (**1**) and villamine (**7**), as well as the addition spectrum of macroline (**9**) and 2,7-dihydropleiocarpamine (**6**), as recorded in 95.5% EtOH*

addition to these, a Me *doublet* with fine structure ($J \approx 8$ Hz) is observed at 1.55 ppm.

The UV spectrum (95% EtOH) of villalstonine is presented in *Fig. 3.6*, suggesting that the compound does not possess a simple known, but rather a composite chromophore.

To clarify the nature of the functional groups, some simple reactions were carried out. If villalstonine is treated with lithium aluminium hydride (LiAlH$_4$) in tetrahydrofuran (THF) over 8 h at 20° C, villalstoninol (**2**, M_r 632) is formed, which displays some characteristic alterations relative to villalstonine. From the difference in molecular masses (−28 mass units = −28 amu), a methoxycarbonyl group (MeOOC) must have been reduced to a primary alcohol (COOMe (59 amu) → CH$_2$OH (31 amu); difference: 28 amu).

2

129

In accord with this, it is possible, under conventional acetylation conditions (pyridine/Ac$_2$O), to transform compound **2** into a monoacetate derivative. Additionally, the UV spectra of **1** and **2** show no significant differences, which implies that the MeOOC group in **1** is not conjugated with the rest of the molecule. The IR spectrum (CCl$_4$) of **2** is notable for the absence of both of the ester carbonyl absorptions mentioned above. Intensive bands are observed only at 3030 (OH) and 1662 cm^{-1}. From this, it can be concluded that the carbonyl bands at 1754 and 1730 cm^{-1} in the villalstonine spectrum must stem from only *one* ester carbonyl group; the mass difference of 28 amu between **1** and the reduced **2** cannot be explained otherwise. This crucial conclusion is also confirmed in the ^{1}H-NMR spectrum of **2**, in which only three of the originally observed four villalstonine Me *singlets* are present (3.70, 2.36, and 1.31 ppm).

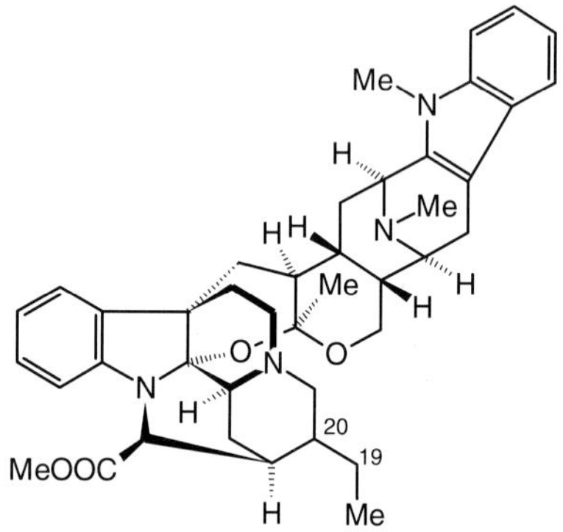

Villalstonine may be catalytically hydrogenated with H$_2$/Pd to give dihydrovillalstonine (**3**), which lacks the exocyclic double bond (the UV spectrum of **3** is almost identical with that of **1**, and the molecular mass has changed by +2). In the ^{1}H-NMR spectrum of dihydrovillalstonine, the *quadruplet* (*q*) at 5.37 ppm is now absent, as is the *doublet* at 1.55 ppm. From this, it follows in turn that villalstonine clearly must contain a trisubstituted ethylidene group (>C=CH–Me). Incidentally, this also allows the *quadruplet* at 5.37 ppm to be used as an integration standard (1 H) for the spectrum of **1**.

Two of the four N-atoms present in villalstonine are tertiary and thus basic. Treatment of **1** with MeI (containing some MeOH) was followed by addition of anhydrous Na$_2$CO$_3$ and refluxing for 24 hours. This gave rise to villalstonine dimethoiodide (**4**) with the empirical formula C$_{43}$H$_{54}$I$_2$N$_4$O$_4$ · 3 H$_2$O, obtained by combustion analysis[31]. However, a molecular mass of only 660 (as for **1**) was obtained for the ammonium salt **4** by mass spectrometry. This result can be explained by thermal production (fragmentation) of the 'di-nor base' villalstonine (**1**) in the mass spectrometer. Were one of the two N-atoms secondary (or primary), than the molecular mass of the 'di-nor base' would have been 14 amu (or 28 amu) higher (N–H → N–*CH$_2$*–H). That apart, the combustion analysis would also have given different results in this case.

The findings thus far obtained for the structural determination of villalstonine may be summarized as follows:

3

Definitely present in villalstonine (**1**) are:

Fragment			**Evidence**
1 double bond	$1 \times$	![C=C with H and Me]	^{1}H-NMR, catalytic hydrogenation
1 COOMe	$1 \times$	COOMe	^{1}H-NMR, LiAlH$_4$ reduction; both are stronger arguments than IR, which suggests two ester groups
8 arom. H			^{1}H-NMR
2 basic N	$2 \times$	N	Evidence of the dimethoiodide **4**
2 (N)–Me	$2 \times$	(N)—Me	^{1}H-NMR, determination of (N)–Me; one of the two Me groups is presumably bound to a basic N-atom
1 (C)–Me	$1 \times$	—C—Me	^{1}H-NMR

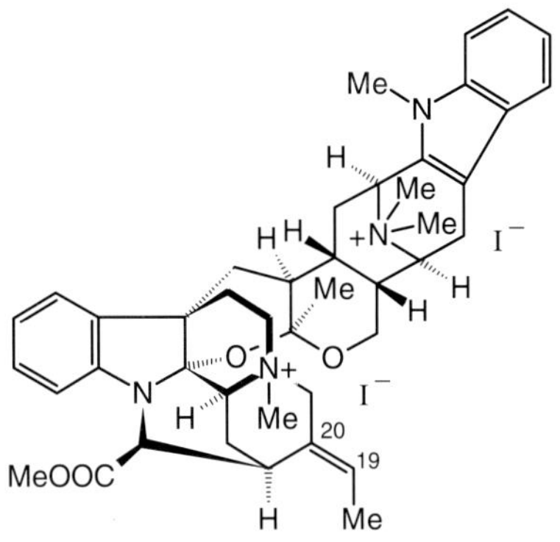

4

Definitely **_not_** present in villalstonine (**1**) are:

OH or NH groups	N—H, OH	acetylation and methylation experiments
additional (C=O)-groups	C=O	IR

What comes out of this inventory is the conclusion that some insights can indeed be gained through the methods of investigation described, but also that no genuine understanding of the molecular structure nor the determination of the alkaloid type is possible solely on the basis of the above informations at hand. To further elucidate the structure of villalstonine, it is necessary, on one hand, to look for chemical reactions that may reveal more about the chemical properties of the substance. Suitable starting points are functional groups (such as amides, esters, ethers) that may be cleaved or otherwise transformed by appropriate reagents. On the other hand, the mass spectrum may also help; it is possible to look for signals that belay the presence of known alkaloids, or at least known structural elements. Only if this is indeed the case, it is then possible to specifically perform appropriate reactions to further check for characteristic structure elements.

As far as villalstonine is concerned, there are, as a matter of fact, some clues of this type present that can be used in this way. It should be stressed though that clues are not facts but rather well-founded suspicions, which may or may not be endorsed experimentally. The following hints appear particularly promising:

1) The mentioned methoxycarbonyl double band in the carbonyl region is a rarely observed phenomenon. It would be conceivable that this double band might also occur in the spectra of known model compounds incorporating similar or even identical chromophores. In fact, this phenomenon is also found in the spectrum of the indole alkaloid pleiocarpamine (**5**)[32]. The possibility, therefore, exists that either pleiocarpamine itself, a derivative of it, or some of its essential structural elements might be present in villalstonine.

2) Before the actual structure elucidation started, it was already known that indole alkaloids occur in *Alstonia* species. The *Alstonia* genus belongs to the Apocynaceae family, also well-known as a plant family containing indole alkaloids (*cf.* also *Chapt. 7*). It would thus not be totally outlandish to assume that villalstonine might also be an indole alkaloid. Further evidence in support of this hypothesis is provided in the IR spectrum of villalstonine: the absorption band at 1610 cm^{-1} can be found in many compounds with an indoline (dihydroindole) chromophore. If villalstonine is indeed a compound incorporating an indoline chromophore[33], it should then also exhibit a specific color reaction behavior on exposure to Ce^{IV}-sulfate reagent. In the case of villalstonine, however, findings were negative, which, in turn, flatly contradicts the IR spectral evidence. There is, however, a let-out in that one possible explanation for the absence of the color reaction might be provided by the strongly acidic character of the reagent (4% aqueous solution of $Ce(SO_4)_2$ in $2 \text{ N } H_2SO_4$)[34–36]. It would thus be imaginable that the molecule undergoes some profound change under strongly acidic conditions and does not react in the expected manner. This consideration might, if correct, provide a further clue as to the structure of villalstonine (**1**).

At this point, attention should once more be paid to the empirical formula of villalstonine: $C_{41}H_{48}N_4O_4$. The substance thus contains four O-atoms, of which two can be attributed to a methoxycarbonyl group. The other two exist – as we know from the earlier experiments – neither as part of an aldehyde moiety, nor as part of a ketone. From this, it follows that only ethers

(+)-Pleiocarpamine (**5**)

and acetals come into question as possible functional groups, since these are known to be hydrolyzed by (strong) acids. On these grounds, it is only logical to examine the effect of acids on villalstonine.

3.4.4. Acid-Catalyzed Cleavage Reactions of Villalstonine

If villalstonine is treated at ambient temperature with 70% perchloric acid (HClO$_4$) for 45 min, the result is a purple solution from which, after neutralization and alkalization, an extract can be isolated, which, after chromatographic purification on aluminum oxide, affords pleiocarpamine (**5**; C$_{20}$H$_{22}$N$_2$O$_2$, M_r 322) in a yield of *ca.* 40%. The pleiocarpamine obtained synthetically displays the same properties (TLC, m.p., $[\alpha]_D$, UV, IR, ^{1}H-NMR, MS, and color reactions) as the natural compound, isolated from the Apocynaceae *Pleiocarpa mutica* BENTH. Despite great efforts, it was not possible to isolate any other compounds from the HClO$_4$ hydrolysate. The greater proportion of the villalstonine used was converted into a resin-like product.

Since the molecular mass of villalstonine (**1**) does not correspond to double the mass of pleiocarpamine, the former cannot be a simple dimer of the latter. There must, therefore, exist another, rather heavy component bound to pleiocarpamine. Of this second component, we know that it is being liberated by HClO$_4$, but immediately destroyed under these strongly acidic conditions. Thus, the next goal must be trying to achieve the same cleavage under yet acidic, but nonetheless gentle conditions. In doing so, both HCl and AcOH proved to be too mild (the starting material was recovered). Upon using sulfuric (H$_2$SO$_4$) and phosphoric acid (H$_3$PO$_4$), pleiocarpamine was formed – albeit in poorer yield than with HClO$_4$ – but it was not possible to obtain any second compound in this way.

An alternative approach was to reduce the obviously highly sensitive product immediately after its formation. Several experiments followed, in which villalstonine was treated with concentrated HCl in the presence of tin, zinc, or zinc^{II} chloride. In none of these experiments, however, it was possible to isolate the unknown component: a dihydro derivative of pleiocarpamine (2,7-dihydropleiocarpamine (**6**), M_r 324) was obtained instead. With the aid of a control experiment, it was possible to ascertain that pure pleiocarpamine is also subject to transformation into **6** under the above (reductive) conditions. This experiment was therefore also a flop.

After a number of futile attempts at acidic hydrolyses, a breakthrough finally came with the application of a mixture of trifluoroacetic acid/trifluoroacetic

6

acid anhydride (TFA/TFAA). The reagent was first allowed to act on crystalline villalstonine bishydrochloride tetrahydrate ($C_{41}H_{48}N_4O_4 \cdot 2\,HCl \cdot 4\,H_2O$) for 5 minutes at room temperature. The excess of TFA/TFAA was removed under a stream of N_2, and the residue was made basic with NH_3 and worked up. Under these conditions, it finally proved possible to isolate two isomers of villalstonine, together with a little starting material. The main product (40.5%) was named *villamine* (**7**), the minor one *villoine* (**8**). No pleiocarpamine (**5**) was produced under these reaction conditions.

As far as the characterization of these two new substances is concerned, the following points are relevant:

1) The mentioned double IR band for the methoxycarbonyl group is found in the spectra of both isomers at 1762 and 1737 cm^{-1} in **7** (CCl_4), and at 1754 and 1731 cm^{-1} in **8** ($CDCl_3$), respectively. From this, it may be concluded that both compounds still incorporate the pleiocarpamine moiety[37].

2) Both villamine and villoine may be converted back into villalstonine by treatment with dilute HCl.

3) A very important finding came from the mass spectrum of villamine (**7**; *Fig. 3.7*), which clearly differs from that of villalstonine (**1**; *Fig. 3.8*), but displays a certain similarity to that of pleiocarpamine (**5**; *Fig. 3.9*).

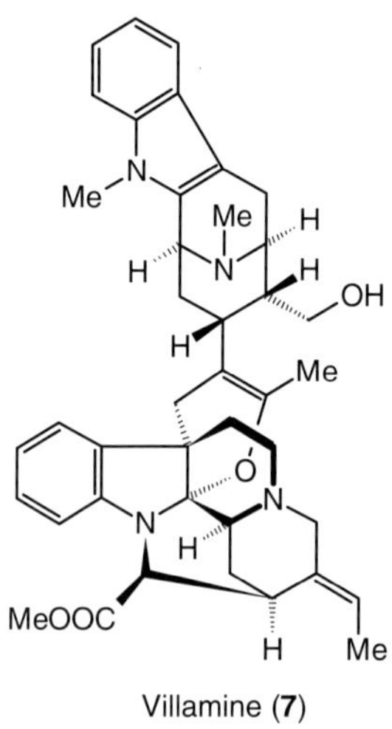

Villamine (**7**)

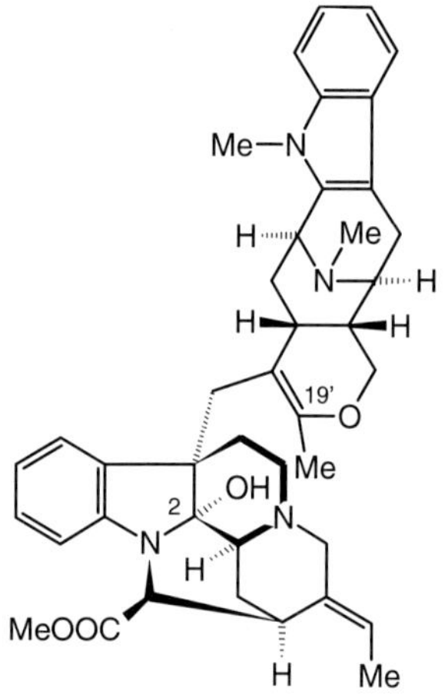

Villoine (**8**)
(assumed structure)

Obviously, the mass spectrum of pleiocarpamine (**5**) partially corresponds to that of villamine (**7**). In other words: if the spectrum of pleiocarpamine is subtracted from that of villamine, then the mathematically generated spectrum might in fact represent that of the unknown component. This assumption is based on simple arithmetic: the molecular mass of the one component (pleiocarpamine) is 322. The putative molecular ion of the other compound would have the mass 338; both added together would give 660, which matches the molecular mass common to villalstonine (**1**), villamine (**7**), and villoine (**8**). Even though this may be only an assumption, perhaps just a piece of circumstantial evidence, then this trail still appeared worth following up, especially because all the earlier attempts to derive other compounds from villalstonine had failed.

Let us briefly think about how an electron-impact (EI) mass spectrum of an organic, N-containing compound is produced. Before the molecules of a sample can be ionized, the probe sample has to be vaporized. While hydro salts of organic bases easily dissociate thermally into their respective components (molecular hydrogen halide and uncharged base)[38], and may be ionized independently during mass spectrometric electron bombardment, no such reaction can take place with the thermally more stable tetraalkylammonium compounds, for obvious reasons.

Transfering a salt into the gas phase under EI-MS conditions is not possible without transforming it into neutral components. This thermal conversion, in turn, takes place according to certain laws, which make it possible to draw certain conclusions about the molecular weight of the salt originally involved. Tetraalkylammonium compounds may be transformed into neutral constituents in

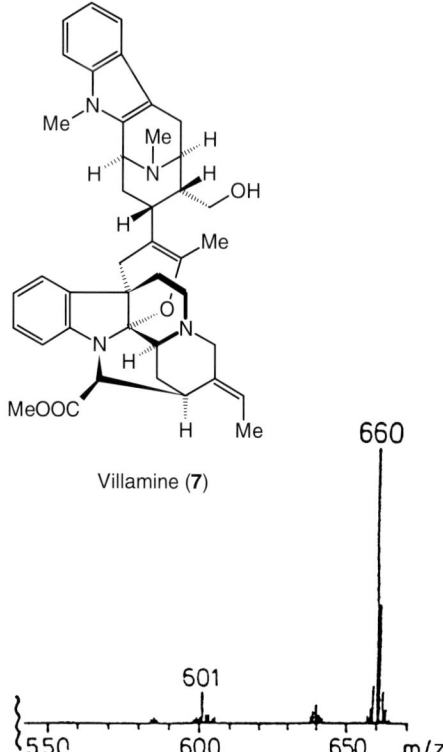

Villamine (**7**)

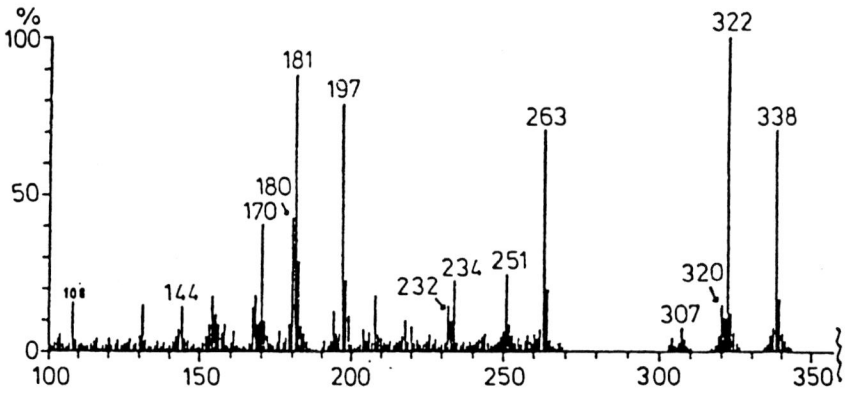

Fig. 3.7. *EI Mass spectrum of villamine (**7**)*

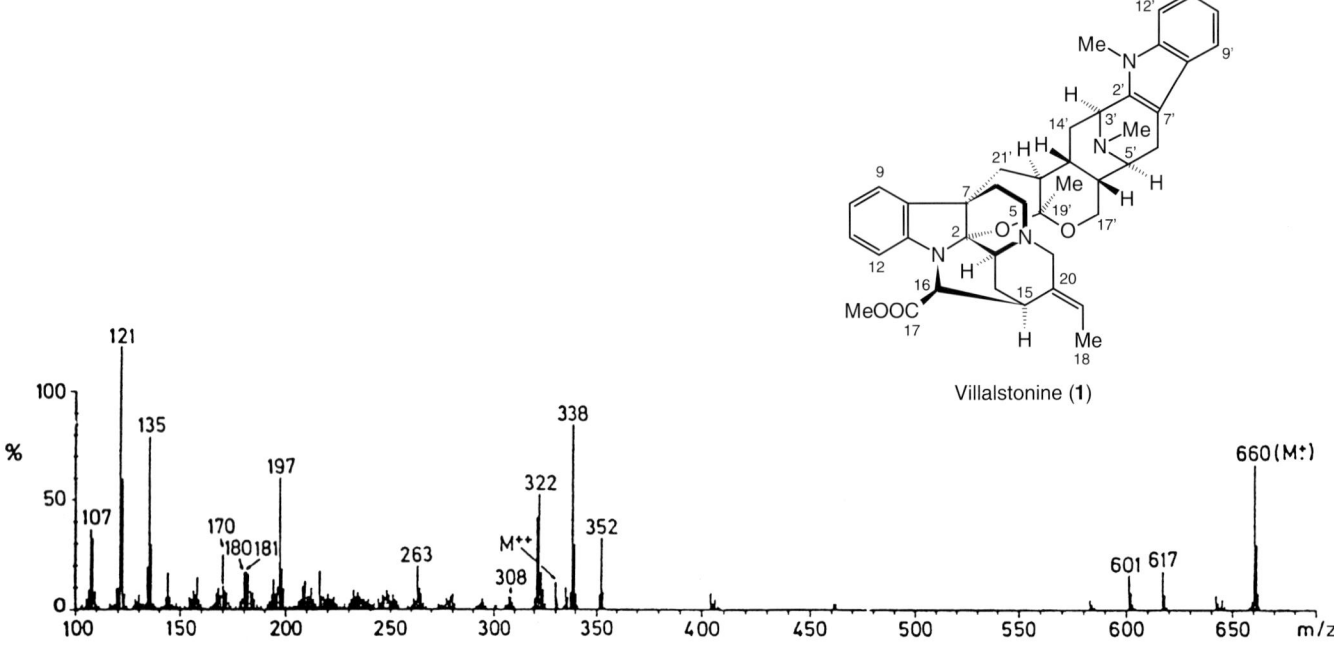

Villalstonine (**1**)

Fig. 3.8. *EI Mass spectrum of villalstonine* (**1**)

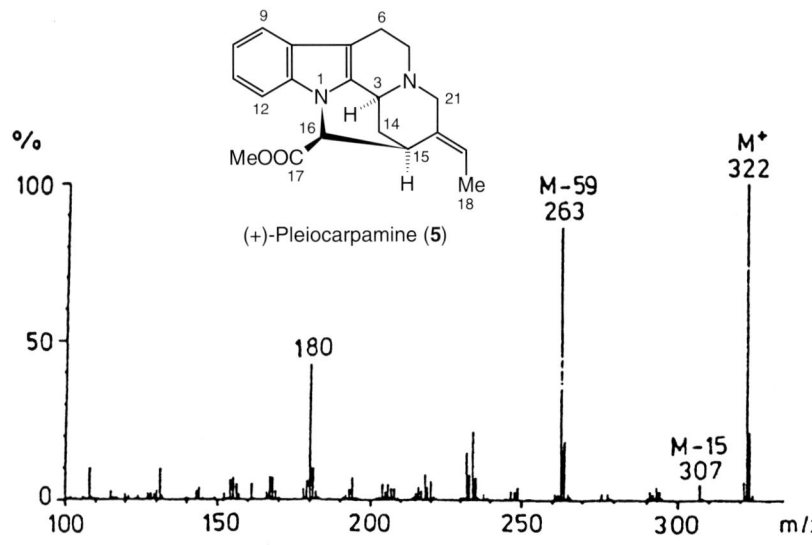

(+)-Pleiocarpamine (**5**)

Fig. 3.9. *EI Mass spectrum of pleiocarpamine* (**5**)

three ways: by dealkylation, by thermal *Hofmann* degradation, and by substitution reactions (with the amine becoming incorporated into the organic moiety, leading to the formation of an uncharged molecule).

But back to the mass spectrum of villamine. As mentioned above, it is necessary to vaporize the molecules of a test sample before ionization; during this procedure a number of thermal reactions can come into play. The components originating from such reactions are ionized independently of one another and lead to discrete, superimposed mass spectra. If some thermal reaction of this kind were also to be active in the vaporization of villamine (**7**), then it would indeed be possible to explain the superposition of the spectra of pleiocarpamine and the other compound in the mass spectrum of villamine in this manner, but not, however, the appearance of signals of $m/z > 400$. As an alternative, therefore, it should be taken into account that villamine itself features in its mass spectrum the signals shown in *Fig. 3.7*. Therefore, to examine the thermal stability of villamine (**7**), the compound was distilled *in vacuo*[39]. From the distillate produced, it was indeed possible to obtain a second compound named *macroline* (**9**; M_r 338) together with pleiocarpamine (**5**; $C_{20}H_{22}N_2O_2$, M_r 322). This was evidence that, in the mass spectrometer, villamine would at least partially decompose thermally. The mass spectrum of pure macroline (*Fig. 3.10*) and the corresponding spectrum derived by subtraction techniques agree with each other astonishingly well.

Macroline (**9**)

Now, with a part of the problem concerning the structure of villalstonine solved, the subsequent steps must include: *a*) determination of the structure of macroline (**9**) as a component base of villamine (**7**); *b*) derivation of the structure of villamine (**7**) as a component of villalstonine (**1**); *c*) determination of the structure of villalstonine (**1**) itself.

3.4.5. Solution of the Molecular Structure of Macroline

(+)-Macroline (**9**; $C_{21}H_{26}N_2O_2$, M_r 338) crystallizes from MeOH/Et$_2$O as colorless needles. This compound always decomposes upon melting at 211–213° C. Specific rotation: $[\alpha]_D = + 19 \pm 5$ ($c = 0.406$, MeOH). Its UV spectrum (*Fig. 3.11*) fits that of an *N*-alkylated indole chromophore: λ_{max} [nm] 231 (log $\varepsilon = 4.56$), 282 (3.67), 288 (3.66); λ_{min} 254 (3.12), 286 (3.64) (measured in EtOH).

This assignment was achieved by comparison of spectra from the literature with that of the 'unknown' macroline (**9**). Its IR spectrum (*Fig. 3.12*) shows the presence of an OH group (weak band at 3205 cm^{-1}). The corresponding

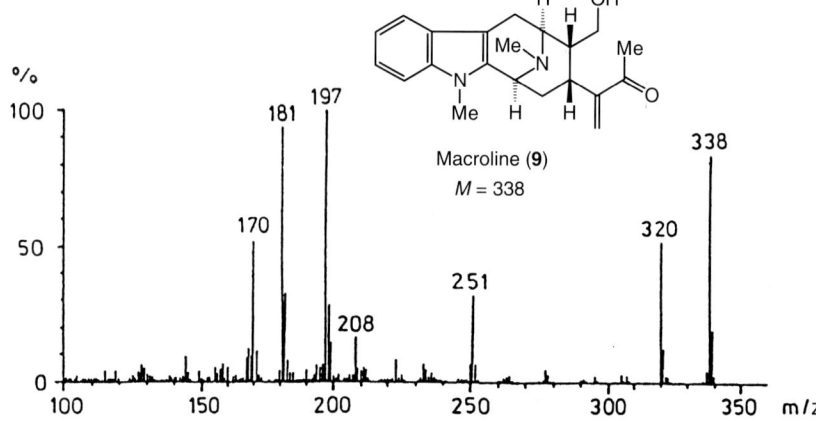

Fig. 3.10. *EI Mass spectrum of macroline* (**9**)

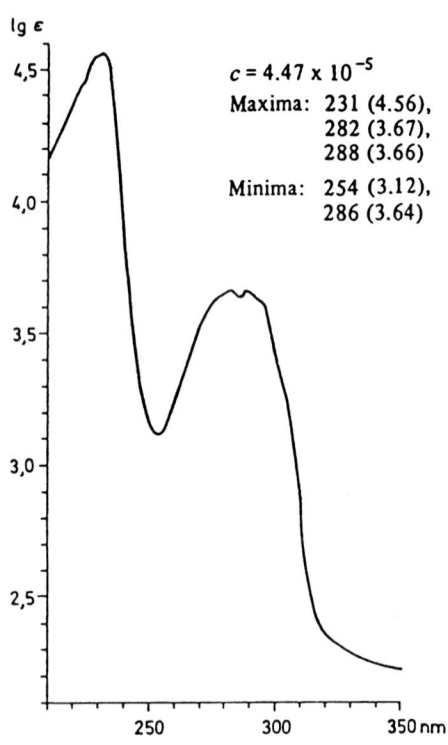

$c = 4.47 \times 10^{-5}$

Maxima: 231 (4.56),
282 (3.67),
288 (3.66)

Minima: 254 (3.12),
286 (3.64)

Fig. 3.11. *UV Spectrum of macroline* (**9**), *recorded in EtOH* (95.5%)

^{1}H-NMR spectrum (60 MHz, CDCl$_3$) confirms the findings from the UV and the IR spectra and, at the same time, provides additional information.

In the region between 6.8 and 7.8 ppm, there appears a *multiplet*, corresponding to four aromatic protons. The *N*-alkyl group on the indole nucleus is a Me residue, as shown by a *singlet* at 3.60 ppm (*cf.* villalstonine; 3.55 or 3.63 ppm). The α,β-unsaturated ketone, anticipated from the corresponding IR spectrum, is easily recognized: the C=C bond corresponds to a vinyl group (*singlet* at 6.18 ppm (1 H) and *singlet* with fine splitting ($J = 1.5$ Hz) at 5.96 ppm). The fine splitting is caused by an allylic proton and can be suppressed by application of a second irradiation at 2337 Hz, typical for the absorption of allylic protons. A terminal Me group (*singlet* at 2.26 ppm) belongs to the α,β-unsaturated ketone. As the natures of both O-functionalities of macroline are known by now (OH and C=O), and so no second oxo group is present, it can be assumed that the original ^{1}H-NMR assignments are definite. Another well-defined NMR signal is to be found at 2.38 ppm, probably attributable to another Me group on a second, non-indolic N-atom (*cf.* villalstonine: 2.25 ppm).

If the spectroscopic findings from UV, IR, and ^{1}H-NMR are combined, then it is possible to formulate the partial structure **I** for macroline (**9**). What still remains unknown is the construction of the ring system. The findings obtained so far thus still require further substantiation.

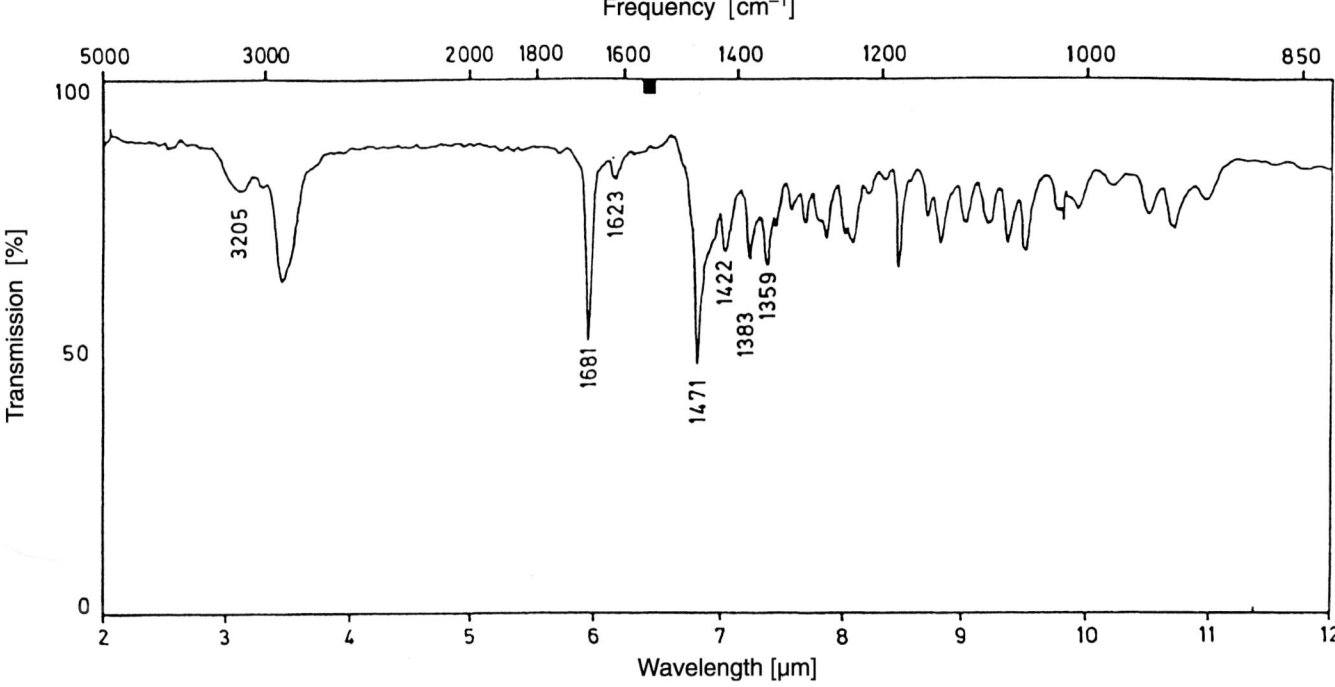

Fig. 3.12. *IR Spectrum of macroline (**9**), recorded in CCl₄.* Microcell, 0.2 mm; region indicated in black: solvent absorption.

Carbonyl Group

Reduction of macroline (**9**) with LiAlH₄ in THF gives macrolinol (**10**) ($C_{21}H_{28}N_2O_2$, M_r 340) as the major product. Its UV spectrum turns out to be very similar to that of the starting material **9**, meaning that the C=O group in the latter is not conjugated with the main chromophore. In the IR spectrum of **10**, recorded in CHCl₃, the carbonyl absorption band is no longer present, while stronger absorptions in the OH region are found. The absorption band corresponding to the C=C bond remains weak, as previously. Clearly identifiable, however, is the absorption of the vinylic C=C bond centered at 913 cm⁻¹. Finally, in the corresponding ¹H-NMR spectrum, the reduction of the C=O group to the corresponding alcohol results in the following major changes: the signals of the two vinyl protons are shifted to higher field by *ca.* 1 ppm to 5.20 (*singlet*) and 5.05 ppm (*singlet*), respectively – a region typical of vinyl protons of allylic alcohols[40,41]. Also, the terminal Me group of the ketone fragment (*cf.* **9**) appears in the ^{1}H-NMR spectrum of **10** as a *doublet* at 1.15 ppm ($J \approx 6.5$ Hz).

Macrolinol (**10**) may be converted into its diacetyl derivative **11** (M_r 424, IR: 1727 cm^{-1}, ester carbonyl group), establishing that there are indeed two OH groups present. This means that macroline (**9**) cannot contain any secondary N-atoms (primary amines have already been ruled out on the basis of the two *N*-Me groups). The formation of *O*-acetylmacroline from macroline proceeds unevenly; *O*-acetylmacroline (**12**) may, however, be obtained in high yield from pyrolysis of *O*-acetylvillamine (**13**).

Chemical evidence for the C=C bond present in macroline (**9**) was provided by catalytic hydrogenation (PtO$_2$, H$_2$, EtOH) yielding 20,21-dihydromacroline (**14**; M_r 340).

II

Taking these more recent findings into account, we can now formulate the partial structure **II**.

For the definitive elucidation of the ring system and the determination of the interconnectivity of the structural elements indicated in **II**, analysis of mass spectra proved particularly valuable. Comparison of the major signals in the mass spectrum of macroline (*Fig. 3.10*) with spectra of some of its derivatives (*Table 3.1*) gave rise to a number of structurally relevant conclusions:

1) The fragments listed in *Table 3.1* all contain the aromatic core, easily recognized in the mass spectrum of (D$_4$)macroline (**15**) due to the shifting of signals by four mass units. The deuterated compound **15** was obtained from villalstonine (**1**) by isomerization to villamine (**7**) in the presence of

Table 3.1. *Characteristic Signals in the Mass Spectra of Macroline and some of its Derivatives*

Compound		*m/z*						
Macroline (**9**)		338 ($M^{+\cdot}$)	320	251 (*g*)	208 (*h*)	197 (*a*)	181 (*i*)	170 (*d*)
	HR ^{a}	C$_{21}$H$_{26}$N$_2$O$_2$	C$_{21}$H$_{24}$N$_2$O	C$_{17}$H$_{17}$N	C$_{15}$H$_{14}$N	C$_{13}$H$_{13}$N$_2$	C$_{13}$H$_{11}$N	C$_{12}$H$_{12}$N
	$m^{*\,b}$	• →		• →				
Macrolinol (**10**)		340	322	253	208	197	181	170
20,21-Dihydromacroline (**14**)		340	322	253	210	197	181	170
O-Acetylmacroline (**12**)		380	320	251	208	197	181	170
(D$_4$)Macroline (**15**)		342	324	255	212	201	185	174
Macroline · CD$_3$I (**16**)		341 + 338	323 + 320	251	208	200 + 197	181	170

^{a} HR = High-resolution mass spectrometry
^{b} m^* = Transition signal

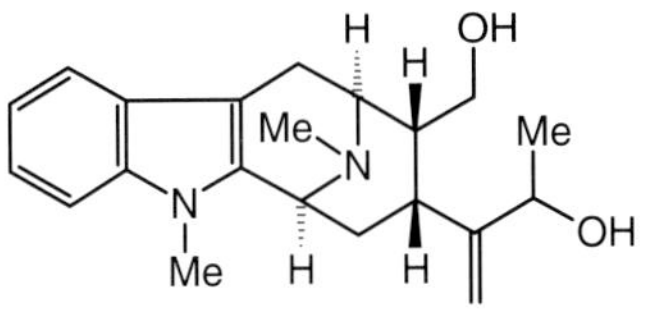

Macrolinol (**10**)

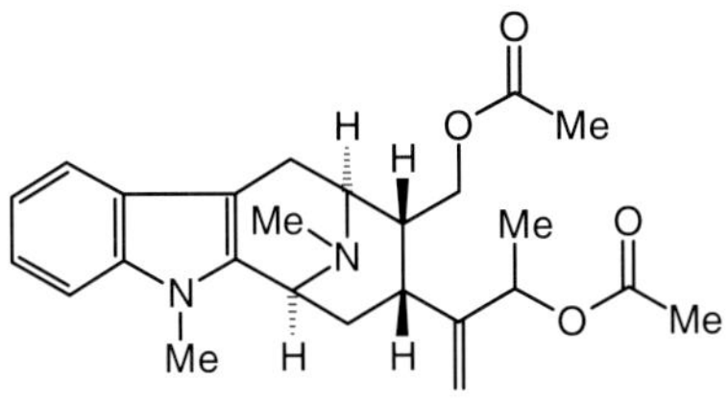

O,O-Diacetylmacrolinol (**11**)

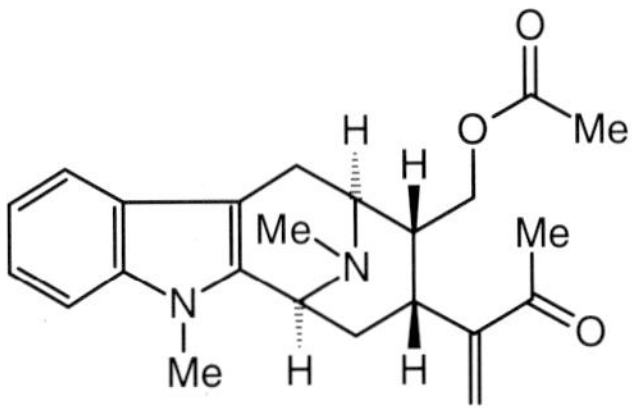

O-Acetylmacroline (**12**)

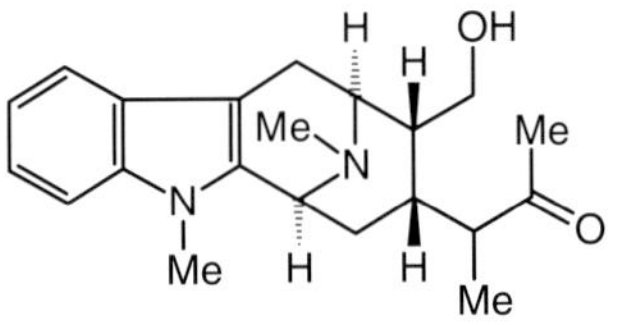

20,21-Dihydromacroline (**14**)

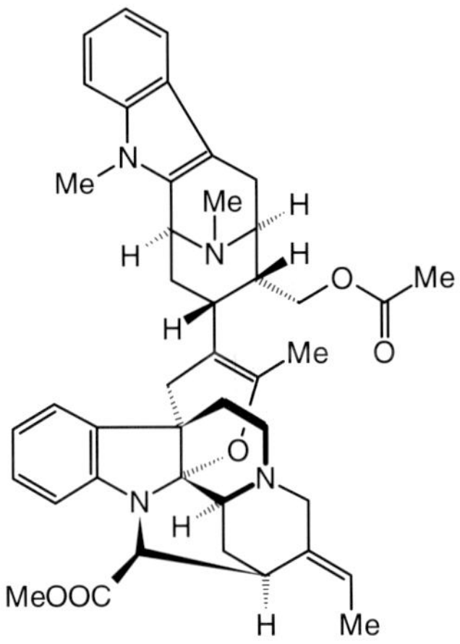

O-Acetylvillamine (**13**)

CF$_3$COOD/(CF$_3$CO)$_2$O, followed by pyrolysis. From the spectrum of (D$_4$)macroline, it follows that the C=O group detected in macroline (**9**) cannot yet be present in villamine (**7**)[42], a finding that we shall revisit later.

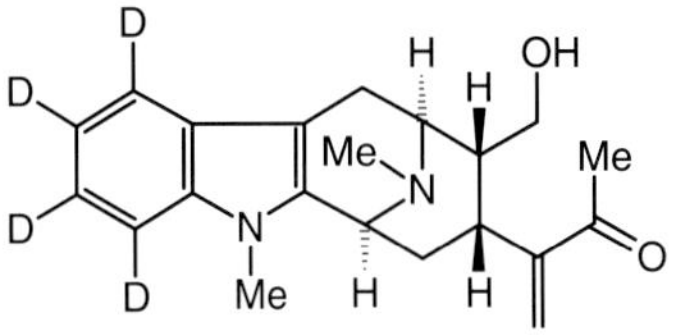

2) All ions possess at least one N-atom; those with *m/z* 338 ($M^{+\bullet}$), 320, and 197 contain two (as determined from high-resolution data). As we have already seen, macroline incorporates two N-atoms, of which one is part of a weakly basic indole chromophore, while the other is significantly more basic in character. Treatment of macroline with MeI produces the quaternary salt 'macroline methoiodide'. If deuterated MeI (CD$_3$I) is used, the corresponding '(D$_3$)macroline methoiodide' (**16**) is formed. In the mass spectrometer, thermal dealkylation of this salt gives rise to CD$_3$I

(D$_4$)Macroline (**15**)

(D₃)Macroline Methoiodide (**16**)

and macroline (**9**), together with CH₃I and (D₃)macroline. The molecular masses of the two compounds **9** and (D₃)-**9** must differ by three mass units. Consequently, discrete, superimposed mass spectra are obtained; the exact data are shown in *Table 3.1*. The three signals at *m/z* 338, 320, and 197 all display 'shadows' *ca.* 3 amu heavier (at *m/z* 341, 323, and 200, respectively), their relative intensities being about equal. The other four signals, though, appear 'alone'. From this finding it may be inferred that the two fragments at *m/z* 320 and 197 contain the basic N-atom with a Me group attached, while none of the others do. It further follows that there must be at least two different fragmentation pathways: one involving signals at *m/z* 338, 320, and 197, the other at *m/z* 338 (or 320, corresponding to 338 – H₂O), 251, 208, 181, and 170, respectively. As well as this, it can be seen that the last two ions listed cannot originate from *m/z* 197 (this ion would have to lose NH₂ in order to produce 181, and this is not reconcilable with the mass differences found for (D₃)-**9**; an analogous situation also applies for *m/z* 170).

To derive their structures, it is thus necessary, with the aid of the observed signals, to interpret the sequence of reactions that produce the individual ions in the mass spectrometer.

3) A transition signal (*m**) indicates that the ion of *m/z* 208 is produced from *m/z* 251. In the course of this fragmentation, C₂H₃O is eliminated (high resolution). Comparison of the spectrum of macrolinol (**10**), in which the C=O group has been reduced to an alcohol, with that of macroline (**9**) indicates that the observed fragment is represented by the acetyl (Ac) group of the α,β-unsaturated ketone fragment. In the spectrum of **9**, the heavier ion is found at *m/z* 253. The lighter ion, though, remains unchanged at *m/z* 208, implying that ·CH(OH)CH₃ has been eliminated in place of ·COCH₃.

4) It is extremely unlikely that the ion at *m/z* 208 is transformed into that of mass 181, since this would require the elimination of a ·C=CH₂ radical (evidence: in the spectrum of 20,21-dihydromacroline (**14**), in which the vinyl C=C bond of **9** has been hydrogenated, the macroline ion at *m/z* 208 is shifted to 210, while *m/z* 181, however, remains unchanged). Consequently, it is to be assumed that the ion at *m/z* 181 originates from ions with a mass of 251, 338, or 320.

5) The fragment observed at *m/z* 251, with the elemental composition C₁₇H₁₇NO, is a radical ion[43] and must, therefore, arise through a 'neutral process'[44].

Signal at m/z *197*

The lightest fragment (with the smallest number of unknown structural elements) corresponds to *m/z* 197, therefore, to first derive the structure of this particular ion seems especially promising. Taking account of the results outlined above, it is possible to establish that the ion features the following structural elements:

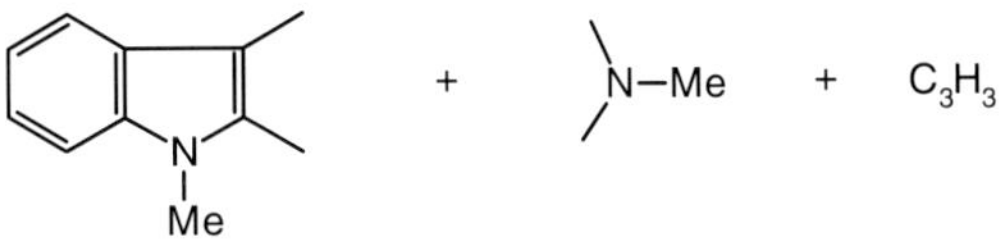

The remaining unknown structural component amounts to a mass of 39 amu, which corresponds to some C_3H_3 moiety.

On the basis of the UV spectrum of macroline (**9**), it is possible to rule out a direct bond between the indole nucleus and the second N-atom (lack of conjugation). Hence, C,C bonds with the indole nucleus can also be assumed. Given these assumptions, the two possible structures **a** or **b** result for *m/z* 197.

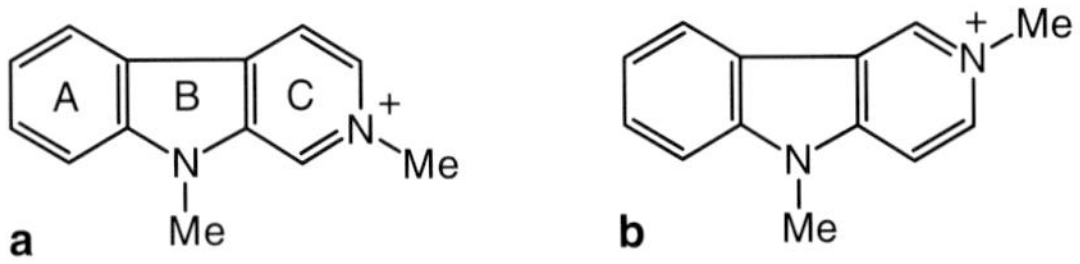

Decision in favor of one or the other structure is not possible yet on the basis of the experimental results so far obtained. However, there is a biogenetic argument speaking for compound **a**, which corresponds to a *β*-carboline system (and hence a tryptamine derivative), while **b** represents a *γ*-carboline. Previously, only *β*-carboline derivatives have been isolated from *Alstonia* species, and of the naturally occurring carbolines, incidentally, 99.9% also are *β*-carboline varieties with hardly any *γ*-derivatives among them. Accordingly, it is highly probable that macroline itself belongs to the *β*-carboline type, although there is as yet no definite proof of this.

Ion **a** is a derivative of the molecular ion of macroline (**9**), and thus of macroline itself. This component, however, cannot be incorporated in macroline in its present form (quaternary N-atom, aromatic C-ring). Therefore, it is much more likely (*cf.* UV spectrum) that the macroline constituent corre-

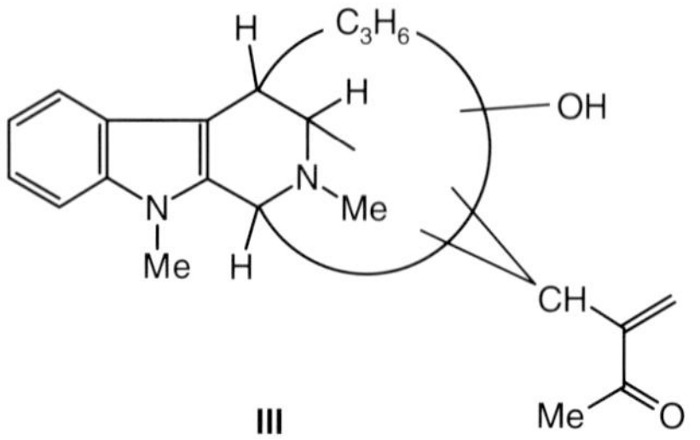

III

sponding to **a** is a tetrahydro-β-carboline. With the aid of this piece of information, the partial structure **II** for macroline can be augmented to **III**.

Signal at m/z *170*

Now, with the three extra C-atoms having been fixed in the structure of macroline, deduction of the next molecule fragment can be attempted.

Compounds that possess a six-membered ring with a C=C bond are susceptible to thermal *retro-Diels-Alder (RDA)* reactions in the mass spectrometer. If we postulate a *RDA* reaction for the C-ring in macroline, then the result is the partial structure **c**, which differs in mass only slightly from *m/z* 170 (if R^1 is taken as H, and R^2 as CH_2, or *vice versa*, the correct mass or atom composition is obtained). Given this constraint, it is possible to deduce two possible structural variants: **d** and **e**.

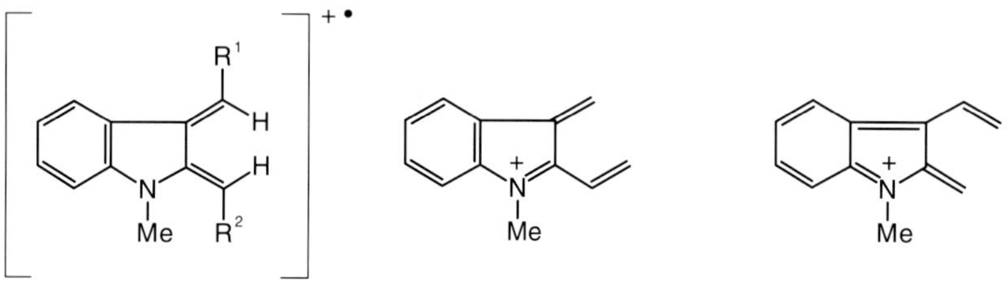

c   d   e

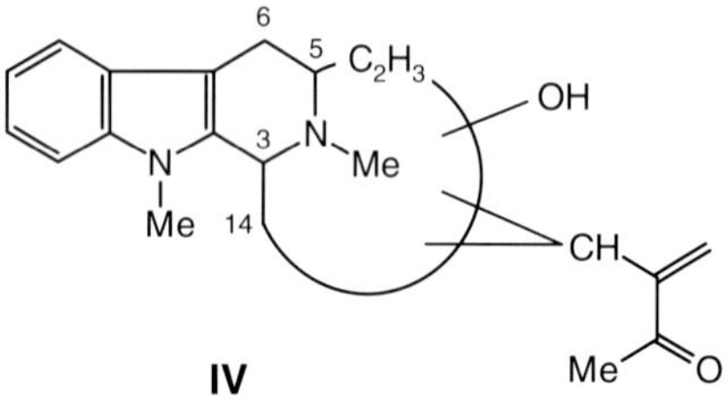

With the chemical and spectroscopic findings available at the moment, it is not possible *a priori* to differentiate between the two structures. However, structural variant **e** can be provisionally ruled out for the time being on the basis of analogy with other indole alkaloids (*cf.* discussion above). Accordingly, the macroline partial structure **III** can once again be improved to give **IV**.

IV

Signal at m/z *251*

As already explained, the basic NMe and the OH group are both absent from the ion with *m/z* 251; on the other hand, both the oxo group and the indole chromophore are present. Therefore, it is useful to consider how the NMe and the OH group are cleaved off from macroline. One possibility for the elimination of the basic NMe group would be another *RDA* reaction. In this, though, in the partial structure **IV**, the bond between C(5) and the rest of the molecule would also be abolished (although the bond from C(14) would be preserved).

If such a *RDA* reaction is indeed the primary pathway to the ion with *m/z* 251, then C(14) has to be connected with the α,β-unsaturated ketone fragment. This, in turn, implies the partial structure **f**, which contains all of the C-, N-, and O-atoms required to explain the signal at *m/z* 251.

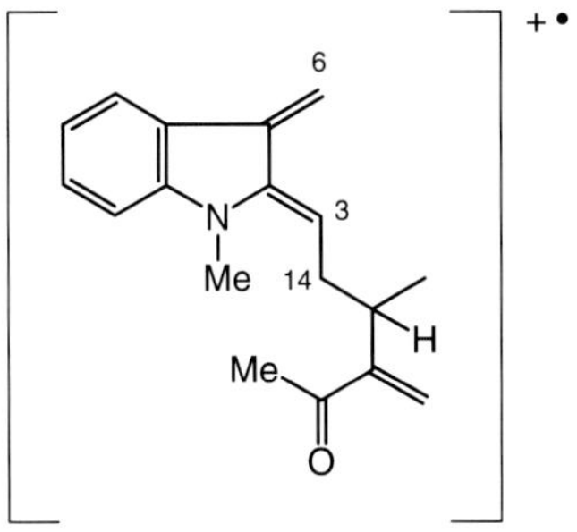

However, the attentive observer is immediately confronted with an additional H-atom plus an unsaturated bond. This should not be the cause of too much disquiet, though, as it appears logical that the H-atom could split off during the fragmentation, thus giving rise to the formation of a C=C bond, which represents part of a multiply unsaturated system. Looking at it this way, one comes up with the revised partial structure **V**.

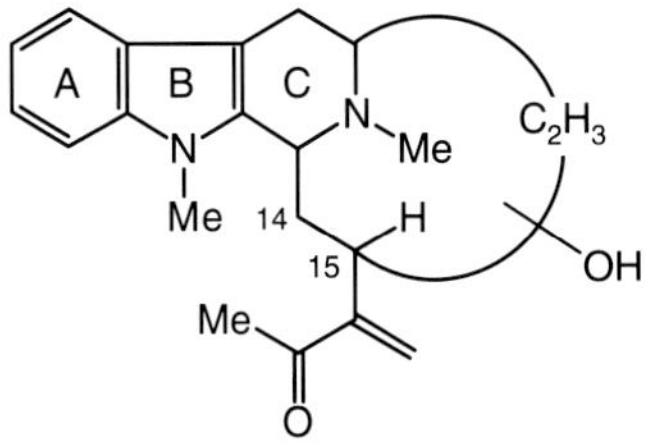

The above results in mind, it is possible to disentangle the 'molecular makeup' of macroline (**9**) a little more: as we have seen, the assumption of an *RDA* reaction in the C-ring of **9** is clearly plausible. A rearrangement of this sort, however, is sufficient to explain neither the removal of a H-atom from C(14), nor the cleavage of the bond between C(15) and the unknown residue. There must, therefore, be another neutral reaction[45] taking place in the primary fragment (arising from the *RDA* reaction), through which H–C(14) is eliminated. A *McLafferty* rearrangement, for example, would be plausible. As a H-acceptor, the MeN=C(5) bond created in the corresponding *RDA* reaction is worth considering.

Since it is well-known that *McLafferty* rearrangements generally proceed *via* a six-membered cyclic transition state, the fragmentation proposed in *Scheme 3.4* is compelling (a seven-membered transition state, by the way,

f

V

Scheme 3.4. *Formation of the Fragment Ion at* m/z 251 *by* McLafferty *Rearrangement of the* retro-Diels-Alder *Product of Macroline* (**9**)

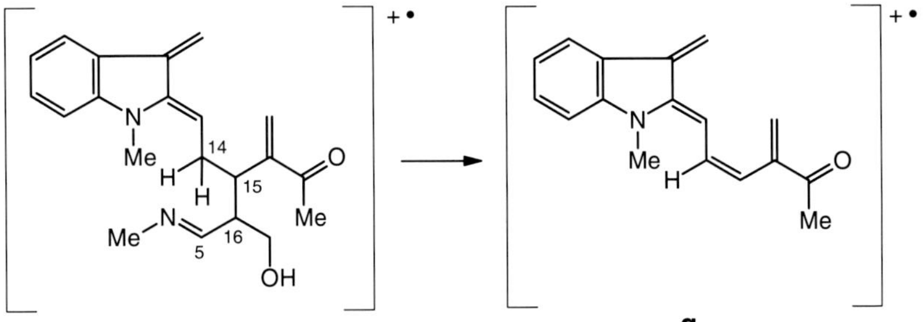

g

(*m/z* 251)

such as would be possible through inclusion of the second, still 'unknown' C-atom, is extremely unlikely). Moreover, because the OH group cannot be situated on C(16) (otherwise an extra Me group would have been visible in the ^{1}H-NMR spectrum), we can finally draw the ('planar') constitutional formula **9** for macroline.

Scheme 3.5 now shows the complete fragmentation mechanism of macroline. Before we can go on further, though, some clarification of the formation of the ions with *m/z* 208, 181, and 320 is necessary.

The ion with *m/z* 208 originates from *m/z* 251 (*m**) by elimination of an acetyl radical. Generally, eliminations adjacent to a C=C bond are only feasible when conjugated (or otherwise stabilized) species are formed as in the case of the neutral acetyl radical (·Ac) as well as the corresponding ion fragment (*m/z* 208).

The formation of the ion of *m/z* 181 can only be explained by the following process pictured in *Scheme 3.5*: sequential *RDA* reaction, *McLafferty* rearrangement, *Diels-Alder* reaction, terminated by a second *McLafferty* rearrangement. There is no doubt that neither the structure nor the formation of this ion constitute any kind of supporting evidence for the macroline structure, with the basis of this ion being made up of a carbazole ring system, and not a β-carboline one.

The elimination of H_2O from the molecular ion (*m/z* 338) and the consequent formation of the ion of *m/z* 320 have yet to be fully explained. The reaction probably starts from the hemiacetal form **9a** though.

The case for the structural elucidation thus far presented, essentially based on spectroscopic arguments, has therefore brought us to formula **9**. This structure is further supported by the spectroscopic properties of synthetic model compounds[46], and other similarly constituted alkaloids occurring naturally in the same plant[47]. This, though, does not completely establish the structure of macroline. Other (chemical) investigations are necessary, especially to clarify the configuration of the compound.

Now that the structures of the two component bases may be taken as known, it is time to move on to the elucidation of the structures of villamine (**7**) and then villalstonine (**1**).

9a

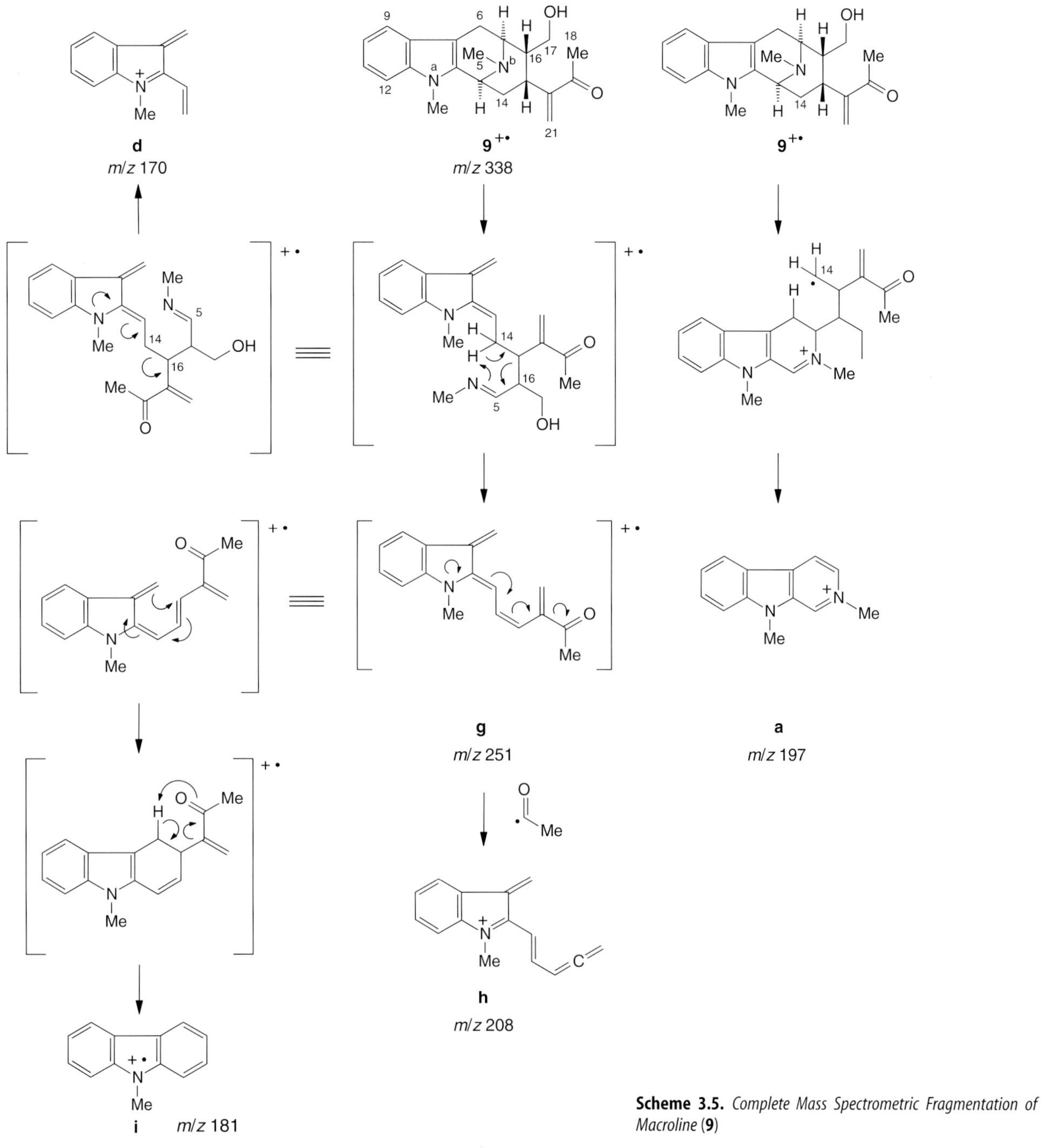

d
m/z 170

9$^{+\bullet}$
m/z 338

9$^{+\bullet}$

g
m/z 251

a
m/z 197

h
m/z 208

i   *m/z* 181

Scheme 3.5. *Complete Mass Spectrometric Fragmentation of Macroline* (**9**)

3.4.6. Structure Elucidation of Villamine and Villalstonine

As explained previously, the pyrolysis of villamine (**7**) produces both pleio-carpamine (**5**) and macroline (**9**) (*cf. Scheme 3.6*). The sum of the molar masses of the two cleavage products is equal to that of villamine. This clearly means that no compounds other than those stated were produced in the course of the pyrolysis, not even H_2O, CO_2, or other small fragments. In other words: for the derivation of the structure of villamine (**7**), only compounds **5** and **9** need to be considered.

Possible connection points between the 'monomer' bases cannot be determined without comparing the functional groups of the two component bases with those of villamine (**7**). Absent, additional, or altered functional groups offer clear clues about the connecting points. *Table 3.2* lists all the functional groups of the three compounds together.

From this list, it can be seen that the connection in the case of macroline (**9**) must involve the α,β-unsaturated ketone, while in pleiocarpamine (**5**) the indole C=C bond is affected. Furthermore, villamine possesses an additional enol ether double bond (IR: 1682 cm^{-1}), an extra (C)Me group (^{1}H-NMR; *singlet* at 1.31 ppm), and an indoline chromophore[48]. It is also known that

Scheme 3.6. *Transformation of Villalstonine (**1**) into Villamine (**7**), Leading to both Pleiocarpamine (**5**) and Macroline (**9**)*

Villalstonine (**1**)
M = 660

(CF$_3$CO)$_2$O /
CF$_3$COOH

2N HCl

Villamine (**7**)
M = 660

Δ

Pleiocarpamine (**5**)
M = 322

Macroline (**9**)
M = 338

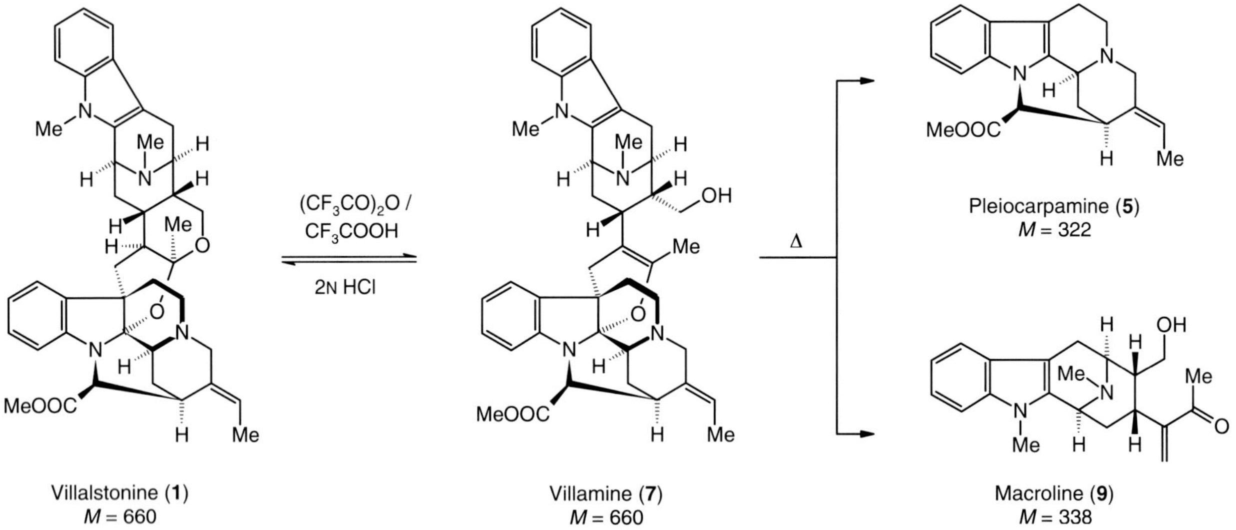

Table 3.2. *Comparison of Functional Groups in the Alkaloids Pleiocarpamine* (**5**) *and Macroline* (**9**) *vs. Villamine* (**7**)

Group	Pleiocarpamine (**5**) + Macroline (**9**)	Villamine (**7**)
COOMe	1 (**5**)	1
CH$_2$OH	1 (**9**)	1
C=O	1 (**9**)	missing
C=C–O–R	missing	1
Basic N-atom	2 (**5** + **9**)	2
Arom. H	4 + 4	8
Chromophores	Indole + Indole	Indole + Indoline
Ethylidene group	1 (**5**)	1
Vinyl group	1 (**9**)	missing
Methyl groups	2 (N) – Me (**9**)	2 (N) – Me
	1 (C(O)) – Me (**5**)	missing
	1 (COO) – Me (**5**)	1 (COO) – Me
	1 (C=C) – Me (**5**)	2 (C=C) – Me

5 and **9** are readily formed by means of a thermal reaction. In the light of these data, there are essentially two options for the thermal cleavage of villamine (*Schemes 3.7 and 3.8*).

It is not possible to distinguish between these two possibilities without additional measures, however. On biogenetic grounds, the former seems the more plausible, though: the most electronegative position in the indole nucleus (apart from the N-atom) is the β-position, while the α-position is appreciably less nucleophilic (*Scheme 3.9*).

From this, structure **7** follows for villamine. The α-*cis*-orientation of the macroline component about C(2) and C(7) of pleiocarpamine is assumed by analogy with the position of the H-atoms in 2,7-dihydropleiocarpamine. Moreover, a β-orientation is not even possible sterically.

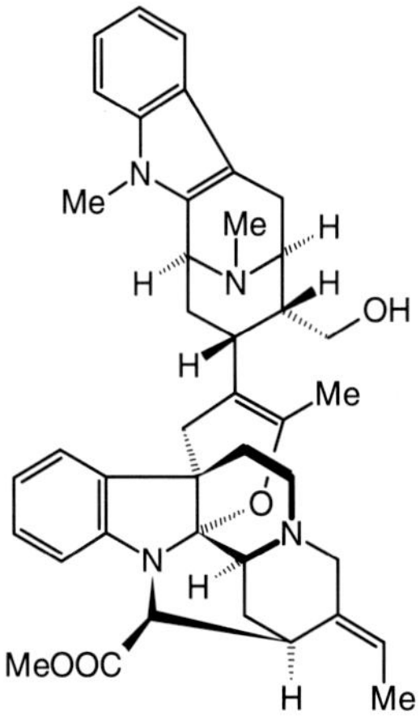

The last step, the deduction of the structure of villalstonine (**1**) itself, is similarly achieved by considering functional groups: whereas villamine possesses a free OH group and shows an enol ether band in its IR spectrum, both these two functional groups are absent in **1**. On the other hand, an extra (C)Me *singlet* is observed in villalstonine (^{1}H-NMR: 1.23 ppm (1.31 ppm in villamine)). From these facts it is possible to formulate the structure **1** for villalstonine[46]. The isomeric villoine (**8**) probably has a similar structure, but with an open C(2)–O–C(19′) bond.

Villamine (**7**)

Scheme 3.7

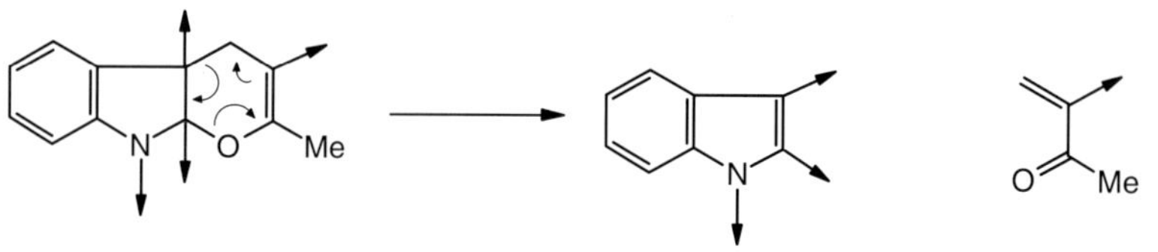

Villamine

Pleiocarpamine

Macroline

Scheme 3.8

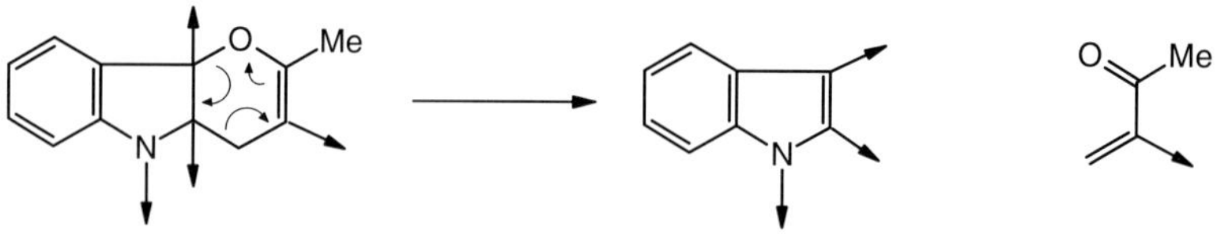

Pleiocarpamine

Macroline

Scheme 3.9

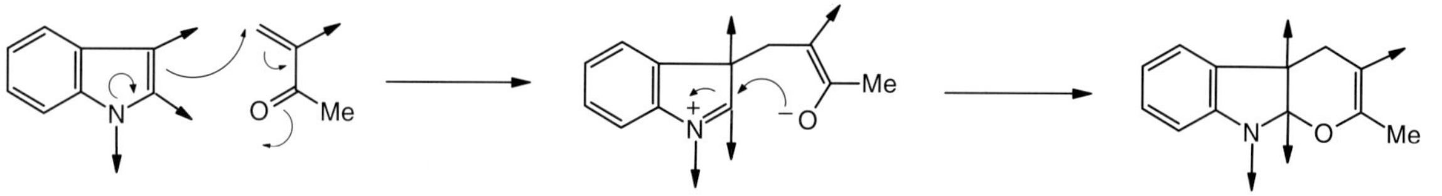

Partial Structure of Villamine

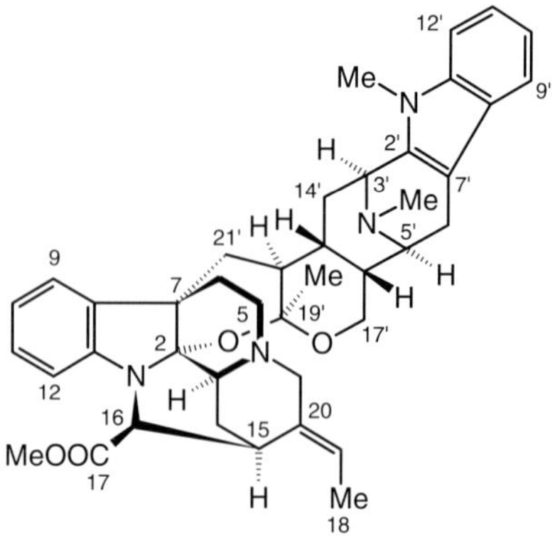

Villalstonine (**1**)

The villalstonine structure deduced in this manner was later confirmed by means of an X-ray crystal-structure analysis (*vide infra*). The configurations at C(3′), C(5′), C(16′), C(19′), and C(20′) of the macroline component were also determined this way.

Comprehensive interpretation of all the spectra has to be refrained from here. It should be stressed, though, that this kind of spectral interpretation is absolutely essential for proper corroboration of the structures. In the context of the structural elucidation of macroline and villalstonine discussed above, however, initial priorities are confined only to the most prominent bands.

The structure elucidation of villalstonine impressively demonstrates the capability of modern spectroscopy. In particular, the crucial success of mass spectrometry in the analysis of villalstonine has to do with the fact that macroline happens to exhibit an exceptionally compelling fragmentation pattern, and that this has its origins in the C-ring of the molecule. It was only thanks to this that there was the opportunity to structurally exploit the three derivatives **10**, **11**, and **14**, which all display completely analogous behavior in their mass spectra.

Structure elucidation of other alkaloids generally involves deployment of similar approaches and techniques. Which spectroscopic method happens to be particularly suitable, entirely depends on the substance class under investigation. If it is not possible to derive the skeleton structure of a particular alkaloid through its spectra alone, then it also becomes necessary to identify one or more products from chemical degradation reactions to permit conclusions about the basic skeleton of that alkaloid.

3.4.7. X-Ray Crystal-Structure Analysis

It is also possible to unravel molecular structures another way, independent of the results of chemical degradation reactions, spectra of degradation products, or partial syntheses. The imperative for this, though, is that either the substance itself or one of its derivatives must be crystallizable. A single crystal, featuring no abnormalities (such as being too small or possessing a twin structure), is required.

Irradiation of a crystal with X-ray beams results in their diffraction by the elements (atoms, ions, molecules of the lattice), giving rise to interference patterns that can be collected photographically or electronically. These phenomena may also be viewed as reflections of the X-rays at the crystal planes. Analysis of the measured interference pattern consequently makes it possible to determine the crystal system and symmetry, together with its lattice constants, and even interatomic and intermolecular spacing and bond angles. As a result, for an organic compound, it is possible to determine the structure itself.

Villalstonine crystallizes from methanolic solution. The crystals are monoclinic and belong to the space group $P21$. The unit cell has the dimensions $a = 13.756$ Å, $b = 13.645$ Å, and $c = 10.045$ Å, with $\beta = 101°41'$. It consists of two molecules each of villalstonine and MeOH. Determination of the reflections obtained after X-ray diffraction[23] produced the structure shown in *Fig. 3.13* in full agreement with the structure derived using chemical and spectroscopic methods[46].

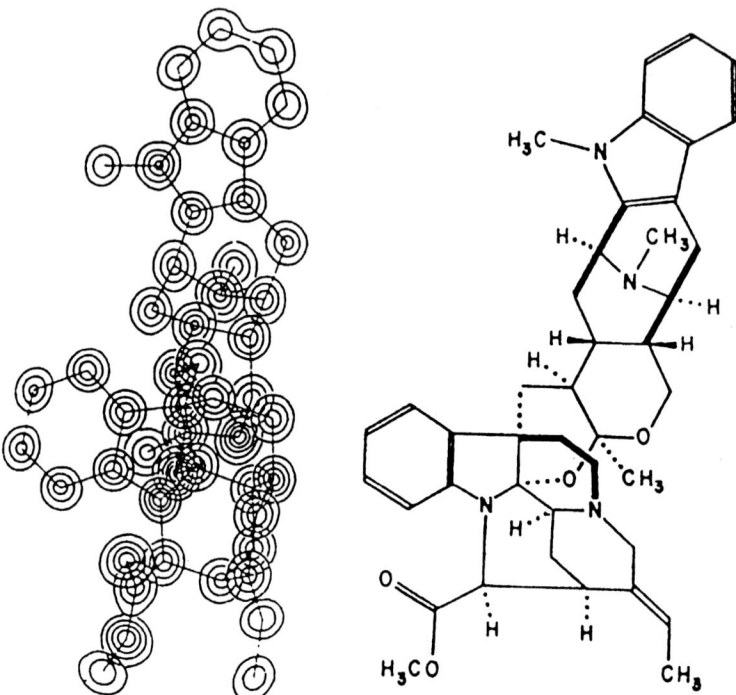

Fig. 3.13. Left: *electron density distribution in the villalstonine molecule.* Right: *structural formula* (with kind permission of the *J. Am. Chem. Soc.*)[23].

X-Ray crystal-structure analysis enables the relative positions of atoms to be determined to one another. It is only possible, however, to determine the absolute configuration if, for example, the chiral sense at one (or more) center(s) in the molecule in question is already known (in this particular case, the absolute configuration of pleiocarpamine was known), or if a heavier atom (sulfur or higher) is present.

3.5. Structure Elucidation of the Components of a Naturally Occurring Mixture of Extremely Small Quantities of Substances: Pollen of *Hippeastrum × hortorum*

During the course of investigations into the pollen of various plant species (from the Acanthaceae, Betulaceae, Fagaceae, and Juglandaceae families, among others), aryl-substituted derivatives of spermidine have frequently been detected or isolated. Hence, for example, four different 4-hydroxycinnamic acid derivatives of spermidine were found in Acanthaceae: three (E,E)-N,N'-bis(4-hydroxycinnamoyl)spermidines plus (E,E,E)-N,N',N''-tris(4-hydroxycinnamoyl)spermidine[49]. Investigations of other plants (such as *Quercus dentata* (Fagaceae)) had, without exception, produced similar results[50]. Pollen of *Hippeastrum* species, however, had not been studied.

The genus *Hippeastrum* belongs to the botanical family of the Amaryllidaceae, found mainly in Central and South America. To date, *ca.* 60 *Hippeastrum* species are known. While some of them have awakened the interest of flower growers, and so are available commercially as popular (house)plants, chemists had previously devoted their attention more to *Hippeastrum* bulbs, which had been intensively studied for alkaloids. Examples of these so-called Amaryllidaceae alkaloids are to be found in *Chapt. 2*. For a deeper examination of the makeup of *Hippeastrum × hortorum* pollen (*Fig. 3.14*), the proceedings undertaken during the investigation of a mixture containing extremely small quantities of substances – structural results supported by current findings[51] – will be illustrated at this point.

Processing of 1 g pollen from *Hippeastrum × hortorum* produced a mere 9 mg of alkaloidal substances. TLC Analysis revealed that the isolated extract constituted a mixture of a great many compounds. Crude chromatographic separation produced three well-distinguished main fractions (TLC on SiO_2; $CHCl_3$/MeOH/25% aq. NH_4OH 78:19:3). The least mobile fraction (2.7 mg in total) was then further studied. This fraction, however, also consisted of a great many substances according to TLC; too many, in any case, to be separated, purified, and analyzed as with coniine and villalstonine (*Sect. 3.3* and *3.4*). Thus, it was clear from the start that any attempt at a classical structure elucidation would hardly be likely to succeed. How could such a challenge be overcome otherwise, though? The answer is: with an 'online' combination of micro high-performance liquid chromatography (μ-HPLC), ultraviolet diode-array detection (UV-DAD), atmospheric pressure chemical ionization (APCI), and tandem mass spectrometry (MS^n), as schematically shown in *Fig. 3.15*.

In the examination of extremely small quantities of substances (down to the femtogram scale, 1 fg = 10^{-15} g), it is, as a rule, necessary to refrain from individual collection of the fractions after chromatographic separation: the losses in the manipulation of these minimal quantities are excessive on one hand, while, on the other, there is also a constant risk that the fractions may become contaminated or decompose on exposure to aerial oxygen. Therefore, the sample is subjected to direct ('on-line') mass spectrometric analy-

Fig. 3.14. *Hippeastrum* × *hortorum* (Amaryllidaceae). *Amaryllis* hybrids, popular as large-flowered, winter-blooming, decorative houseplants, contain *Amaryllis* alkaloids in the bulb and spermidine derivatives of cinnamic acids in the pollen (photo *C. Werner*).

sis directly after chromatographic separation (by means of HPLC, μ-HPLC, or capillary electrophoresis (CE)). We should note in passing that the technical requirements for efficient execution of this procedure have only been met in the last few years, especially thanks to the development of enormously effective high-vacuum pumps capable of fast removal of large quantities of gas produced on evaporation of the solvent. Thus, after HPLC separation, the sample is injected into the MS and converted into the corresponding $[M + H]^+$ ion by means of an especially gentle ionization technique[52], allowing to immediately determine the molecular mass of the substance. At the same time, the presence or absence of signals corresponding to other molecular ions also enables the homogeneity of the HPLC fraction to be assessed. Thanks to the mild ionization technique, we do not normally obtain any information concerning the substance's fragmentation pattern at this point, however. A fragmentation of the $[M + H]^+$ ion can be induced if desired,

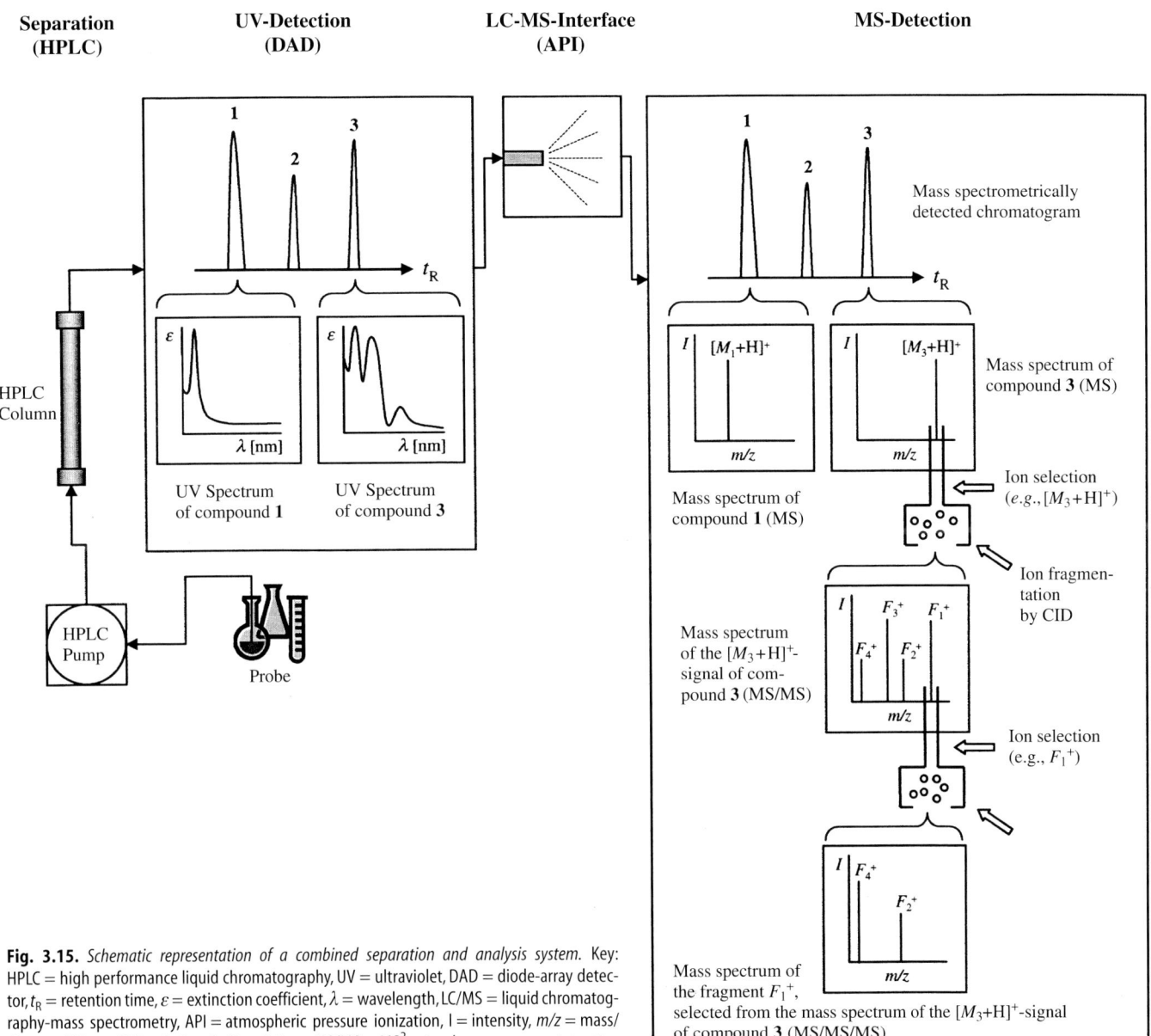

Separation (HPLC)   **UV-Detection (DAD)**   **LC-MS-Interface (API)**   **MS-Detection**

Fig. 3.15. *Schematic representation of a combined separation and analysis system.* Key: HPLC = high performance liquid chromatography, UV = ultraviolet, DAD = diode-array detector, t_R = retention time, ε = extinction coefficient, λ = wavelength, LC/MS = liquid chromatography-mass spectrometry, API = atmospheric pressure ionization, I = intensity, m/z = mass/charge, CID = collision-induced dissociation, MS/MS = MS^2 = tandem mass spectrometry, MS/MS/MS = MS^3 = 'third generation' mass spectrum.

though: use of electrical acceleration can increase the kinetic energy of the molecular ion to such a degree that collision with neutral gas molecules (in the so-called collision chamber) can give rise to fragmentation ($\rightarrow$ 'collision-induced decomposition' (CID)). A new spectrum is thus obtained, with the $[M + H]^+$ signal frequently absent, due to the complete nature of fragmentation. With the aid of an 'ion trap' as a mass filter plus a second, down-line mass spectrometer (tandem MS), individual ions may then be further fragmented in a controlled fashion, which is enormously helpful for structure elucidation of unknown compounds.

The technique, as outlined in broad brush, is less successful, though, if the substances in a mixture are not fully separable by means of HPLC. This was sadly the case on investigation of fraction I of the basic *Hippeastrum* extract. *Fig. 3.16,g* shows the HPLC chromatogram of fraction I, recorded by means of an UV diode-array detector. It displays significant signal overlap, and hence incomplete separation. Apart from intensity differences, *Fig. 3.16,g* agrees with the mass spectrometrically measured chromatogram depicted in *Fig. 3.16,f.* In mass spectrometric recording of HPLC samples, practically all of the pertinent ions are detected. Integration over all ion intensities of all completely scanned spectra enables the so-called 'reconstructed ion current' (RIC) to be obtained. Since all the information is stored electronically, the data may later be sorted, selected, and compiled according to whatever criteria are desired.

In our case, the masses of the molecular ions lay between *m/z* 438 and 558. Consequently, the separate spectra of the compounds with $[M + H]^+$ 438 (*Fig. 3.16,a*), 468 (*b*), 498 (*c*), 528 (*d*), and 558 (*e*) were printed out sequentially and compared with one another: since the signals of the different molecular ions differ by 30 amu, it can be assumed that this is due to sequential loss of a MeO group in each case. Summation of the spectra (*cf. Fig. 3.17,a*) produced a curve that – small intensity differences aside – showed extensive similarities to that obtained in the RIC chromatogram (*Fig. 3.16,f*). This, in turn, implied that all the essential information lay in the spectra (*Figs. 3.16,a–3.16,e*) without any signals being absent. It is important to point out, though, that the intensities represented in *Figs. 3.16,a–e* do not in any way correspond to the actual distribution in the RIC chromatogram. The signal intensities are each normalized to 100%, and hence merely represent relative, substance-specific values. In fact, the quantities of the individual compounds differ from one another appreciably.

With the above mass chromatograms at hand, it is possible to start measuring the fragmentation mass spectra of the molecular ions. As an illustration of the methodology, spectra *a* and *b* from *Fig. 3.16* shall be considered more

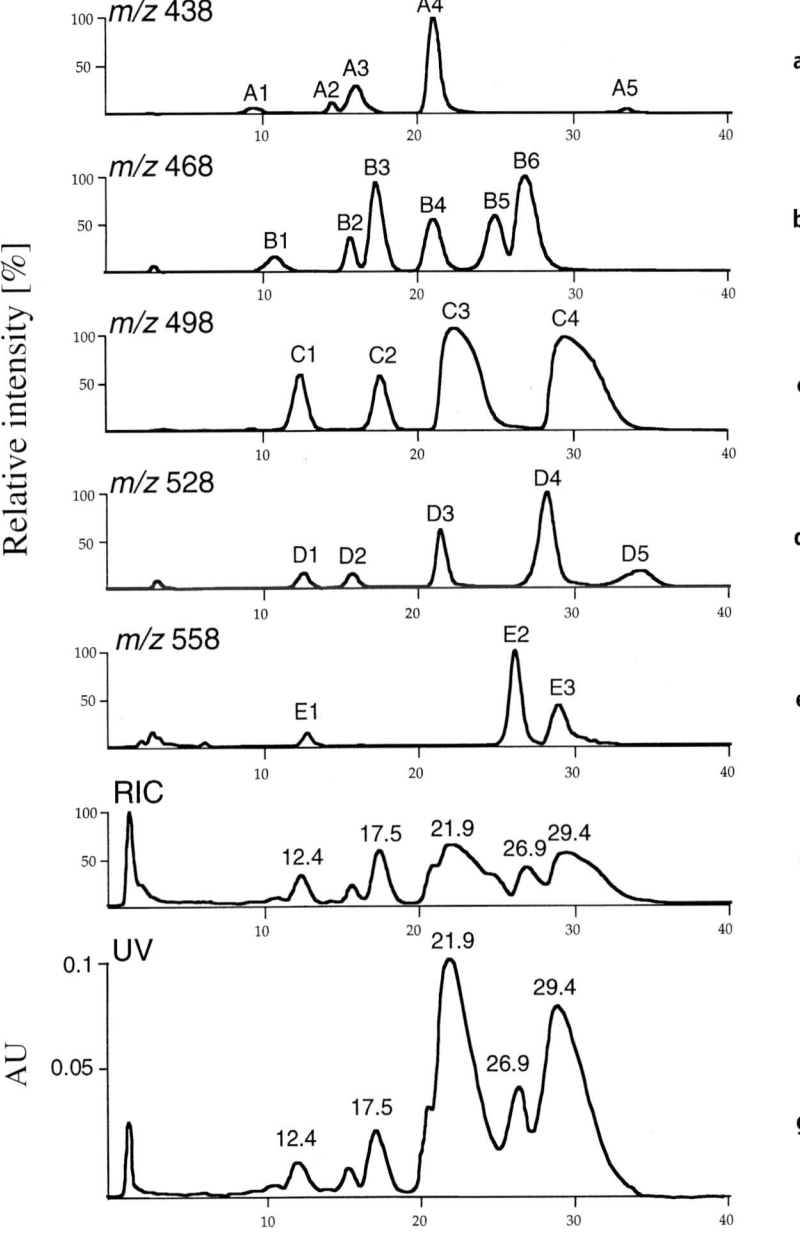

Fig. 3.16. *Chromatographic resolution of the basic Hippeastrum pollen fraction I. g*: overall HPLC chromatogram. Shown: absorption intensities in the UV (recorded by UV-DAD), plotted against retention time (min). *f*: Same chromatogram as in *g*, but with mass spectrometric detection using 'reconstructed ion current' (RIC; see text) in place of UV detection. *a*–*e*: *Mass chromatograms*. All peaks with masses 438 (*a*), 468 (*b*), 498 (*c*), 528 (*d*), and 558 (*e*) in spectrum *f* were filtered out and reproduced individually. Intensities are normalized to 100% in each mass chromatogram, thus preventing absolute mass comparison.

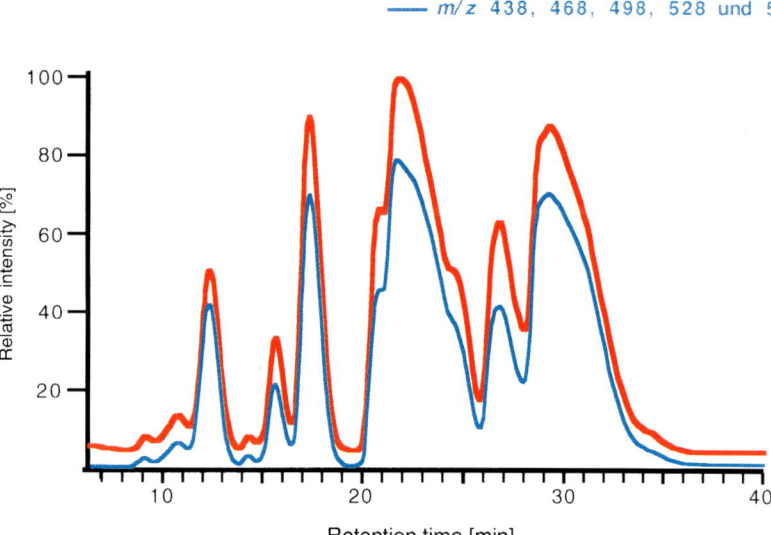

Fig. 3.17. *RIC Chromatograms.* The RIC chromatogram from *Fig. 3.16,f* (——) is shown in comparison with the curve produced by summation (——) of the mass chromatograms of *Figs. 3.16,a–e* in the correct intensity ratios. Since the two curves run almost parallel to one another, it follows that all molecular ions of the original chromatogram (red) have been taken into account by the other five chromatograms (blue).

closely. The corresponding fragmentation spectra are shown in *Figs. 3.18* and *3.19.*

The five daughter-ion spectra in *Fig. 3.18* are conspicuously similar to one another. We are, therefore, dealing with structurally very closely related compounds such as stereoisomers or constitutional isomers with the same underlying skeleton. From the tandem mass spectra shown in *Fig. 3.19*, it is clear that substances of very similar structures are also present, with some signals (*m/z* 147, 204, 218, 221, 275, 292) even having exactly the same mass as those given in *Fig. 3.18*. Other signals are shifted by 30 amu to higher masses (($147 \rightarrow 177$), ($204 \rightarrow 234$), ($218 \rightarrow 248$), ($221 \rightarrow 251$), ($275 \rightarrow 305$), and ($292 \rightarrow 322$)).

On the basis of analogy with other pollen classes, it may legitimately be suspected that the whole substance class is uniform, and that the substances are basically different substituted dicumaroyl-spermidines. To test this suspicion, an assortment of studies was carried out:

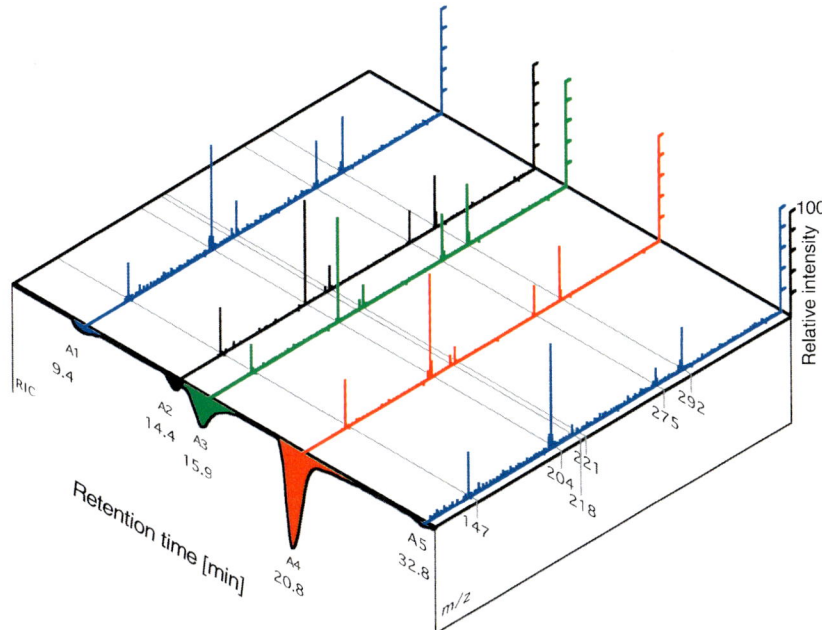

Fig. 3.18. *Three-dimensional representation of the HPLC-APCI-MS/MS experiment.* The chromatogram of *Fig. 3.16,a* is shown on the left with peaks A1, A2, A3, A4, and A5. Also given are the corresponding retention times in min. The daughter-ion mass spectra ($[M + H]^+$ is absent, obviously) are shown perpendicularly in the same color as their chromatogram peaks.

1) Rentention Times (t_R)

Under identical chromatographic conditions, synthetic N^1,N^5-dicumaroylspermidines displayed the same t_R values as substances of the series A (*cf. Fig. 3.16,a* and *Fig. 3.18*). The compounds examined were N^1,N^5-bis(4-hydroxycinnamoyl)spermidine (**17**), as well as the corresponding N^5,N^{10}-isomer (**24**) and the N^1,N^{10}-isomer (**31**) (see *Table 3.3*). The t_R values of both the synthetic and the natural compounds are given in *Table 3.4*.

2) Fragmentation Patterns

As mentioned, the exceptional similarity of all the mass spectra justifies the assumption of a great deal of structural similarity. The structures of the fragment ions were deduced with the aid of synthetic compounds and are represented in *Scheme 3.10*. The results obtained are wholly in accord with the structural types depicted in *Table 3.3*.

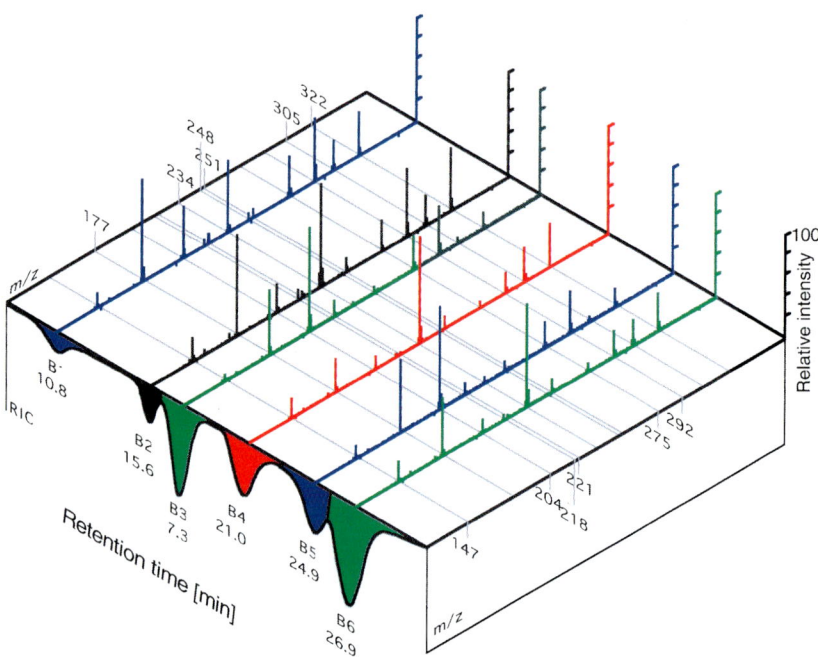

Fig. 3.19. *This three-dimensional representation corresponds to the one in* Fig. 3.18, *but it relates to ions of mass 468* (from *Fig. 3.16,b*)

3) *Hydrogenation Experiments*

To test for the C=C bonds in the compounds in pollen fraction I, a small amount (*ca.* 0.5 mg) of the substance was catalytically hydrogenated (H$_2$, 10% Pd/C, MeOH/AcOH). HPLC-APCI-MS analysis carried out subsequently resulted – in place of the [*M* + H]$^+$ ions at *m/z* 438, 468, 498, 528, and 558 – in the following counterparts at *m/z* 442, 472, 502, 532, and 562, which corresponds to an uptake of four H-atoms per molecule. Thus, each molecule must contain two C=C bonds, which fits exactly for the expected two cinnamoyl residues.

4) *Photoreactivity*

Finally, a sample of the basic *Hippeastrum* pollen fraction I was irradiated in methanolic solution with an UV lamp (low-pressure Hg-vapor lamp, 366 nm, 1 min). The product mixture was examined in turn by HPLC-UV(DAD)-APCI-MS, and mass chromatograms were recorded similar to those shown in *Fig. 3.16*. The result was unequivocal: UV irradiation

Table 3.3. *Presumed Structures of the Substances from Fraction I of the Basic Hippeastrum Pollen Extract.* There are a maximum of 21 compounds of (*E,E*)-, 21 of (*E,Z*)-, 21 of (*Z,E*)-, and 21 of (*Z,Z*)-configuration: hence a maximum of 84 compounds.

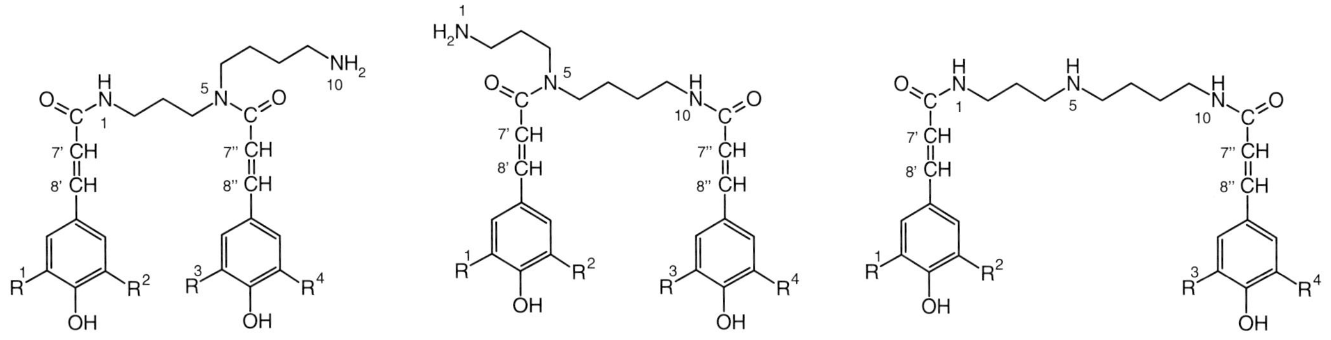

17 – 23   24 – 30   31 – 37

Compound	R^1	R^2	C(7')=C(8')	R^3	R^4	C(7'')=C(8'')	[M+H]$^+$
17, 24, 31	H	H	(*E*) or (*Z*)	H	H	(*E*) or (*Z*)	438
18, 25, 32	MeO	H	(*E*) or (*Z*)	H	H	(*E*) or (*Z*)	468
19, 26, 33	H	H	(*E*) or (*Z*)	MeO	H	(*E*) or (*Z*)	468
20, 27, 34	MeO	H	(*E*) or (*Z*)	MeO	H	(*E*) or (*Z*)	498
21, 28, 35	MeO	MeO	(*E*) or (*Z*)	MeO	H	(*E*) or (*Z*)	528
22, 29, 36	MeO	H	(*E*) or (*Z*)	MeO	MeO	(*E*) or (*Z*)	528
23, 30, 37	MeO	MeO	(*E*) or (*Z*)	MeO	MeO	(*E*) or (*Z*)	558

caused isomerization of the compounds. Thus, *Fig. 3.20,a* displays two primary signals at retention times ($t_R \approx 10$ and 15 min) known to be attributable to (*Z,Z*)- and (*E,Z*)/(*Z,E*)-isomers, but not to the corresponding (*E,E*)-isomers.

In the pollen fraction under discussion, therefore, there are at most 4 × 21 = 84 different compounds present (*cf. Table 3.3*), and these may fundamentally be described as differently substituted (*E,Z*)-isomers of *N,N'*-dicinnamoylspermidine.

Table 3.4. *Retention Times of All the Synthetic* N,N-*Bis(4-hydroxycinnamoyl)spermidines of (E,E)-, (E,Z)-, (Z,E)-, and (Z,Z)-Configuration, Together with Those of Fraction A (cf. Fig. 3.18).* The chromatographic conditions for both the synthetic and natural compounds were the same (HPLC: *Waters Symmetry* C_8 column (5 μm, 2.1 × 150 mm; flow rate 0.22 ml/min; mobile phase MeCN/H$_2$O 9:91 (*v/v*) + 1% AcOH). (*E,Z*)- and (*Z,E*)-isomers cannot be differentiated under these conditions.

Compound	Retention times of the synthetic compounds [min]			Retention times of the pollen fraction A [min]
	(*E,E*)	(*E,Z*) + (*Z,E*)	(*Z,Z*)	*Fig. 3.16, a*
17	20.8	15.8	9.3	9.4
				14.4
24	32.8	22.3	15.7	15.9
				20.8
31	20.4	14.4	9.7	32.8

It remains to comment that, so far, only spectra and chromatograms have been at hand, not the pure substances. Comparison of different substances or proof of identity can only be achieved, however, by chromatographic separation and measurement of individual spectra. In this case, a small measure of uncertainty concerning the location of the MeO group in the aromatic moieties of the methoxylated compounds still remains. Isomers other than those depicted (*e.g.*, 3-OH instead of 4-OH) would also produce very similar mass spectra, to a first approximation. In this particular case, however, it was possible to make a distinction with the aid of the UV spectra (not analyzed here). NMR Spectra, however, would certainly be the most valuable technique for structural analysis. Systems of the HPLC-UV(DAD)-NMR type are known and have already proved themselves in structure elucidation, although these require significantly larger quantities, as well as full separation of the compounds involved.

Scheme 3.10. *Structures of the Fragment Ions of Substituted N,N'-Dicinnamoylspermidines.*
Distinction between (*E*)- and (*Z*)-isomers cannot be made solely on the basis of mass spectrometry.

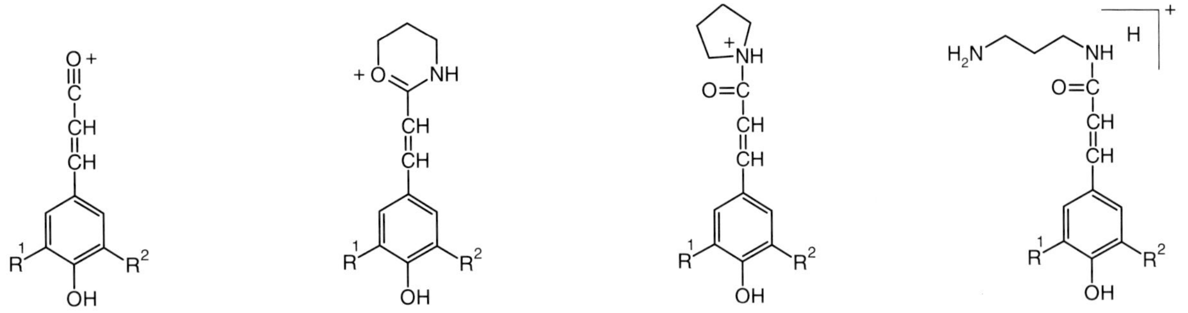

a'	$R^1 = R^2 = H$	*m/z* 147
a''	$R^1 = MeO, R^2 = H$	*m/z* 177
a'''	$R^1 = R^2 = MeO$	*m/z* 207

b'	$R^1 = R^2 = H$	*m/z* 204
b''	$R^1 = MeO, R^2 = H$	*m/z* 234
b'''	$R^1 = R^2 = MeO$	*m/z* 264

c'	$R^1 = R^2 = H$	*m/z* 218
c''	$R^1 = MeO, R^2 = H$	*m/z* 248
c'''	$R^1 = R^2 = MeO$	*m/z* 278

d'	$R^1 = R^2 = H$	*m/z* 221
d''	$R^1 = MeO, R^2 = H$	*m/z* 251
d'''	$R^1 = R^2 = MeO$	*m/z* 281

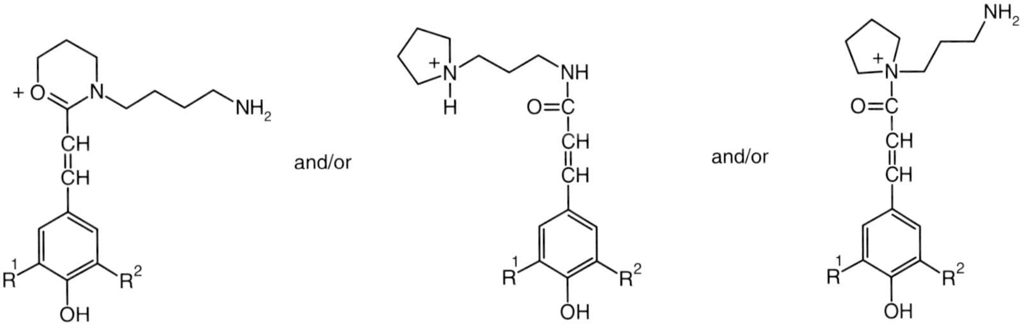

and/or   and/or

e'	$R^1 = R^2 = H$	*m/z* 275
e''	$R^1 = MeO, R^2 = H$	*m/z* 305
e'''	$R^1 = R^2 = MeO$	*m/z* 335

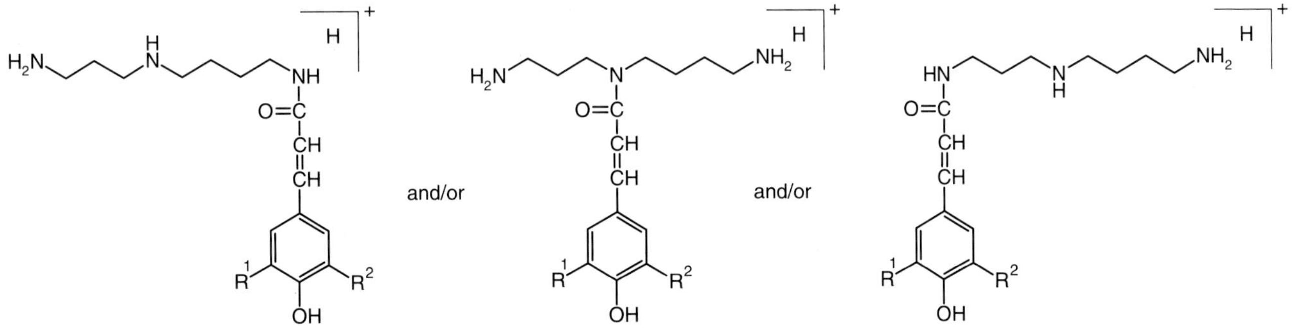

and/or   and/or

f'	$R^1 = R^2 = H$	*m/z* 292
f''	$R^1 = MeO, R^2 = H$	*m/z* 322
f'''	$R^1 = R^2 = MeO$	*m/z* 352

163

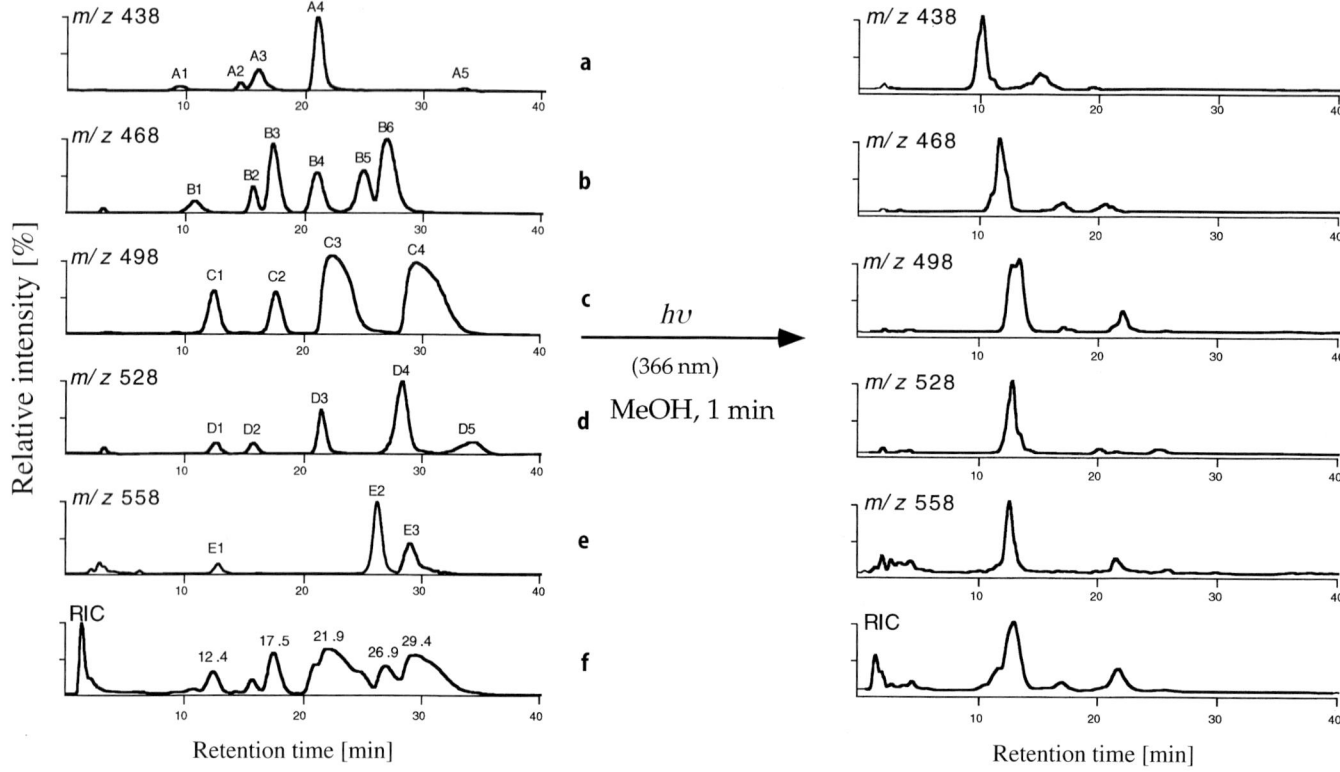

Fig. 3.20. *Chromatographic separation of the basic pollen fraction I of Hippeastrum after irradiation with UV light* (low-pressure Hg-vapor lamp, 366 nm, MeOH, 1 min). *Apart from that, the same characteristics apply as given in Fig. 3.16.*

References and Notes

1. J. Blyth, '*Über die Zusammensetzung und die Zersetzungsproducte des Coniins*', *Ann. Chem. Pharm.* **1849**, *70*, 73.
2. *Annalen der Pharmacie* **1837**, *21*, 344.
3. G. P. Schiemenz, *Der Schierlingsbecher, die Zauberflöte und der Druckfehlerteufel*, Lecture 2nd Int. Conf. on the History of Chemistry, Eger, Hungary Aug. 15–19, 1995.
4. P. L. Geiger, J. Liebig, '*Über die Zusammensetzung des Coniins*', *Magazin der Pharmacie* **1831**, *9*, 159.
5. J. J. Berzelius, *Lehrbuch der Chemie*, German translation by *F. Woehler*, 10 Vols., Arnoldische Buchhandlung, 3rd edn., Dresden, p. 1835–1841.
6. V. Ortigosa, '*Zusammensetzung des Coniins*', *Ann. Chem. Pharm.* **1842**, *92*, 313.

7. I. Guareschi, *Einführung in das Studium der Alkaloide mit besonderer Berücksichtigung der vegetabilischen Alkaloide und der Ptomaine*, Gaertner's Verlagsbuchhandlung, Berlin, 1896.

8. A. v. Planta, A. Kekulé, '*Beiträge zur Kenntniss einiger flüchtiger Basen*', Ann. Chem. Pharm. **1854**, *89*, 129.

9. A. W. Hofmann, '*Einwirkung der Wärme auf die Ammoniumbasen*', Ber. Dtsch. Chem. Ges. **1881**, *14*, 705.

10. E. Winterstein, G. Trier, *Die Alkaloide*, Verlag Bornträger, Berlin, 1910.

11. A. W. Hofmann, '*Noch einige Beobachtungen über Piperidin und Pyridin*', Ber. Dtsch. Chem. Ges. **1883**, *16*, 586.

12. A. W. Hofmann, '*Über die Einwirkung des Broms in alkalischer Lösung auf die Amine*', Ber. Dtsch. Chem. Ges. **1883**, *16*, 558.

13. A. W. Hofmann, '*Zur Kenntnis der Coniin-Gruppe*', Ber. Dtsch. Chem. Ges. **1885**, *18*, 5.

14. Conydrine was isolated by *T. Wertheim* ('*Ueber ein neues Alkaloid in Conium maculatum*', *Liebigs Ann. Chem.* **1856**, *100*, 328) and structurally elucidated by *E. Späth and E. Adler* ('*Zur Konstitution des Konhydrins*', *Monatshefte Chem.* **1933**, *63*, 127).

15. A. W. Hofmann, '*Zur Geschichte des Conydrins*', Ber. Dtsch. Chem. Ges. **1882**, *15*, 2313.

16. A. W. Hofmann, '*Zur Kenntnis des Coniin-Gruppe*', Ber. Dtsch. Chem. Ges. **1885**, *18*, 109.

17. A. W. Hofmann, '*Zur Kenntnis des Coniins*', Ber. Dtsch. Chem. Ges. **1884**, *17*, 825.

18. A. Cahours, A. Etard, *Compt. rend.* **1881**, *92*, 1079.

19. R. Laiblin, '*Über Nicotin und Nicotinsäure*', *Liebigs Ann. Chem.* **1879**, *196*, 129.

20. H. Weidel, '*Studium über Verbindungen aus dem animalischen Theer*', Ber. Dtsch. Chem. Ges. **1879**, *12*, 1989.

21. T. A. Henry, *The Plant Alkaloids*, 4th edn., Churchill Ltd., London, 1949.

22. T. M. Sharp, '*The Alkaloids of Alstonia Barks. Part II. A. macrophylla WALL, A. somersetensis F. M. BAILEY, A. verticillosa F. MUELL, A. villosa BLUM*', J. Chem. Soc. **1934**, 1227.

23. C. E. Nordman, S. K. Kumra, '*The Structure of Villalstonine*', J. Am. Chem. Soc. **1965**, *87*, 2059.

24. N. K. Hart, S. R. Johns, J. A. Lamberton, '*Tertiary Alkaloids of Alstonia spectabilis and Alstonia glabriflora (Apocynaceae)*', Austral. J. Chem. **1972**, *25*, 2739.

25. The term *drug* is used here in the older sense: '*An original, simple, medicinal substance, organic or inorganic, used by itself, or as an ingredient in medicine, or, formerly, in the arts generally*' (Shorter Oxford English Dictionary, 3rd edn., 1955). The restriction to euphorics and intoxicant drugs is a more recent development.

26. All mass spectra, mentioned in this section, relate to electron-impact ionization mass spectrometry (amu: atomic mass unit).

27. In a *Zeisel* determination, the substance is refluxed with HI, forming alkyl iodides, which are treated with alcoholic $AgNO_3$ solution. The AgI separating out is then determined either gravimetrically or by mass analysis.

28. SRB test = 'Sheep-Red-Blood-Cell' test.

29. N. Keawpradub, P. J. Houghton, E. Eno-Amooquaye, P. J. Burke, '*Activity of Extracts and Alkaloids of Thai Alstonia Species Against Human Lung Cancer Cell Lines*', *Planta Medica* **1997**, *63*, 97.

30. ^{1}H-NMR Spectra were recorded at 100 MHz in $CDCl_3$ solution (unless stated otherwise) with tetramethylsilane (TMS) as the internal standard.

31. Perchlorate (ClO_4^-), picrate ($C_6H_2N_3O_7^-$), and iodide (I^-) are particularly well suited as anions for tetraalkylammonium compounds (quaternary nitrogen salts). These salts may usually be crystallized from acetone/H_2O. Generally, the content of crys-

tal water is low (between 0 and 3 equivalents of H_2O). Other salts such as bromide (Br^-), chloride (Cl^-), fluoride (F^-), or sulfate (SO_4^{2-}) crystallize less easily and may frequently incorporate much more water in the crystal lattice. Their crystallization mostly requires higher alcohols (propanol, isopropanol *etc.*) or mixtures of lower alcohols and ethers. For anion exchange, precipitation reactions between either aqueous alkaloid chloride or alkaloid fluoride solutions and saturated sodium perchlorate, picrate, or iodide solutions have proven ideal. Otherwise, it is possible to convert alkaloid picrates or alkaloid iodides (in acetone/H_2O 1 : 1) into the corresponding chlorides or fluorides by ion exchange on exchange resin columns (Cl^- or F^- forms). Commercially available ion-exchange resins of the chloride form may be converted into the fluoride analogs by washing with an excess of saturated aqueous KF.

32. In the IR spectrum of pleiocarpamine (**5**), the double band corresponding to the MeOOC group is found at 1736 and 1770 cm^{-1} (CS_2), at 1724 cm^{-1} (KBr), and at 1730 cm^{-1} (nujol), respectively.

33. The UV spectrum would also have to indicate the presence of the indoline chromophore, of course. As mentioned, though, the spectrum is complex in nature, representing the superposition of at least two chromophores. No interpretation was, therefore, possible at that stage of the investigation.

34. Other color reagents commonly used in alkaloid chemistry are: *a*) potassium iodoplatinate reagent[35]: two solutions are prepared (*soln. 1*: 1 g $PtCl_4 \cdot 2\,H_2O$ + 6 ml H_2O + 20 ml 1N HCl; *soln. 2*: 9 g KI + 90 ml H_2O) and mixed in the ratio of 1 ml (*soln. 1*) : 9 ml (*soln. 2*) with 20 ml H_2O. *b*) *Dragendorff* reagent[36]: 8 g $Bi(NO_3)_3 \cdot 5\,H_2O$ in 20 ml HNO_3 (spec. grav. 1.18) + 27.2 g KI in 50 ml H_2O. After mixing, KNO_3 crystallizes out, the solution is decanted and diluted with H_2O to a total volume of 100 ml. *c*) *Mayer* reagent: 1.36 g $HgCl_2$ in 60 ml H_2O + 5 g KI in 10 ml H_2O, made up to 100 ml with H_2O.

35. E. Schlittler, J. Hohl, '*Über die Alkaloide aus Strychnos melinoniana BAILLON*', *Helv. Chim. Acta* **1952**, *35*, 29.

36. B. T. Cromwell, '*The Alkaloids*', in K. Peach, M. V. Tracey (Eds.), *Moderne Methoden der Pflanzenanalyse*, Springer-Verlag, Berlin, **1955**, *4*, 367.

37. In their IR spectra, pleiocarpamine as well as 2,7-dihydropleiocarpamine and 16-*epi*-pleiocarpamine all display double bands for the ester carbonyl group.

38. In adsorption mass spectrometry, some cases, in which the 'molecular ion' of the hydro salt appears, are known (papaverine is one example). These signals are probably attributable not to the actual molecular ion of the hydro salt, but of some addition product of base and acid. Addition products of this kind are found fairly commonly.

39. Distillations of this type are preferably carried out in a small flask consisting of a glass tube blown into a small bulb at one end. A test sample of the dissolved substance is directly introduced into the flask by means of a long funnel (pulled-out test tube). The solvent is then removed by swirling the flask in warm water under vacuum. The result should be the thinnest possible film on the walls of the flask. The actual pyrolysis is then carried out by briefly dipping the flask into a heated metal bath, with the end of the tube connected to a vacuum pump. The substance then condenses inside the tube above the metal bath as a slightly colored coating.

40. According to *Silverstein* and *Bassler*'s tables[41], the following vinyl proton signals are to be expected for macroline (**9**) and macrolinol (**10**): 5.72 and 6.20 ppm (**9**); 4.90 ppm (**10**).

41. R. M. Silverstein, G. C. Bassler, *Spectrometric Identification of Organic Compounds*, 2nd edn., John Wiley & Sons, New York, 1967.

42. Aromatic protons (H^+) may be exchanged with deuterons (D^+) under acid-catalyzed conditions. In addition, acidic protons in α-positions to electron-withdrawing functional groups such as CO, CN, or NO_2 (*etc.*) can also be exchanged. Protons linked

directly to heteroatoms (OH, NH *etc.*), are indeed exchanged with deuterons, but under normal aqueous workup conditions, without rigorous exclusion of protons (water), the reverse exchange usually proceeds so rapidly that only OH, NH *etc.* are observed afterwards.

43. Molecules consisting exclusively of C, H, N, and O and of odd molecular mass contain an odd number of N-atoms. With fragment ions containing an odd number of N-atoms, those with an odd mass are radical cations (+•), those with an even mass are regular cations (+).

44. Mass spectrometric neutral processes are taken to be those reactions that give rise to neutral components, such as CO, H_2O, C_2H_4, ketenes *etc.* Some examples of such reactions are *McLafferty* rearrangements, retro-*Diels-Alder* reactions, onium reactions, and $S_N i$-type reactions.

45. See note 43; the fragment corresponding to *m/z* 251 is a radical ion; it contains an odd number of N-atoms, and its mass is also odd.

46. M. Hesse, H. Hürzeler, C. W. Gemenden, B. S. Joshi, W. I. Taylor, H. Schmid, '*Die Struktur des Alstonia-Alkaloides Villalstonin*', *Helv. Chim. Acta* **1965**, *48*, 689.

47. T. Kishi, M. Hesse, W. Vetter, C. W. Gemenden, W. I. Taylor, H. Schmid, '*Macralstonin*', *Helv. Chim. Acta* **1966**, *49*, 946.

48. Subtraction of the UV spectrum of macroline from that of villamine (*Fig. 3.6*) generates the spectrum of an indoline chromophore (*e.g.*, that of 2,7-dihydropleiocarpamine (**6**)). The same result is obtained by adding the UV spectra of macroline and 2,7-dihydropleiocarpamine (*Fig. 3.6*). Evidence that the indole chromophore of pleiocarpamine is involved, and not that of macroline, comes from the corresponding ¹H-NMR spectra; in both spectra (**9** and **7**), the signal of the Me group of the indole moiety appears at the same position, which would not be expected in the presence of a 2,7-dihydromacroline chromophore.

49. C. Werner, W. Hu, A. Lorenzi-Riatsch, M. Hesse, '*Dicoumaroylspermidines and Tricoumaroylspermidines in Anthers of Different Species of the Genus Aphelandra*', *Phytochemistry* **1995**, *40*, 461.

50. M. Nimtz, M. Bokern, B. Meurer-Grimes, '*Minor Hydroxycinnamic Acid Spermidines from Pollen of Quercus dentata*', *Phytochemistry* **1996**, *43*, 487.

51. N. Youhnovsky, L. Bigler, C. Werner, M. Hesse, '*On-Line Coupling of High-Performance Liquid Chromatography to Atmospheric Pressure Chemical Ionization Mass Spectrometry (HPLC/APCI-MS and MS/MS). The Pollen Analysis of Hippeastrum × hortorum (Amaryllidaceae)*', *Helv. Chim. Acta* **1998**, *81*, 1654.

52. For a detailed discussion on this topic, see: M. Hesse, H. Meyer, B. Zeeh, *Spektroskopische Methoden in der Organischen Chemie*, Thieme-Verlag, Stuttgart, 6th edn., 2001.

4. Artifacts

4.1. Introduction

In chemistry, by 'artifact' we mean an 'artificial' compound not found as such in nature. Both alkaloids and N-free compounds can equally well come into play as precursors for alkaloid-like artifacts. Structurally, it is not possible to recognize artifacts *a priori*. Their structures are frequently entirely plausible and indistinguishable from naturally occurring analogs, while the palette of chemical reactions that may contribute to the formation of artifacts is exceedingly broad.

During processing of biological materials (plants, fungi, animal products *etc.*) with a view to obtaining alkaloids, a whole series of purification steps is often carried out. These inevitably bring solvents, acids, and bases into contact with the active substance. As well as this, work is only rarely carried out under exclusion of air. Given the great variety of alkaloid structures, it is, therefore, not too surprising that chemical reactions may easily take place between the alkaloids and the reagents necessary to isolate them. To illustrate this, let us pick out a few representative examples from the great number of reactions of this type.

4.2. Acetone

The two compounds 1,2-dehydrobeninine (**1**) and its derivative **2** were isolated, together with a great number of others alkaloids, from the plant *Hedranthera barteri* (HOOK. f.) PICHON, a member of the Apocynaceae family[1] (*Scheme 4.1*).

Scheme 4.1

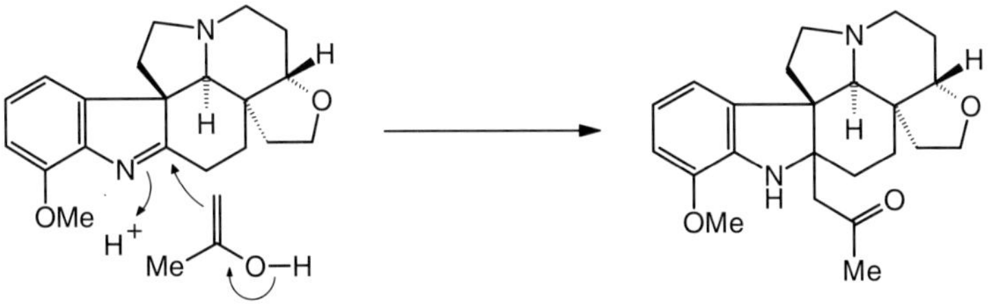

1,2-Dehydrobeninine (**1**)   **2**

Compound **2** differs from **1** in containing one additional molecule of acetone. When treated with 2N aqueous HCl solution, 1,2-dehydrobeninine (**1**) is formed. Conversely, when a model compound that resembles **1** in incorporating a methoxyindolenine chromophore is exposed to acetone under acid catalysis, it is possible to prepare the corresponding acetonyl compound without any problems. From this experiment, it may be concluded that the acetone used in the extraction and isolation of the drug might have reacted with 1,2-dehydrobeninine (**1**) to produce compound **2** as an artifact. To test this assumption, the isolation of the drug was also performed under exclusion of acetone; as expected, it was no longer possible to isolate the acetonyl derivative **2**.

Acetone has often been suspected to cause artifacts during alkaloid isolation. A more recent example in this area is karachine (**3**), isolated from *Berberis aristata* DC (Berberidaceae). Its structure[2] was unequivocally established at the beginning of the 1980s, revealing that the compound is made out of two molecules of acetone and one molecule of berberine (**4**), the principal alkaloid of *B. aristata* (*Scheme 4.2*). In an attempt to allay the suspicion that karachine (**3**) represents an artifact, the plant, indigenous to India and Pakistan, was extracted once more, but this time with acetone-free solvents (probably EtOH). Karachine, however, was (surprisingly) also found under these controlled conditions. Nonetheless, it still has to be assumed that acetone incorporation into berberine is an unselective *chemical* rather than a *biological* process, especially because karachine (**3**) was found to be optically non-active ($[\alpha]_D$ close to zero; sadly, the authors did not mention the solvent used). In other words: karachine – with four stereogenic centers, all of which come into being during the process of acetone incorporation into the achiral berberine (**4**) – is most likely a racemate. Also, *post hoc* racemization of karachine may likewise be ruled out on structural grounds. Consequently, the source of the acetone remains to be explained.

Hydroxypropyllycodine (**5**) was isolated, together with lycodine (**6**) and other *Lycopodium* alkaloids, from *Lycopodium obscurum* L. (Lycopodiaceae). It differs from the principal alkaloid **6** in that it contains one additional molecule of acetone. It is also the first compound of this type with a C_{19}-skeleton. To explain its formation, a biogenetic hypothesis has been developed, although it has not yet been supported by tracer[3] experiments. According to this model, hydroxypropyllycodine (**5**) may be formed from two subunits of pelletierine (**5a**) and from one component of 3-oxobutanoic acid (*Scheme 4.3*)[4]. However, as shown in *Scheme 4.4*, acetone would also merit consideration as a source of the bridging C_3-unit: in this case, acid-catalyzed incorporation of acetone into the lycopodinium ion would be followed

Scheme 4.2

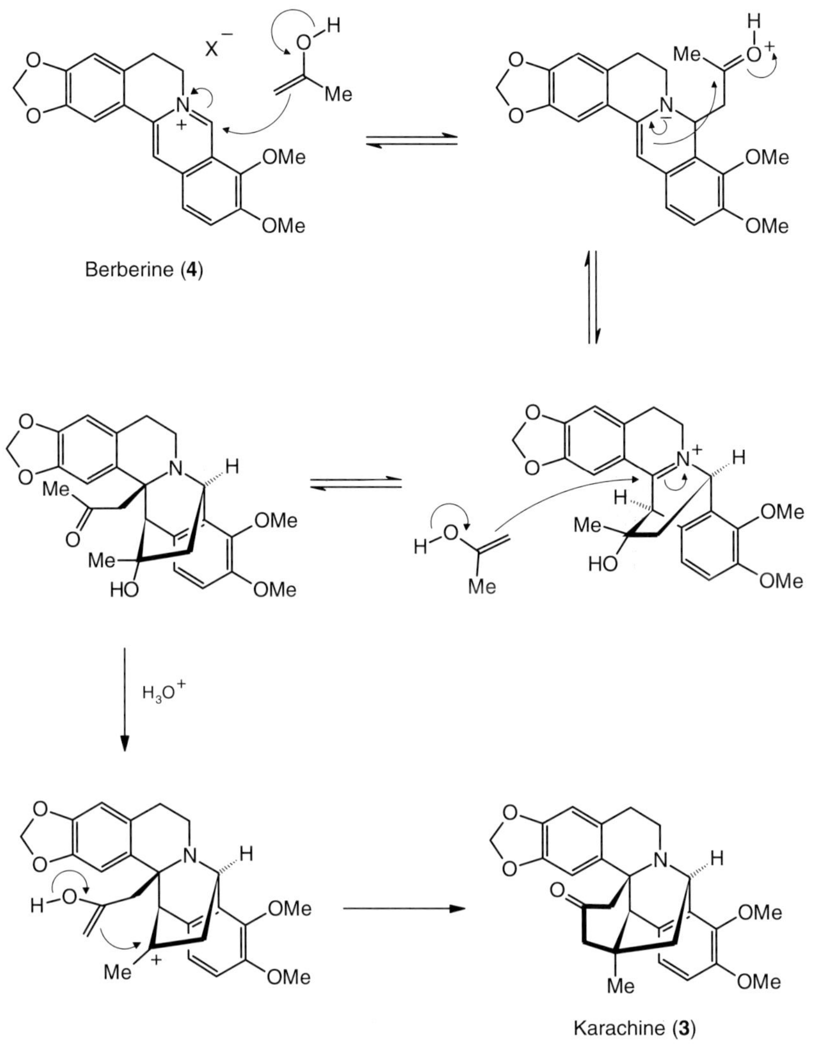

Berberine (**4**)

H_3O^+

Karachine (**3**)

by a hydride transfer to the C=O group upon re-aromatization of the substituted pyridine ring.

However, the presence of an acetonyl side chain in an alkaloid does not by any means dictate that the compound is necessarily an artifact. Pelletierine

Scheme 4.3

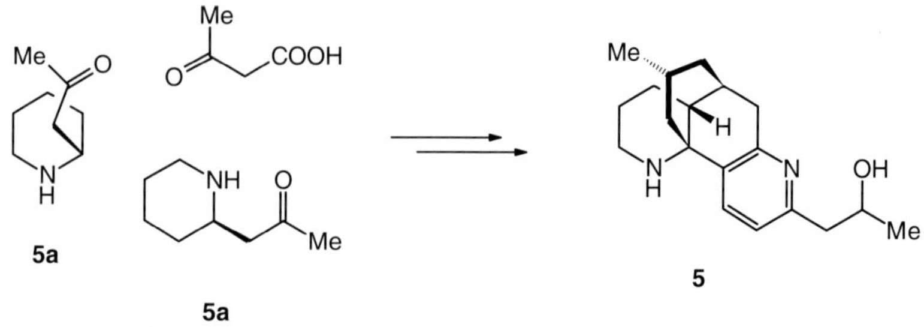

5a

5a

5

Scheme 4.4

6

5

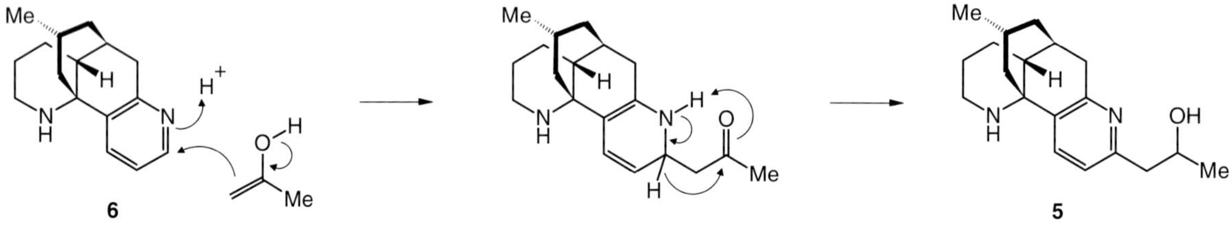

(**7**), isolated for instance from *Punica granatum* L. (Punicaceae), is one such example. Its biogenesis has been established to proceed *via* lysine (**8**) and 3-oxobutanoic acid (**9**) (*Scheme 4.5*). In contrast, the plant *Conium maculatum* L. produces coniine (**10**), which possesses the same carbon skeleton as pelletierine (**7**), by the polyketide pathway[5, 6] (*Scheme 4.6*).

Scheme 4.5

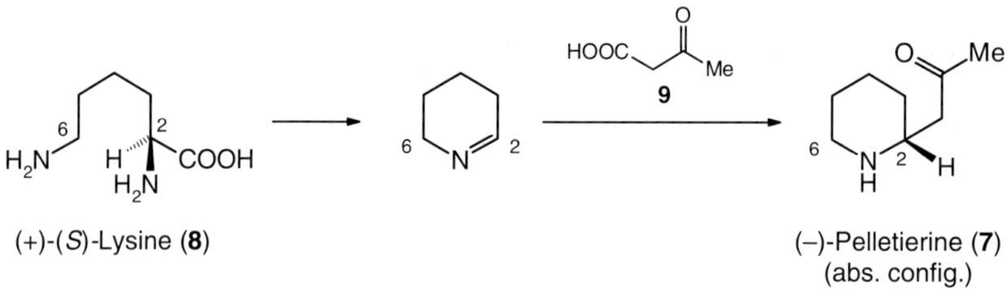

(+)-(*S*)-Lysine (**8**)

9

(–)-Pelletierine (**7**)
(abs. config.)

Scheme 4.6

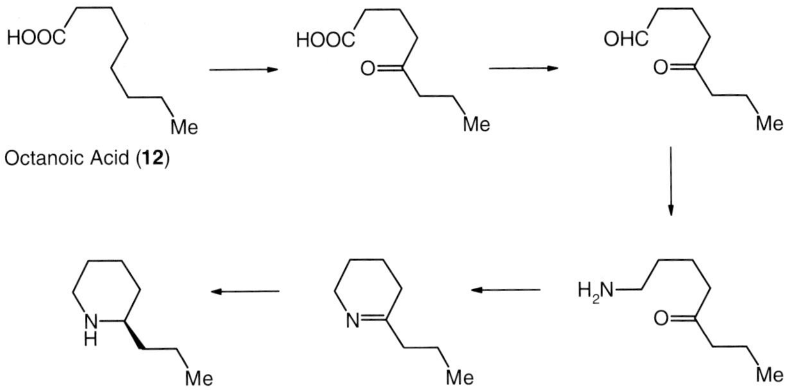

Octanoic Acid (**12**)

(+)-Coniine (**10**)
(abs. config.)

γ-Coniceine (**11**)

4.3. Ammonia

The base gentianine (**13**) has been isolated from a number of plants of the families Gentianaceae and Loganiaceae. A biogenetically related compound also found in Gentianaceae is gentiopicroside (**14**), a N-free iridoid. Swertiamarin (**15**), from plants of the *Swertia* (Gentianaceae) genus, is also closely related to **13**. Compounds **14** and **15** both contain a labile cyclic acetal function.

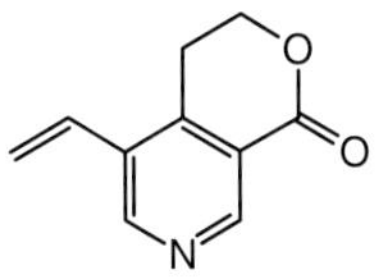

On treatment with ammonia (NH_3) under mild conditions, both **14** and **15** can be converted into gentianine (**13**). This very easily occurring transformation process provides grounds for assuming that gentianine could be formed from **14** and/or **15**, on treatment of the active substance with NH_3 (which is very often used for alkalization of plant extracts). Accordingly, a number of plants of the Gentianaceae family, reported to contain gentianine, were processed once more, but this time with sodium carbonate (Na_2CO_3) as the base and avoiding NH_3. The results of this study were as follows: only *Gentiana fetisowii* was found to contain gentianine in large amounts, all the other plants, in contrast to earlier findings, were found either not to contain the base or to do so only in trace amounts[7,8]. Therefore, it may be concluded with reasonable certainty that most of the isolated gentianine (**13**) was an artifact indeed, formed only on processing with NH_3. Whether the active material, isolated in the absence of NH_3, did originally contain 'natural' gentianine, is not clear, however. It is conceivable that the action of Na_2CO_3 might

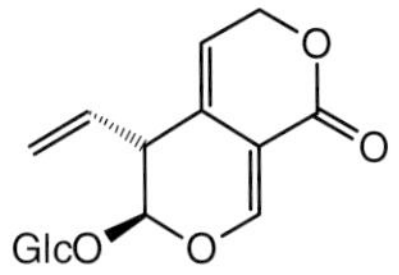

Gentianine (**13**)

Gentiopicroside (**14**)

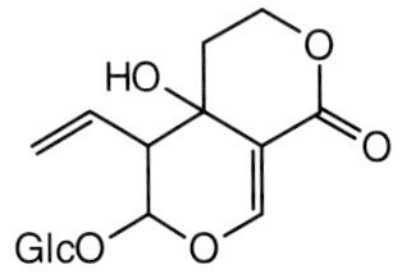

Swertiamarin (**15**)

173

bring about the release of NH_3 from certain plant materials. This similarly 'non-biological' NH_3 might then be responsible for the formation of gentianine, as discussed.

4.4. Formaldehyde

Formaldehyde (CH_2O) as such plays hardly any role as a reagent in the isolation of natural chemical products. When simple chemical reactions are carried out on alkaloids, its use is deliberately avoided, and so to natural product chemists – concerned primarily with simple extraction, chromatography, or recrystallization – it tends not to come to mind as a cause of side reactions. Nevertheless, formaldehyde is present whenever methanol (MeOH) is used, entering into reactions with ease.

Methanol is cheerfully and frequently used as a solvent for the extraction of plant materials, in chromatography, in chemical reactions, and also for measuring spectra. The molar ratio of solvent to substance is usually generous. As well as this, occasionally prolonged refluxing is necessary, or a sample might be left standing at room temperature in a particular solvent. For all these reasons, even small amounts of solvent impurities may suddenly turn out to be present in quantities equimolar to the substance. Hence, if a reaction can take place, then a considerable proportion of the alkaloid might be transformed. Despite the 'super-pure' chemicals on the market today, considerations of this sort are still always important. In particular, problems arise with CH_2O impurities whenever MeOH comes into contact with light and air. Freshly distilled MeOH, stored under inert atmosphere (N_2, Ar), is mostly pure; with ingress of O_2, however, CH_2O is rapidly produced. Just repeatedly opening a bottle of MeOH causes the CH_2O content to increase significantly; and in old, half empty bottles there is plenty of CH_2O to be found. In contrast, EtOH, the next higher homologous alcohol, is stable under comparable conditions.

$$CH_3OH + [O] \rightarrow CH_2O + H_2O$$

Once produced, CH_2O may react with primary or secondary amines, which often occur in alkaloids. Two typical examples from alkaloid chemistry are given here.

In the course of structural investigations, the indole alkaloid talbotine was catalytically hydrogenated in MeOH (*Scheme 4.7*). The major project, together with a very small quantity of the desired 19,20-dihydrotalbotine, was its *N*-methyl derivative most probably arising from the reduction of the corresponding (intermediate) *N*-methyl-'iminium'-talbotine. In the presence on-ly of CH$_2$O – under non-reducing conditions – talbotine had not, however, undergone any permanent modification: on contact with H$_2$O, the iminium salt would have hydrolyzed back into talbotine and CH$_2$O.

Scheme 4.7

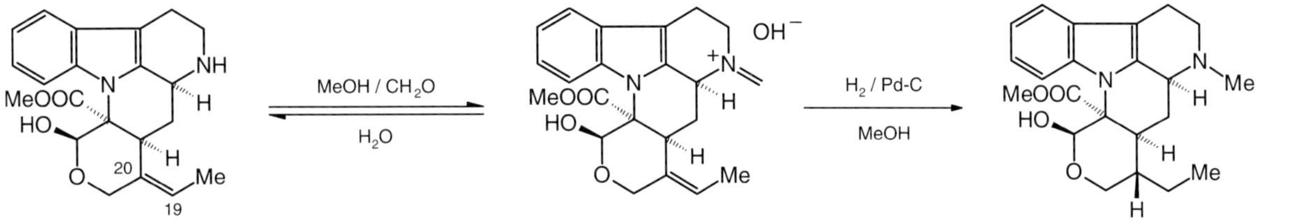

(–)-Talbotine
(abs. config.)

19,20-Dihydro-*N*-methyltalbotine

While the above conversion is rather easy to understand, things are some-what more complicated in reactions between CH$_2$O and 1,3-diamines. 1,3-Diamines are found in derivatives of spermidine, spermine, and other (also naturally occurring) polyamines. They react with CH$_2$O (through neighbor-ing group participation) to yield stable hexahydropyrimidines. Interestingly, no analogous reaction has to date been observed in the case of natural 1,4-diamines. Also, less reactive aldehydes (other than CH$_2$O) seem not to react to the corresponding 2-substituted hexahydropyrimidines.

By electrospray-ionization mass spectrometry (ESI-MS), hexahydropyrimi-dines can be easily identified by their additional mass of +12 amu (plus one C-atom) relative to the corresponding 1,3-diamines. If electron-impact-ion-ization mass spectrometry (EI-MS) is used, a fragmentation signal 11 amu heavier ($[M + 11]^+$) is observed, in which the signal intensities are consider-ably higher than those of the corresponding 1,3-diamines[9] (*Scheme 4.8*).

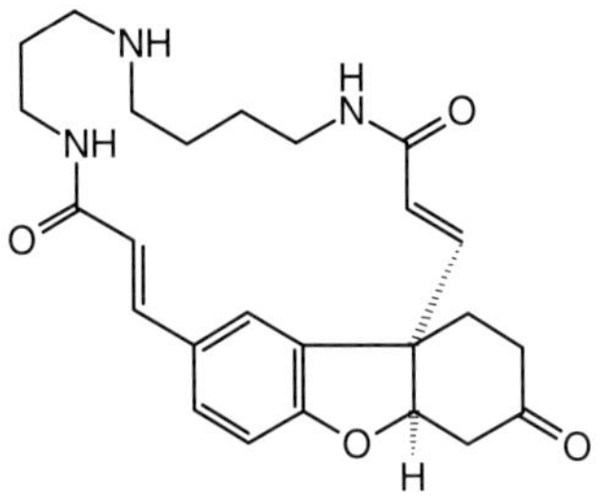

As an example of 'natural' polyamine alkaloids incorporating a hexahydro-pyrimidine substructure, lunarine and the so-called LBX alkaloid should be mentioned (*cf. Lunaria* alkaloids from *Lunaria biennis* MOENCH (Crucifera-ceae)).

Scheme 4.8

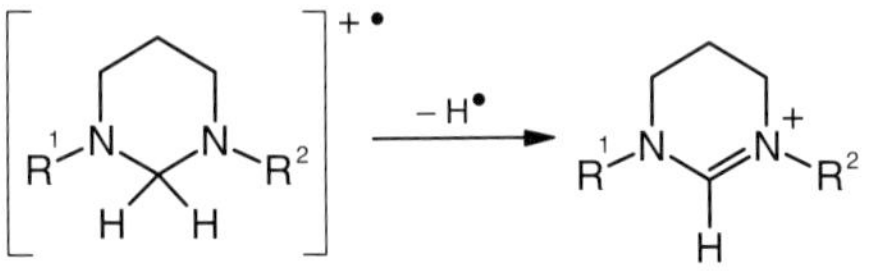

R^1, R^2 = alkyl

(+)-Lunarine
(abs. config.)

(+)-Alkaloid LBX

19-Oxocoronaridine

19-Hydroxycoronaridine

19-Ethoxycoronaridine

4.5. Chloroform

Despite some serious drawbacks, chloroform (CHCl$_3$) is still very popular as a solvent (*inter alia* in deuterated form (CDCl$_3$) for NMR spectroscopy). Its disadvantages mostly concern the disposal of waste solvent, since HCl is formed on incineration. For the chemist working in the laboratory, however, one other inconvenient property is that CHCl$_3$, in the presence of O$_2$ and light, turns into phosgene and HCl.

$$CHCl_3 + [O] \rightarrow COCl_2 + HCl$$

This process may be prevented by addition of 1–2% EtOH. Commercially, CHCl$_3$ is therefore mostly offered with added EtOH. If the stabilizer upsets certain reactions or investigations (*e.g.*, recording NMR), it can easily be removed. In the same way, though, EtOH gets removed when CHCl$_3$ is used to extract aqueous layers. Consequently, the oxidation reaction can take place during aqueous extractions, and phosgene and/or HCl may enter into reactions with aldehydes, esters, enamines, and other functional groups resulting in artifacts. For illustration, a few of these possible reaction products are presented here.

From *Tabernaemontana crassa* BENTH. (Apocynaceae), together with 19-oxocoronaridine, there were also isolated both 19-hydroxy- and 19-ethoxy-coronaridine. In the latter two compounds, C(19) is formally at the oxidation level of an aldehyde (rather than of a carboxylic acid). Given the condition that alcohol is present, an acid-catalyzed acetalization or transacetalization seems perfectly plausible. In the case of 19-ethoxycoronaridine, we are thus dealing with great certainty with an artifact; whether 19-hydroxycoronaridine is of natural origin, however, cannot be established without renewed examination of the plant material. This problem, incidentally, tends to arise with all acetals isolated from natural sources.

Rather rarer are cases in which alkaloids containing secondary N-functionalities react with phosgene. Not so very long ago, the spermine alkaloid verbaskine was isolated from *Verbascum nobile* (Scrophulariaceae) and structurally elucidated (*Scheme 4.9*). Compared with previously known spermine alkaloids, it possesses an additional C=O group. The two isomeric alkaloids verbacine and verballocine, also recently isolated from *Verbascum pseudonobile* STOJ. et STEF., can be converted into verbaskine by treatment with phosgene (*Scheme 4.9*), which indicates that verbascine is likely to be an artifact[10].

Scheme 4.9

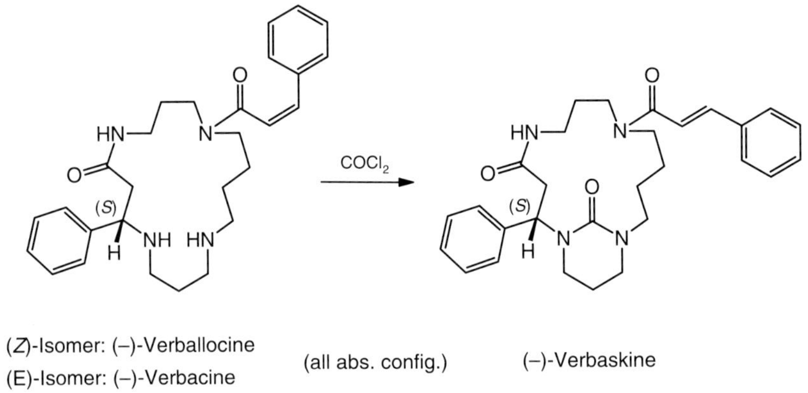

(Z)-Isomer: (–)-Verballocine
(E)-Isomer: (–)-Verbacine

(all abs. config.)

(–)-Verbaskine

The 'dimeric' aporphine alkaloid ovihernangerine (*Hernandia nymphaeifolia* (PRESL.) KUBITZKI, Hernandiaceae)[11] is very probably also an artificial product, although experimental evidence (treatment of the monomeric base with phosgene) is still awaited.

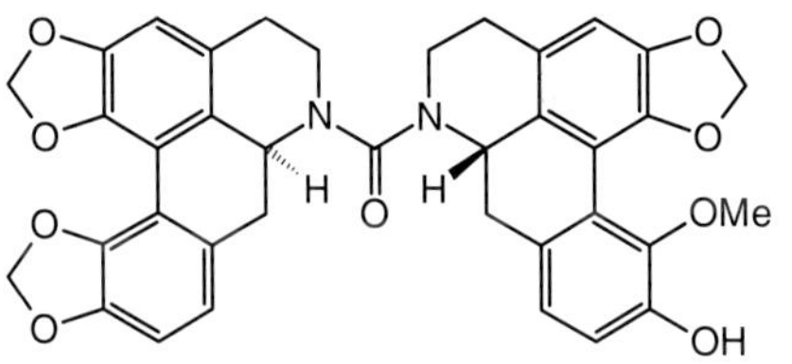

Ovihernangerine

4.6. Atmospheric Oxygen

Not uncommonly, the presence of oxygen (O_2) during plant extraction or product purification also results in the formation of artifacts.

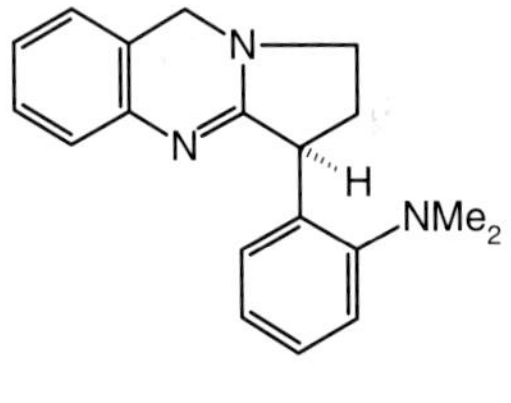

From young plants of *Adhatoda vasica* NEES, belonging to the Acanthaceae (acanthus) family, the bases vasicoline (**16**) and vasicolinone (**17**) were isolated together with other alkaloids[12]. During the course of processing vasicoline-containing fractions, especially if allowed to stand unprotected for a few hours, vasicolinone (**17**) is formed.

Vasicoline (**16**)

The relative quantities of the two bases isolated from the plant are about equal. When the plant material was gently processed under inert atmosphere, it was possible to show that both compounds **16** and **17** are being produced by the plant. However, it is perfectly conceivable that improper processing of *A. vasica* would result exclusively in the isolation of the oxidized product **17**.

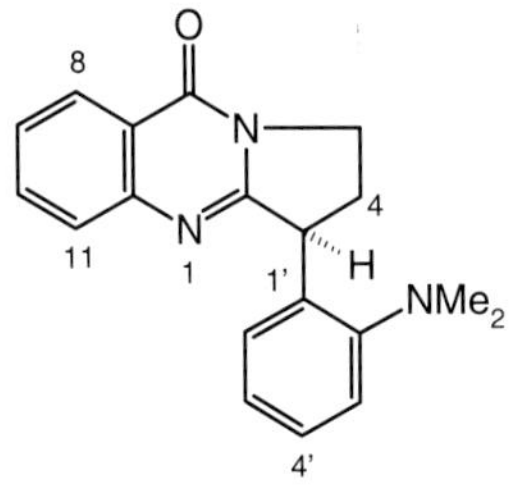

An example of an enzyme-catalyzed oxidation has been observed during the processing of the roots of *Aphelandra tetragona* (VAHL) NEES[13]. Aphelandrine (**18**) is the major alkaloid of this plant. When suitable raw materials

Vasicolinone (**17**)

R = H   : (+)-Aphelandrine (**18**)

R = OH : **19**

(lyophilizates of fresh roots) are processed rapidly, aphelandrine can be isolated in relatively large quantities (0.9% based on dry weight). Air-dried roots, on the other hand, showed an aphelandrine content of only *ca.* 0.1%. Finally, if the crushed plant material was exposed to air, after a few hours no more aphelandrine could be detected at all. Systematic investigations led to the conclusion that aphelandrine (**18**) is being hydroxylated enzymatically resulting in compound **19**, which incorporates an air-sensitive 1,2-dihydroxybenzene moiety. The latter is presumably oxidized rapidly to the corresponding *ortho*-quinone and so is no longer detectable after the aphelandrine-specific workup.

4.7. Atmospheric Oxygen and Light

Curare is a drug made and used by indigenous peoples of northern South America. Depending on tribe and local circumstances (availability), various different plants are used in its preparation. The plant components are first extracted, and the juice is then concentrated. This constitutes a process that – carried out with the most primitive means – promotes artifact formation in an absolutely exemplary manner.

In the record of his journey to Venezuela (July 16, 1799 – November 24, 1800), *Alexander von Humboldt* vividly described the process, which he was able to observe as performed by natives of the Orinoco region[14].

> '*As luck would have it, we met an old Indian. […] The man was a local chemist. At his dwelling we found great clay pots for boiling plant juices, flat vessels well suited, thanks to their large surface areas, for performing evaporation, conical rolled-up banana leaves for straining liquids contained in the more or less fibrous substances. […] 'I know', he told us 'that the Whites understand the arts of making soap and black powder with the evil in it, that makes noise and frightens off the animals when you miss. Curare, the making of which is handed down from father to son, is better than anything that you people over there (over the sea) know how to make. It is the juice of a plant, that kills very quietly (without anyone knowing where the shot came from).*
>
> *This chemical operation, upon which the Master of Curare laid so much importance, seemed very simple to us. The creeper (bejuco) that they make the poison from in Esmeralda […] is the 'Bejuco de Mavacure', and is to be found in great quantity east of the mission on the left bank of the Orinoco, on the other side of the Rio Amaguaca in the granite mountain country of Guanaya and Yumariquin. The fresh juice of the*

liana is not poisonous; perhaps it only becomes active when it is highly concentrated. The terrible poison is contained in the bark and partly in the sapwood. With a knife, one scrapes 4–5 lines (a line is one twelfth of an inch; hence 9–11 mm) *thick mavacure branches and pounds the scraped off bark on a stone, such as is used for grinding manioc flour, into very thin fibers. Because the poisonous juice is yellow, the whole fibrous mass takes on this color. This is all put in a wide funnel. […] It was a conical, rolled-up banana leaf, mounted in another, stronger cone of palm leaves; the whole contraption rested on a light frame of leafstalks and fruit-spindles of a palm. First, one makes a cold infusion by pouring water onto the fibrous, pounded bark of the mavacure. A yellow liquid drips out of the Embudo, the leaf funnel, for several hours. This water trickling through is the poisonous liquid; it only acquires its proper potency, though, after being concentrated like molasses in a large clay container. The Indian demanded from time to time that we taste the liquid; its greater or lesser bitterness is the test of whether the juice has been thickened enough. There is no danger in this, since curare is only deadly when it comes directly into contact with the blood.*

The juice of the mavacure, even though boiled down, is not thick enough to stick to arrows. So, merely to give more body to the poison, the concentrated infusion is added to another, very sticky plant sap that comes from a big-leafed tree called kiracaguero. […] As soon as the sticky sap of the kiracaguero tree is immersed in the thickened, boiling, poisonous juice, it blackens and curdles into a mass of the consistency of tar or thick syrup. This mass is now curare'.

The name 'mavacure' is not completely unambiguous from a modern viewpoint. However, the use of *Strychnos toxifera* SCHOMBURGK (Loganiaceae) as a major component of calabash curare (*vide infra*) is reasonably certain.

Also instructive are the accounts of the two brothers *Robert H.* and *Richard Schomburgk*[15], regarding their expedition (1840–1844) to Guyana, in which they describe the Macusi people, '*who live on the banks of the Rio Branco in Brazilian Guyana and into British Guyana, on the Rupununi and elsewhere, mostly on the inaccessible river courses*'.

The Macusi also produce curare, which they obtain by processing bark infusions of *Strychnos toxifera*, *S. schomburgkii* KLOTZSCH, and *S. congens* BENTH.

'*[…] After the mass had been boiled down to about a seventh of its original volume, it was poured through a strainer made of palm leaves and*

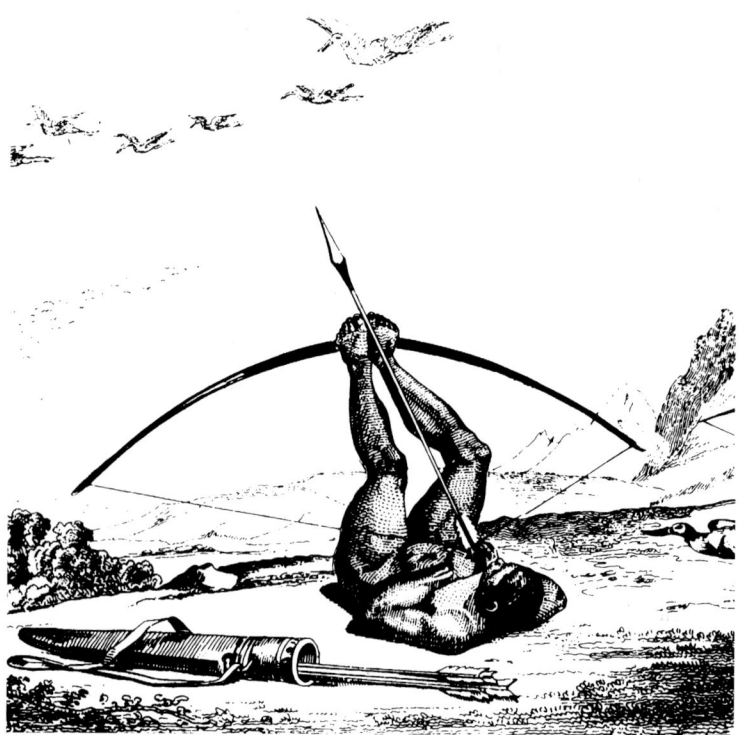

Fig. 4.1. *South American native shooting for birds with poisoned arrows* (Bildarchiv Stiftung Preussischer Schlösser und Gärten Berlin Brandenburg).

containing silk-grass. The strained liquid was left in the sun in flat vessels for three hours and the slimy sap of the Muramu root, which had previously been soaked a little and then squeezed out, was added. The poison immediately curdled into a gelatinous mass. On further drying in the sun, it reached the right consistency to be put into the calabashes[16] or earthenware pots. The poison-maker also gauged the efficacy of the poison straightaway. A lizard that had some of the poison rubbed into a small needle prick died after ten minutes, a rat after four, a hen after three. […]'.

'According to the maker of the poison, the poison should only remain effective for two years. How wrong this is can be seen from the fact that I [L. Lewin, 1923] have found that the original poison, acquired by Richard Schomburgk from the Macusi almost one hundred years ago, and of which some is in the [Berlin] Anthropological Museum, is still effective'.

Fig. 4.2. *Tubocurare is sealed in sections of bamboo, using a piece of banana leaf, and offered by the natives of Amazonia.* Chemically, it consists of bisisoquinoline alkaloids (such as tubocurarine dichloride), but Menispermaceae plants may also play a role, depending on the source. Calabash curare is packed in bottle gourds, sealed with a wooden peg. Pot curare, the third sort, is to be found in small, gray or black earthenware pots (photo to *C. Hesse*).

In the eyes of a natural products chemist, the curare makers were committing a fatal error in exposing the plant juice to air, sun, and heat: ideal conditions for triggering off a variety of reactions 'in the field'. Chemical evidence for the operation of photochemical oxidation reactions on warming the plant juice in the sun was obtained in laboratory tests between 1955 and 1965[17].

From diverse *Strychnos* varieties (especially *S. toxifera, S. froesii* DUCKE, *S. solimoesana* KRUK., and *S. trinervis* (VELL.) MART.), there were isolated numerous bisindole alkaloids, the structures of which were established as, among others, *C*-dihydrotoxiferine (**20**), *C*-alkaloid H (**21**), and *C*-toxiferine I (**22**). Their oxidation products – namely *C*-curarine I (**23**), *C*-alkaloid G (**24**), and *C*-alkaloid E (**25**), as well as *C*-calebassine (**26**), *C*-alkaloid F (**27**), and *C*-alkaloid A (**28**) – were also isolated (in smaller quantities) from the same plants (*Scheme 4.10*). The same compounds were also isolated from calabash curare and pot curare[18], but in these cases the oxidation products were always found to predominate (*Fig 4.2*). In the isolation of pure *C*-dihydrotoxiferine (**20**), the two oxidation products *C*-curarine I (**23**)

Scheme 4.10

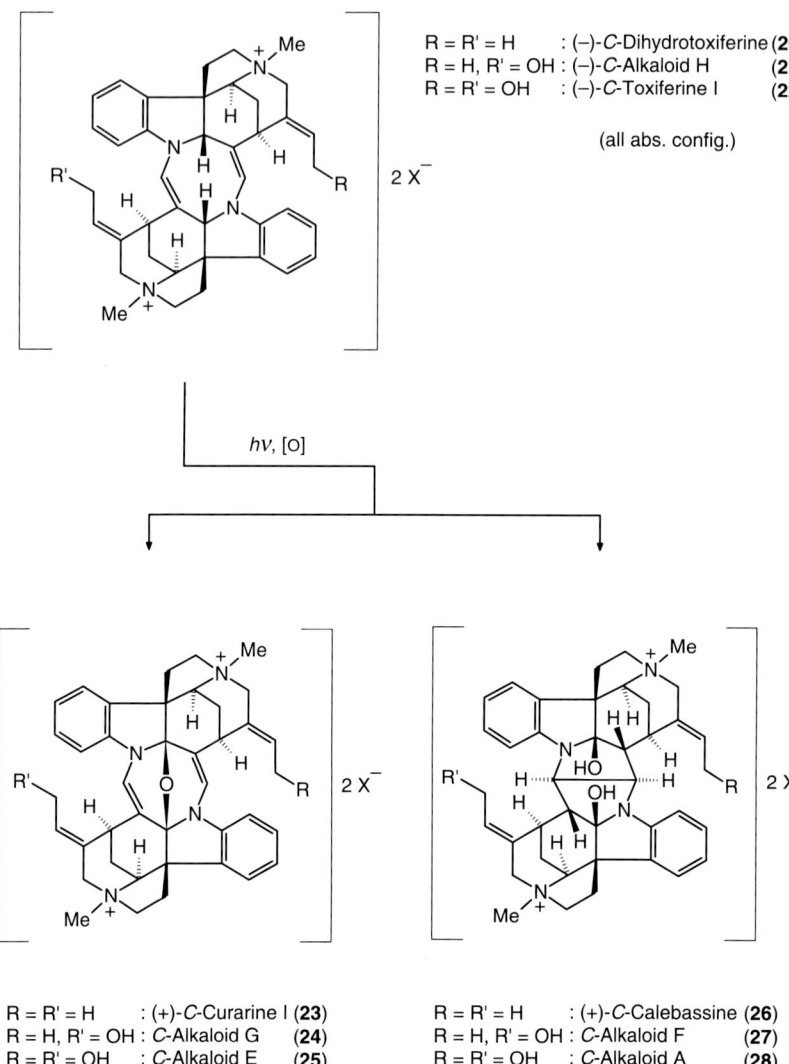

R = R' = H : (–)-*C*-Dihydrotoxiferine (**20**)
R = H, R' = OH : (–)-*C*-Alkaloid H (**21**)
R = R' = OH : (–)-*C*-Toxiferine I (**22**)

(all abs. config.)

hv, [O]

R = R' = H : (+)-*C*-Curarine I (**23**)
R = H, R' = OH : *C*-Alkaloid G (**24**)
R = R' = OH : *C*-Alkaloid E (**25**)

R = R' = H : (+)-*C*-Calebassine (**26**)
R = H, R' = OH : *C*-Alkaloid F (**27**)
R = R' = OH : *C*-Alkaloid A (**28**)

and *C*-calebassine (**26**) were always found in the mother liquors. In oxidation experiments (O$_2$, PtO$_2$), the two compounds **21** and **22** proved convertible into the analogous curarine and calebassine derivatives **24** and **25,** and **27** and **28**, respectively[17]. It is, therefore, not too surprising that these oxidation products are readily found when insufficient care is taken.

Table 4.1. *Toxicities of Some Curare Alkaloids* (intravenous, mouse)[a]

Alkaloid	No.	HD[b]	DML[c]	Duration of Paralysis [min]
C-Dihydrotoxiferine	**20**	30	60	5.5
C-Curarine I	**23**	30	50	4
C-Calebassine	**26**	240	320	3
C-Alkaloid H	**21**	16	24	3.7
C-Alkaloid G	**24**	0.6	12	7
C-Alkaloid F	**27**	75	120	1.3
C-Toxiferine I	**22**	9	23	12
C-Alkaloid E	**25**	0.3	8	18
C-Alkaloid A	**28**	7	150	2

[a] Values in $\mu g/kg$ mouse (1 μg = 10^{-6} g). [b] HD = Headdrop dose. [c] DML = *Dosis minimalis letalis.*

In total, some sixty different indole alkaloids have been isolated from curare.

Certainly, it is not exactly advantageous to have an excess of oxidation products in curare, but neither is it a grave drawback. Toxiferine I produces the longest lasting paralysis in frogs, while its oxidation product *C*-alkaloid E (**25**) represents the most toxic curare alkaloid of all. The major components, however, are *C*-dihydrotoxiferine, *C*-calebassine, and *C*-curarine (*cf. Chapt. 6.7*). The toxic action (as listed in *Table 4.1*) of different curare alkaloids was determined as follows[19]:

> '*We injected mice intravenously with increasing quantities of curare components and ascertained the time of its onset of action through several symptoms of paralysis (head drop, lying on side, death from respiratory paralysis, or recovery with raising of the head and standing upright), using diagrams. On average, we used 20–50 mice (!) for the investigation of a single alkaloid, so that good limiting doses for the characterization of the substance might be obtained from the measurements*'.

The list of artifacts may be extended almost as far as one likes. As well as the reactions already mentioned, transesterifications with EtOH (as a stabilizer in $CHCl_3$), acid- or base-catalyzed epimerizations, or quaternizations (with $CHCl_3$ or its degradation products) are also regularly encountered.

Artifact formation cannot be avoided in every eventuality. For precisely this reason, one must always keep clear in mind that the reactions described above may come into play during extraction and processing of natural products in general. It is, therefore, always essential to keep back a sufficient quantity of the material to be extracted, so as to be in a position, if necessary, to repeat all the experiments under different conditions.

References and Notes

1. V. Agwada, M. B. Patel, M. Hesse, H. Schmid, 'Die Alkaloide aus Hedranthera barteri (HOOK. f.) Pichon', Helv. Chim. Acta **1970**, *53*, 1567.
2. G. Blaskó, N. Murugesan, A. J. Freyer, M. Shamma, A. A. Ansari, Atta-ur Rachman, 'Karachine: An Unusual Protoberine Alkaloid', J. Am. Chem. Soc. **1982**, *104*, 2039.
3. A 'tracer' is a radioactive marker compound with the aid of which biological processes (among other things) in plants and animals can be tracked.
4. W. A. Ayer, G. C. Kasitu, 'Some New Lycopodium Alkaloids', Can. J. Chem. **1989**, *67*, 1077.
5. E. Leete, J. C. Lechleiter, R. A. Carver, 'Determination of the Starter Acetate Unit in the Biosynthesis of Pinidine', Tetrahedron Lett. **1975**, 3779.
6. M. F. Keogh, D. G. O'Donovan, 'Biosynthesis of Some Alkaloids of Punica granatum and Withania somnifera', J. Chem. Soc. **1970**, 1792.
7. H.-G. Floss, U. Mothes, A. Rettig, 'Die Beziehung zwischen Gentianin und Gentiopikrosid', Z. Naturforsch. **1964**, *19b*, 1106.
8. T. R. Govindachari, S. S. Sathe, N. Viswanathan, 'Gentianine, an Artifact in Enicostemma littorale', Indian J. Chem. **1966**, *4*, 201.
9. H. Kühne, M. Hesse, 'Der [M + 11]⁺-Peak in den Massenspektren von Diaminen', Helv. Chim. Acta **1982**, *65*, 1470.
10. K. Drandarov, 'Verbacine and Verballocine, Novel Macrocyclic Spermine Alkaloids from Verbascum pseudonobile STOJ. et STEF. (Scrophulariaceae)', Tetrahedron Lett. **1995**, *36*, 617.
11. J.-J. Chen, T. Ishikawa, C.-Y. Duh, I.-L. Tsai, I.-S. Chen, 'New Dimeric Aporphine Alkaloids and Cytotoxic Constituents of Hernandia nymphaeifolia', Planta Medica **1996**, *62*, 528.
12. S. Johne, D. Gröger, M. Hesse, 'Neue Alkaloide aus Adhatoda vasica NEES', Helv. Chim. Acta **1971**, *54*, 826.
13. M. Todorova, C. Werner, M. Hesse, 'Metabolism of Polyamine Alkaloids in Plants: Enzymatic Phenol Oxidation and Polymerization of the Spermine Alkaloid Aphelandrine', Phytochemistry **1994**, *37*, 1251.
14. E. Lehmann, G. Alschner, H. Münnich, R. Ogrissek, 'Die amerikanische Reise', in Alexander von Humboldt, G. Harig (Ed.), Urania-Verlag, Leipzig, 1959, p. 174.
15. L. Lewin, Die Pfeilgifte, J. A. Barth-Verlag, Leipzig, 1923, p. 442.
16. Calabash: a bulbous, long-necked flask made out of a bottle gourd, or the fruit of the calabash tree.
17. M. Hesse, Indolalkaloide in Tabellen, Springer-Verlag, Berlin, 1964, Suppl. Vol. 1968.
18. The chemical composition of the calabashs is subject to great fluctuations, depending on their various origins. Tubocurare, being made by natives of Amazonia, is sealed in sections of bamboo using a peace of banana leaf. Chemically, it consists of bis(isoquinoline) alkaloids (such as tubocurarine dichloride), but Menispermaceae plants may also play a role, depending on the source. Calabash curare is offered in bottle gourds, sealed with a wooden peg. Pot curare, the third sort, is kept in small, gray or black earthenware pots.
19. P. G. Waser, 'Die Pharmakologie der Calebassenalkaloide', in Curare, Bull. Schweiz. Akad. Med. Wiss. **1966**, *22*; ibid. **1967**, *23*.

5. Chiroptical Properties of Alkaloids

5.1. General

The structural diversity of alkaloids is strongly reflected in their broad range of chiroptical properties. A few achiral representatives – *e.g.* the aromatic compounds papaverine (**1**) and olivacine (**2**), or the aliphatic spermidine derivative inandenin-12-one (**3**) – are hugely outnumbered by a multitude of compounds possessing from one stereogenic center (such as (–)-coniine (**4**)) up to fourteen stereogenic centers (such as *C*-alkaloid D dichloride (**5**)). Accordingly, we shall discuss important stereochemical aspects and illustrate the kind of (often surprising) discoveries that have been made with the help of chiroptical studies.

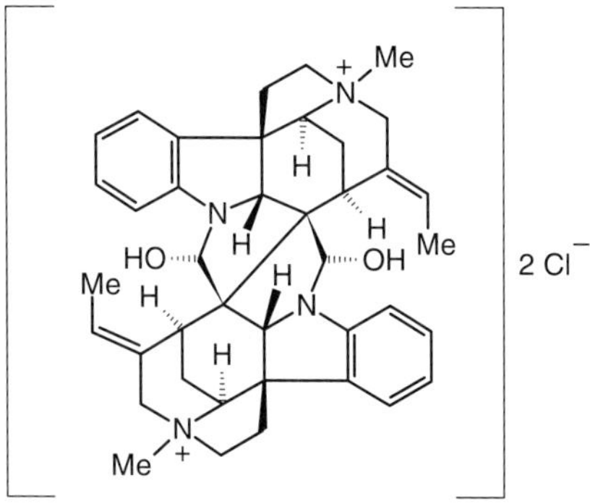

(–)-*C*-Alkaloid D dichloride (**5**) (abs. config.)

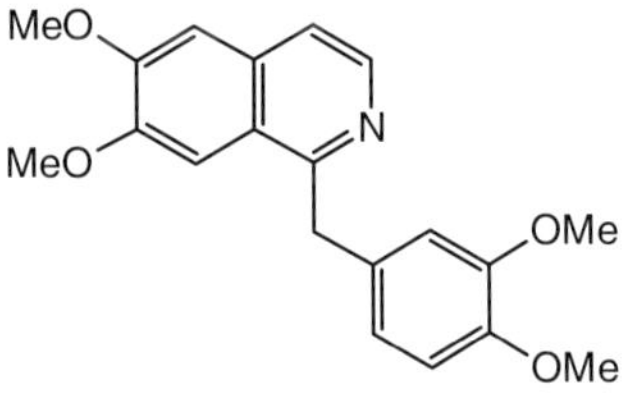

Papaverine (**1**)

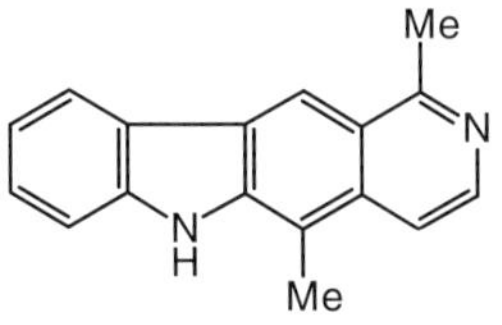

Olivacine (**2**)

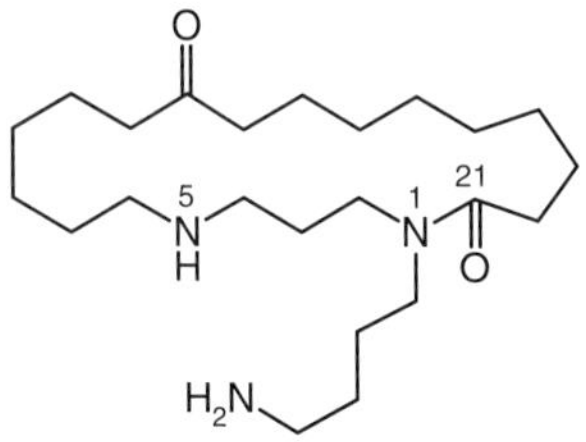

Inandenin-12-one (**3**)

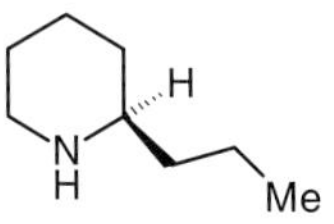

(–)-Coniine (**4**)

5.2. Influence of Solvents

Optical activities are frequently reported in the form of $[\alpha]_D$ values, referring to the 'specific rotation' of a substance. Particularly in the early period of modern organic chemistry (before *ca.* 1963), measurement of optical rotation was a very popular technique for characterizing substances. This would not be a problem, were the $[\alpha]_D$ value not dependent on a large variety of factors and, *per definitionem*, only valid for one specific wavelength (the sodium-D-line: $\lambda = 589.3$ nm). One particularly important factor influencing the specific rotation is the solvent. Kopsine (**6**), for example, displays

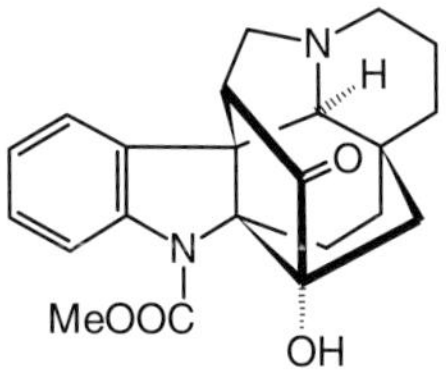

(–)-Kopsine (**6**) (abs. config.)

185

an $[\alpha]_D$ value of $+16$ in 95% EtOH[1], which, according to the literature[2], changes to -18 in CHCl$_3$. When a finding of this kind presents itself, there is a natural tendency to assume that the two samples under investigation must have been antipodes. Antipodal kopsines, however, have yet to be isolated from plants. Consequently, the specific rotation of the substance was indeed solvent dependent, or, accidentally, both the sample and reference cuvettes had been switched.

Examples of alkaloids with highly solvent-dependent specific rotations are polyneuridine (7) ($[\alpha]_D = -69$ in pyridine[3], ± 0 in MeOH[3], $+1$ in CHCl$_3$[4]), akuammidine (8) ($[\alpha]_D = +21$ in EtOH[5], ± 0 in CHCl$_3$[6]), and akuammigine (9) ($[\alpha]_D = -44$ in EtOH[5], -1 in pyridine[7]). Such extreme solvent-dependencies, however, are the exception rather than the rule. Nevertheless, it does mean that the commonly used prefixes '(+)' and '(–)', which denote the sign of the specific rotation, are not too meaningful, since the solvent to which the direction of rotation refers is often not known. Hence, under the right circumstances, (–)-polyneuridine can eventually become (±)-polyneuridine or even (–)-polyneuridine. Since such examples are rare though, serious consideration is scarcely afforded to these borderline cases, despite occasional ambiguities encountered. If these are to be avoided altogether, the sign of the $[\alpha]_D$ value should either not be given, or, if so, then only with concomitant listing of the solvent. If known, then the absolute configurations ((R)/(S)) and their respective locants should be stated before the name of the compound in addition to the direction of rotation.

Significantly more informative than the optical rotation, though, is the measurement of (likewise solvent-dependent) optical rotatory dispersion (ORD) or circular dichroism (CD) spectra, which do provide clear information about whether racemic mixtures or single antipodes are present.

Polyneuridine (7) (abs. config.)

Akuammidine (8) (abs. config.)

Akuammigine (9) (abs. config.)

5.3.   Free Base *vs.* Hydrochloride: Mayfoline

The 13-membered spermidine alkaloid mayfoline was isolated from *Maytenus buxifolia* (A. RICH.) GRISEB. (Celastraceae). It possesses only one stereogenic center. In the literature, the natural product was described as the (+)-compound ($[\alpha]_D = +10.6$ in CHCl$_3$), with the (2S) absolute configuration[8]. Some years later, when synthetic (2S)-mayfoline was examined, prepared from methyl (–)-(S)-3-amino-3-phenylpropanoate (*cf. Scheme 5.1*), the latter was interestingly found to exhibit a negative specific rotation ($[\alpha]_D = -52.3$ in CHCl$_3$)[9].

Scheme 5.1

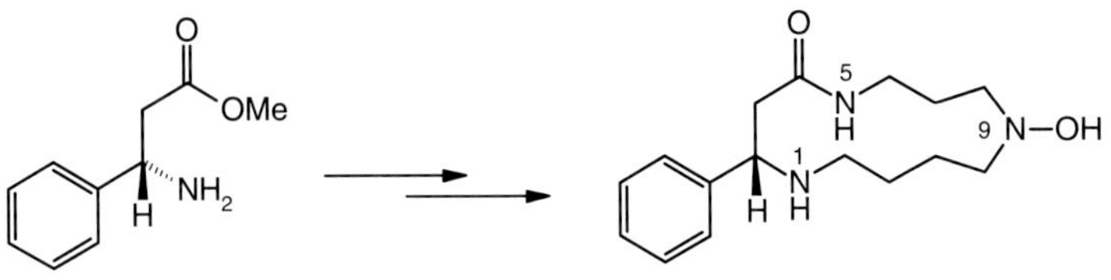

(–)-**10**   (–)-(2*S*)-Mayfoline (**11**)

To check the absolute configuration, an X-ray crystal-structure analysis was carried out on the synthetic product; this confirmed that the assumed structure was correct. At this point, therefore, there were two putative mayfoline structures – the naturally occurring (+)-(2*S*)-mayfoline and the synthetic (–)-(2*S*)-mayfoline – differing both in the direction and magnitude of their specific rotation, thus ruling out the involvement of antipodes. The solution to the problem was surprisingly simple: the author, who had first described the natural product, had measured the specific rotation not of mayfoline itself but of *mayfoline hydrochloride*, assuming that it was the free base. However, the compound had been protonated *in situ* by HCl, produced by the decomposition of CHCl$_3$, the solvent used for the measurement (*cf. Chapt. 4.5*). The natural product in its free-base form, prepared after the event, and synthetic mayfoline finally displayed parallel CD curves, thanks to which it was possible to verify the identities of the two samples beyond doubt (*Fig. 5.1*). Natural mayfoline, expressed precisely, is thus (–)-(2*S*)-mayfoline.

5.4. Racemic and Partially Racemic Mixtures

The biosynthesis of alkaloids almost always involves optically pure α-amino acids, and so it might reasonably be assumed that bases possessing at least one stereogenic center should also be optically active. However, racemic or partially racemic alkaloid mixtures have occasionally been isolated from natural sources.

5.4.1. Narwedine

The alkaloid narwedine (**12**) has been isolated from various Amaryllidaceae species. As the compilation in *Table 5.1* shows, both the racemic mixture and the (+)-rotating form have been found.

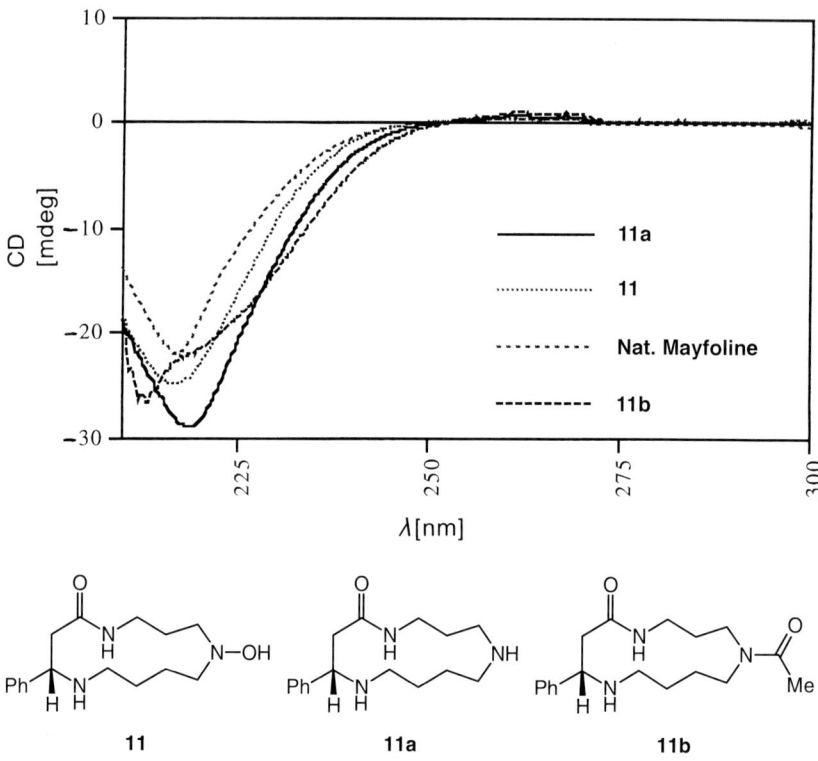

Fig. 5.1. *Circular dichroism spectra of deoxymayfoline (**11a**) and N¹-deoxy-N¹-acetylmayfoline (**11b**) in EtOH as compared with natural and synthetic (–)-(2 S)-mayfoline (**11**)*

Table 5.1. *Specific Rotations of Narwedine (**12**) of Various Provenance*

Origin	$[\alpha]_D$ (Solvent)
Galanthus elwesii Hook f. (cf. Fig. 5.2)	± 0 (MeOH)[10]
Galanthus nivalis L.	not available[11]
Lycoris guangxiensis Y. Hsu & Q. J. Fan	+ 112 (CHCl₃)[12]
Narcissus incomparabilis var.	+ 100 (CHCl₃)[13]

A number of related alkaloids, with absolute configurations corresponding to that of (+)-**12**, are also found in *Galanthus* species. Since no (–)-rotating narwedine has yet been isolated, it is reasonable to assume that the racemic mixture was produced from the (+)-antipode (*cf. Scheme 5.2*). The reaction may proceed under base (or acid) catalysis and might even be triggered by narwedine itself. The base induces the removal of an α-proton relative to the carbonyl group, giving rise to the dienone **12a**. Since the latter is symmetrical, the chiral sense of (+)-**12** is naturally lost. In the subsequent ring-closure, the two antipodes (+)-**12** and (–)-**12** are formed in a 1 : 1 ratio. It remains unclear, though, whether in the plant the (+)-enantiomer occurs exclusively, or whether the racemic mixture is present, since acids and bases were routinely used in the plant extraction.

Scheme 5.2

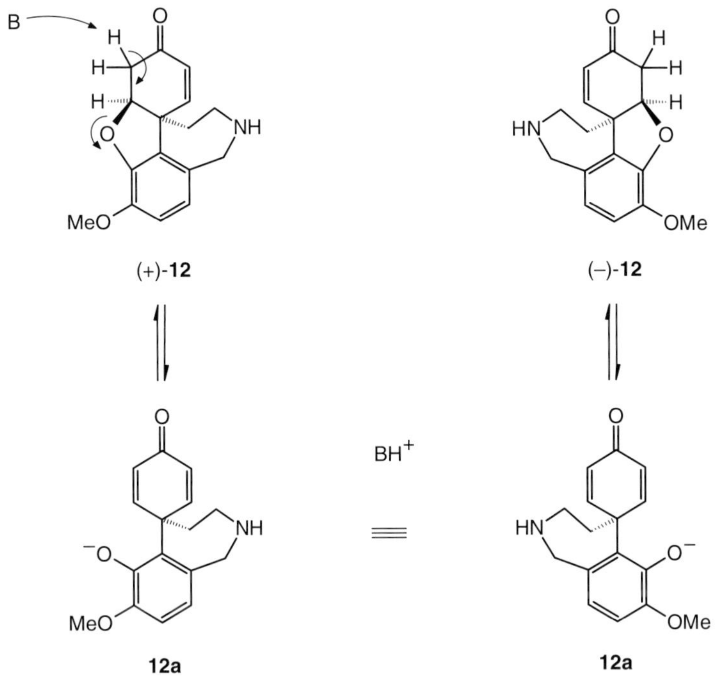

(+)-**12**   (−)-**12**

BH⁺

12a   ≡   **12a**

5.4.2. Vincadifformine

Vincadifformine (**13**) is an indole alkaloid of the plumeran type, isolated from many species of the Apocynaceae family. Depending on its origin, the alkaloid may exist in the (+)-, (−)-, or (±)-form. (+)-Vincadifformine has the (7*S*,20*R*,21*R*) absolute configuration (biogenetic atom-numbering; *vide infra*). The preparation of (−)-vincadifformine from the (+)-antipode requires complete inversion of configuration at all the three stereogenic centers C(7), C(20), and C(21).

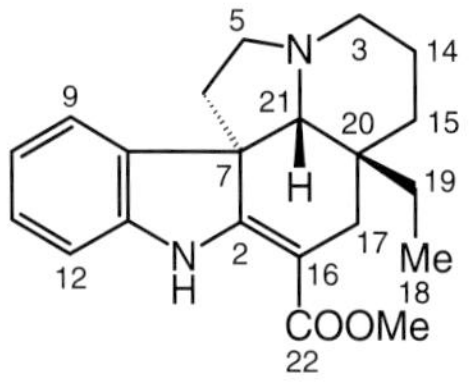

A very informative experiment was accomplished with (±)-vincadifformine obtained from *Vinca difformis*: that a racemic mixture had to be present was conclusively demonstrated by successfully resolving it. As is evident from *Table 5.2*, vincadifformine clearly occurs largely in racemic or partially racemic form[25]; only the sample from *Amsonia tabernaemontana* together with the semisynthetic vincadifformine (prepared from (−)-tabersonine) seem to have been optically pure. Disregarding speculation about the accu-

(+)-Vincadifformine ((+)-**13**)

Table 5.2. *Specific Rotations of Vincadifformine Samples of Different Origins*

Origin	Solvent	$[\alpha]_D$
Aspidosperma album	EtOH	$(+)^{a\,14}$
Amsonia tabernaemontana	EtOH	$+600^{15}$
Melodinus aeneus	EtOH	$+535^{16}$
Melodinus scandens	EtOH	$+526^{17}$
Rhazya stricta	EtOH	$+402^{18}$
Tabernaemontana riedelii	EtOH	$+185^{19}$
Vinca difformis, V. minor	MeOH	$\pm0^{20,\,21}$
Vinca minor	EtOH	-540^{22}
Synthesis from Tabersonine[b]	EtOH	-600^{23}
Hunteria elliotii	EtOH	$(-)^{a\,24}$

[a] Only the sign of rotation was determined.
[b] Tabersonine = 14,15-didehydrovincadifformine.

racy of the values given in *Table 5.2*, it may be assumed that all three optical forms of vincadifformine (even partial racemates) naturally occur in the Apocynaceae family. As far as we know, other alkaloids of the vincadifformine type always occur in enantiomerically pure form, and so it seems that we are dealing here with a unique case (at least for the time being). How, though, can this finding be explained? Discussion of possible causes brings up the question of the thermal stability of the compound. As found experimentally by ourselves, neither rearrangement nor decomposition reactions take place at room temperature, not even over a period of several years; but this only means that vincadifformine is not in a position to undergo spontaneous thermal isomerization. On the other hand, it cannot be ruled out that, at higher temperatures, the combination of *retro-Diels–Alder* plus subsequent *Diels–Alder* reaction might be responsible for the racemization (*cf. Scheme 5.3*). This reaction sequence was established for the mass spectrometric decomposition of vincadifformine (EI-MS)[20]. It is not known, whether the same transformation is feasible also under plant physiological conditions. Also, since the occurrence of both enantiomers cannot be explained by subsequent chemical reactions by means of 'wet chemistry' during the isolation or purification of vincadifformine, it is necessary to seek alternative means of explanation.

A fundamentally different way to account for the occurrence of the two vincadifformine enantiomers is offered by a biogenetic-chemotaxonomic interpretation. Compound **13** belongs to those indole alkaloids that incorporate a rearranged secologanin skeleton (*cf. Chapt. 7.2*). The direct biogenetic precursors of vincadifformine incorporate the skeleton of quebrachamine (**14**) as shown in *Scheme 5.4*.

Quebrachamine possesses one stereogenic center, C(20), and both isomers may be isolated from natural sources. In the (−)-form, it is found, *e.g.*, in *Pleiocarpa pycnantha*[26] and *Vinca minor*[27], while (+)-quebrachamine is

Scheme 5.3

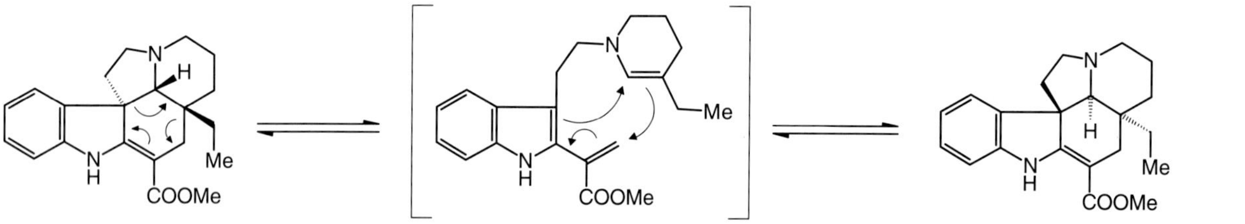

(+)-Vincadifformine ((+)-**13**)

(−)-Vincadifformine ((−)-**13**)

Scheme 5.4. *Probable Biogenetic Pathway Producing the Two Enantiomers of Vincadifformine ((+)-**13** vs. (−)-**13**) from the Two Methoxycarbonyl Derivatives of the Similarly Naturally Occurring Antipodes of Quebrachamine (**14**). Oxidation at C(21) gives rise to an immonium ion in a nine-membered, strained ring, which reacts via* **16** *to yield* **13**. *Cyclization reactions of this type are well-known in vitro.*

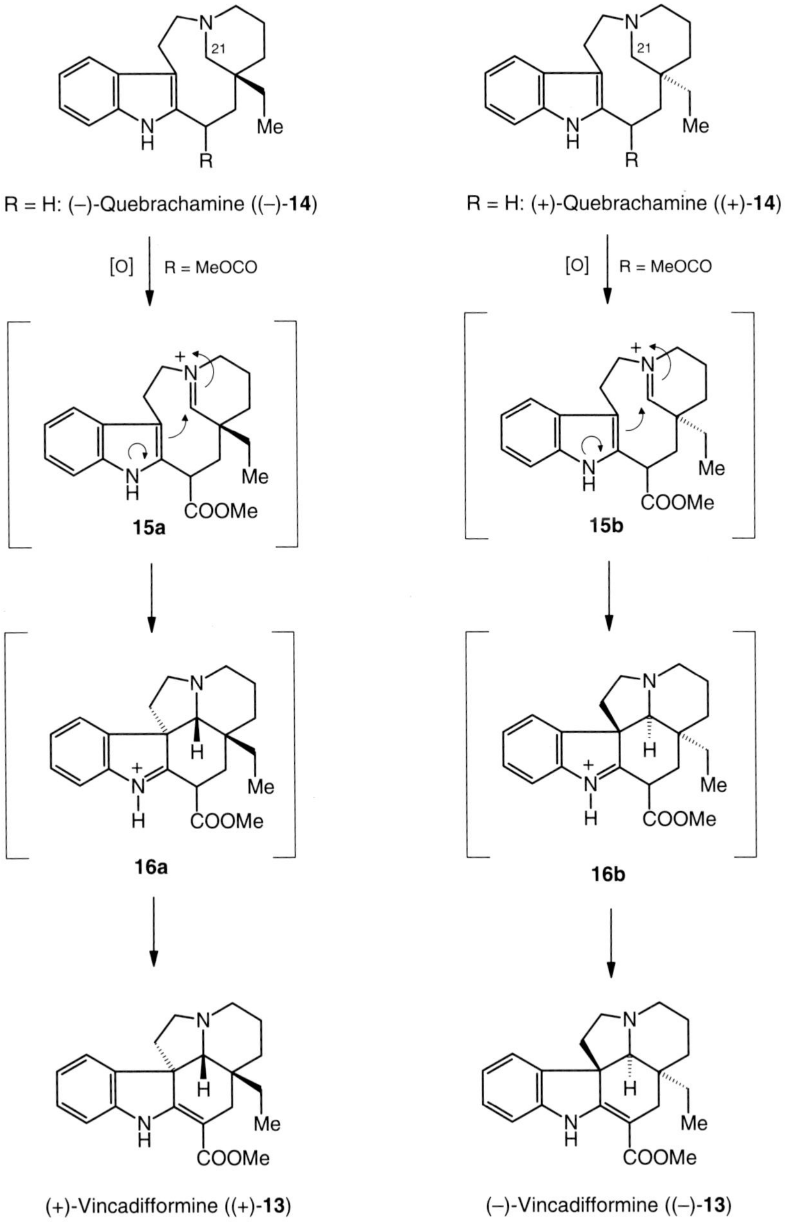

R = H: (−)-Quebrachamine ((−)-**14**)   R = H: (+)-Quebrachamine ((+)-**14**)

[O] | R = MeOCO   [O] | R = MeOCO

15a   **15b**

16a   **16b**

(+)-Vincadifformine ((+)-**13**)   (−)-Vincadifformine ((−)-**13**)

191

found, *inter alia*, in *Aspidosperma album*[28] and *Rhazya stricta*[29]. If it is now assumed that there are two different enzyme systems present, each capable, respectively, of converting only the methoxy derivative of either (+)- or (–)-quebrachamine into (+)- or (–)-vincadifformine (*via* **15** and **16**; *cf. Scheme 5.4*), then it would also be conceivable that there might exist some Plumerioideae plants in which *both* enzyme systems are active *simultaneously*. Were the activities of the two enzyme systems identical, then a racemate would result; otherwise there would be a partial racemate, with one enantiomer in excess. Of course, the activities of the participating enzymes may also vary in different parts of the plant, explaining why both (+)- and (–)-vincadifformine (or even (±)-vincadifformine) have been isolated from the same plant (*e.g.*, *Vinca minor*). Some detailed investigations into the chemotaxonomy of the Plumerioideae (Apocynaceae)[30] led to some other interesting discoveries besides these. Alkaloids derived from vincadifformine (**13**), all of which possess at least one additional ring, all belong exclusively to one enantiomerically pure skeleton series (*Scheme 5.5*). The (+)-vincadifformine structural type (+)-**P5** gives rise to the haplophytine type **P6d** and to the obscurinervine type **P7d.** On the other hand, (–)-vincadifformine (corresponding to (–)-**P5**) produces the venalstonine type **P6a**, the hedrantherine type **P6e**, and the kopsan type **P7a**[30].

Hence, while two different enzyme systems clearly must exist, synthesizing (+)- and (–)-vincadifformine, respectively, there is only one such system responsible for the subsequent transformations in either case, operating in a stereochemically homogenous fashion. In stereochemical terms, the **P6** and **P7** types are uniform in themselves. If the (+)-form is now preferentially consumed from a (±)-vincadifformine pool, with the formation of **P6d** alkaloids, then enrichment of the other form will automatically take place, to give a negatively rotating, partially racemic mixture (and *vice versa*).

Still closer examination of the biogenetic processes has uncovered further remarkable relationships. The vincadifformine skeleton **P5** is very closely related to the eburnan type **E5** (*cf. Chapt. 7.2*) through the quebrachamine type **P4**. Examples of alkaloids of the eburnan type include (+)-eburnamonine (**17**) and (+)-vincamine (**18**) (*cf. Table 5.3*). In nature, alkaloids **17** and **18**, like vincadifformine, occur in both enantiomeric forms. No direct derivatives of the two compounds appear to exist; their production removes only one enantiomer from the racemic pool.

Scheme 5.5. *Biosynthetic Pathways Involving the Plumeran Alkaloids, Isolated from Plants of the Apocynaceae Family.* The prefixes α, β denote absolute configurations; R = H or MeOCO; the descriptors **P4** – **P7** are explained in *Chapt. 7.2.*

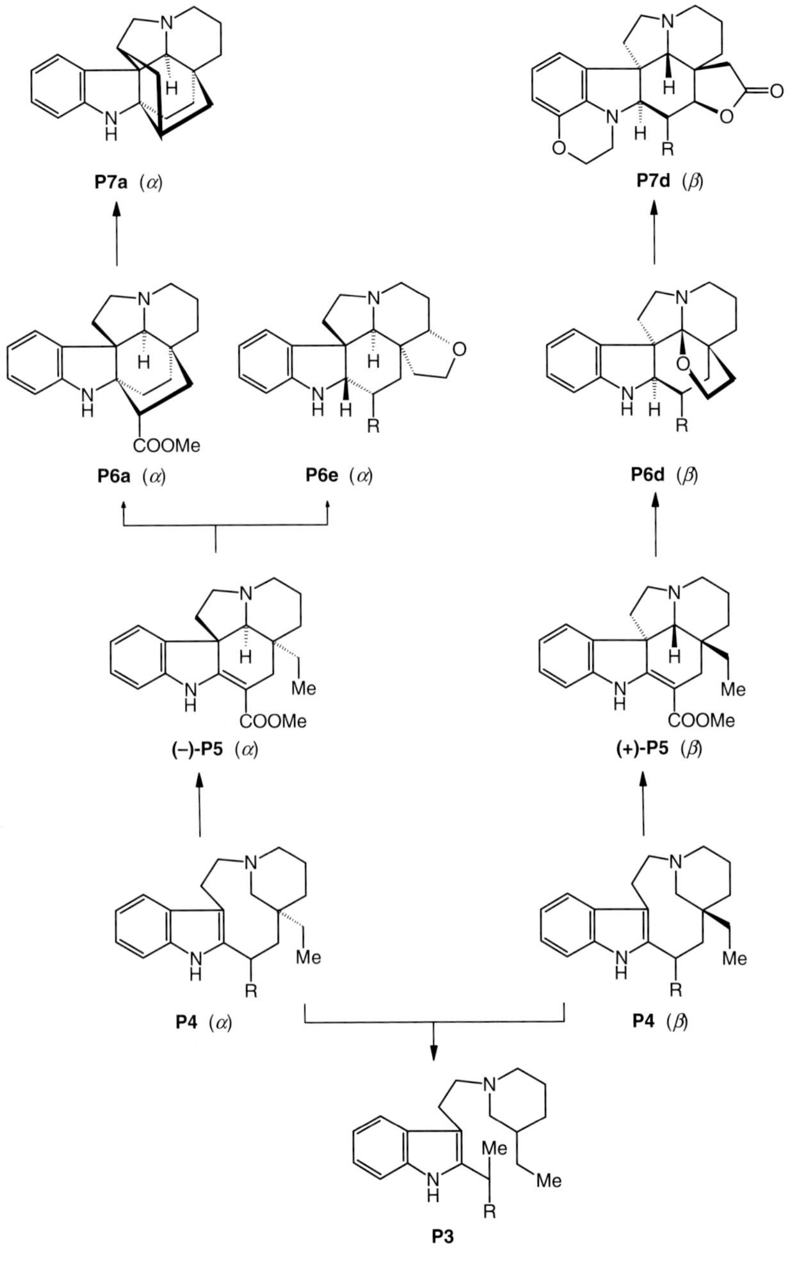

P7a (α)

P7d (β)

P6a (α)

P6e (α)

P6d (β)

(−)-**P5** (α)

(+)-**P5** (β)

P4 (α)

P4 (β)

P3

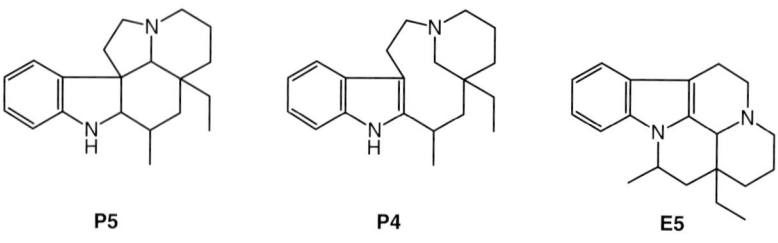

P5   P4   E5

Table 5.3. *Examples of Alkaloids with Enantiomeric Skeletons of the* **E5a** *Structure Type.* In the cases of the two compounds (+)-eburnamonine (**17**) and (+)-vincamine (**18**), the opposite enantiomers have also been found.

(+)-Eburnamonine (**17**)   (+)-Vincamine (**18**)

Plant	$[\alpha]_D$ (Solvent)		Plant	$[\alpha]_D$ (Solvent)
Hunteria eburnea	+89 (CHCl$_3$)[33]		Vinca difformis, V. erecta, V. major, V. minor, Tabernaemontana rigida	+41 (Pyridine)[19,34]
Vinca minor	±0 (CHCl$_3$ or EtOH)[27]			
Vinca minor	−105 (CHCl$_3$)[27]		Aspidosperma album, Tabernaemontana rigida	±0[14,19]

5.5 The Unexpected Inversion at the N-Atom of (+)-9-*O*-Demethyl-homolycorine

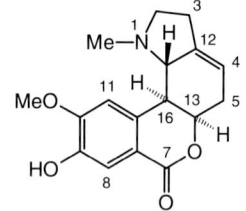

(+)-*O*-Demethylhomolycorine (**19**)

(+)-9-*O*-Demethylhomolycorine (**19**), with the absolute (13*R*,16*S*,17*S*)-configuration, has been isolated from various *Galanthus* species (Amaryllidaceae)[10]. Interestingly, five different sorts of crystals were obtained from five fractions separated during processing of *Galanthus elwesii* HOOK. f. (*Fig. 5.2*). Assuming that five different compounds had been isolated, researchers were highly surprised when they could not find any significant difference between the samples – neither chromatographically (TLC) nor spectroscopically (CD (MeOH), *Fig. 5.3*; UV (MeOH); ^{1}H- (400 MHz, CDCl$_3$) and ^{13}C-NMR (50 MHz, CD$_3$OD)), nor spectrometrically (EI- or CI-MS (NH$_3$)). Nevertheless, the crystals were clearly distinct by visual inspection,

Fig. 5.2. *Galanthus elwesii* Hook. f. Many Amaryllidaceae alkaloids have been isolated from snowdrop bulbs, (+)-9-O-demethylhomolycorine among them (Karakorum, Turkey; photo *T. Gözler*).

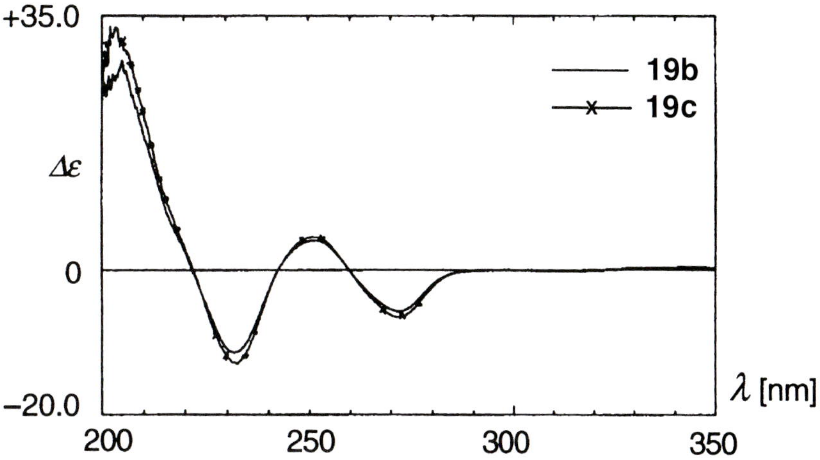

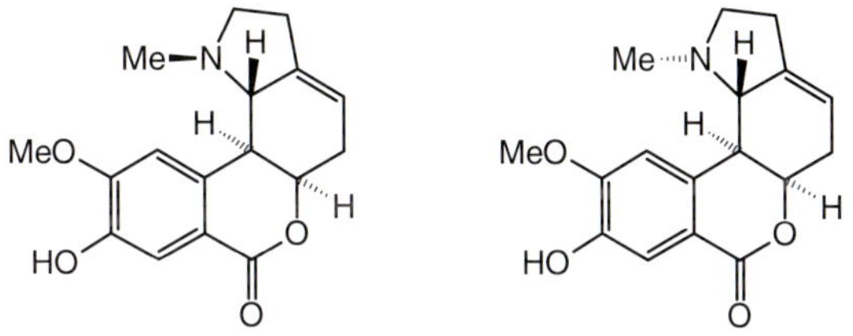

Fig. 5.3. *Circular dichroism (CD) spectra of (+)-9-O-demethylhomolycorine* (**19**) *in MeOH.* Continuous curve: preparation **19b**; broken curve: preparation **19c**.

by different melting points, and (slightly) by differing specific rotations (*Table 5.4*).

Recrystallization of the five samples under identical conditions from a single solvent produced crystals with melting points that no longer differed.

On the basis of these data, it might be suspected that the five crystalline samples originally merely differed in their solvent content; this, at least, could explain both the different melting points and the small variation in specific rotations[35]. It was indeed found that H_2O was present in the crystal lattice in one case, and ethyl acetate in another. However, the quantity of these impurities was too low to be detected by UV or NMR. In the hope of reconciling

Table 5.4. *Selected Physical Data of Five Different Sorts of Crystals of (+)-9-O-Demethylhomolycorine (**19**),* *Designated as **19a** – **19e***

Crystal (Type)	From	M.p.	$[\alpha]_D$
19a	MeOH	118–119	+99.0 (MeOH; $c = 0.22$)
19b (*Fig. 5.3*)	AcOEt	214–216	+101.8 (MeOH; $c = 0.33$)
19·2H$_2$O	MeOH/CHCl$_3$	207–210	
19c (*Fig. 5.3*)	AcOEt	183–185	+115.6 (MeOH; $c = 0.85$)
19d (**19**·0.5 AcOEt)	MeOH	181–184	+125.0 (MeOH; $c = 1.18$)
	AcOEt	217–219	+99.4 (MeOH; $c = 0.50$)
19e	AcOEt	211–213	

the remaining contradictions, three of the five samples were subjected to X-ray crystal-structure analysis, from which it was finally established that the methyl group and the free electron pair on the N-atom had obviously inverted (*cf. Fig. 5.4*).

Amines are known to possess a pyramidal geometry. Provided that there are no opposing steric factors, in solution they invert very quickly at ambient temperature. This means that (potential) optical antipodes are normally not observable or to be isolated under these conditions. This explains why at room temperature the NMR spectra of the two diastereoisomers **19b** and **19c** (*Fig. 5.4*) turned out to be identical; the signals simply averaged out. On crystallization, though, the two diastereoisomers might very well display different behavior: in solution, the two isomers are in a true equilibrium due to the rapid inversion at the N-atom. Should one of the isomers be less soluble than the other, however, then the former begins to crystallize out. This in turn perturbs the solution equilibrium; more of the less soluble isomer is then produced to compensate, but this then straightaway crystallizes out again, and so on. In this manner, it may even be possible to obtain crystals of the less soluble diastereoisomer quantitatively[56].

In retrospect, therefore, it was evident that both the diastereoisomers **19b** and **19c** in crystalline form together with diastereoisomerically pure mixed crystals (with varying quantities of solvent) had been isolated.

5.6. Morphines

As already discussed elsewhere, (–)-morphine was the first alkaloid to be isolated in pure form. The principal component of opium[37] (*cf. Chapt. 11.5*),

a)

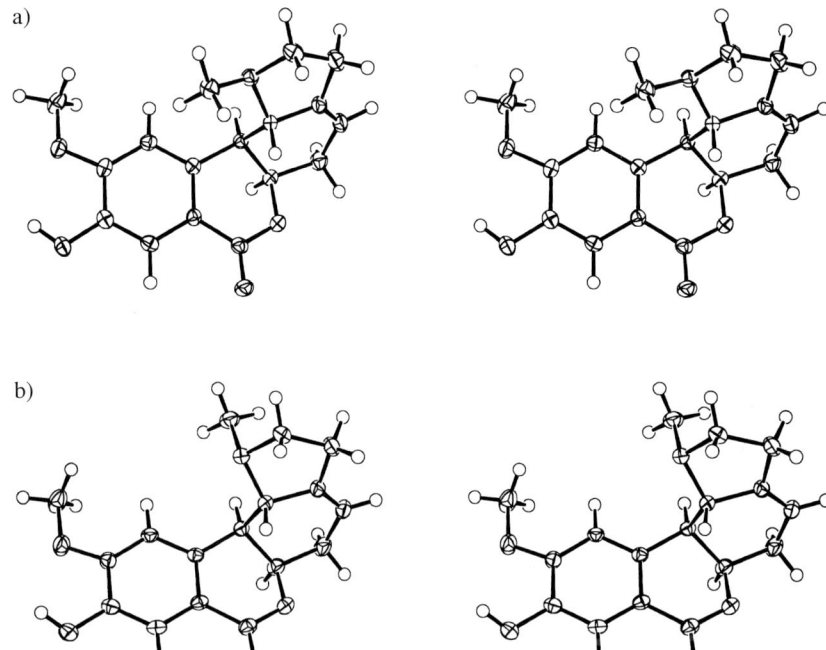

b)

Fig. 5.4. *Stereoscopic representation of the two (+)-9-O-demethylhomolycorines* **19b** *(a) and* **19c** *(b) arising from inversion of the Me group at the N-atom.* H-Atoms are shown with arbitrary radii, the remaining atoms with 50% probability thermal ellipsoids.

it was discovered by *Sertürner*, who, incidentally, also coined the term '*principium somniferum*' (Lat. 'sleep-inducing principle')[38]. Administered in water-soluble form as the hydrochloride, (–)-morphine acts in the first instance on the central nervous system (CNS); more precisely, on the cortex. Here, regions specifically adapted to pain sensation are modulated in such a way that even the most intense pain can be made tolerable. (–)-Morphine, therefore, is an analgesic. Areas in the CNS other than the cortex are also affected, however. The spectrum of activity of the substance is very complex: in small doses, a feeling of well-being (euphoria) is produced with abolition of listlessness, accompanied by a generally quieting effect (sedation). Higher doses, however, can result in strong excitation and, ultimately, even in death by respiratory paralysis. Medicinally, (–)-morphine is an ideal agent against coughing, far outperforming (–)-codeine in this respect[39]. For relief of extreme pain (intensive burns, open fractures *etc.*), morphine still constitutes the sole effective treatment; because of its enormously addictive character (morphinism), however, its use is restricted to absolute emergencies.

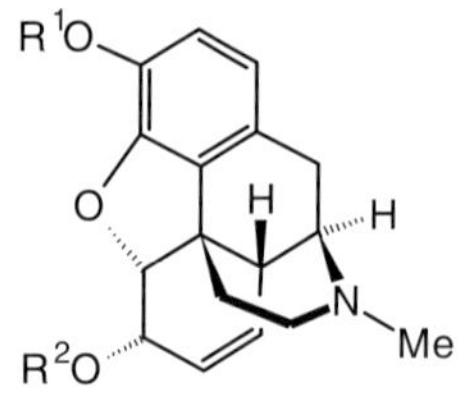

Natural Series	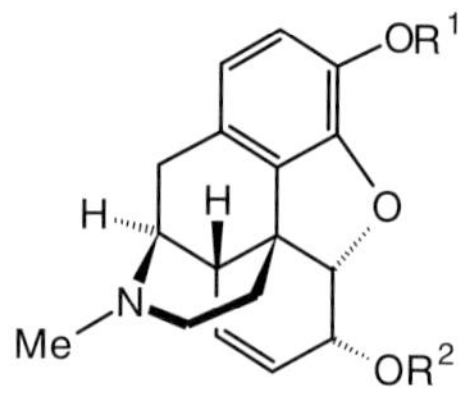	Synthetic Series	
R^1 = R^2 = H	: (–)-Morphine	R^1 = R^2 = H	: (+)-Morphine
R^1 = Me, R^2 = H	: (–)-Codeine	R^1 = Me, R^2 = H	: (+)-Codeine
R^1 = R^2 = Acetyl	: (–)-Heroin	R^1 = R^2 = Acetyl	: (+)-Heroin

Finally, (–)-heroin[40], obtained by double acetylation of (–)-morphine, should be mentioned here. Qualitatively, it displays the same effects as morphine does, but heroin is *ca.* six times more powerful. In glaring contrast, thebaine (= *O,O*-dimethylmorphine = *O*-methylcodeine) induces muscle spasm in a manner reminiscent of strychnine, and so is not used therapeutically.

Structure and Activity

Towards the end of the 1970s, interest began to grow in how the pharmaco-logical activities of natural (–)-morphine and (–)-codeine relate to those of the respective mirror images: (+)-morphine and (+)-codeine.

Naturally occurring (–)-sinomenine (**20**) (isolated from *Sinomenium* and *Menispermum* species (Menispermaceae)) served as the precursor for the synthesis of the artificial opiates (*Scheme 5.6*). Catalytic hydrogenation of the cyclohexenone ring of **20** furnished an initial diastereoisomeric mixture, which could be transformed into the pentacyclic structure **21** with the aid of polyphosphoric acid[41]. This compound was then converted *via* an acetal in-to the enol ether **22**. For the synthesis of (+)-codeine methyl ether (**23**), it was necessary to shift the double bond from the 6,7- to the 7,8-position, which was achieved with methyl hypobromite, followed by treatment with base (*t*-BuOK in DMSO). From **23**, it was possible to obtain (+)-morphine,

Scheme 5.6. *Transformation of the Naturally Occurring (–)-Sinomenine (**20**) into Codeine Methyl Ether (**23**), from which the Nonnatural (+)-Morphine (**23a**), (+)-Codeine (**23b**), and (+)-Heroin (**23c**) Were Obtained*[41]

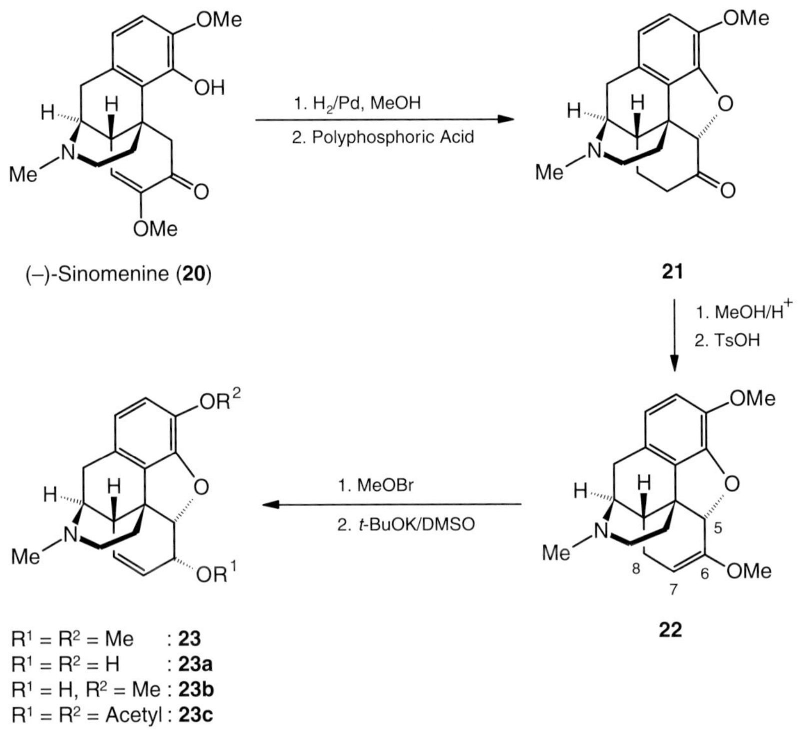

(–)-Sinomenine (**20**)

1. H₂/Pd, MeOH
2. Polyphosphoric Acid

21

1. MeOH/H⁺
2. TsOH

22

1. MeOBr
2. *t*-BuOK/DMSO

R¹ = R² = Me : **23**
R¹ = R² = H : **23a**
R¹ = H, R² = Me : **23b**
R¹ = R² = Acetyl : **23c**

(+)-codeine, and (+)-heroin by already known procedures. Apart from their (opposite) optical rotations, the three compounds displayed the same physical properties as their natural counterparts.

However, '*(+)-morphine, (+)-codeine, and (+)-heroin showed no analgesic activity on subcutaneous injection in mice in routine screening for centrally active analgesics*'[41]. In other words: the morphine receptor is able to selectively bind (–)-morphine, but not (+)-morphine.

Similar notable properties are also demonstrated by the synthetically produced methorphans, close cousins to morphine. Levomethorphan (**24**) is active as an analgesic, the enantiomeric dextromethorphan (**25**) is not. The latter does, however, exhibit sustained antitussive activity and so is used (under the name *Romilar*™) as an anticoughing agent.

Levomethorphan (**24**)   Dextromethorphan (**25**)

The biological activities of enantiomers of chiral, active substances are quite often different, and indeed not only in the case of alkaloids. There are also cases, though, in which two enantiomers are found to be equally effective, such as in the case of nicotine. To determine precise biological-activity profiles of individual enantiomers, detailed *in vivo* studies are always required

References and Notes

1. A. Bhattacharya, A. Chatterjee, P. K. Bose, '*On an Alkaloid of Kopsia fruticosa*', *J. Am. Chem. Soc.* **1949**, *71*, 3370.
2. A. R. Battersby, H. Gregory, '*Alkaloids of Kopsia Species. Part I. Kopsine, Fruticosamine, and Fruticosine*', *J. Chem. Soc.* **1963**, 22.
3. M.-M. Janot, J. Le Men, J. Gosset, J. Lévy, '*Dégradation de la vincamédine et configuration absolue des alcaloïdes apparentés: vincamajine, akuammidine, polyneuridine, voachalotine et macusine A. Alcaloïdes des pervenches*', *Bull. Soc. Chim. Fr.* **1962**, 1079.
4. L. D. Antonaccio, N. A. Pereira, B. Gilbert, H. Vorbrueggen, H. Budzikiewicz, J. M. Wilson, L. J. Durham, C. Djerassi, '*Alkaloid Studies. XXXIII. Mass Spectrometry in Structural and Stereochemical Problems. VI. Polyneuridine, a New Alkaloid from Aspidosperma polyneuron and Some Observations on Mass Spectra of Indole Alkaloids*', *J. Am. Chem. Soc.* **1962**, *84*, 2161.
5. T. A. Henry, '*The Alkaloids of Picralima klaineana, PIERRE. Part II.*', *J. Chem. Soc.* **1932**, 2759.
6. A. Chatterjee, C. R. Ghosal, N. Adityachaudhury, S. Ghosal, '*Alkaloids of Rhazya stricta DECAISNE*', *Chem. Ind. (London)* **1961**, 1034.
7. R. Robinson, A. F. Thomas, '*The Alkaloids of Picralima nitid, STAPF, TH. et H. DURAND, Part I. The Structure of Akuammigine*', *J. Chem. Soc.* **1954**, 3479.
8. H. Ripperger, '*Mayfoline, a Novel Alkaloid from Maytenus buxifolia*', *Phytochemistry* **1980**, *19*, 162.
9. P. Kuehne, M. Hesse, '*Asymmetric Synthesis of the Alkaloids Mayfoline and N(1)-Acetyl-N(1)-deoxymayfoline*', *Helv. Chim. Acta* **1996**, *79*, 1085.
10. A. Latvala, M. A. Önür, T. Gözler, A. Linden, B. Kivçak, M. Hesse, '*Alkaloids of Galanthus elwesii*', *Phytochemistry* **1995**, *39*, 1229.
11. H.-G. Boit, W. Döpke, '*Alkaloide aus Hippeastrum aulicum var. robustum*', *Naturwissenschaften* **1960**, *47*, 109.

12. H.-Y. Li, G.-E. Ma, Y. Xu, S.-H. Hong, 'Alkaloids of Lycoris guangxiensis', *Planta Medica* **1987**, *53*, 259.

13. H.-G. Boit, W. Döpke, A. Beitner, 'Alkaloide aus Trompeten-Narcissen, Schalen-Narcissen und gefüllten Narcissen', *Chem. Ber.* **1957**, *90*, 2197.

14. M. Urrea, A. Ahond, H. Jacquemin, S.-K. Kan, C. Poupat, P. Potier, M.-M. Janot, 'Nouveaux alcaloïdes extraits des grains de Aspidosperma album (VAHL) R. BENTH. (Apocyanacées)', *Compt. Rend. Acad. Sci. Ser. C.* **1978**, *287*, 63.

15. B. Zsadon, P. Kaposi, '(+)-Vincadifformine from Amsonia tabernaemontana WALT. (Apocynaceae)', *Tetrahedron Lett.* **1970**, 4615.

16. K. Biemann, 'Mass Spectrometry of Selected Natural Products', *Fortschr. Chem. organ. Naturstoffe* **1966**, *24*, 1.

17. H. Mehri, M. Plat, P. Potier, 'Plantes de Nouvelle-Calédonie. V. Melodinus scandens FORST. Isolement de dix Alcaloïdes Monomères. Description de deux Alcaloïdes Nouveaux: N-Oxy-épiméloscine et Méloscandonin', *Ann. Pharm. Franc.* **1971**, *29*, 291.

18. G. F. Smith, M. A. Wahid, 'The Isolation of (±)- and (+)-Vincadifformine and of (+)-1,2-Dihydroaspidospermidine from Rhazya stricta', *J. Chem. Soc.* **1963**, 4002.

19. M. P. Cava, S.S. Tjoa, Q. A. Ahmed, A. I. Da Rocha, 'The Alkaloids of Tabernaemontana riedelii and T. rigida', *J. Org. Chem.* **1968**, *33*, 1055.

20. C. Djerassi, H. Budzikiewicz, J. M. Wilson, J. Gosset, L. Le Men, M.-M. Janot, 'Mass Spectrometry in Structural and Stereochemical Problems. Vincadifformine (Alcaloïdes des Pervenches)', *Tetrahedron Lett.* **1962**, 235.

21. J. Mokrý, I. Kompiš, L. Dúbravková, P. Šefcović, 'Vincadifformin und Monovincin, zwei weitere razemische Alkaloide aus Vinca minor L.', *Experientia* **1963**, *19*, 311.

22. M. Plat, J. Le Men, M.-M. Janot, 'Structure de quatre alcaloïdes de la petite Pervenche (Vinca minor L.): (–)-vincadifformine, minovincine, minovincinine et methoxy-16-minovincine', *Bull. Soc. Chim. Fr.* **1962**, 2237.

23. B. Zsadon, M. Rákly, R. Hubay, 'Indole and Indole Derivatives, VII. Tabersonine form the Seeds of Amsonia tabernaemontana WALT. (Apocynaceae) Grown in Hungary', *Acta Chim. Acad. Scient. Hung.* **1971**, *67*, 71.

24. A.-M. Morfaux, J. Vercauteren, J. Kerharo, L. Le Men-Olivier, J. Le Men, 'Alcaloïdes des Feuilles de l'Hunteria elliottii', *Phytochemistry* **1978**, *17*, 167.

25. Since the measurements involve different samples of which little is known as far as the degrees of purity are concerned, the formulation must remain vague.

26. B. W. Bycroft, D. Schumann, M. B. Patel, H. Schmid, 'Weitere Alkaloide aus den Blättern von Pleiocarpa tubicina; Umwandlung von (–)-Quebrachamin in (+)-1,2-Dehydroaspidospermidin', *Helv. Chim. Acta* **1964**, *47*, 1447.

27. J. Mokrý, I. Kompiš, G. Spiteller, 'Alkaloids from Vinca minor L. XX. Further Minor Alkaloids', *Coll. Czech. Chem. Commun.* **1967**, *32*, 2523; W. Doepke, H. Meisel, 'Über die Isolierung von Vincatin und Vincesin, zweier neuer Alkaloide aus Vinca minor L.', *Pharmazie* **1966**, *21*, 444.

28. C. Djerassi, L. D. Antonaccio, H. Budzikiewicz, J. M. Wilson, 'Mass Spectrometry in Structural and Stereochemical Problems: The Structure of the Aspidosperma Alkaloid Aspidoalbine', *Tetrahedron Lett.* **1962**, 1001.

29. H. K. Schnoes, A. L. Burlingame, K. Biemann, 'Application of Mass Spectrometry to Structure Problems: The Occurrence of Eburnamenine and Related Alkaloids in Rhazya stricta and Aspidosperma quebracho blanco', *Tetrahedron Lett.* **1962**, 993.

30. D. Ganzinger, M. Hesse, 'A Chemotaxonomic Study of the Subfamily Plumerioideae of the Apocynaceae', *Lloydia* **1976**, *39*, 326.

31. Alkaloids with the ring system type **4a** have yet to be found in nature.

32. Alkaloids with the ring system types **E5a** and **P5** naturally occur in both 'enantiomeric skeleton' forms. Thereby, 'enantiomeric skeleton' refers to compounds, the ring systems of which exist in the same relationship as (true) enantiomers, but which

differ from one another due to different substitution patterns and functional groups. These compounds belong to the group of indole alkaloids with rearranged secologanin skeletons **3a**, for which the starting material is a vincoside precursor with only one center of absolute configuration at C(15).

33. M. F. Bartlett, R. Sklar, A. F. Smith, W. I. Taylor, 'The Alkaloids of Hunteria eburnea PICHON. III. The Tertiary Bases', J. Org. Chem. **1963**, 28, 2197.

34. E. Schlittler, A. Furlenmeier, 'Vincamin, ein Alkaloid aus Vinca minor L. (Apocynaceae)', Helv. Chim. Acta **1953**, 36, 2017.

35. The specific optical rotation $[\alpha]_D$ is a concentration-dependent quantity. Since the relative proportion of solvent molecules present in the crystal influences the effective mass of substance, the $[\alpha]_D$ value may also vary.

36. A. Latvala, M. A. Önür, T. Gözler, A. Linden, B. Kivçak, M. Hesse, 'Nitrogen Inversion in 9-O-Demethylhomolycorine', Tetrahedron: Asymm. **1995**, 6, 361.

37. The alkaloid content of raw opium varies according to situation, climate etc. The total amount comprises ca. 20–30% (over thirty alkaloids). The content of (–)-morphine typically amounts to 12% (3–23%), that of narcotine, papaverine, thebaine, and codeine to ca. 5, 0.8, 0.4, and 0.3%, respectively.

38. R. Schmitz, F.-J. Kuhlen, 'Schmerz- und Betäubungsmittel vor 1600', Pharmazie in unserer Zeit **1989**, 18, 11.

39. Codeine is one of the main active agents in powerful anti-cough drugs. It specifically inhibits the center of cough in the central nervous system (CNS).

40. Ironically, the 'heroically' effective heroin was originally lauded as a nonaddictive alternative to morphine; mistakenly, as we now know.

41. I. Iijima, J.-i. Minamikawa, A. E. Jacobsen, A. Brossi, K. C. Rice, 'Studies in the (+)-Morphinan Series. 4. A Markedly Improved Synthesis of (+)-Morphine', J. Org. Chem. **1978**, 43, 1462.

6. Alkaloid Synthesis

6.1. General

In a book looking into all aspects of alkaloids, a section about the important field of alkaloid synthesis naturally cannot be omitted. Because of the great diversity of alkaloid structures, however, it is extraordinarily difficult, if not utterly impossible, to hit upon a representative selection of typical examples. For this reason, it is necessary to restrict ourselves to a few appetizers.

Ladenburg's synthesis of (+)-coniine is described as an introduction. This example sheds some light on those problems that confront chemists at any point in history, such as how to identify (especially noncrystalline) synthetic products with the means currently available and to make meaningful comparisons with pertinent natural substances. This is followed by the synthesis of mesembrine, illustrating how a simple tricyclic system is being constructed. As an example of a biomimetic synthesis, the preparation of porantherine is addressed next. An examination of the syntheses of the isomeric spermidine alkaloids cyclocelabenzine and isocyclocelebenzine, in turn, provides information about the procedures involved in producing macrocyclic compounds, while also demonstrating the significance of synthetic approaches in the elucidation of absolute configurations of natural products. This is followed by an investigation of the total synthesis of the indole alkaloid vincamine and completed with one example of the partial synthesis of the bisindole alkaloid alloferin.

The term 'total synthesis' formally refers to the preparation of a target molecule from its component elements. In practice, this means that the starting materials used are structurally simple molecules that can either be purchased or easily synthesized out of their component elements. 'Partial synthesis', in contrast, involves the construction of the target molecule from a comparatively complex precursor, often an optically pure natural product structurally similar to the target molecule.

6.2. The First Total Synthesis of an Alkaloid: (+)-Coniine

The structure of the alkaloid coniine was first proposed by *Hofmann* in 1885, on the basis of a great many experimentally observed details. This was at a time in which scarcely anything was known about the nature of the great majority of coniine degradation products (*cf. Chapt. 3.3*). *Ladenburg* (*Fig. 6.1*) described many attempts[1] to produce 2-propylpyridine by heating pyridine with propyl iodide or isopropyl iodide. He was undoubtedly attempting to synthesize the skeleton of coniine in this simple way, but this plan was doomed to fail. After that, he turned his attention to the condensation reactions between quinaldine (= 2-methylquinoline) and benzaldehyde or chloral (= 2,2,2-trichloroacetaldehyde), which had been accomplished by *Reimer, Einhorn*, as well as by *Miller* and *Spady*. *Ladenburg* studied the '*effect of paraldehyde on α-picoline*'[2]. The two compounds react together only at high temperatures, and even at 250° C he only succeeded in isolating small quantities of the desired product[1]. Finally, by heating the compounds for ten hours at 250–260° C in a sealed tube, *Ladenburg* was able to obtain 2-allylpyridine in a yield of 9% (*Scheme 6.1*)[3]. The substance was removed from the reaction mixture by solvent extraction and distillation, on which he commented that '*it possesses a perceptible coniine aroma*'[3]. To rule out side-chain migration during the synthesis, *Ladenburg* oxidized the allyl group and identified the reaction product as 2-picolinic acid (= pyridine-2-carboxylic acid). Experiments of this kind were, naturally, much more labor-intensive than they are today.

'*Reduction of the α-allylpyridine to α-propylpiperidine was successfully accomplished by means of sodium in alcoholic solution at the boiling point, giving rise to a nearly quantitative yield.* [...] *The base displayed the closest resemblance to coniine and may be regarded as chemically identical to that substance*'[3].

So as to be quite sure, he converted the synthetic propylpiperidine into conyrine (zinc-dust distillation), which, as expected, matched the original compound in both melting point and elemental analysis of its salts. On administration to white mice, α-propylpiperidine also exhibited the same physiological effects as coniine. Only in its behavior towards '*the polarized ray of light*' did the synthetic (optically inactive) α-propylpiperidine differ from coniine. It was, therefore, necessary to carry out the separation of the two enantiomers. *Ladenburg* first attempted this with the aid of *Penicillium glaucum*, but without success. Finally, the racemate could be separated with the help of tartaric acid:

'*Schorm has prepared and studied dextrotartaric coniine, and that gentleman was so kind as to send me some crystals of this beautiful salt some while ago. If a fragment of this is added to a highly concentrated solution of α-propylpiperidine bitartrate, prepared by rapid evaporation, then the substance begins to crystallize; this can be assisted by stirring. After 5–6 days, a crystalline mass embedded in a syrup was obtained; this was transferred onto a blotting paper, which slowly drew off the syr-*

Fig. 6.1. *Albert Ladenburg* (July 2, 1842 – August 15, 1911). Ph. D. from the University of Heidelberg, professor at Kiel 1874 – 1889, Breslau 1889 – 1909. Highly reputed organic chemist, who first succeeded in the synthesis of an alkaloid: coniine.

*up. The crystals were then strongly pressed repeatedly, until a complete-
ly dry, white crystal mass was left behind. This was decomposed by the
action of potash and the base was distilled off. After it had been dried, its
optical behavior was examined in a Laurent half-shade instrument. It
was optically active, dextrorotatory, and displayed almost the same
angle of rotation (11°46' at 1 dm cell length) as the natural substance
under the same conditions (11°40' at 1 dm cell length). This corresponds
to $\alpha_D = 13°87'$ for the synthetic base, and $\alpha_D = 13°79'$ for coniine*[3].

Ladenburg's synthesis, therefore, made it possible to establish the identity of
natural coniine and dextrorotatory α-propylpiperidine.

Scheme 6.1. *First Synthesis of an Alkaloid:* Ladenburg's Preparation of (+)-Coniine[3]

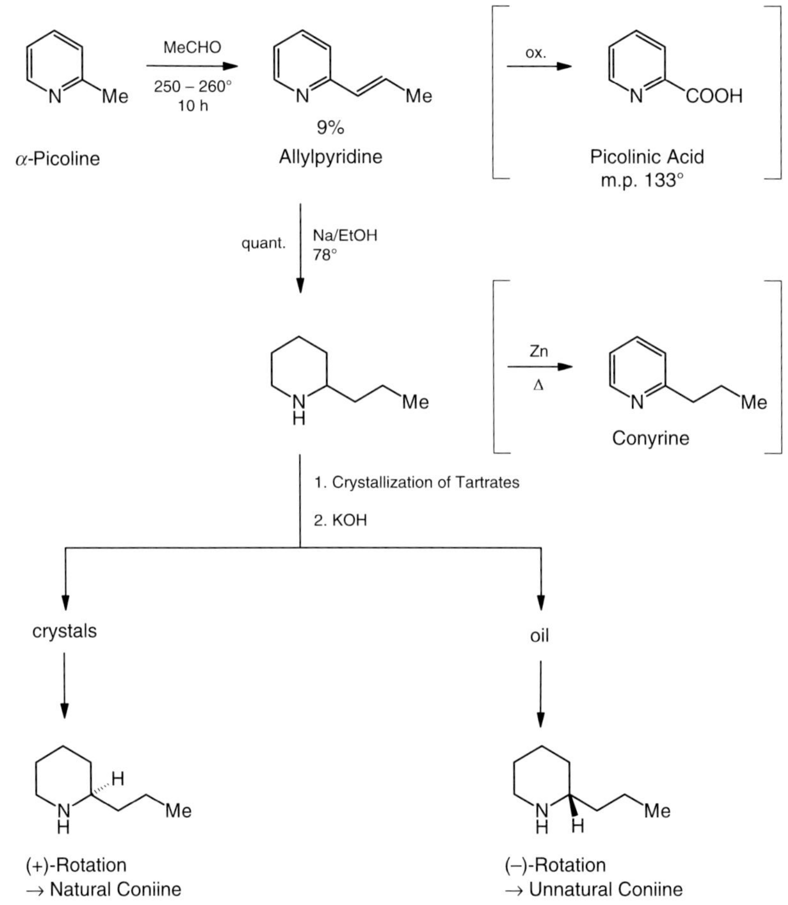

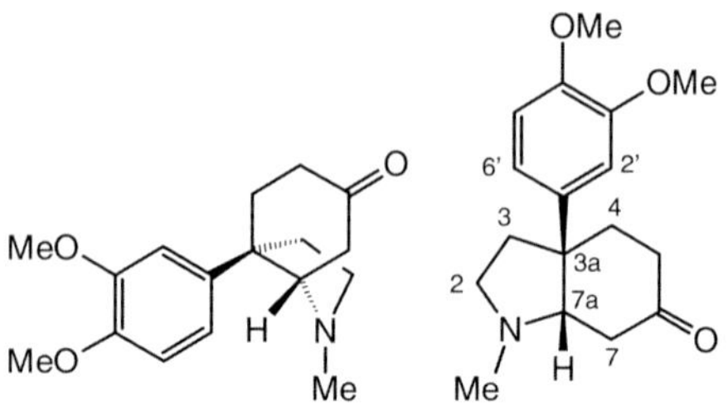

(–)-Mesembrine (**1**)

6.3. The Synthesis of Mesembrine

(–)-Mesembrine (**1**) was isolated for the first time in 1957 from *Mesembryanthemum tortuosum*, a species belonging to the Aizoaceae[4]. The structure of natural (–)-mesembrine was elucidated in 1960[5]. The compound possesses three rings: one benzene, one pyrrolidine, and one cyclohexanone ring fused to the heterocycle. The two fused rings are joined in a *cis*-configuration.

While the formula of mesembrine shown on the left-hand side better illustrates the biogenetic relationship to the Amaryllidaceae alkaloids of the ambelline type, the representation on the right-hand side has the advantage that the mutual arrangement of the three rings can be seen more easily.

As far as the synthesis of this alkaloid is concerned, the first question to be addressed is how the substituted cyclohexanone ring should be constructed. Looking at the right-hand formula, it can be seen that the oxo group is situated in the same position relative to the fused ring as in the 3-oxosteroids. How is this carbonyl group introduced in steroid synthesis though? Here, one extremely effective and frequently used method is the so-called *Robinson* anellation. In this transformation – a combination of a *Michael* addition and an aldol reaction – a cyclohexanone is treated with but-3-en-2-one (methyl vinyl ketone) or with one of its equivalents under basic reaction conditions, which gives rise to the formation of the desired ring (*Scheme 6.2*).

In the final step, though, H_2O is eliminated producing the α,β-unsaturated derivative instead of the saturated ketone. This type of reaction is hence not directly applicable to the synthesis of the corresponding ring system in mesembrine, since the same procedure using pyrrolidin-2-one (in place of cyclohexanone) would not lead to the target compound but rather yield an unsaturated system. To suppress the elimination of H_2O, it would, therefore, be better to begin not with the ketone, but (formally) with the corresponding alcohol, and so the starting material of choice would be compound **3**. The latter can be readily dehydrated to yield the enamine **4**, which is ideally suited for reactions with the α,β-unsaturated carbonyl compound (*Scheme 6.3*).

The intermediates **5** (from enamine **4**) and **2** (from cyclohexanone) correspond to one another: while, in the case of **2**, the nucleophilic enolate anion attacks the C=O group, the electrophilic C(2)-atom is attacked in the case of compound **5**. Species **6**, resulting from **5**, does not have the option of eliminating H_2O, since no OH group is present. Furthermore, compound **6** features all the required structural elements present in the 'lower' half of the me-

Scheme 6.2

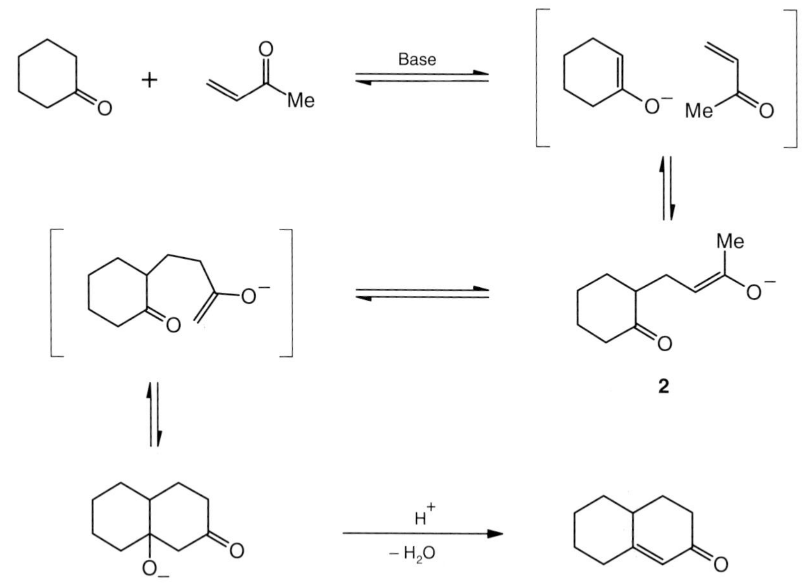

Scheme 6.3

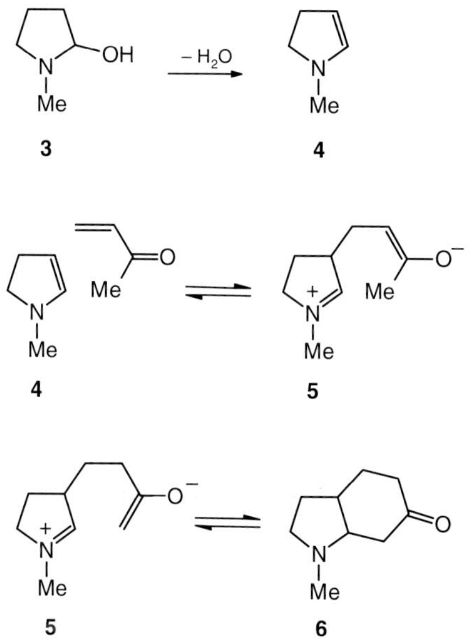

sembrine formula, and so the first part of the problem is solved. The next step is the synthesis of enamine **7**, which may also be regarded as a dihydropyrrole. In general, dihydropyrroles can be obtained by H_2O elimination from pyrrolidin-2-ols or pyrrolidin-3-ols[6]. 1-Methylpyrrolidin-2-ol (**3**) itself corresponds to the cyclic aminoacetal of 4-(methylamino)butanal. Pyrrolidin-3-ol might be attainable through reduction of 1-methylpyrrolidin-3-one. Formally, though, the same effect may be achieved by letting 1-methylpyrrolidin-3-one react at C(3) with a nucleophile (*Scheme 6.4*).

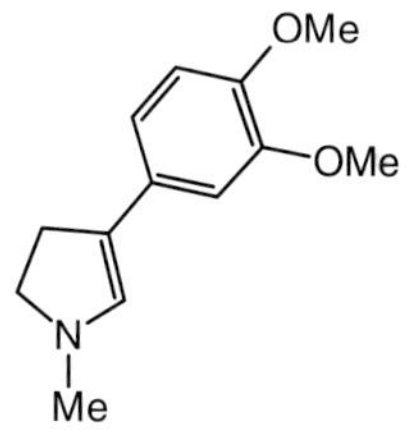

The choice of the residue R is directly at hand; it has to be the appropriately substituted aromatic ring. For a nucleophilic reaction with an aromatic ring to be feasible, it is necessary to use organometallic Li or Mg reagents.

Altogether, the synthesis of mesembrine (**1**) is now roughly outlined (*Scheme 6.5*). As a matter of fact, the compound could indeed be synthesized in 37.4% overall yield by the way depicted. Thereby, the *cis*-configuration of the fused ring system was formed automatically, since, for stereoelectronic reasons (maximum π-orbital overlap), the nucleophilic attack in **8** proceeds *trans* to the phenyl nucleus.

7

Scheme 6.4

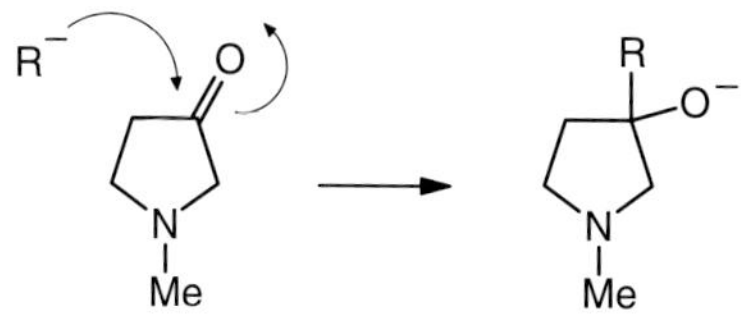

Scheme 6.5. *Synthesis of (±)-Mesembrine* (**1**)[9]

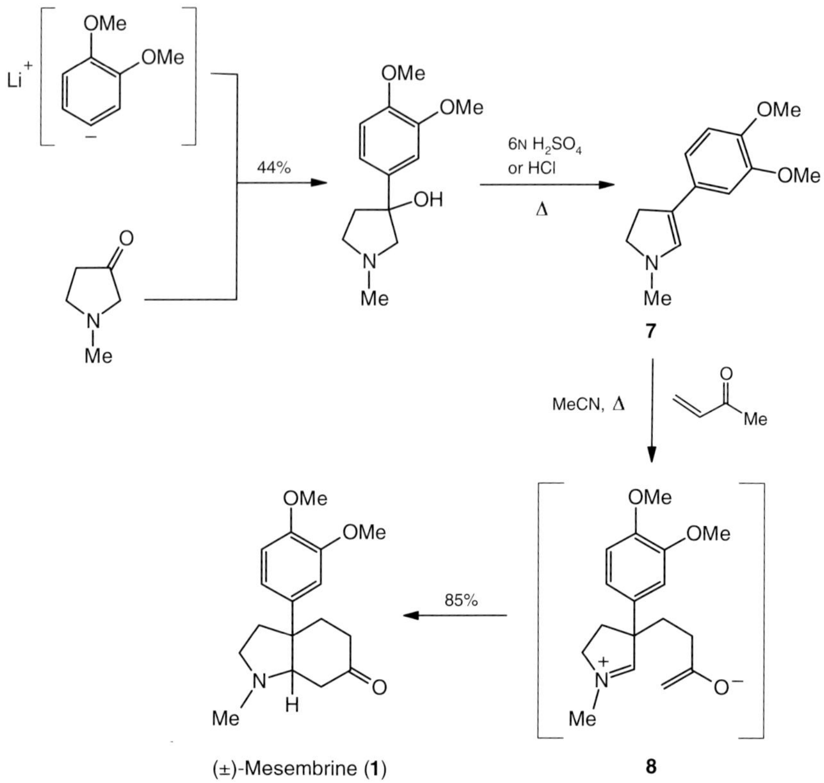

6.4. A Biomimetic Synthesis of Porantherine

The plant *Poranthera corymbosa*, indigenous to Australia and belonging to the Euphorbiaceae family, contains a number of different alkaloids. The alkaloid extraction yield, relative to the dry weight of the extract, was 0.4% with the bases **9 – 14** constituting the major components.

The structures of the alkaloids **9 – 12** were elucidated by X-ray crystallography (*cf.* **9**[10]; **10**[11,12]; **11**[11,13]; **12**[14]); those of **13** and **14** were resolved later by chemical correlation with the main bases[15].

The ring system of porantherine (**9**), 9b-azaphenalene, was for a long time unknown to be present in plants. Alkaloids possessing this ring system, how-

Porantherine (**9**)
(abs. config.)

Porantheridine (**10**)

R = H : Poranthericine (**11**)
R = Ac: *O*-Acetylporanthericine (**14**)

Porantherilidine (**12**)

Porantheriline (**13**)

Propyleine (**15**)

ever, had been isolated from beetles of the Coccinellidae family; one example is propyleine (**15**), found in *Propylaea quatuordecimpunctata*[16]. Coccinellins from *Coccinella septempunctata*[17] (*Fig. 6.2*) are similarly constituted, as are hippodamine and convergine from *Hippodamia convergens*[18].

The biogenetic interrelationship of the *Poranthera* alkaloids **9**–**14** is clearly recognizable. Porantherine (**9**) itself possibly results from ring closure between a suitable 2-substituted Et group and C(6a) in poranthericine (**11**), accompanied by elimination of H_2O. Poranthericine, for its part, may be derived from porantherilidine (**12**) by a selective cyclization between the appropriately functionalized atoms C(3′) and C(4) and epimerization at C(9a). Ring closure between the OH groups at C(2′) and C(4) in porantherilidine (**12**), in turn, produces porantheridine (**10**).

6.4.1. Retrosynthetic Analysis

If a retrosynthetic (antithetic) analysis[19] is performed on porantherine (**9**) (*Scheme 6.6*), then a potential synthetic pathway is created[20].

As a starting point, the C(2)=C(3) bond of porantherine is worthwhile considering. A C=C bond of this type may be produced by the elimination of

Coccinelline

Fig. 6.2. *Coccinella pentapunctata* L. (Coccinellidae) *on a four-leafed clover*

Scheme 6.6

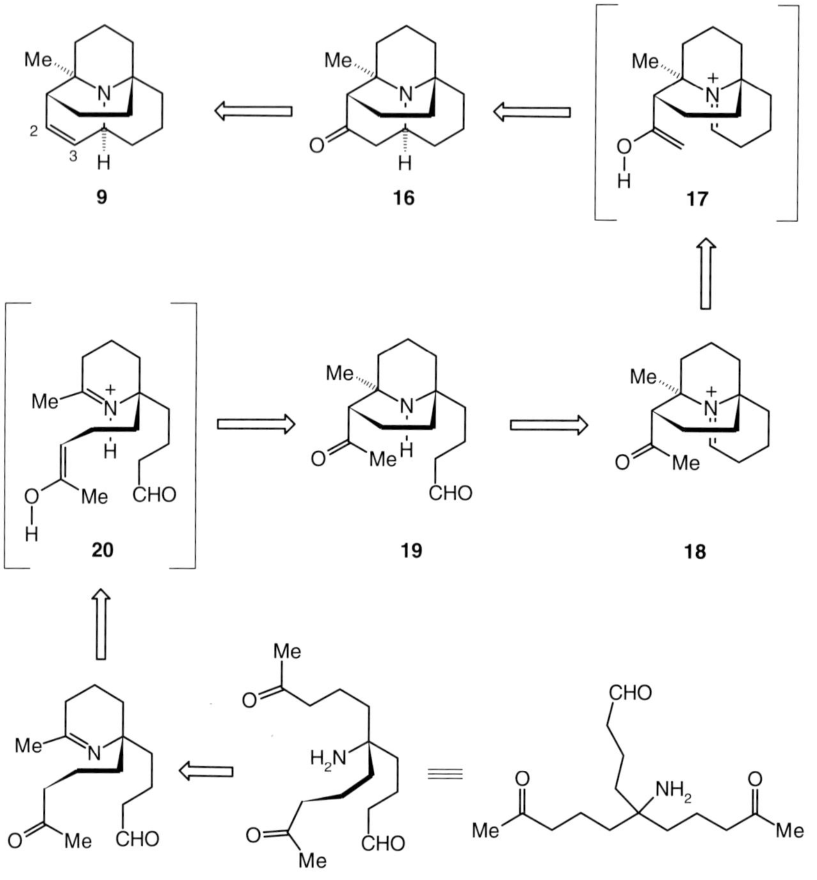

Scheme 6.7

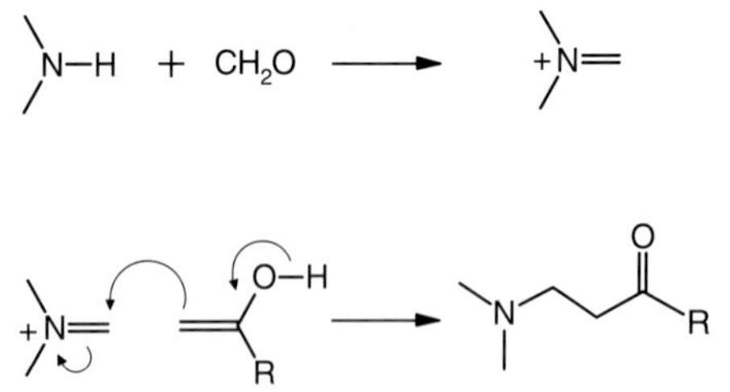

H_2O from a corresponding alcohol. A C(2)-alcohol, for example, might be prepared by the reduction of the 2-oxo group of compound **16**. The latter features a (synthetically) interesting characteristic: the N-atom is in (double) β-position relative to the C=O group. This immediately makes an experienced chemist think of a *Mannich* reaction, in which the condensation product of an amine and an aldehyde (*i.e.*, an imine) reacts with an enolizable ketone (*Scheme 6.7*). Compound **16**, therefore, represents a so-called *Mannich* base that might have been produced from compound **18** (*via* **17**) by means of bond formation between C(3) and C(3a). Compound **18**, in turn, may be derived from the aldehyde **19**, which once again features a β-aminocarbonyl moiety, suggesting another *Mannich* reaction. Hence, one finally arrives at the aliphatic, symmetrical compound **22** as a possible starting material for the synthesis of porantherine (**9**) according to the pathway given in *Scheme 6.6*. Such a transformation would involve two *Mannich* reactions and one reduction, followed by the elimination of H_2O. It is less probable, though, that compound **22**, which contains two oxo groups and one formyl group, can be converted directly into **16**: there would simply be too many competing reactions. For a laboratory synthesis, therefore, it is necessary to synthesize the compound in protected form and to selectively and stepwise induce the above reactions.

Since it is assumed that the (intermediate) products outlined in the synthesis of porantherine also arise during its biogenesis, the above process can be described as a biomimetic synthesis. The term 'biomimetic', though, should not be taken as implying anything about the actual proceedings involved; while *biosynthesis* of natural products takes place at ambient temperatures in an aqueous environment, *biomimetic syntheses* largely proceed under appreciably more drastic conditions in the laboratory.

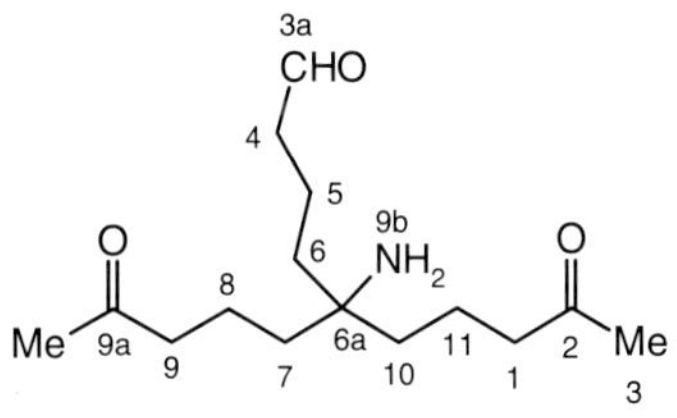

6.4.2. Synthesis of Porantherine

Compound **22** is a symmetrical amine. Its protected counterpart **27** can be synthesized as follows. If two equivalents of the *Grignard* compound **23** are heated under reflux with one equivalent of ethyl formate, then the secondary alcohol **24** results, in which the two original C=O groups (C(9a) and C(2)) are protected as acetals (*Scheme 6.8*).

22

Scheme 6.8

23   **24**

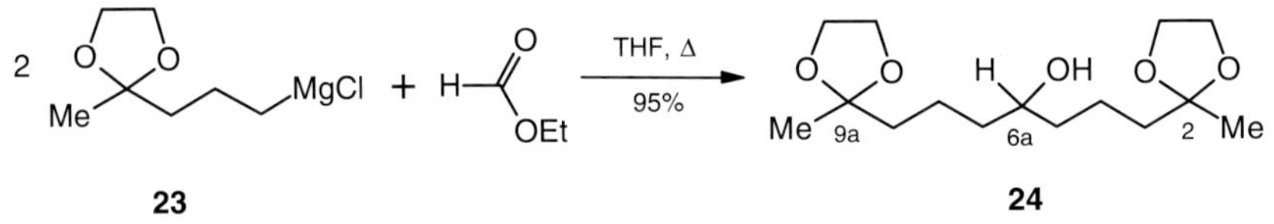

To introduce both the amino function and the butanal residue at C(6a) later, **24** is oxidized with *Collins* reagent (CrO_3 in pyridine, 6 equiv.) at 20° C in CH_2Cl_2 to give the intermediate **25** (93% yield). The latter is treated with methylamine in toluene in a sealed bomb to produce compound **26** (*Scheme 6.9*). Addition of molecular sieves (4 Å) enables the H_2O liberated during the course of the reaction to be removed, which irreversibly shifts the equilibrium towards the target compound.

Scheme 6.9

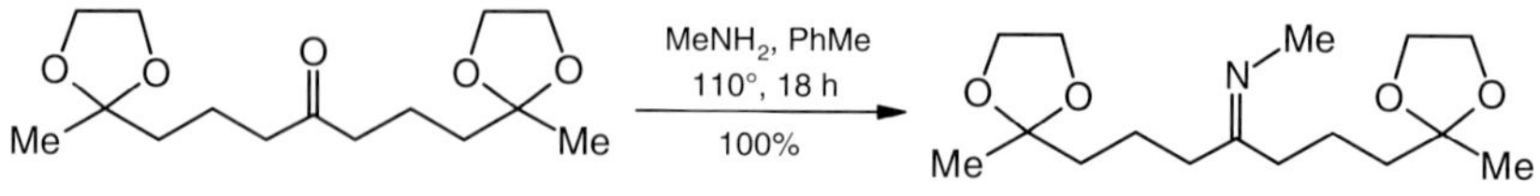

25   **26**

Nucleophilic alkylation of the imino group at C(6) with 1-lithiopent-4-ene, prepared from 5-bromopent-1-ene in ether, affords compound **27**, in which all the reactive groups are protected (*Scheme 6.10*). Oxidative cleavage of the C=C bond will later reconstitute the aldehyde.

Scheme 6.10

26

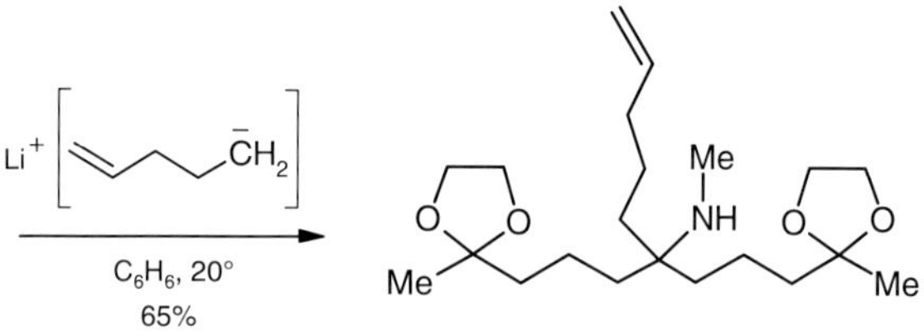

27

The two acetal groups in **27** are now carefully cleaved with 10% aqueous HCl solution, upon which one out of the two C=O groups promptly reacts with the amine to furnish **28** (*Scheme 6.11*; *cf.* also **21** in *Scheme 6.6*). The bicyclic compound **29** is formed by another *Mannich* reaction in the presence of toluene-4-sulfonic acid/isopropenyl acetate. Under these conditions, the transformation proceeds *via* an intermediary enol acetate. The next step involves the ring closure between the N-atom and the formyl group at C(3a), but both the oxidation of the *N*-Me group and the reconstitution of the aldehyde are required first. The *N*-Me group may be oxidized to an *N*-formyl group by using *Collins* reagent (10 equiv., CH$_2$Cl$_2$, 72 h), to yield **30**. Lastly, cleavage of the C=C bond is effected with osmium(VIII) tetroxide and sodium metaperiodate in *t*-BuOH/H$_2$O solution.

The intermediate **31** is directly transformed into the acetal **32** with an excess of ethylene glycol in benzene (*Scheme 6.12*). This is necessary in order to suppress the formation of undesired side products in the ensuing deformylation with 3N aqueous KOH solution, resulting in **33**. The latter compound may, in turn, be converted into the unstable enamine **34** (*cf.* also **18** in *Scheme 6.6*) by treatment with 10% aqueous HCl solution, and this compound then affords **35** under acid catalysis in refluxing toluene. Reduction of the

Scheme 6.11

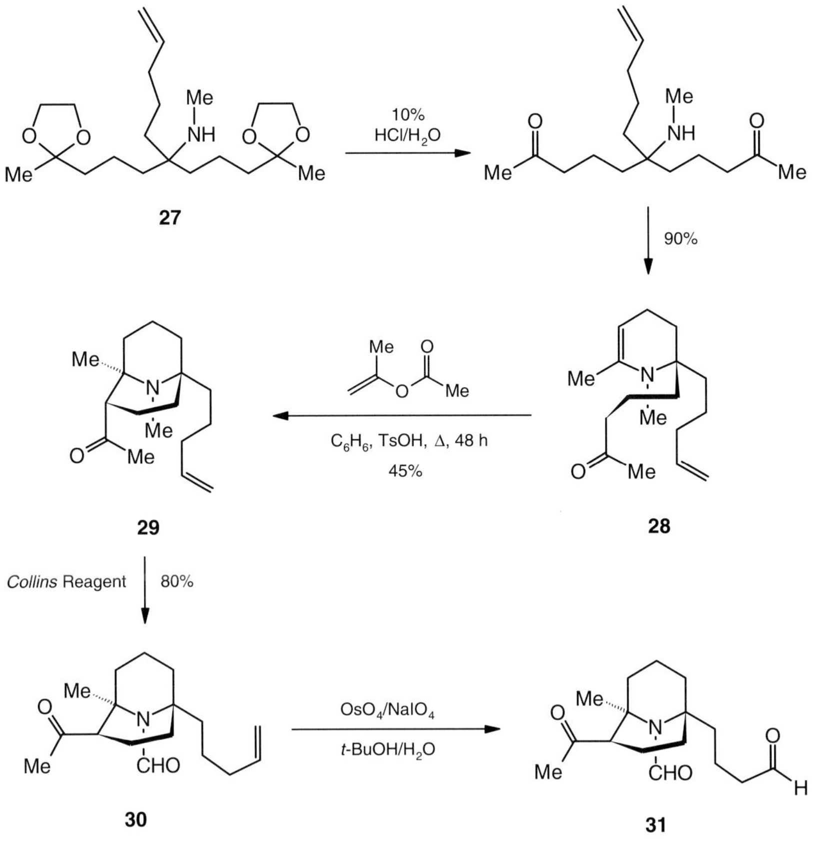

C=O group with NaBH$_4$ and subsequent elimination of H$_2$O with SOCl$_2$ in pyridine finally affords (±)-porantherine (**9**). The overall yield of the pathway shown was around 2.5% in practice.

6.5. Syntheses of Macrocyclic Spermidine Alkaloids: (+)-Cyclocelabenzine and (+)-Isocyclocelabenzine

In the course of intensive investigations of the constituent substances of the genus *Maytenus* (Celastraceae) during the 1970s, maytansine was isolated from *Maytenus ovatus* in 1972. A maytansine content of 0.00015% of dry weight was found. Since this compound was a very potent cytostatic agent[21,22], the alkaloid contents of other Celastraceae species were later

213

Scheme 6.12

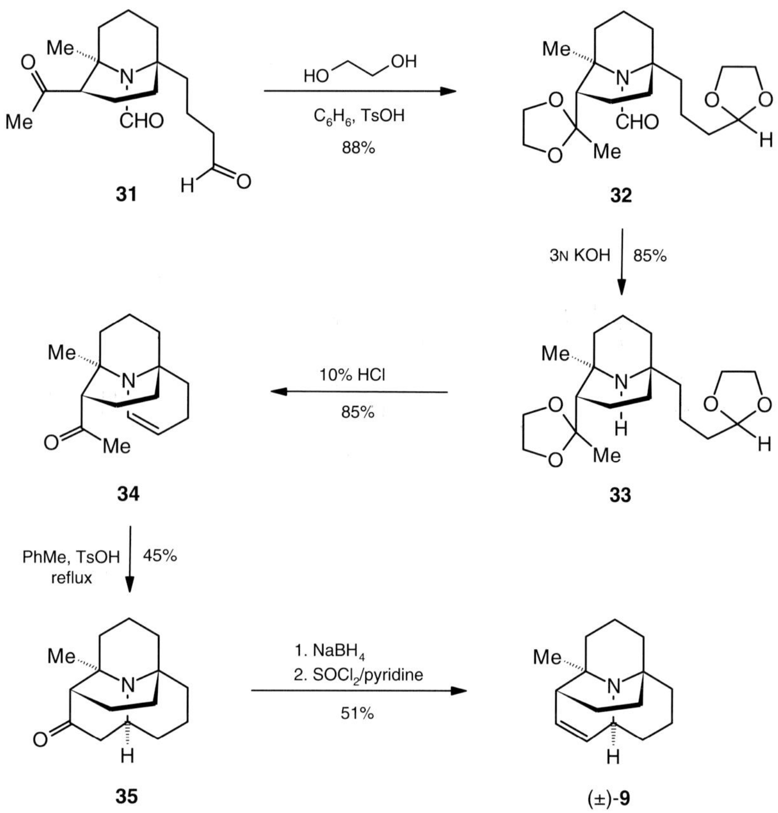

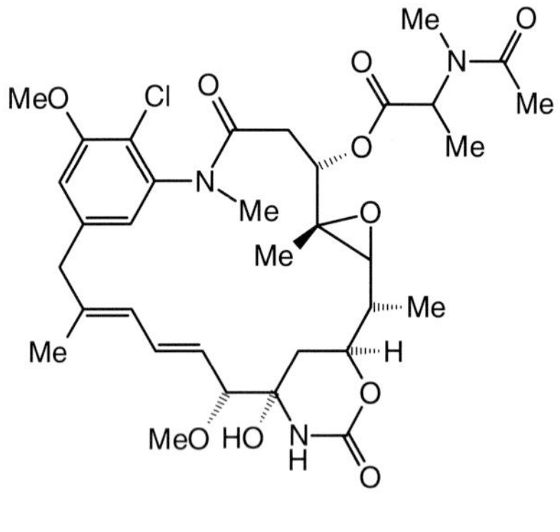

studied in the hope of discovering richer sources of maytansine or similarly effective substances. Except for a few spermidine alkaloids (especially macrocyclic compounds), however, no further bases were isolated. It was later realized that maytansine is not actually a product of the plant, but of a parasitic fungus (*Nocardia* sp.) that had infected the plants (*Maytenus rothiana*, *M. ovatus*[23], *M. buchananii*, *Trewia nudiflora*, *Putterlickia verrucosa*, etc.), which also explained, why it had only been possible to isolate such small quantities of the substances. During investigations of *Maytenus mossambicensis* var. *mossambicensis*, however, three compounds were isolated and named as (0)-celabenzine (**36**)[24], (+)-cyclocelabenzine (**37**), and (+)-isocyclocelabenzine (**38**)[25] (*cf. Schemes 6.13* and *6.15*). It soon came to light that there was one stereogenic center present in **36**, and two in **37** and **38**, respectively. Neither the relative nor the absolute configuration was known,

Maytansine

however, and so it was decided to deduce the configuration by total synthesis followed by structural comparison[26].

The biogenetic interrelationship of the three structurally related compounds is not yet known. It is presumed, though, that some cinnamic acid ester (3-phenylprop-2-enoic acid ester) and spermidine are joined through a β-addition in such a manner that the terminal NH_2 group of the longer methylene chain initiates a nucleophilic attack on the C=C bond, while the other NH_2 group reacts with the acid moiety to an amide, producing **A** (*cf. Scheme 6.13*). From biogenetic considerations, it may be assumed that the stereogenic center present – as in all spermidine alkaloids currently known – has the (*S*)-configuration. This mode of incorporation of spermidine is very often observed. 'Reversed' spermidine incorporation also happens, though; in (–)-periphylline and (–)-isoperiphylline (from *Peripterygia marginata* (Celastraceae)), for example. Benzoylation of **A** provides celabenzine (**36**), from which cyclocelabenzine (**37**) is produced (*via* **B**) by an additional ring closure. Isocyclocelabenzine (**38**) is made in an analogous manner (*via* 'iso-**B**') by means of dehydrogenation of the spermidine tetramethylene moiety in the second step (**36**→**B**).

Essentially, the cyclization of the dehydrogenated precursor **B** may take place in two different ways: either photochemically[27] or through acid catalysis

Scheme 6.13

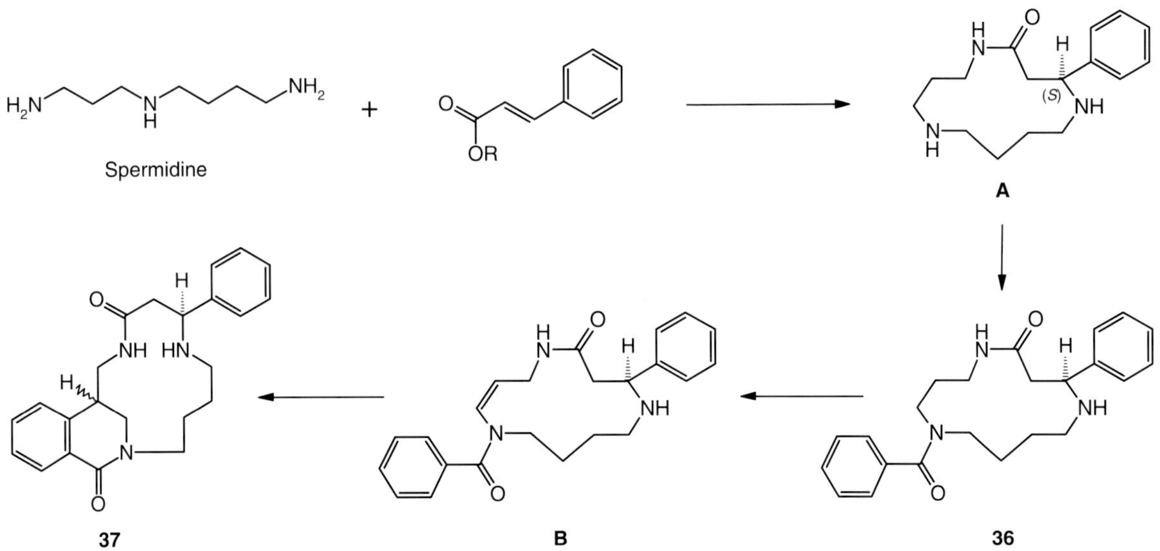

Scheme 6.14. *Possible Mechanisms for the Formation of Cyclocelabenzine* (**37**) *and Isocyclocelabenzine* (**38**) *from Dehydro(iso)cyclocelabenzine* (**B** or iso-**B**)

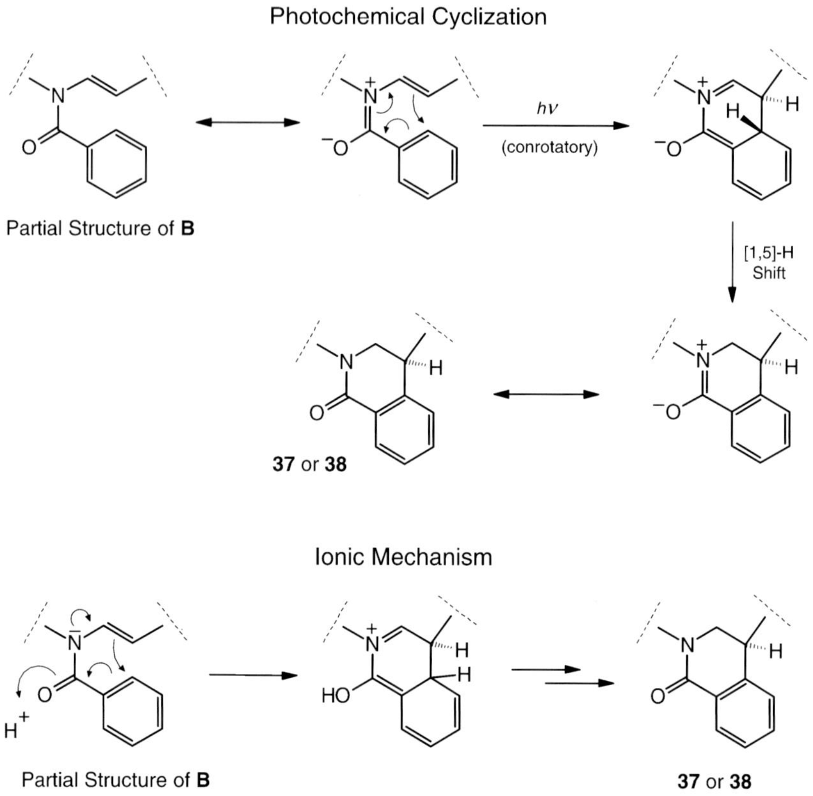
Photochemical Cyclization
Partial Structure of B
hν
(conrotatory)
[1,5]-H
Shift
37 or 38
Ionic Mechanism
Partial Structure of B
37 or 38

(*cf. Scheme 6.14*). Should the need arise, though, a radical mechanism or a transitory epoxidation of the intermediate might also be feasible.

The reactions shown in *Scheme 6.14* could, in principle, give rise to two diastereoisomers, epimeric at C(13). Only one epimer, however, was isolated from the plant. From this, though, it may not be concluded with absolute certainty that the cyclization is enzymatically catalyzed, since explicit examination of the extracts for the presence of the absent diastereoisomer was not carried out.

After a careful retrosynthetic analysis of the target molecules (*Scheme 6.15*), the preparation of cyclocelabenzine and isocyclocelabenzine was finally begun (*vide infra*).

6.5.1. Synthesis of (+)-Cyclocelabenzine[28] (37)

The goal of the first stage of the synthesis of (+)-cyclocelabenzine (**37**) is the preparation of the ethyl carbamate **44** (*cf. Scheme 6.17*). The symmetrical 3-phenylglutaric acid dinitrile (**39**) is obtained by treatment of benzaldehyde with two equivalents of cyanoacetic acid in the presence of piperidine and pyridine (both *double* addition and decarboxylation; *Scheme 6.16*). The two carbonitriles are transformed into amines by hydrolysis to the corresponding carboxamides followed by *Hofmann* degradation. The reaction between the diamine **41** and one equivalent of phthalic anhydride furnishes the carboxylic acid **42**. Since the latter is hardly soluble in the solvent system selected (tetrahydrofuran/toluene) and immediately precipitates out, the yield at this step is especially high. Ethoxycarbonylation of the (remaining) free NH_2 group finally yields **43**, which can be converted into the target compound **44** by (intramolecular) introduction of a phthalimide protecting group (important for the subsequent cyclization).

The transformation of the ester **44** into a 1,2,3,4-tetrahydroisoquinolin-1-one derivative of type **47** can, in principle, be effected by means of a so-called *Bischler–Napieralski* reaction (*Scheme 6.17*). Effective cyclization, though, requires either an electron-rich aromatic moiety[29] or a strong electrophile (or both). Because of strategic considerations, however, a chlorobutyl group (as part of the cyclocelabenzine ring) is introduced for the moment, producing compound **45**. After this, the desired ring closure (*via* **46**) to compound **47** proceeds under typical *Bischler–Napieralski* conditions ($POCl_3$, Δ). Removal of the phthaloyl protecting group with hydrazine hydrate (*Gabriel* synthesis) readily affords the intermediate product **48**.

To complete the ring system, a 3-amino-3-phenylpropionic acid moiety should be introduced. This is accomplished by reacting the *N*-protected, activated (–)-(*S*)-3-{*N*-[(*tert*-butoxy)carbonyl]amino}-3-phenylpropionic acid with **48** – with 2-chloro-1-methylpyridinium iodide as a coupling reagent (*cf. Scheme 6.18*) – resulting in **49** (*Scheme 6.19*).

The obvious attempt to perform the final ring closure by means of intramolecular *N*-alkylation (leaving group Cl^- or I^- in compounds **49** or **50**) is, however, unsuccessful. Therefore, it is necessary to work out another way: hydrolysis of the iodide **50** provides the corresponding hydroxy compound **51**, which is oxidatively (pyridinium chlorochromate) converted directly into the aldehyde **52**. The *tert*-butoxycarbonyl (Boc) group is then removed (trifluoroacetic acid), and the aldehyde is allowed to react with the liberated amine to the corresponding macrocyclic *Schiff* base (imine), which can finally be reduced to cyclocelabenzine (**37**) with $NaBH_3CN/MeOH$.

Scheme 6.15

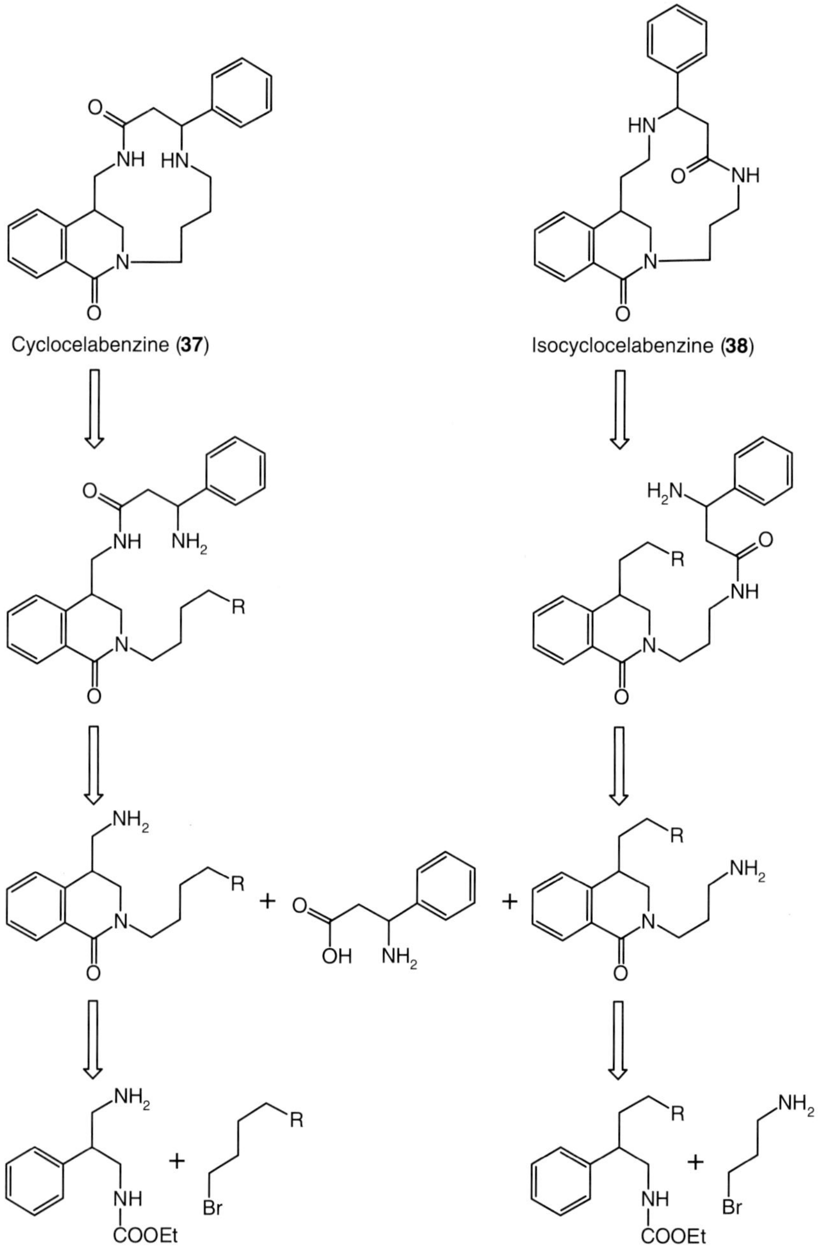

Cyclocelabenzine (**37**)

Isocyclocelabenzine (**38**)

Scheme 6.16

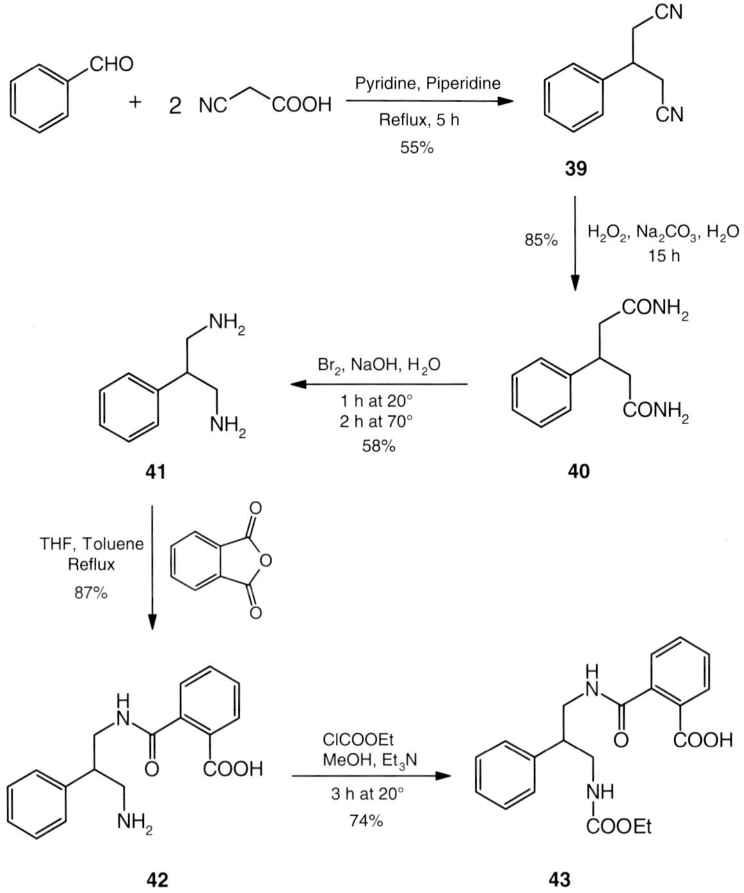

The above reaction sequence provides a 4:1 mixture of the two C(13)-epimers of cyclocelabenzine, with the predominant diastereoisomer **37a** matching the natural product in all its physical and chemical properties. How, though, could the absolute configuration of the (oily) **37a** now be determined? Fortunately, the other diastereoisomer, **37b**, obtained in lower yield, was crystalline in nature. Hence, it was possible to perform an X-ray crystal-structure analysis and to determine its relative configuration. Since the configuration at C(8) in **37b** was known from the synthetic approach, the overall absolute configurations of both isomers could be derived automatically. The naturally occurring, noncrystalline epimer **37a**, therefore, corresponds to (+)-(8S,13R)-, the crystalline compound **37b** to (−)-(8S,13S)-cyclocelabenzine.

Scheme 6.17

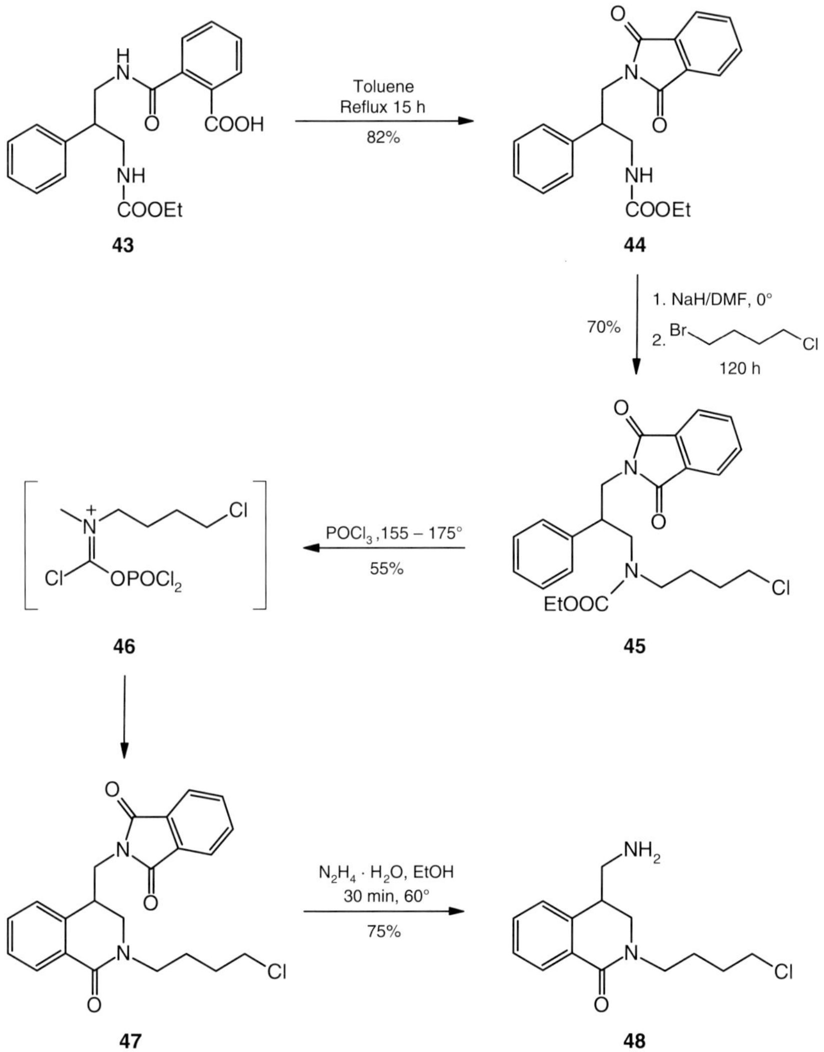

6.5.2. Synthesis of (+)-Isocyclocelabenzine[30] (38)

For the synthesis of (+)-isocyclocelabenzine, it has to be borne in mind that the benzoyl residue is bound to a *tri*methylene rather than to a *tetra*methylene moiety, found in cyclocelabenzine. Some essential alterations in the

Scheme 6.18

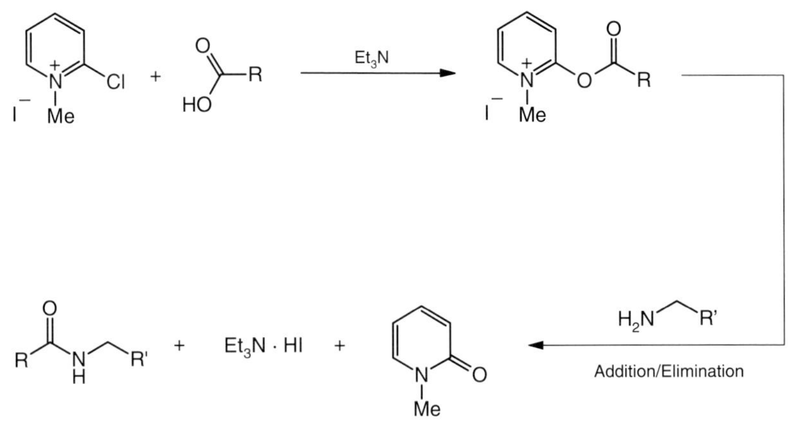

planning of the synthesis are, therefore, necessary, although the individual steps can be performed using the same methodology.

In a first step, phenylacetonitrile is alkylated under basic conditions with 2-bromoethanol and, to inhibit side reactions, directly acetylated to provide the (acetoxy)butyronitrile **53** (*Scheme 6.20*). The low yield of 43% is essentially due to the formation of the dialkylated product. Catalytic reduction (H_2/Pt) in the presence of a little $CHCl_3$ provids the ammonium salt **54**, into which it is possible to incorporate the trimethylene moiety found in **55**. The transformation of the latter into 1,2,3,4-tetrahydroisoquinolone **57** is carried out with ethyl chloroformate and $POCl_3$ (*cf. Scheme 6.17*) . The subsequent treatment of **57** is shown in *Scheme 6.21* and largely corresponds to the cyclocelabenzine synthesis. Upon *in situ* reduction of the *Schiff* base, obtained from **58**, the two C(13) epimers of isocyclocelabenzine are isolated in *ca.* 1:1 ratio.

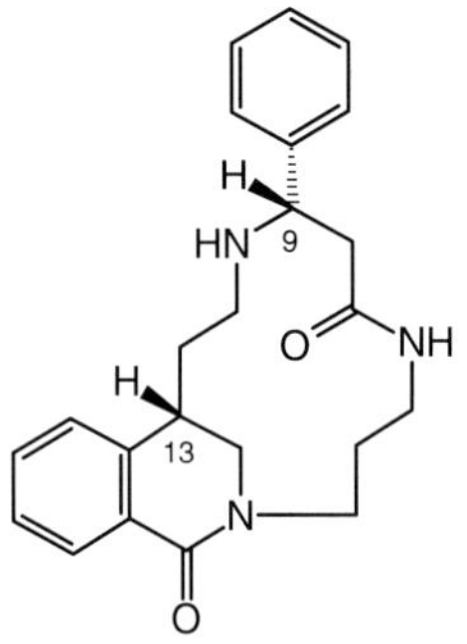

Again, the absolute configuration of the naturally occurring base could be determined by X-ray crystal-structure analysis: here, the natural product is (+)-(9*S*,13*S*)-isocyclocelabenzine (**38a**).

The synthetic work described above underpins the hypothesis that all macrocyclic alkaloids containing a spermine or spermidine ring system possess the (*S*)-configuration[31] at the C-atom to which fatty or cinnamic acid moieties usually are bound (corresponding to C(8) in **37** and C(9) in **38**, respec-

(+)-(9*S*,13*S*)-Isocyclocelabenzine (**38a**)

Scheme 6.19

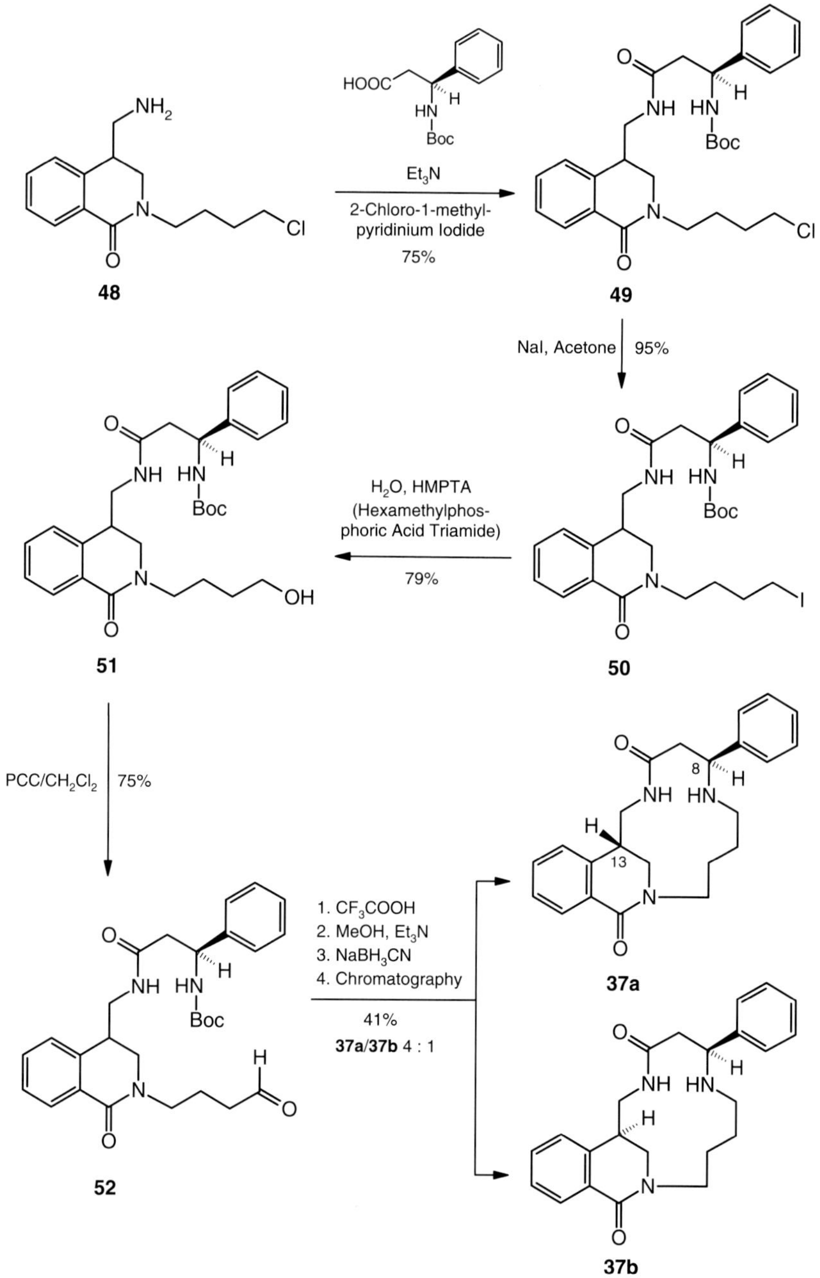

48

HOOC

HN
Boc

Et₃N

2-Chloro-1-methyl-
pyridinium Iodide

75%

49

NaI, Acetone | 95%

H₂O, HMPTA
(Hexamethylphos-
phoric Acid Triamide)

79%

51

50

PCC/CH₂Cl₂ | 75%

1. CF₃COOH
2. MeOH, Et₃N
3. NaBH₃CN
4. Chromatography

41%
37a/37b 4 : 1

52

37a

37b

Scheme 6.20

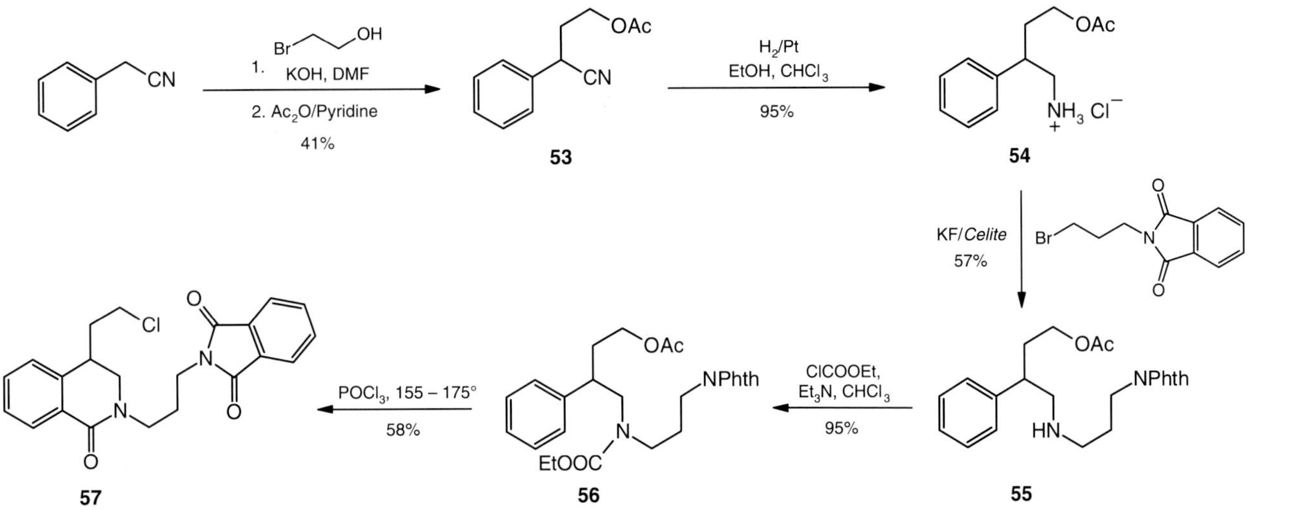

Scheme 6.21

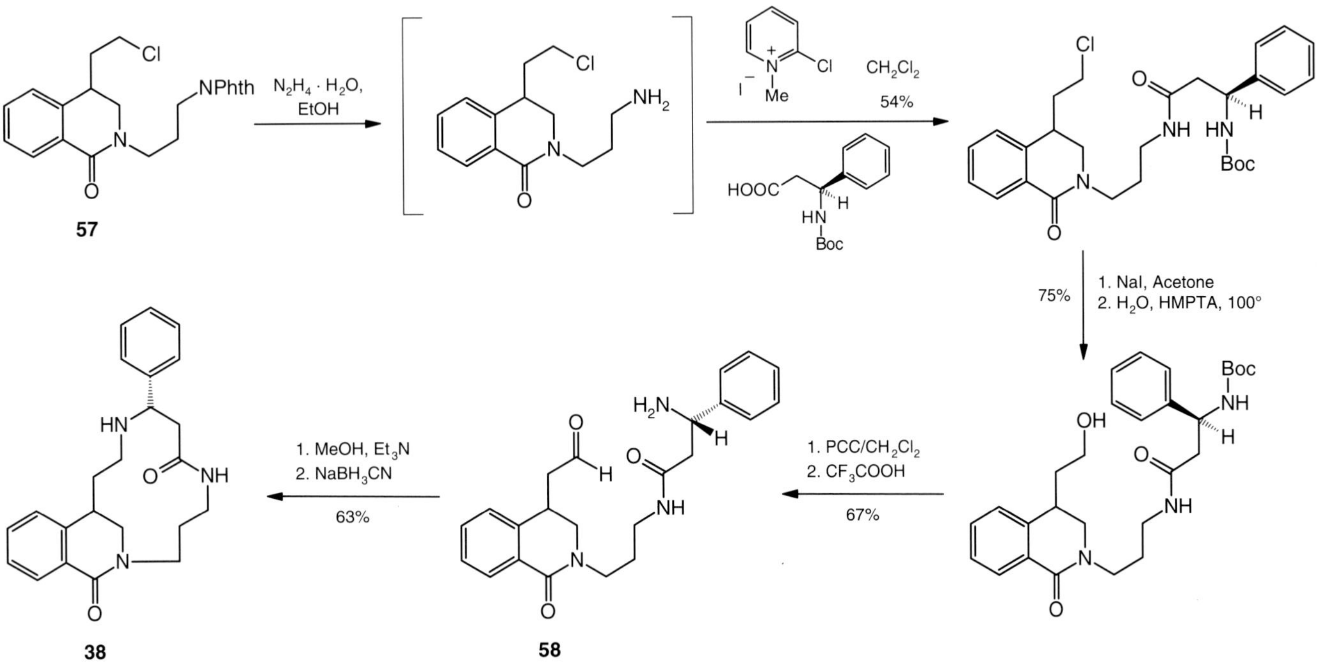

Scheme 6.22

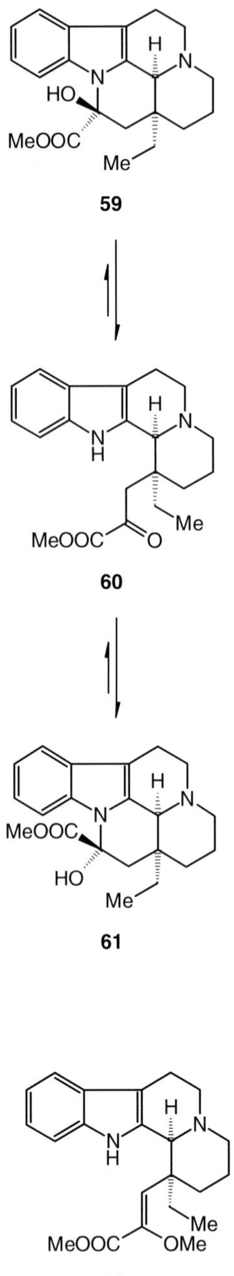

59

60

61

62

tively). The biomolecular mechanism behind this phenomenon has not as yet been elucidated; and so exceptions to this 'rule' cannot be completely dismissed.

6.6. The Synthesis of (3S,14S,16S)-Vincamine

Vincamine (**59**)[32], an indole alkaloid from various *Vinca* species and other genera of the Apocynaceae, is especially interesting, since it acts as a regulator of cerebral blood circulation. Hence, a great number of publications relating to both nonstereoselective[33,34] and stereoselective syntheses[35–40] of vincamine exist. Here, the variant by *Oppolzer* (*Fig. 6.3*) and co-workers[40] (overall yield 4.5% before optical resolution) is introduced and discussed as a representative example.

6.6.1. Retrosynthetic Analysis

As mentioned, vincamine (**59**) possesses three stereogenic centers[41]. Thereby, C(14), in α-position to the indole N-atom, has the oxidation level of a ketone. Hence, vincamine might be in equilibrium with compound **60**, which, hardly stable in itself, can cyclize to yield 14-epivincamine (**61**), a compound that has also been isolated from *Vinca minor*, the lesser periwinkle (*Scheme 6.22*).

At the equilibrium, the concentration of compound **59** relative to **61** essentially depends on the absolute configuration at C(3) and C(16). However, meticulous stereochemical examination is not necessary here, since it is possible to prepare the intermediate **60** *via* the more stable compound **62**, in which the original α-oxo group is protected as an enol ether. The choice of **62** as a target molecule would largely prevent side reactions occurring at the oxo group during the course of the synthesis; also, its ring system is partly made up of a tetrahydro-β-carboline (*cf.* **63**), which is relatively easy to prepare.

Treatment of the *N*-acetyl model compound **67** with dehydrating reagents such as P_2O_5 or $POCl_3$ in benzene or tetralin results in a *Bischler–Napieralski*-type ring closure, which leads to **65** (*Scheme 6.23*). Subsequent imine reduction, carried out with $NaBH_4$, for example, proceeds smoothly providing **63** as a racemate. With compound **69** as the starting material, analogous treatment leads to the vincamine precursor **62**. Thereby, the reduction of the primary *Bischler–Napieralski* adduct with $NaBH_4$ proceeds preferentially from the less sterically hindered side; in other words: the configuration at

Scheme 6.23

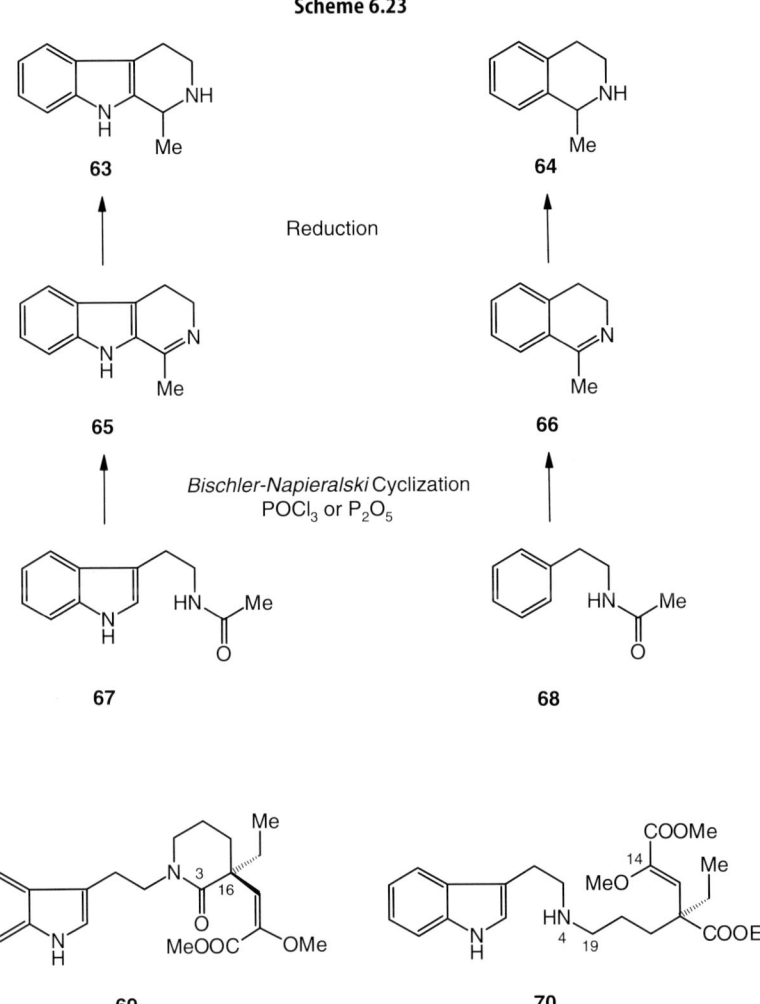

Reduction

Bischler-Napieralski Cyclization
POCl₃ or P₂O₅

63

64

65

66

67

68

69

70

Fig. 6.3. *Wolfgang Oppolzer* (August 4, 1937–March 15, 1996). Ph.D. from the Swiss Federal Institute of Technology (ETH Zurich) with *V. Prelog*, since 1975 professor of organic chemistry at the University of Geneva. Developed new synthetic methods that have found wide application in the stereoselective total syntheses of numerous natural products (terpenes, steroids, alkaloids).

C(3) is dictated by that at C(16). Since epimerization at C(16) is not possible in **69** (quaternary center), the 'open chain' structure of **70** is certainly worth consideration as a precursor. The ester is preferred to the free acid in this case, avoiding the usual preparative problems in dealing with free amino acids. The synthon **70** may be constructed out of commercially available tryptamine (**71**) and an ester such as **72**. Formation of the bond between N(4) and C(19) can, in this case, be accomplished by means of an S_N2 reaction with a suitable leaving group X. (Reductive alkylation of the amine, by way of the

Vincamine (**59**)

225

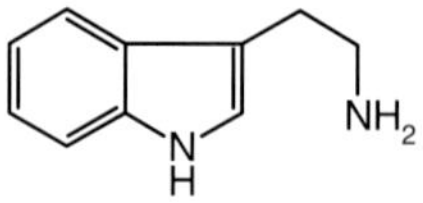

71

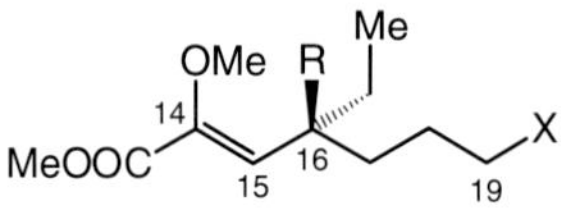

72

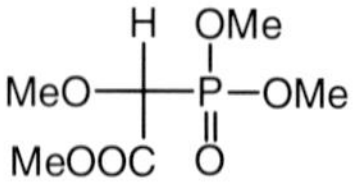

73

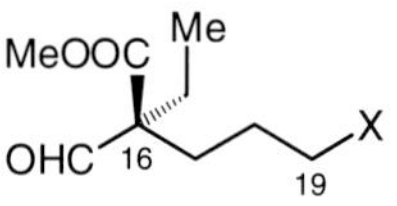

74

Schiff base formed between N(4) and a formyl group at C(19), would also be conceivable, but might pose unnecessary difficulties[42]). But how might the chiral building block **72** now be prepared? As a matter of fact, its structure can be simplified by cleaving the C(14)=C(15) bond, originally formed by means of a *Wittig–Horner* reaction. Thus, it should be possible to synthesize **72** by treatment of methyl (dimethylphosphono)(methoxy)acetate (**73**) with the formylated ester **74**. While compound **73** is commercially available, **74** needs to be further simplified: the side chain containing C(17), C(18), and C(19) may be introduced by alkylation, exploiting the CH acidity of **75**. Superficially, the use of a compound of the following type would seem suitable:

$$Y-C(17)H_2-C(18)H_2-C(19)H_2-X$$

That solution is not ideal though, since the treatment of **75** with such a bifunctional type of synthon would unavoidably result in the formation of 'bis-adducts'.

The preparation of **74**, thus, has to be performed over several steps with a single allyl residue attached as a blocking group, such as in **76**. The latter compound would be suitable also for optical resolution, since racemization at C(16) is no longer possible at this stage (*cf. Scheme 6.24*). Introduction of the leaving group X at C(19) can be finally effected by anti-*Markovnikov* functionalization of the C=C bond with HX.

Scheme 6.24

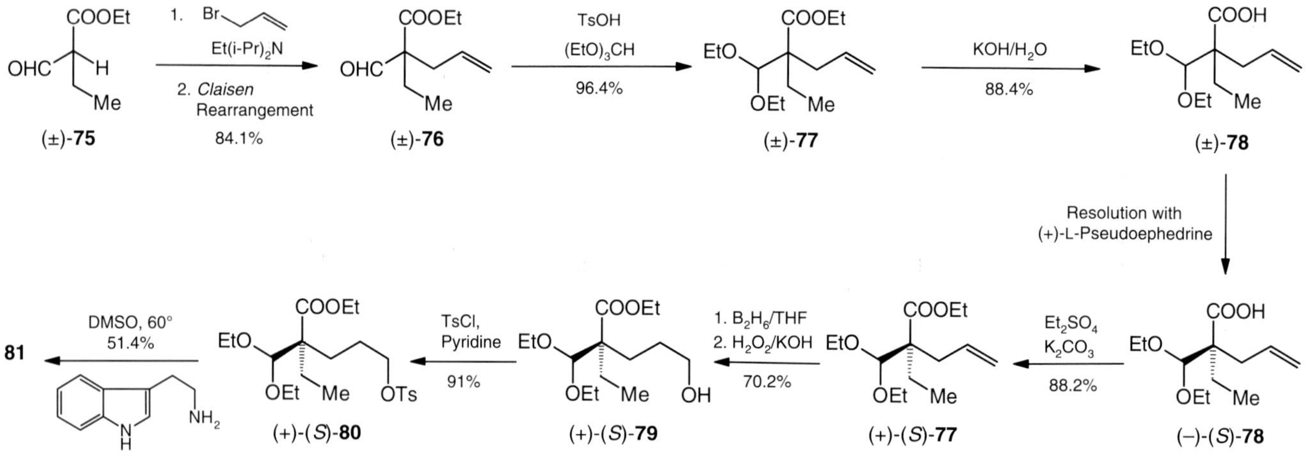

Conceptually, the overall synthetic pathway is complete now. However, one shortcoming that might easily lead to complications should be pointed out at this stage: during the conversion of **72** into **69**, a hydrolytic step will be required. If this proceeds under acidic conditions, then concomitant cleavage of the remaining ester group and hydrolysis of the enol ether are both very likely. It is, therefore, advisable to carry out the *Wittig–Horner* reaction at a stage at which C(3) is already lactamized.

6.6.2. Synthesis of Vincamine

The individual steps in the synthesis of vincamine (**59**) are depicted in *Schemes 6.24–6.26*. The starting material, racemic ethyl 2-formylbutanoate ((±)-**75**), is treated with allyl bromide and ethyldiisopropylamine (*Hünig*'s base), producing a mixture of *O*- and *C*-alkylated products that are directly converted to (±)-**76** by *in situ Claisen* rearrangement. Acetalization of the CHO group affords (±)-**77**, which is hydrolyzed under basic conditions to the carboxylic acid (±)-**78**. The latter turns out to be suitable for resolution with an optically active base such as L-pseudoephedrine, providing optically pure (–)-(*S*)-**78**. Alkylation with diethyl sulfate, in turn, allows (+)-(*S*)-**77** to be prepared, which is treated with diborane/H_2O_2 to afford the hydroxylated ester (+)-(*S*)-**79**. To transform the OH group into a suitable leaving group, the latter is tosylated to yield (+)-(*S*)-**80**, which, finally, upon treatment with tryptamine (**71**), furnishes (+)-(*S*)-**81**.

Interestingly, the conversion of **81** to **82** is successfully accomplished in the presence of imidazole at 130° C in the molten state (74% yield, *Scheme 6.25*). Next, the acetal group in (+)-(*S*)-**82** is cleaved upon treatment with concentrated acetic acid, producing **83**, which, after *Wittig–Horner* reaction with **73**, provids compound **69**. The latter is cyclized under *Bischler–Napieralski* conditions to yield **84**, which, upon $NaBH_4$ reduction, exclusively provids the isomer **85**, possessing the 'wrong' configuration at C(3). However, catalytic hydrogenation (H_2, Pd/C) under special reaction conditions (Et_3N plus an empirically determined solvent combination) finally affords the desired **62** as the major component with a ratio **62**/**85** of 81:19. Each of the three compounds **69**, **84**, and **62** are obtained as mixtures of the corresponding (*E/Z*)-isomers, analytical samples of which are separated only for the purposes of characterization. This makes sense insofar, as double-bond isomerism is irrelevant as far as the end product is concerned.

For the cyclization to vincamine (**59**), the masked oxo group in **62** (enol ether) needs to be deprotected first, which is only possible under drastic reaction conditions (HBr/glacial acetic acid), providing 81% yield of the de-

Scheme 6.25

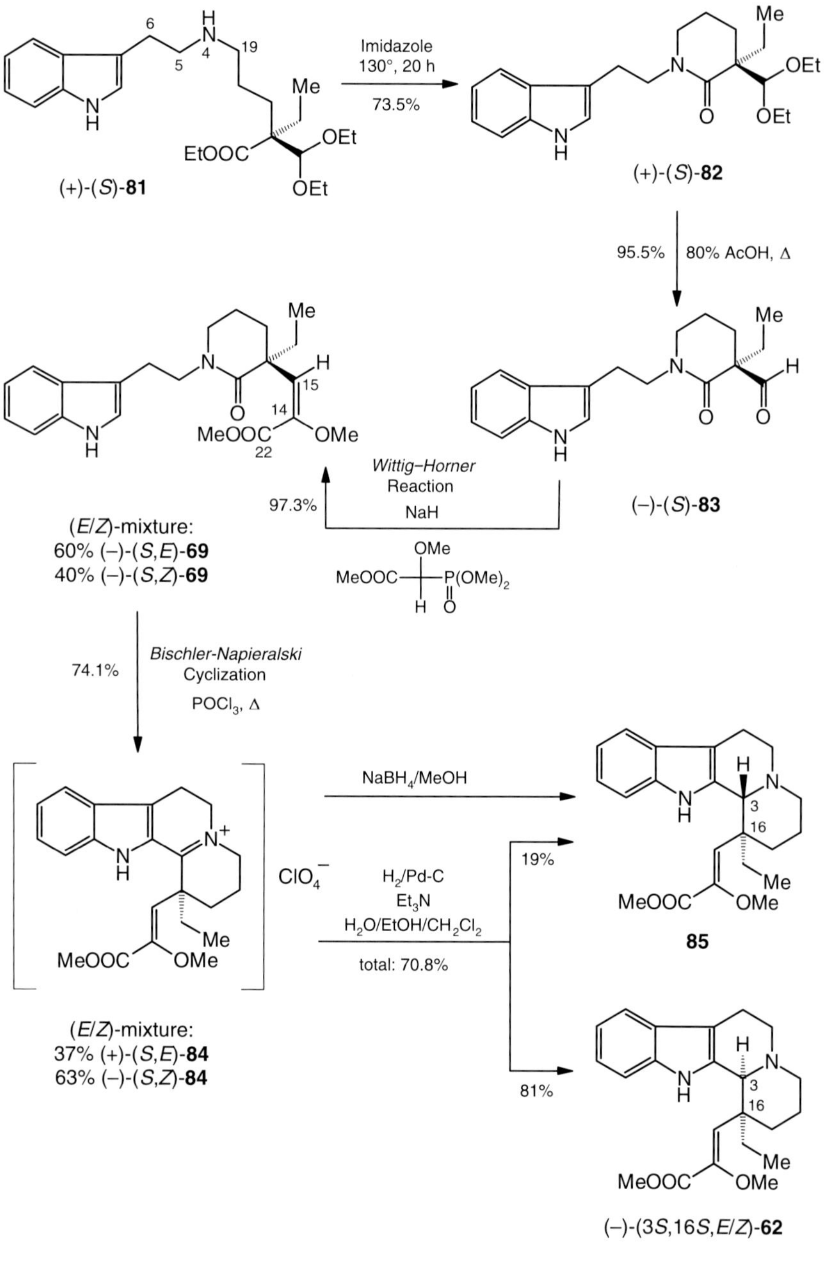

(+)-(S)-**81**

Imidazole
130°, 20 h

73.5%

(+)-(S)-**82**

95.5% | 80% AcOH, Δ

(E/Z)-mixture:
60% (−)-(S,E)-**69**
40% (−)-(S,Z)-**69**

Wittig–Horner
Reaction
NaH

97.3%

(−)-(S)-**83**

MeOOC—P(OMe)$_2$
H Ö
OMe

74.1%

Bischler-Napieralski
Cyclization
POCl$_3$, Δ

ClO$_4^-$

NaBH$_4$/MeOH

H$_2$/Pd-C
Et$_3$N
H$_2$O/EtOH/CH$_2$Cl$_2$

total: 70.8%

19%

85

(E/Z)-mixture:
37% (+)-(S,E)-**84**
63% (−)-(S,Z)-**84**

81%

(−)-(3S,16S,E/Z)-**62**

Scheme 6.26

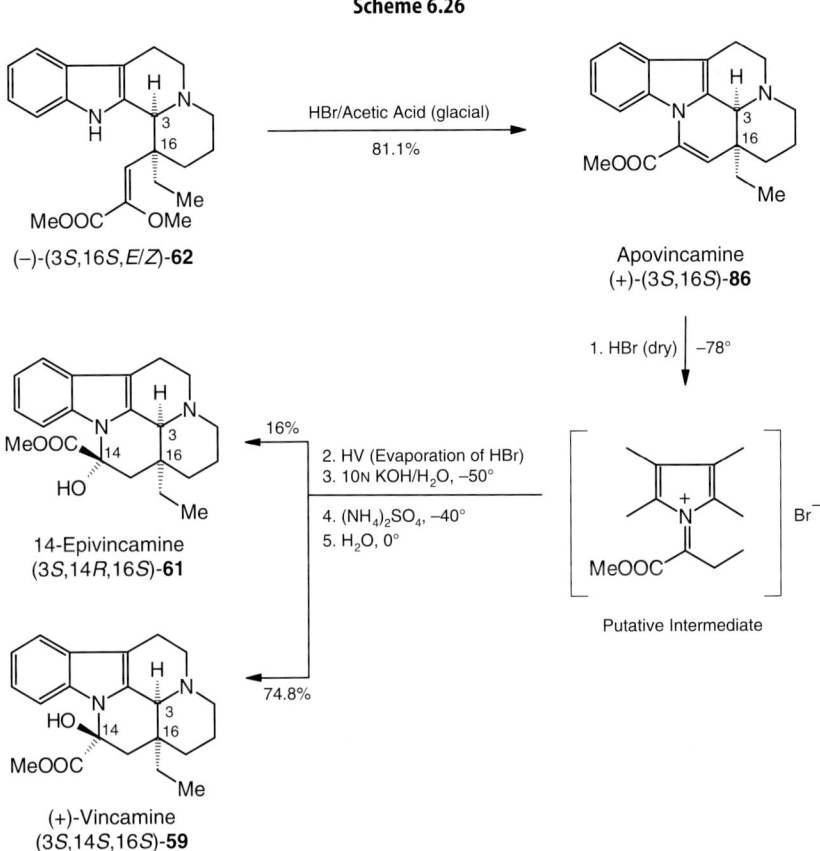

(−)-(3*S*,16*S*,*E*/*Z*)-**62**

HBr/Acetic Acid (glacial)

81.1%

Apovincamine
(+)-(3*S*,16*S*)-**86**

1. HBr (dry) │ −78°

MeOOC
HO
14-Epivincamine
(3*S*,14*R*,16*S*)-**61**

16%

2. HV (Evaporation of HBr)
3. 10N KOH/H₂O, −50°

4. (NH₄)₂SO₄, −40°
5. H₂O, 0°

Putative Intermediate

74.8%

(+)-Vincamine
(3*S*,14*S*,16*S*)-**59**

hydrated apovincamine ((+)-**86**), while the C(14) epimer of vincamine (**61**) is obtained only as a side-product (*Scheme 6.26*). The transformation of apo-vincamine ((+)-**86**) into vincamine (**59**) is finally achieved – clearly inspired by the preceding transformation – under fairly 'special' conditions, not nec-essarily applicable to other, analogous reactions: apovincamine is dissolved in anhydrous HBr at −78° C and dried at that temperature under high vacu-um. A 10N solution of KOH is then added at −50° C (an NaOH solution would be solid at this temperature), followed, with slow warming to −40° C, by (NH₄)₂SO₄. Addition of H₂O at *ca.* −5° C finally results in the formation of vincamine (**59**) together with the epimer **61** in a total yield of 91%. The ratio **59/61** is *ca.* 5:1.

Fig. 6.4. *Hans Eduard Schmid* (March 24, 1917 – December 19, 1976). Studies and Ph. D. in Vienna (*E. Späth*), 1947 assistant professor, 1959 full professor at the University of Zurich. A natural-product chemist, particularly notable with regard to the structural elucidation and syntheses of curare and other indole alkaloids (photo *Photo Peyer*, 1975).

As can clearly be seen from this example, it is best to perform the resolution of a racemate as early as possible in order to reduce the amount of substance needed – an advantage not to be underestimated in terms either of time or of cost.

6.7. Partial Synthesis of the Bisindole Alkaloid *Alloferin*®

Isolation of the dimeric curare alkaloids from natural sources (*e.g.*, from curare itself or from *Strychnos* species) for the preparation of medicinal products, unlike the *Catharanthus* alkaloids, is completely uneconomic (*cf. Chapt. 4.7*). Total synthesis would also mean excessively high costs. However, for the industrial production of the muscle relaxant *Alloferin*® (= diallylbis(nortoxiferine) dichloride[43,44]) a partial synthesis has been developed, starting from strychnine, readily obtained in high purity from *Strychnos nux-vomica*. The strategy consists of removing C(22) and C(23) of strychnine, followed by dimerization of the so-called *Wieland–Gumlich* aldehyde. The first step of the synthesis is based on the work of *Wieland*, *Gumlich*, and *Kaziro*, who discovered this degradation reaction in 1932/33 in the course of the structure elucidation of strychnine[45,46]. The methods used for the dimerization of the *Wieland–Gumlich* aldehyde and those for the preparation of *Alloferin*® were developed by *Schmid* (*Fig. 6.4*), *Karrer* (*Fig. 6.5*), and co-workers during their structural investigations of dimeric curare alkaloids.

6.7.1. Synthesis of *Alloferin*®

In the style of an ester condensation, first a NO group is introduced in α-position to the C=O group of strychnine under base catalysis, with amyl nitrite as a reagent (*Scheme 6.27*). This reaction, leading to isonitrosostrychnine, which exists mainly in the oxime form, proceeds in very high yield. A type-II *Beckmann* rearrangement with $SOCl_2$ provides the carbamic acid hydrochloride as an intermediate. Decarboxylation and nucleophilic replacement of HCN by an OH group finally affords the *Wieland–Gumlich* aldehyde, existing predominantly in the hemiacetal form, in moderate yield. Dimerization of the latter takes place in the presence of glacial acetic acid/ AcONa, with each of the two secondary amino groups reacting with one of the two formyl groups of a partner molecule. This gives rise to caracurine V, which also occurs naturally, and which can be isomerized to bis(nortoxiferine) in the presence of sulfuric acid (*Scheme 6.28*). Quaternization of the two peripheral (tertiary) amino groups with allyl chloride finally produces the diallylbis(nortoxiferine) dichloride. Reaction sequences in which the

Scheme 6.27

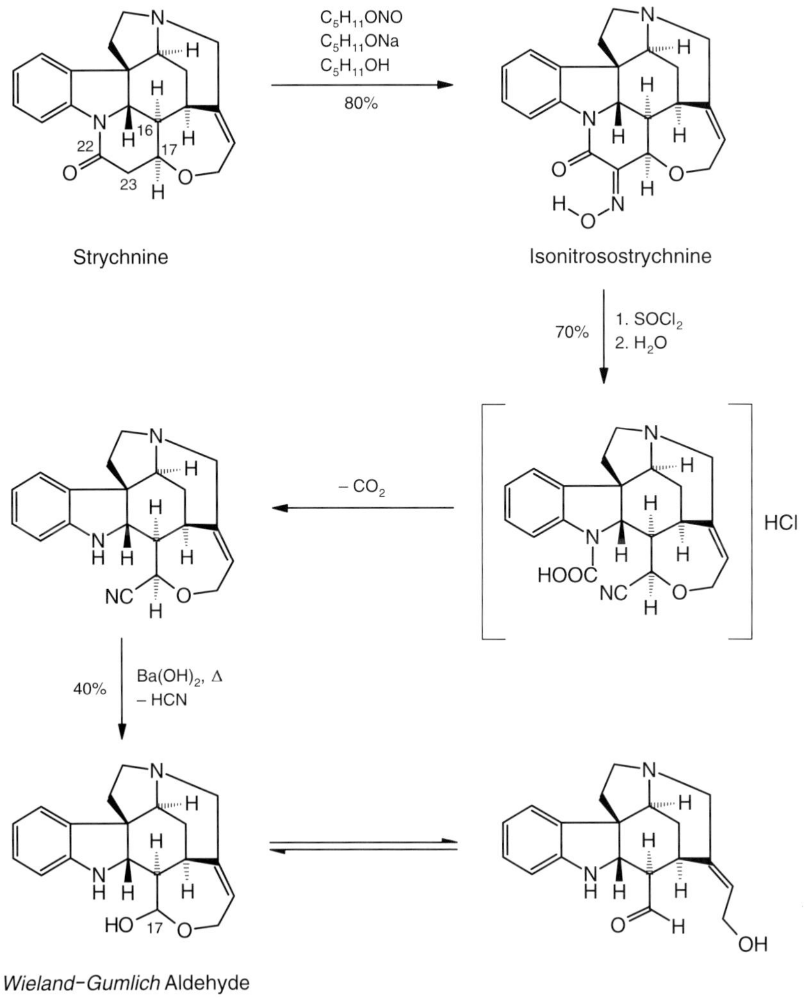

Strychnine

$C_5H_{11}ONO$
$C_5H_{11}ONa$
$C_5H_{11}OH$
80%

Isonitrosostrychnine

70% | 1. $SOCl_2$
2. H_2O

HCl

$- CO_2$

40% | $Ba(OH)_2$, Δ
$- HCN$

Wieland–Gumlich Aldehyde

Fig. 6.5. *Paul Karrer* (April 21, 1889 – June 18, 1971). Studied in Zurich and Frankfurt am Main 1912 – 1918 (*P. Ehrlich*), Ph. D. with *A. Werner*, since 1918 professor of organic chemistry at the University of Zurich, *Nobel* prize in 1937. Natural-product chemist (vitamins, carotenoids, flavines). Isolation and structural elucidation especially of the curare alkaloids (photo *R. Gfeller*, 1968).

Wieland-Gumlich aldehyde is already quaternized prior to the dimerization give the same result.

At this point it should be commented that the originally reported yields relate to small-scale, nonoptimized conditions. It is, thus, entirely plausible that significantly higher yields are obtained in the industrial partial synthesis of *Alloferin*®.

231

Scheme 6.28

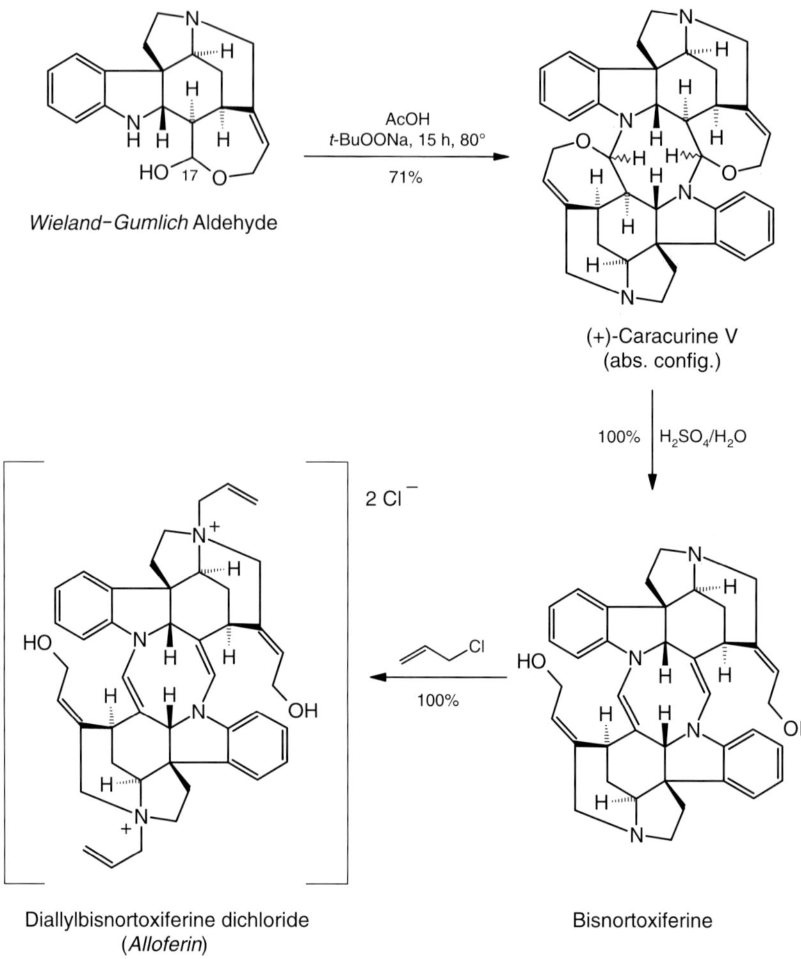

Wieland–Gumlich Aldehyde

AcOH
t-BuOONa, 15 h, 80°
71%

(+)-Caracurine V
(abs. config.)

100% | H_2SO_4/H_2O

Diallylbisnortoxiferine dichloride
(*Alloferin*)

2 Cl⁻

100%

Bisnortoxiferine

References and Notes

1. Ladenburg, '*Versuche zur Synthese des Coniins*', *Ber. Dtsch. Chem. Ges.* **1886**, *19*, 439.
2. 'Paraldehyde' is a trimeric acetaldehyde (($CH_3CHO)_3$) with the systematic name 2,4,6-trimethyl[1.3.5]trioxane. α-Picoline is the trivial name for 2-methylpyridine.
3. Ladenburg, '*Synthese der activen Coniine*', *Ber. Dtsch. Chem. Ges.* **1886**, *19*, 2578.
4. K. Bodendorf, W. Krieger, '*Über die Alkaloide von Mesembryanthemum tortuosum L.*', *Arch. Pharm.* **1957**, *290*, 441.

5. Popelak, G. Lettenbauer, E. Haack, H. Spingler, '*Die Struktur des Mesembrins und Mesembrinins*', *Naturwissenschaften* **1960**, *47*, 231.

6. The preparation of **7** has also been accomplished by other approaches[7,8].

7. S. L. Keely, F. C. Tahk, '*The 3-Arylpyrrolidine Alkaloid Synthon. A New Synthesis of dl-Mesembrine*', *J. Am. Chem. Soc.* **1968**, *90*, 5584.

8. R. V. Stevens, M. P. Wentland, '*Thermal Rearrangement of Cyclopropyl Imines. IV. Total Synthesis of dl-Mesembrine*', *J. Am. Chem. Soc.* **1968**, *90*, 5580.

9. T. J. Curphey, H. L. Kim, '*A New Synthetic Approach to the Amaryllidaceae Alkaloids. Application to the Synthesis of Mesembrine and Mesembrinine*', *Tetrahedron Lett.* **1968**, 1441.

10. W. A. Denne, S. R. Johns, J. A. Lamberton, A. McL. Mathieson, '*The Absolute Structure of Porantherine*', *Tetrahedron Lett.* **1971**, 3107; W. A. Denne, A. McL. Mathieson, '*Poranthera corymbosa Alkaloids. I. Crystal and Absolute Molecular Structure of Porantherine Hydrobromide*', *J. Cryst. Mol. Struct.* **1973**, *3*, 79.

11. W. A. Denne, S. R. Johns, J. A. Lamberton, A. McL. Mathieson, H. Suares, '*The Molecular Structure and Absolute Configuration of Poranthericine and Porantheridine*', *Tetrahedron Lett.* **1972**, 1767.

12. W. A. Denne, A. McL. Mathieson, '*Poranthera corymbosa Alkaloids. II. Molecular Structure and Absolute Configuration of Porantheridine Hydrobromide*', *J. Cryst. Mol. Struct.* **1973**, *3*, 87.

13. W. A. Denne, A. McL. Mathieson, '*Poranthera corymbosa Alkaloids. III. Molecular Structure and Absolute Configuration of Porantheridine Hydrobromide*', *J. Cryst. Mol. Struct.* **1973**, *3*, 139.

14. W. A. Denne, '*Poranthera corymbosa Alkaloids. IV. Crystal and Absolute Structure of 4-Methyl-6-(2'-benzoyloxypentyl)quinolizidine*', *J. Cryst. Mol. Struct.* **1973**, *3*, 367.

15. S. R. Johns, J. A. Lamberton, A. A. Sioumis, H. Suares, '*The Alkaloids of Poranthera corymbosa (Euphorbiaceae)*', *Austral. J. Chem.* **1974**, *27*, 2025.

16. B. Tursch, D. Daloze, C. Hootele, '*The Alkaloid of Propylaea quatuordecimpunctata L. (Coleoptera, Coccinellidae)*', *Chimia* **1972**, *26*, 74.

17. B. Tursch, D. Daloze, M. Dupont, C. Hootele, M. Kaisin, J. M. Pasteels, D. Zimmermann, '*Coccinelline, the Defensive Alkaloid of the Beetle Coccinella septempunctata*', *Chimia* **1971**, *25*, 307.

18. B. Tursch, D. Daloze, J. C. Braekman, C. Hootele, A. Cravador, D. Losman, R. Karlson, '*Chemical Ecology of Arthropods, IX. Structure and Absolute Configuration of Hippodamine and Convergine, Two Novel Alkaloids from the American Ladybug Hippodamia convergens (Coleoptera – Coccinellidae)*', *Tetrahedron Lett.* **1974**, 409.

19. The locants used here are those prescribed for the porantherine system.

20. E. J. Corey, R. D. Balanson, '*A Total Synthesis of (±)-Porantherine*', *J. Am. Chem. Soc.* **1974**, *96*, 6516.

21. Active against *sarcoma 180*, *Lewis* lung carcinoma, *L-1210* and *P-388* leukemia in mice, and *Walker 256* intramuscular carcinoma in rats.

22. M. Suffness, G. A. Cordell, '*Antitumor Alkaloids*', *The Alkaloids* **1985**, *25*, 1.

23. S. M. Kupchan, H. P. J. Hintz, R. M. Smith, A. Karim, M. W. Cass, W. A. Court, M. Yatagai, '*Macrocyclic Spermidine Alkaloids from Maytenus serrata and Tripterygium wilfordii*', *J. Org. Chem.* **1977**, *42*, 3660; S. M. Kupchan, Y. Komoda, W. A. Court, G. J. Thomas, R. M. Smith, A. Karim, C. J. Gilmore, R. C. Haltiwanger, R. F. Bryan, '*Maytansine, a Novel Antileukemic Ansa Macrolide from Maytenus ovatus*', *J. Am. Chem. Soc.* **1972**, *94*, 1354.

24. Celabenzine (from *Tripterygium wilfordii* (Celastraceae)) had already been isolated and structurally elucidated.

25. H. Wagner, J. Burghart, W. A. Hull, '*Neue Macrocyclische Spermidinalkaloide aus Maytenus mossambicensis (KLOTZSCH) BLAKELOCK*', *Tetrahedron Lett.* **1978**, 3893; H. Wagner, J. Burghart, '*Macrocyclische Spermidinalkaloide aus Maytenus mossambicensis (KLOTZSCH) BLAKELOCK*', *Helv. Chim. Acta* **1982**, *65*, 739.

26. K. Schultz, M. Hesse, '*Total Synthesis of (+)-(8S,13R)-Cyclocelabenzine*', *Helv. Chim. Acta* **1996**, *79*, 1295.

27. Compounds **37** and **38** have largely been found in leaves and stems, which suggests a photochemical origin.

28. Systematic name of (+)-cyclocelabenzine: (+)-(8*S*,13*R*)-4,5,6,7,8,9,12,13-octahydro-8-phenyl-2,13-methano-2*H*-[2,7,11]benzotriazacyclopentadecine-1,10(3*H*,11*H*)-dione.

29. Hydroxy-, methoxy-, and methylenedioxy-substituted arenes are especially well suited.

30. Systematic name of (+)-isocyclocelabenzine: (+)-(9*S*,13*S*)-3,4,5,6,8,9,10,11,12,13-decahydro-9-phenyl-2,13-methano-2*H*-[2,6,10]benzotriazacyclopentadecine-1,7-dione.

31. K. Schultz, P. Kuehne, U. A. Häusermann, M. Hesse, '*The Absolute Configuration of Macrocyclic Spermidine Alkaloids*', *Chirality* **1997**, *9*, 523.

32. Vincamine possesses the (3*S*,14*S*,16*S*)-configuration. Its specific optical rotation, $[\alpha]_D$, amounts to +41 as measured in pyridine (*cf. Chapt. 5.4.2*).

33. K. H. Gibson, J. E. Saxton, '*Total Synthesis of (±)-Vincamine*', *J. Chem. Soc., Chem. Commun.* **1969**, 1490.

34. M. E. Kuehne, '*The Total Synthesis of Vincamine*', *Lloydia* **1964**, *27*, 435; *J. Am. Chem. Soc.* **1964**, *86*, 2946.

35. J. L. Herrmann, R. J. Cregge, J. E. Richman, C. L. Semmelhack, R. H. Schlessinger, '*A High Yield Stereospecific Total Synthesis of Vincamine*', *J. Am. Chem. Soc.* **1974**, *96*, 3702.

36. G. Hugel, J. Lévy, J. Le Men, '*Méthylène-Indolines, Indolénines, et Indoléniniums VI. Action de Réactif Oxidants. Hemisynthèse de la Vincamine*', *Compt. Rend. Acad. Sci.* **1972**, *274*, 1350.

37. C. Szántay, L. Szabó, G. Kalaus, '*Stereoselective Total Synthesis of (+)-Vincamine*', *Tetrahedron Lett.* **1973**, 191.

38. C. Thal, T. Sevenet, H.-P. Husson, P. Potier, '*Synthèses de la (±)-deséthylvincamine*', *Compt. Rend. Acad. Sci.* **1972**, *275 C*, 1295.

39. E. Wenkert, B. Wickberg, '*General Methods of Synthesis of Indole Alkaloids, IV. A Synthesis of dl-Eburnamonine*', *J. Am. Chem. Soc.* **1965**, *87*, 1580.

40. P. Pfäffli, W. Oppolzer, R. Wenger, H. Hauth, '*Stereoselektive Synthese von optisch aktivem Vincamin*', *Helv. Chim. Acta* **1975**, *58*, 1131.

41. The vincamine numbering system is used here.

42. It would be extremely difficult to selectively modify only one out of the two C=O groups at C(14) and C(19).

43. Roche, *Alloferin*, F. Hoffmann-La-Roche AG, Basel, in-house publication.

44. *Alloferin*® is a medium duration depolarization inhibitor muscle relaxant (onset *ca.* 2–3 minutes after intravenous injection and lasting *ca.* 30 minutes), which causes total or partial paralysis of skeletal muscle. A competitive acetylcholine agonist, it blocks the acetylcholine receptor that normally modulates signal transfer at the synapse. In this way, the flow of sodium ions into the cell, leading to depolarization, is suppressed, and with it signal transduction to striated muscle fibers. The relative sensitivity of individual muscle groups towards relaxants decreases in the following order: Mm. oculomotorii > lid muscles > facial muscles > finger flexors > tongue and pharynx muscles > jaw muscles > greater extremity muscles > pelvic floor > lower abdominal musculature > upper abdominal musculature > intercostal and auxiliary respiratory musculature > larynx musculature > diaphragm.

The sequence in which muscle paralysis takes effect follows the above order and ceases, correspondingly, in reversed order. Muscle relaxants are used in operations to inhibit involuntary muscle spasm. The use of *Alloferin*® has proved extremely helpful especially in surgery on limbs, the abdomen, the thorax, and the brain, in gynecology (including cesarean section), and in ophthalmology.

45. H. Wieland, W. Gumlich, '*Über einige Neue Reaktionen der Strychnos-Alkaloide. XL*', *Liebigs Ann. Chem.* **1932**, *494*, 191.
46. H. Wieland, K. Kaziro, '*Abbauversuche vom Isonitrosostrychnin aus. Über Strychnos-Alkaloide. XIII*', *Liebigs Ann. Chem.* **1933**, *506*, 60.

7. Alkaloids and Chemotaxonomy

7.1 General

The goal of chemotaxonomy is the drawing up of phylogenetic trees for the classification of plants (or other living organisms) on the basis of their component chemical substances. Chemotaxonomy is, therefore, a complementary science of systematic botany, together with morphology, anatomy, cytology, and molecular biology. The findings of chemotaxonomic investigations should, of course, concur with the other features used in classification in the plant kingdom. Hence, chemotaxonomy always has the opportunity to prove its merits if the other criteria of classical systematic grouping contradict one another. There have been cases known in which the correct assignment of a particular plant has only been possible with the aid of chemotaxonomy.

Before an overview of chemotaxonomic analysis with specific examples is given, however, the exact meaning of 'component chemical substance' has to be made clear, especially in the light of natural products isolation.

In addition to alkaloids, the component chemical substances of a plant also encompass all the other plant 'ingredients', such as flavonoids, coumarins, terpenes, saccharides, biogenetic amines *etc*. This is, of course, an enormous range of diverse chemical substances. There are known examples – gone into more deeply in *Chapt. 8.1* – in which the biosynthesis of one and the same component substance can take place in two different ways. It is, therefore, quite understandable that classification can sometimes be assisted not only by knowledge of the chemical substances in themselves, but also by information about the biogenesis of these natural products in the plant concerned.

Thanks to the huge number of plants on Earth, it is impossible, both for individual workers and for groups, to carry out comprehensive investigations into all component substances, be that in the best interests of the subject or not. Consequently, it is necessary to rely on reports from the literature, which is not wholly unproblematic. The main drawback is that usually only 'positive' results are published, or, put another way, it is only reported that a particular compound has been isolated from a plant, and not that, during the course of examination of a given plant, *no* component substances *of a particular sort* could be detected. This, however, is precisely the information that would greatly assist the meaningfulness of chemotaxonomic studies.

Fig. 7.1. *Catharanthus roseus* G. Don (*Vinca rosea*) (from C. Rätsch, *Enzyklopädie der psychoaktiven Stoffe*, AT Verlag, Aarau, 1998)

Verification of whether a given substance is present or absent in a plant is often dependent on the quantity of plant material available. It is likely that trace compounds might not be detected at all, when only small quantities of plant materials are processed. If, though, plants under investigation are available in large quantities – as is the case, for example, with the Apocynaceae *Catharanthus roseus* (*Vinca rosea*; *Fig. 7.1*), from which the active antileukemic bisindole alkaloids vinblastine and vincristine are obtained – then even trace compounds may be isolated and structurally elucidated.

Quantitative aspects aside, still other factors play a role in the evaluation of plant component substances. The substance content can sometimes vary profoundly in different parts of a plant – bark, leaves, roots, blossoms *etc.* Also, many original studies lack precise information about which part of a given plant was used. Even more rarely are details stated of the time of year during which the plants were gathered, or the stage of development they were at. This information is also important, since the substance spectrum of a plant alters during the course of its vegetation period (*cf. Chapt. 9.7*). In conclusion: reported descriptions of the chemical characteristics of isolated (basic) component substances are usually not sufficient for chemotaxonomic ends.

If only small quantities of plant material are available, it is often impossible to characterize all the compounds isolated, which, at the same time, means that investigations of this type are only of very limited use chemotaxonomically. Comparative chromatography and mass spectrometry alone are, as a rule, not enough to guarantee the identity of two substances. Consequently, declaration of chiroptical properties (such as $[\alpha]_D$ values or, better, CD spectra) is essential for positive identification. Examination of the chiroptical properties of a compound is, in turn, only possible if very pure samples are at hand. The slightest impurity (in the form of a strongly rotating substance) in a sample of a weakly optically active alkaloid may lead to fatally wrong conclusions (*cf. Chapt. 5*).

If it is necessary to rely on literature findings regarding the isolation of a component plant substance, then one has to assume that the plant involved has been systematized correctly: checking the result is either very difficult or impossible for third parties. Furthermore, errors can also creep in thanks to carelessness during the isolation itself, which happens much more commonly than generally acknowledged. Several examples on this theme can be found in *Chapt. 4*.

7.2. Chemotaxonomy and Biogenesis of Indole Alkaloids

7.2.1. Introduction

Having now briefly outlined the fundamentals of chemotaxonomy, it is appropriate to examine a special group of alkaloids more closely through the chemotaxonomic lens. The following observations, though, also rely solely on 'positive' findings from plant investigations, since – as commented above – only these generally tend to be reported in the specialist literature[1].

The indole alkaloids represent the biggest single class of all alkaloids. There are *ca.* 1,500 currently known compounds that have appeared *ca.* 5,000 times in the literature, whether in the contexts of their isolation, their chemical identification or modification. As far as their constitution is concerned, it is possible to make a distinction between two large groups: those alkaloids that derive from tryptamine (or tryptophan) and incorporate a C_9- or C_{10}-monoterpene moiety, and those that incorporate the same base component(s) but possess some other (nonbasic) moiety other than a C_9- or C_{10}-monoterpene unit. Compounds such as simple carbazole derivatives, ergot and *Aristotelia* alkaloids, or alkaloids of the eserine type all belong to the latter group (*cf. Chapt. 2*).

The major group of substances under investigation is the one with a C_9- or C_{10}-monoterpene moiety joined to tryptamine. The monoterpene unit usually is (–)-secologanin or a derivative of it, in other words: an iridoid derived biogenetically from geraniol (*Scheme 7.1*). Geraniol itself is composed of two isoprene units.

(–)-Secologanin, the production of which takes place through a number of sequential, stereospecific reactions, occurs in plants only in the (–)-rotatory form. It constitutes a polyfunctional compound with three (partially protected) aldehyde groups, one carboxylic acid moiety, and a terminal C=C bond (*Scheme 7.1*).

As mentioned above, of the three aldehyde groups in secologanin, two are protected. Accordingly, only one CHO group (at C(6)) is capable to react spontaneously. Nevertheless, a total of five functional groups in a relatively small compound open up a wealth of possible reactions for secologanin and its 'hydrolysis' product shown. In the latter, since there is a β-oxo-carboxylic acid moiety present, it can easily be imagined that decarboxylation (extrusion of $C(10)O_2$) might take place under suitable reaction conditions. The existence of both C_9- and C_{10}-monoterpene components in indole alkaloids depends on whether or not such a decarboxylation has oc-

(–)-Secologanin
(abs. config.)

Reaction Equivalent
of Secologanin

Scheme 7.1. *Putative Biogenesis of the C_{10}-Monoterpene Moiety (–)-Secologanin from Geraniol*[2]. *All configurations shown are absolute.*

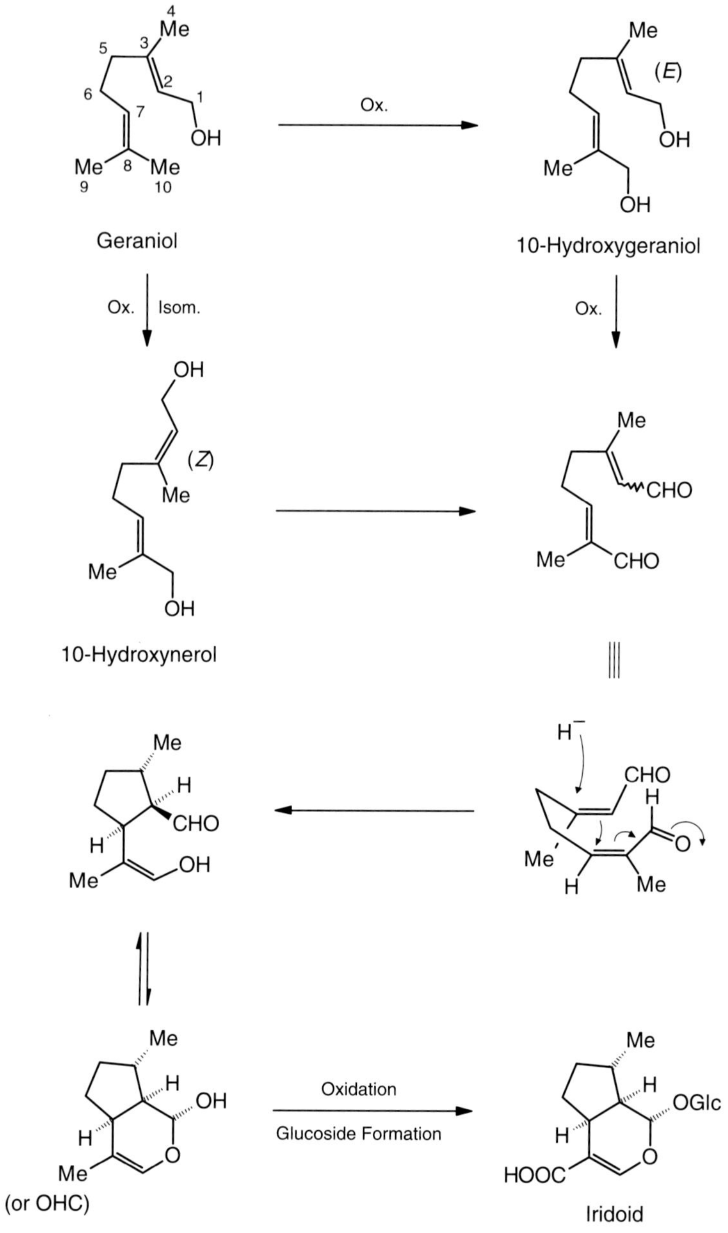

Geraniol

10-Hydroxygeraniol

10-Hydroxynerol

Iridoid

Scheme 7.1 (*cont.*)

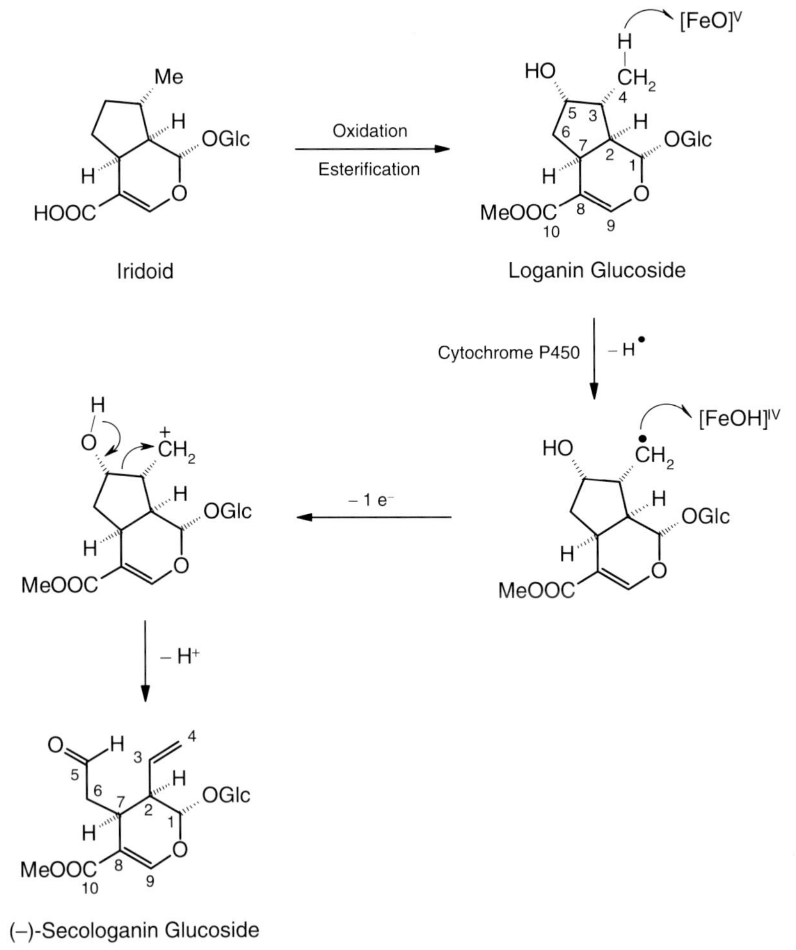

Iridoid

Loganin Glucoside

Oxidation
Esterification

Cytochrome P450 | − H•

[FeO]^V

[FeOH]^{IV}

− 1 e⁻

− H⁺

(−)-Secologanin Glucoside

curred, but biogenetically the two entities may be treated as effectively the same.

Lastly, one additional stereochemical aspect of hydrolyzed (−)-secologanin should be noted in passing. The compound possesses three stereogenic centers, *i.e.*, C(2), C(7), and C(8). Centers C(2) and C(8) are both in α-positions to C=O groups and so are prone to epimerization under acidic or basic conditions. Center C(7), however, can be regarded as stable.

7.2.2. Structure Types

Indole alkaloids with C_9- and C_{10}-monoterpene components may be structurally subdivided into eight main skeleton types shown in *Scheme 7.2*. An example of a typical representative of each type is given, together with an explanation of the structure of the non-tryptamine component and its relation

Scheme 7.2. *The Main Ring System Types* (C, D, V, S, A, E, P, J) *of Indole Alkaloids Consisting of Tryptamine* (or Tryptophan) *and a C_9- and C_{10}-Monoterpene Component* (Secologanin)

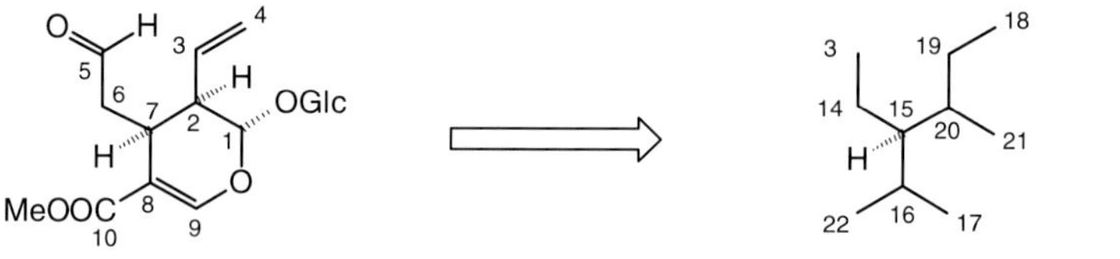

Secologanin Numbering   Indole-Alkaloid Numbering

Main Skeleton Type	Specific Example	Skeleton of the C_{10}-Monoterpene Unit
Corynanthean Type 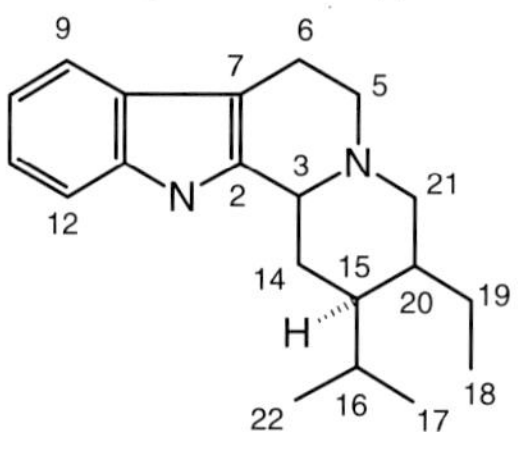 C-Type	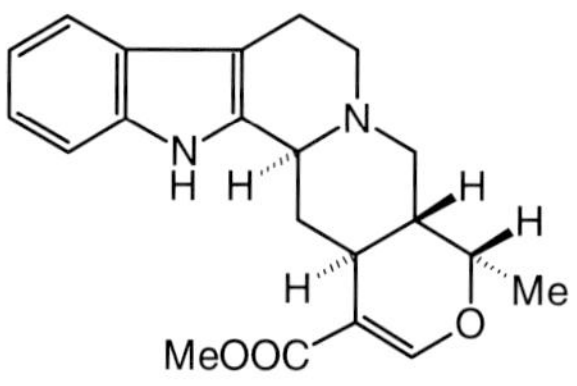 Ajmalicine (**C4a**)	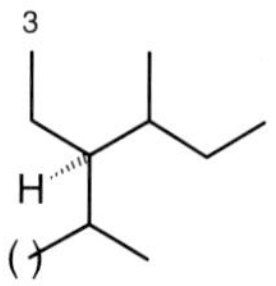 (?)
Vincosan Type 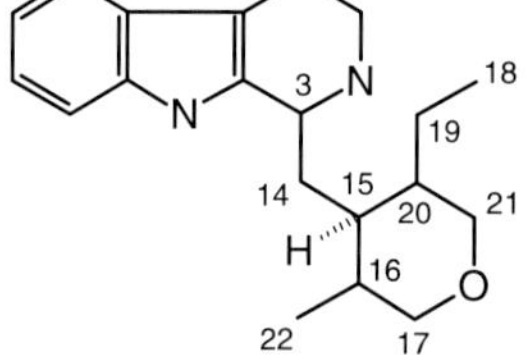 D-Type	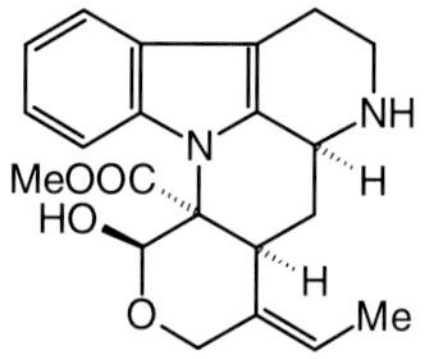 Talbotine (**D4f**)	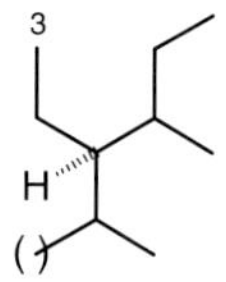 (?)

Scheme 7.2 *(cont.)*

Main Skeleton Type	Specific Example	Skeleton of the C_{10}-Monoterpene Unit
Vallesiachotaman Type 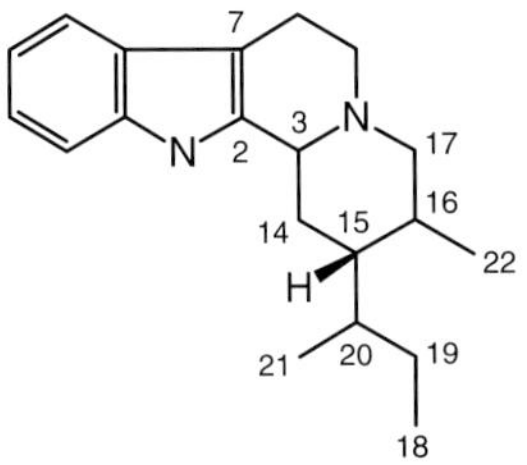 V-Type	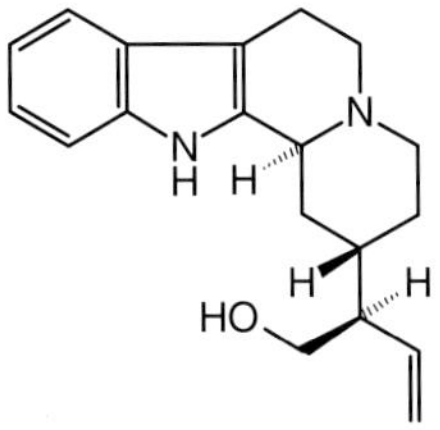 Antirhine (**V3**)	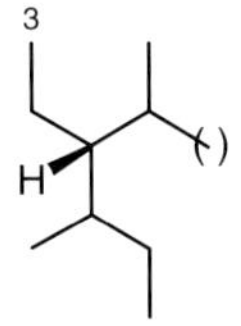
Strychnan Type 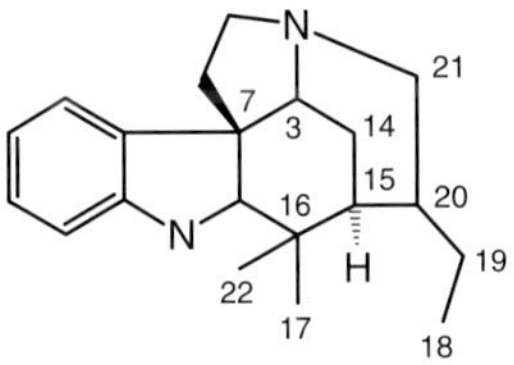 S-Type	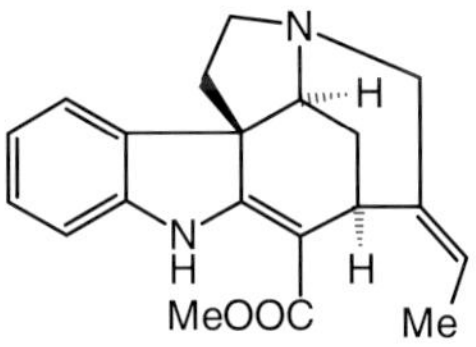 Akuammicine (**S4**)	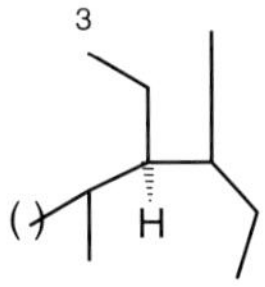
Aspidospermatan Type 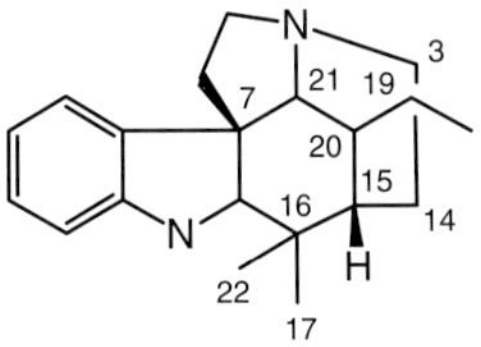 A-Type	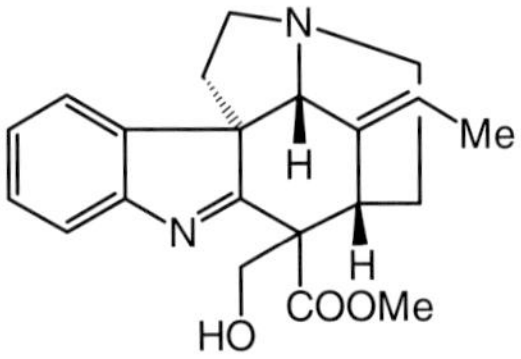 Precondylocarpine (**A4b**)	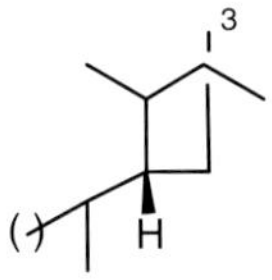

Scheme 7.2 (*cont.*)

Main Skeleton Type	Specific Example	Skeleton of the C_{10}-Monoterpene Unit
Eburnan Type 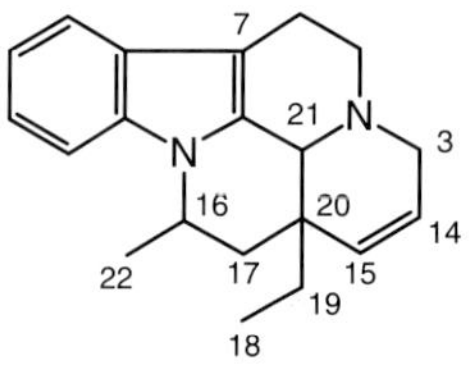 E-Type	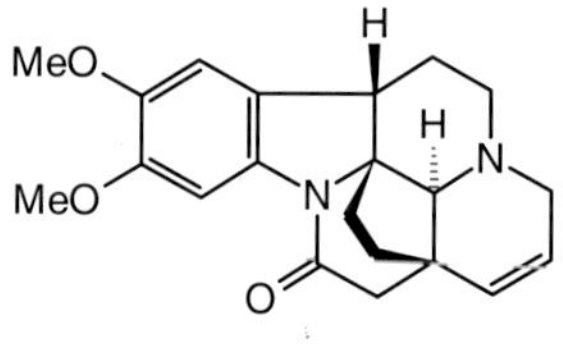 Schizogamine (**E6a**)	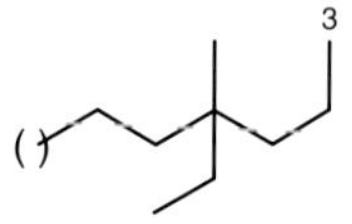
Plumeran Type 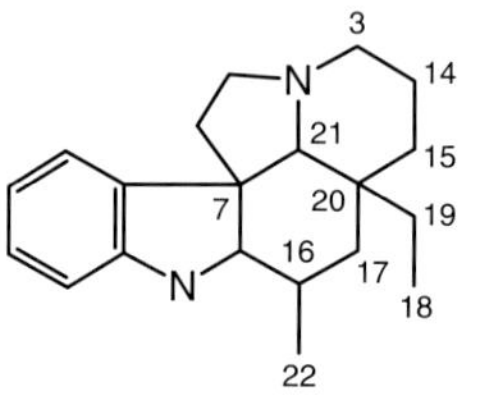 P-Type	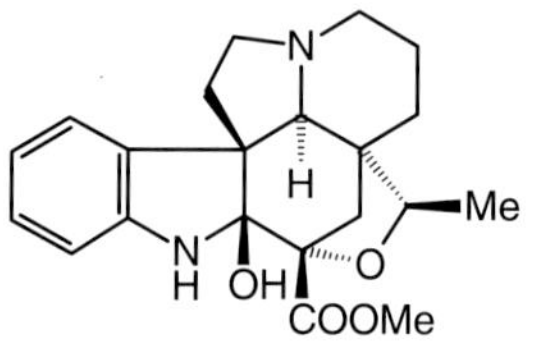 Melobaline (**P6f**)	
Ibogan Type 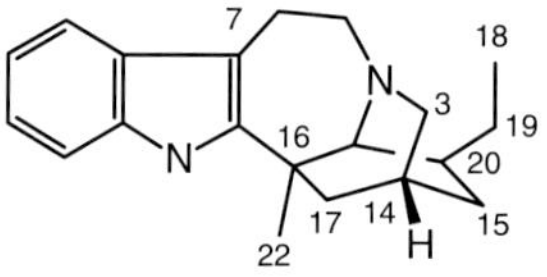 J-Type	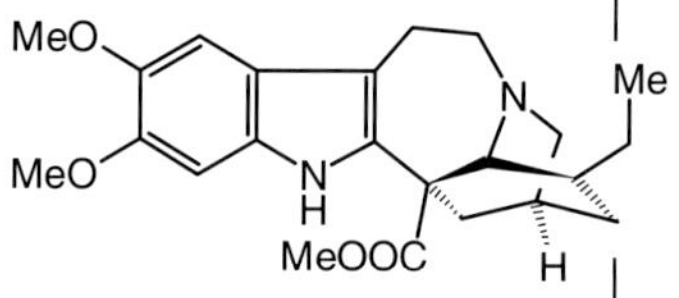 Conopharyngine (**J7a**)	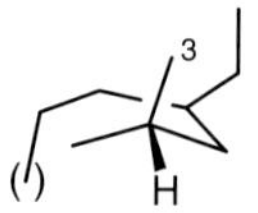

to (−)-secologanin. A comparison of these eight monoterpene skeleton types shows that the pattern of the C-atoms is identical in the first five representatives. Differences only arise from additional O- and N-bonds in the alkaloid. These five monoterpene residues can easily be interconverted by rotation about C–C bonds.

The first important chemotaxonomic point now follows. The absolute configurations at C(15) (or at C(7) by secologanin numbering) of all currently known indole alkaloids of the indicated structures – corynanthean (C), vincosan (D), vallesiachotaman (V), strychnan (S), and aspidospermatan (A) – are always the same and identical to that in (−)-secologanin. This implies that the same enzyme is involved in the synthesis of all these compounds. The above structural types are, therefore, known as the 'alkaloids with nonrearranged secologanin skeleton'. The remaining three types contain structurally diverse C_9- or C_{10}-monoterpene components, and in all these cases, the C(15)-atom is no longer a stereogenic center[3]. Consequently, stereochemical arguments in these three classes are more complicated, as we shall see later.

Thanks to findings and deductions from biogenetic experiments, the corynanthean-type atom-numbering system is also used for all other structural types (biogenetic atom numbering). Atom C(3) is indicated in each case. Atom C(22) (in brackets) is sometimes lost through decarboxylation. The meaning of the parenthetic expressions (in bold) after the names of specific alkaloids will be explained later.

7.2.3. Plant Families

Indole alkaloids with C_9- or C_{10}-monoterpene components occur in three different plant families: the Loganiaceae, the Rubiaceae, and the Apocynaceae. There are very few publications in which the occurrence of C-type indole alkaloids in other plant families is reported, and these exceptions amount to less than 0.1% compared to the main families. Whether the botanical classification of those plants was performed correctly in these few cases is open to question.

Botanically, Loganiaceae, Rubiaceae, and Apocynaceae are three closely related families in which the Rubiaceae and Apocynaceae, viewed in developmental biology terms, have emerged from the Loganiaceae.

The three plant families are subdivided further: the Apocynaceae into the three subfamilies Plumerioideae, Cerberioideae, and Echitoideae (or Apo-

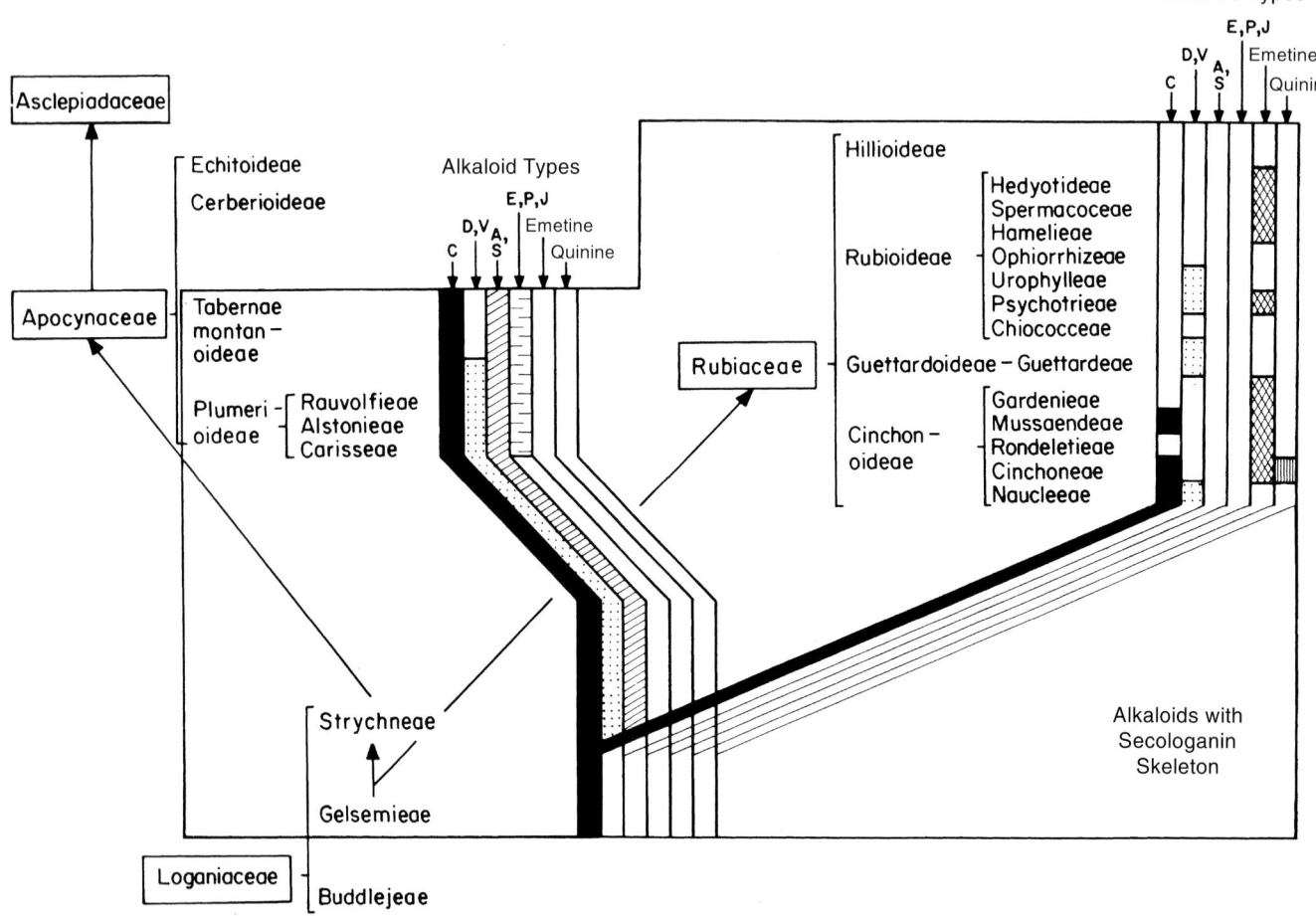

Fig. 7.2. *Botanical classification of the plant families Apocynaceae, Loganiaceae, and Rubiaceae on the basis of chemotaxonomic criteria involving the distribution of indole alkaloids with C_9- or C_{10}-monoterpene components.* The eight alkaloid classes examined are Corynanthean (C), Vincosan (D), Vallesiachotaman (V), Strychnan (S), and Aspidospermatan (A), all with *nonrearranged* (–)-secologanin components, plus Eburnan (E), Plumeran (P), and Ibogan (J) with *rearranged* (–)-secologanin skeletons.

cynoideae), the Rubiaceae into the four subfamilies Cinchonoideae, Guettardoideae, Rubioideae, and Hillioideae. The Loganiaceae are subdivided not into subfamilies, but into so-called *tribus*. The same applies for all subfamilies of the other two families, as follows from the botanical-morphological and developmental-biological subclassification shown in *Fig. 7.2*. The tribus Strychneae (Loganiaceae) is the direct precursor of the Apocynaceae or – put more precisely – the Plumerioideae subfamily. The Rubiaceae family, which represents a very large plant family, difficult to subclassify, branches off from the Gelsemieae tribus.

7.2.4. Structural Hierarchy of Alkaloids

Unquestionably, not all main skeleton types are equal in biogenetic terms; on the contrary, they often differ significantly. For example, the products of the coupling of tryptamine and secologanin, α- and β-vincoside (D-type), are closer to the 'start' than iboluteine (J-type), featuring the same components, but being the product of several skeletal rearrangements and oxidations. In the following discussion of such transformations, these processes will be termed *stages*. 3α-Vincoside is assigned the stage designator **D3**, iboluteine **J8**. This denotes that five different transformations are required to go from **D3** to **J8**.

Viewed in this way, iboluteine is thus further away from the 'start' than 3α-vincoside, because more enzyme systems are needed for its production. In a 'more highly developed' plant, these enzymes *have to be* present. As a rule, such plants are always capable of also synthesizing simpler alkaloid types. Their production might also be caused by other factors, though, such as the seasonally dependent disappearance of certain enzymes or their extreme low concentration *in vivo*. The skeletal rearrangement reactions are summarized in *Scheme 7.3*. The (hypothetical) precursor **1** is converted through ring closure and rearrangement reactions into **2a**, this into **3a**, and this further into **P4**. (Structure designators without capital letters denote that no natural representative alkaloid with the particular skeleton is yet known).

From **P4**, the J-type ring system – more precisely, **J7** – is reached in three steps. The ring system of iboluteine is modified from the **J7** template as the result of an oxidative rearrangement, and so it is assigned the higher hierarchical designator **J8**. (Lower case letters a, b *etc.* serve only to differentiate between hierarchically equivalent alkaloid skeletons).

According to the principle outlined, it is now possible to classify all indole alkaloids with C_9- or C_{10}-monoterpene components in chemotaxonomic rank. Assignment of rank numbers to specific alkaloid examples is shown in *Scheme 7.3*. The numbers indicate, as it were, the distance between each particular alkaloid skeleton and the hypothetical starting compound **1**, but do not characterize the structure completely (*e.g.*, MeO groups and *O*- or *N*-substituents are disregarded).

3α-Vincoside
Type **D3**

Iboluteine
Type **J8**

Scheme 7.3. *Biogenetic Interrelationship of the Eight Main Skeleton Types of Indole Alkaloids with C₉- or C₁₀-Monoterpene Components (cf. Scheme 7.2 for the meanings of individual capital letters)*

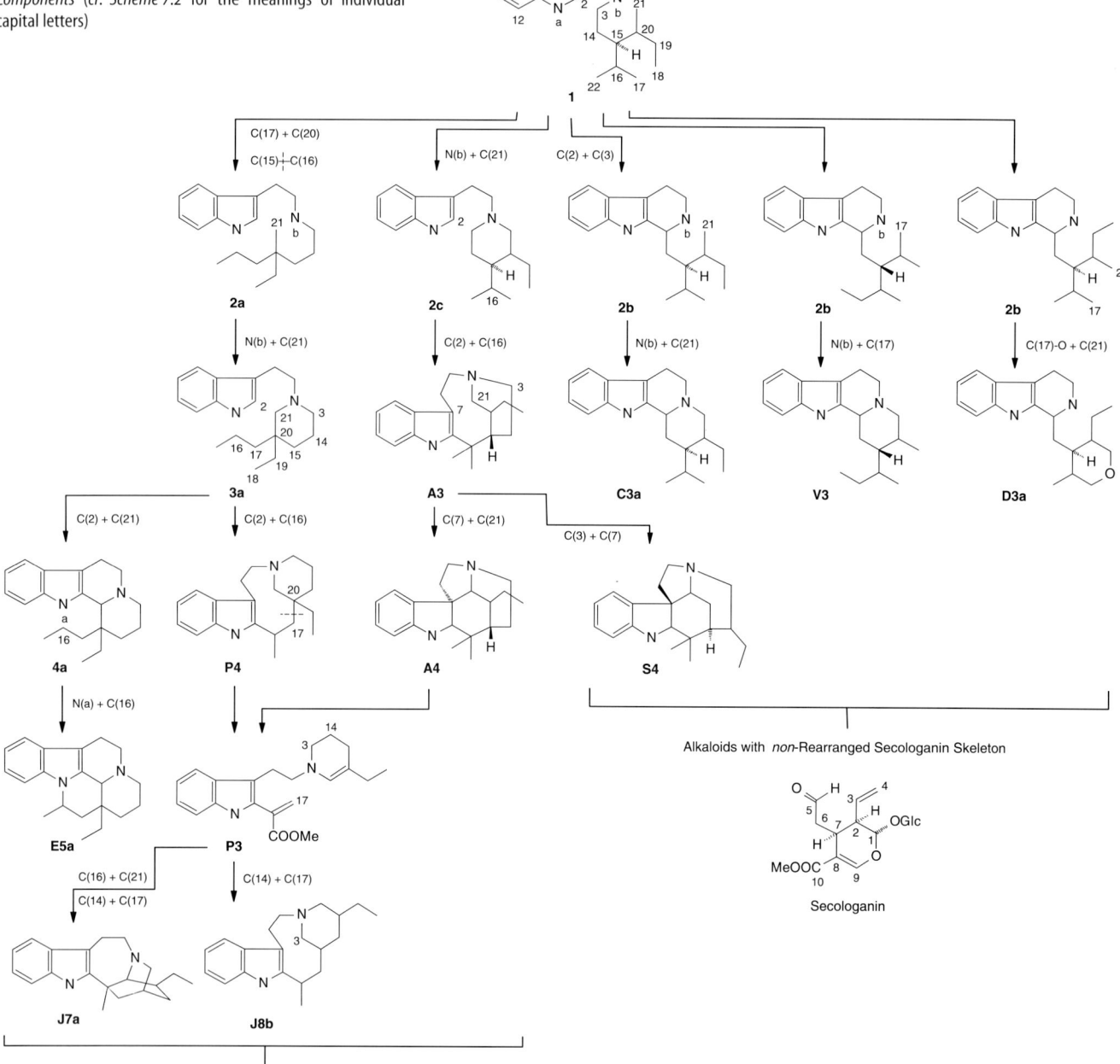

Alkaloids with *non*-Rearranged Secologanin Skeleton

Secologanin

Alkaloids with Rearranged Secologanin Skeleton

7.2.5. Chemotaxonomic Analysis

Before we go on to actual chemotaxonomic appraisal of indole alkaloid studies, we should briefly summarize what has been discussed so far:

Indole alkaloids are found in three plant families that are closely related botanically. The structures of the alkaloids may be divided into eight classes, which have been derived biogenetically. A large number of ring systems are based on these eight main skeleton types, for which it is possible to construct a hierarchy. This hierarchy basically is a 'complexity scale' correlating with the number of (bio)chemical transformations required to reach a compound, starting from the product of the reaction between tryptamine and secologanin.

It is to be assumed that all plants of the Loganiaceae, Rubiaceae, and Apocynaceae can also synthesize the simplest representatives of the C-, V-, and D-types. Such compounds have indeed been found in many – but not in all – of these plants. This, however, is not an argument against the general occurrence of these alkaloid representatives; their stationary concentration in the plants might be very low, since the primary products are presumably quickly converted into more complicated compounds.

Evaluation of all available data on isolation of indole alkaloids allows two generally valid conclusions to be drawn:

1) The lower the rank number of an alkaloid skeleton, the more widely distributed its occurrence. Conversely, the higher the rank number, the fewer the plants that produce the compound.

2) The occurrence of certain main skeleton types is family- and even tribus-specific.

A clear example is shown in the biogenetic series in *Scheme 7.4*. Five C-type alkaloid skeletons of common biogenetic origin are introduced. While alkaloids with the **C3a** ring-system have been found in all three plant families in seven tribus, the close derivative of type **C4e** (arising from ring closure between C(7) and C(16)) is only found in the Apocynaceae, although in plants from all four tribus. Additional ring-closure and ring-opening reactions result in even greater specialization; type **C7a**, for instance, is now only found in a single plant species.

Many biogenetic series of this kind have been discovered. At the same time, it also follows from *Scheme 7.4* that **C4e** to **C7a** specifically belong to the corynanthean alkaloid types of the Apocynaceae.

Scheme 7.4. *A Biogenetic Series.* The structurally simplest alkaloid of the C-group (**C3a**) is much more widely distributed than related alkaloids with a skeleton that has been altered by several rearrangements. The ring-system type **C7a** is found only in one *Hunteria* species. Carisseae (CAR), Tabernaemontaneae (TAB), Alstonieae or Plumerieae (PLU) and Rauwolfieae (RAU) make up the tribus of the Plumerioideae subfamily of the Apocynaceae (APO) family. The tribus Strychneae (STR) belongs to the Loganiaceae (LOG) family, while the tribus Naucleae (NAU) and Cinchoneae (CIN) are subdivisions of the Cinchonoideae subfamily of the Rubiaceae (RUB).

Skeleton Type	Occurring in the Species						
	APO				LOG	RUB	
	CAR	TAB	PLU	RAU	STR	NAU	CIN
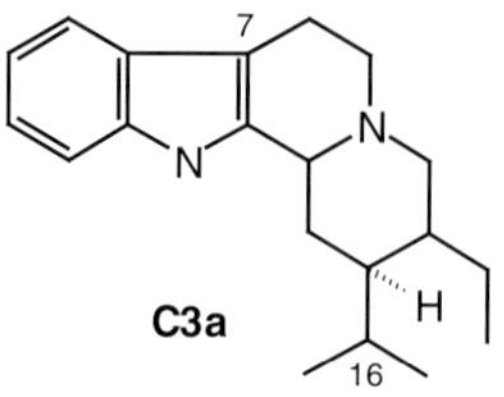 **C3a**	+	+	+	+	+	+	+
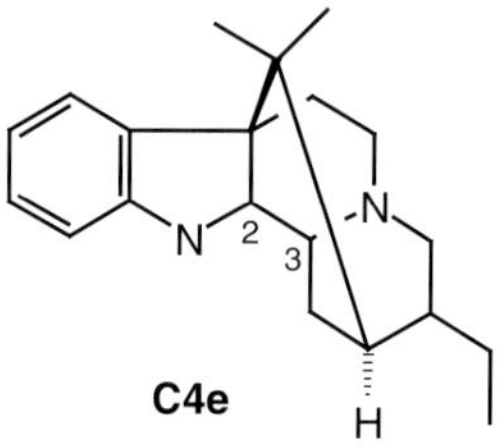 **C4e**	+	+	+	+			
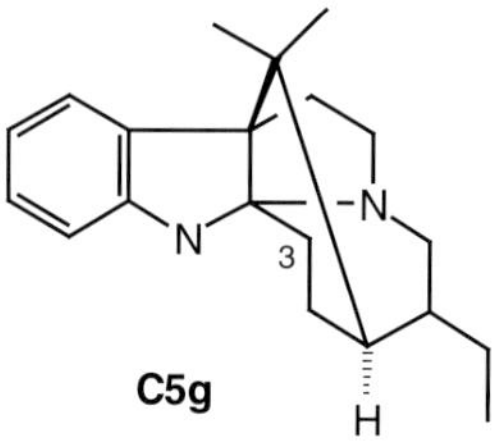 **C5g**	+		+	+			
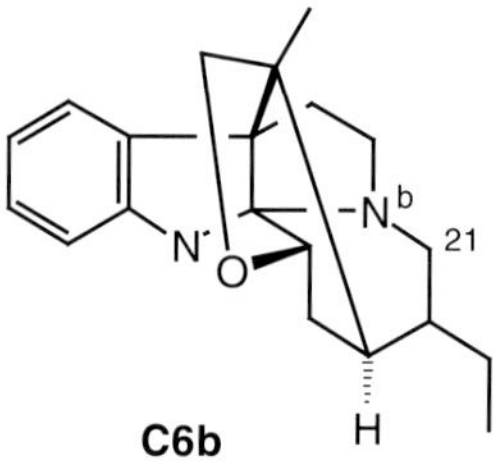 **C6b**	+ Only Hunteria, 4 Species						
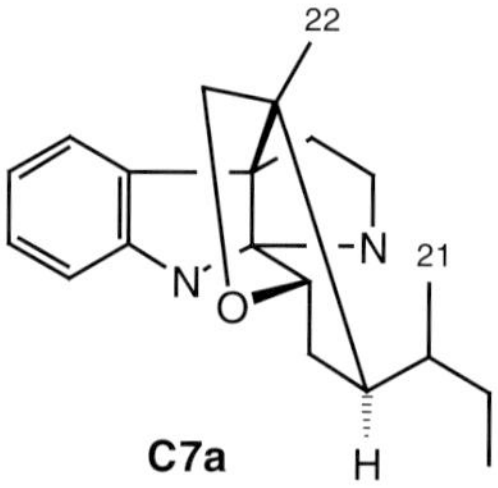 **C7a**	+ Hunteria, 1 Species						

To clarify the second conclusion, it may be commented that alkaloids of the C-, D-, and V-types are found in plants of all three families, whereas the S- and A-types occur only in the Apocynaceae and Loganiaceae. Those alkaloids that incorporate the rearranged secologanin skeleton are exclusively restricted to the Apocynaceae, with J-type alkaloids being found only in three of the four tribus (Carisseae, Alstonieae, and Tabernaemontaneae (*Fig. 7.3*)). In the same way that this is a specific feature of the Apocynaceae alkaloids with rearranged secologanin skeletons, the other two plant families also display some 'species-specific' peculiarities. The corresponding feature of the Loganiaceae in the Strychneae tribus is the occurrence of strychnine and similar substances. The plants distinguish themselves by the experimentally established biogenetic incorporation of an acetic acid moiety, together with tryptamine and secologanin, in the S-type ring skeleton (*Scheme 7.5*).

Corynanthean, vincosan, and vallesiachotaman alkaloids are found particularly often in Rubiaceae genera. An exceptional feature of this plant family, however, is the occurrence of corresponding oxidation products. The C-atoms adjacent to a N-atom seem to be especially readily oxidizable.

Fig. 7.3. *Tabernaemontana alba* Mill. (Apocynaceae) (Photo J. Valverde)

Scheme 7.5. *Strychnine and its Derivatives as Specifics Peculiar to the Loganiaceae*

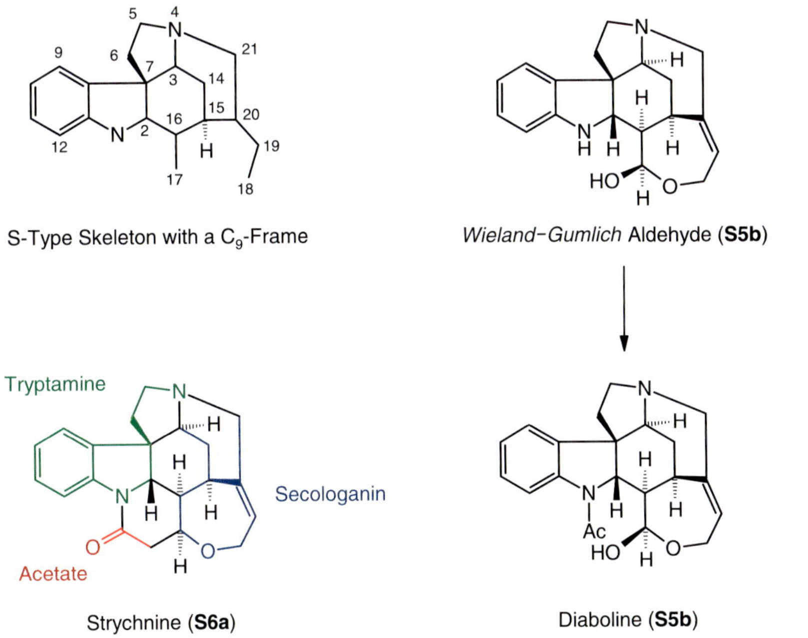

S-Type Skeleton with a C_9-Frame

Wieland–Gumlich Aldehyde (**S5b**)

Strychnine (**S6a**)

Diaboline (**S5b**)

Tryptamine

Secologanin

Acetate

Scheme 7.6. *Oxidation of the C-Type Main Skeleton in the Rubiaceae Family: Formation of Oxindole and Quinine Alkaloids.* * = ^{14}C-labeling in biogenesis experiments. The atom numbering given for cinchonine corresponds to that used for indole alkaloids.

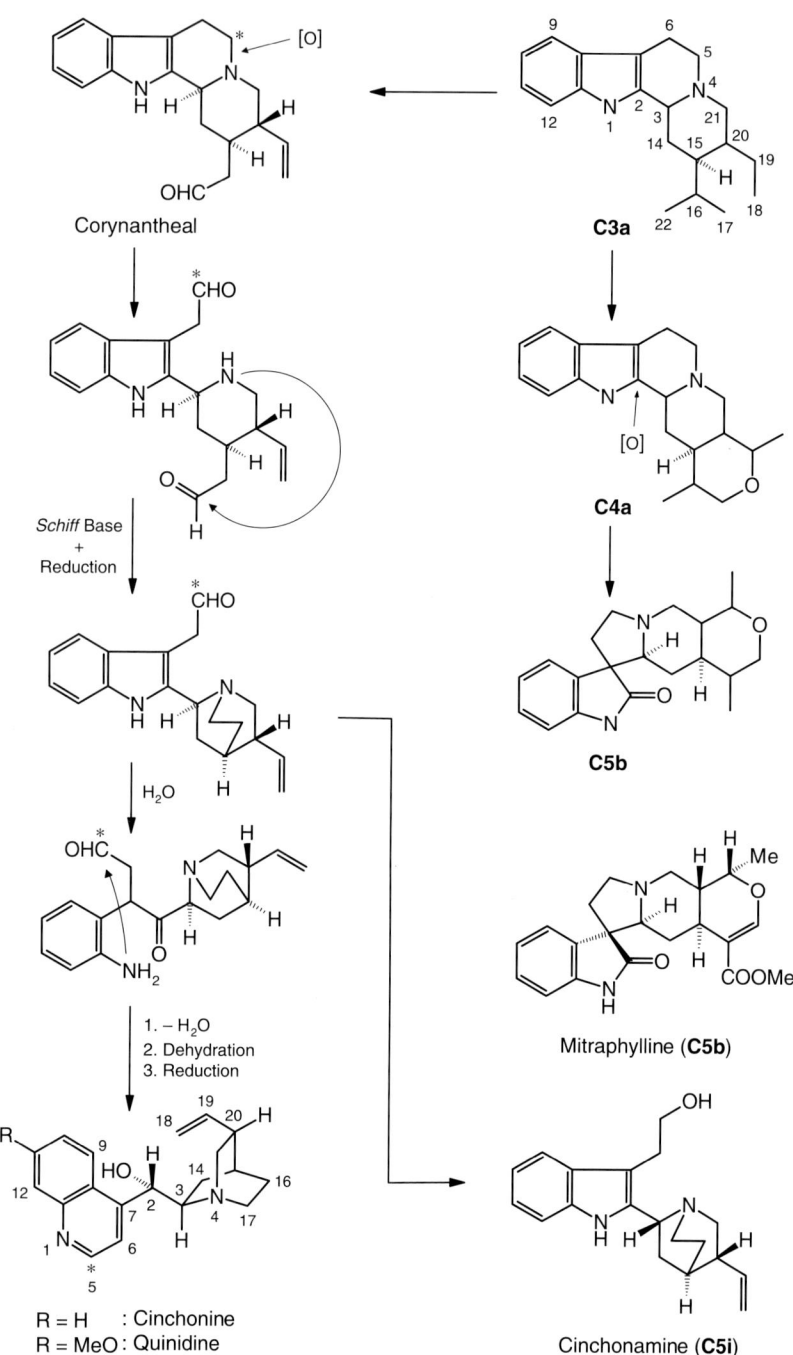

Corynantheal

Schiff Base + Reduction

H_2O

1. $- H_2O$
2. Dehydration
3. Reduction

R = H : Cinchonine
R = MeO : Quinidine

C3a

[O]

C4a

C5b

Mitraphylline (**C5b**)

Cinchonamine (**C5i**)

In one case, the oxindole alkaloids are produced; here, mitraphylline is given as a representative example (*Scheme 7.6*). Particularly prominent in this respect are the genera *Cephalanthus*, *Mitragyna*, and, especially, *Uncaria* (*Fig. 7.4*), all of which belong to the Naucleeae tribus, in which especially many of these compounds have been found. Oxidation at C(5) in α-position to the basic N-atom leads to the quinoline alkaloids quinidine and cinchonine. As can be seen, no reaction has taken place at C(15) of the corynanthean system, and so quinidine consequently possesses the same absolute configuration as its C-type indolic precursor. Quinine alkaloids have been isolated from plants of the Cinchoneae tribus.

The Rubiaceae are distinguished by another special feature: secologanin is capable of reacting not only with tryptamine, but also with two molecules of tyramine (or dihydroxytyramine), producing the isoquinoline alkaloids known as ipecacuanha alkaloids (emetine, cephaleine, psychotrine *etc.*). These have been isolated from sources including the *Psychotria*, *Bothiospora*, *Ferdinandusa*, and *Tocoyena* genera of the Cinchonoideae subfamily (*cf. Scheme 7.7*). Tubulosine has been isolated *inter alia* from *Pogonopus*, also belonging to the Cinchonoideae.

Especially interesting in the context of chemotaxonomy of indole alkaloids with C_9- or C_{10}-monoterpene components are those plants that are to be considered as 'borderline' cases. Such plants still synthesize indole alkaloids 'according to instructions', but at the same time also possess enzymes usually associated with plants lacking them. One such example is, among others, *Kopsia dasyrachis*, an Apocynaceae species assigned to the Rauwolfieae tribus (*Fig. 7.5*) of the Plumerioideae subfamily.

From this plant there were isolated two fundamentally different compounds. The plumeran alkaloid kopsidasine (type **P6a**), on one hand, is an alkaloid that can reasonably be expected, as discussed. Like many other indole alkaloids from *Kopsia* species, this is a compound with a rearranged secologanin moiety (*cf. Scheme 7.8*). On the other hand, a wholly different system was also found in the form of kopsirachine[4], which is composed out of catechin (a flavonoid) and two molecules of the piperidine derivative skytanthine, comprising itself a C_{10}-monoterpene unit. For the coupling of the two types of compounds, it is reasonable to assume a dehydrogenation to an immonium ion, which is attacked by the flavonoid in a *Mannich* reaction (*ortho*-position to the phenolic OH group). In nature, skytanthine had previously only been found in the genus *Skytanthus*. This genus, also belonging to the Apocynaceae, is one of the Cerberioideae, a subfamily directly related to the Plumerioideae. *Kopsia dasyrachis* must, therefore, be a plant possessing characteristic features of both subfamilies (Cerberioideae and Plu-

Fig. 7.4. *Uncaria tomentosa* A. DC. (Rubiaceae) (Photo *J. Valverde*)

Fig. 7.5. *Rauwolfia tetraphylla* L. (Apocynaceae) (Photo *J. Valverde*)

Scheme 7.7. *Borderline Cases: Emetine and Tubulosine from Rubiaceae Plants.* Their production involves tryptamine and dihydroxytyramine as well as secologanin.

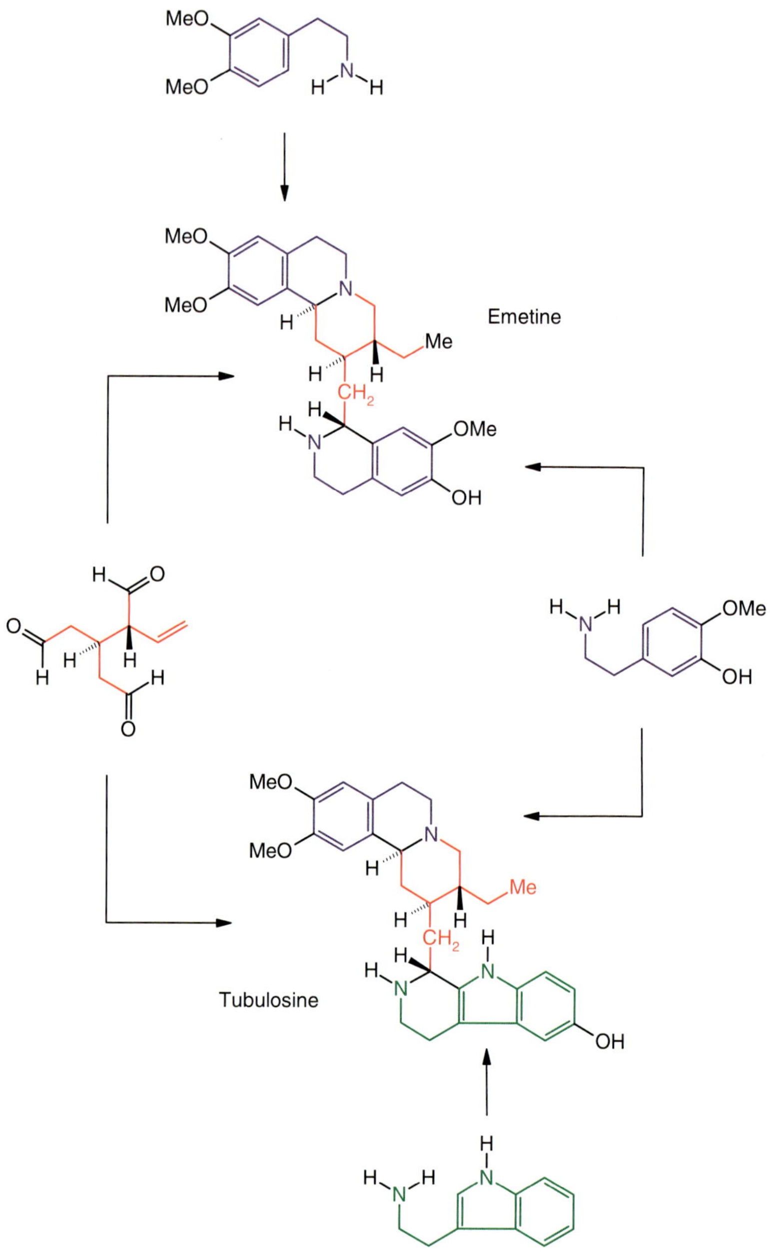

merioideae). These chemotaxonomic findings were also confirmed by classical taxonomic evidence[5].

P6a

Kopsidasine

δ-Skytanthine

Deoxyloganinc Acid

MeNH₂

Catechin

Kopsirachine

The chemotaxonomic investigations discussed in this chapter are highly significant. They are suited for botanical classification of plants on the structural basis of their component chemical substances, or also for reappraisal of their botanical categorization into family, subfamily, and tribus[6,7].

In summary, the more complicated the structure of an indole alkaloid – *i.e.*, the greater the number of biochemical steps from its root compounds (tryptamine and secologanin) – the more specifically it does occur, sometimes with restriction to one genus or even to a single plant species. Examples of this are the so-called 'biogenetic series', the occurrence of alkaloids with rearranged secologanin skeletons (confined to the Plumerioideae), and also the production of quinine/quinidine systems through oxidation reactions (restricted to the Rubiaceae). Particularly noteworthy is the alkaloid spectrum of 'borderline' plants bearing characteristic features of two subfamilies, as demonstrated by the example of *Kopsia*.

References and Notes

1. D. Ganzinger, M. Hesse, 'A Chemotaxonomic Study of the Subfamily Plumerioideae of the Apocynaceae', Lloydia **1976**, 39, 326.
2. H. Yamamoto, N. Katano, A. Ooi, K. Inoue, 'Secologanin Synthase which Catalyzes the Oxidative Cleavage of Loganin into Secologanin is a Cytochrome P450', Phytochemistry **2000**, 53, 7.
3. The abolition of the stereogenic center C(15) is the result of a skeletal rearrangement for which the term 'alkaloids with rearranged secologanin skeletons' is used. The identity of the C(15)-atom has been confirmed by means of biogenetic labeling experiments on appropriate plants.
4. K. Homberger, M. Hesse. 'Kopsirachin, ein ungewöhnliches Alkaloid aus der Apocynaceae Kopsia dasyrachis RIDL.', Helv. Chim. Acta **1984**, 67, 237; P. J. Houghton, 'Chromane Alkaloids', The Alkaloids **1987**, 31, 67.
5. M. E. Fallen, Ph.D. Thesis, University of Zurich, 1983; M. E. Endress, M. Hesse, S. Nilssen, A. Guggisberg, J.-P. Zhu, 'The Systematic Position of the Holarrheninae (Apocynaceae)', Plant Systematics and Evolution **1990,** 171, 157.

6. M. V. Kisakürek, Ph.D. Thesis, University of Zurich, 1980; M. V. Kisakürek, M. Hesse, *'Chemotaxonomic Studies of the Apocynaceae, Loganiaceae and Rubiaceae, with Reference to Indole Alkaloids'*, in *Indole and Biogenetically Related Alkaloids*, Eds. J. D. Philipson and M. H. Zenk, Academic Press, London, 1980; M. V. Kisakürek, A. J. M. Leeuwenberg, M. Hesse, *'A Chemotaxonomic Investigation of the Plant Families of Apocynaceae, Loganiaceae, and Rubiaceae by Their Indole Alkaloid Content'*, in *Alkaloids: Chemical and Biological Perspectives*, Ed. S. W. Pelletier, Wiley, New York, 1983, Vol. 1, 211; J.-P. Zhu, A. Guggisberg, M. Kalt-Hadamowsky, M. Hesse, *'Chemotaxonomic Study of the Genus Tabernaemontana (Apocynaceae) Based on Their Indole Alkaloid Content'*, *Plant Systematics and Evolution* **1990,** *172,* 13.
7. B. Danieli, G. Palmisano, *'Alkaloids from Tabernaemontana'*, *The Alkaloids* **1986,** *27,* 1.

8. Aspects of Alkaloid Biogenesis

8.1 General

The term 'biogenesis' refers to the ways and means by which bacterial, vegetable, or animal organisms produce, modify, or dismantle particular (organic) substances. In the context of the compounds dealt with in this book, the question of alkaloid biogenesis is naturally of particular interest.

If the same alkaloid occurs in different organisms, this does not in any way mean that the corresponding biochemical reactions have to be the same. A well-known example of this phenomenon is nicotinic acid (**1**). One established pathway for its production starts from tryptophan (*Scheme 8.1*). In the course of the degradation of this amino acid through a number of intermediates, such as kynurenine (**2**), nicotinic acid is produced. On the other hand, it can also be synthesized from glycerin and aspartic acid. This example is especially noteworthy in that evidence has been found that some organisms are capable of using both approaches for the synthesis of nicotinic acid. An experimentally verified mechanism of formation of a particular alkaloid, taken strictly, is valid only for that compound in the organism that was the object of investigation. Generalizations to similar compounds and other organisms should be made only with reservation.

Examples of some alkaloids and their accepted N-sources are listed in *Scheme 8.2*.

The absolutely correct approach to alkaloid biogenesis would be to describe results obtained to date for all alkaloids in the organisms that were the objects of investigation. This would go beyond the scope of this book, however. We are, therefore, compelled to choose another method, presenting and explaining investigations carried out on *one specific* alkaloid class. Some typical isoquinoline alkaloids are thus presented here by example. To help better understand the principles involved in the complex biogenesis of alkaloids, we will treat in detail 1-benzylisoquinolines and some of its derivatives.

If alkaloids of similar structures are found in the same plant species or genus, or even just in the same plant family, then this is a sign of similar (or at least related) biogenetic pathways. To provide an idea of the structural similarity of alkaloids, it is helpful to compare the ring systems and, particularly, their substituents with one another. As we shall see, statistical methods can suc-

Scheme 8.1. *Biogenesis of Nicotinic Acid* (**1**). Its production can proceed in two ways. In the fungus *Neurospora crassa*, it is done by degradation of tryptophan. *Mycobacterium tuberculosis*, though, can synthesize nicotinic acid from glycerin and aspartic acid, with pyridine-2,3-dicarboxylic acid as an intermediate.

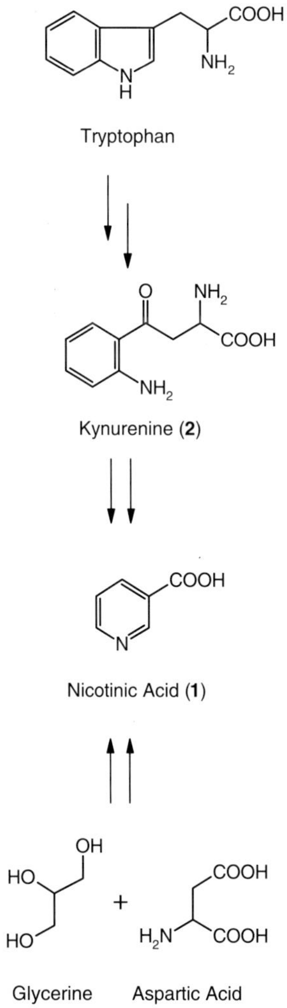

Tryptophan

Kynurenine (**2**)

Nicotinic Acid (**1**)

Glycerine   Aspartic Acid

Scheme 8.2. *Examples of Alkaloids and Their Amino Acid Precursors.* In brackets: formation of other alkaloid classes.

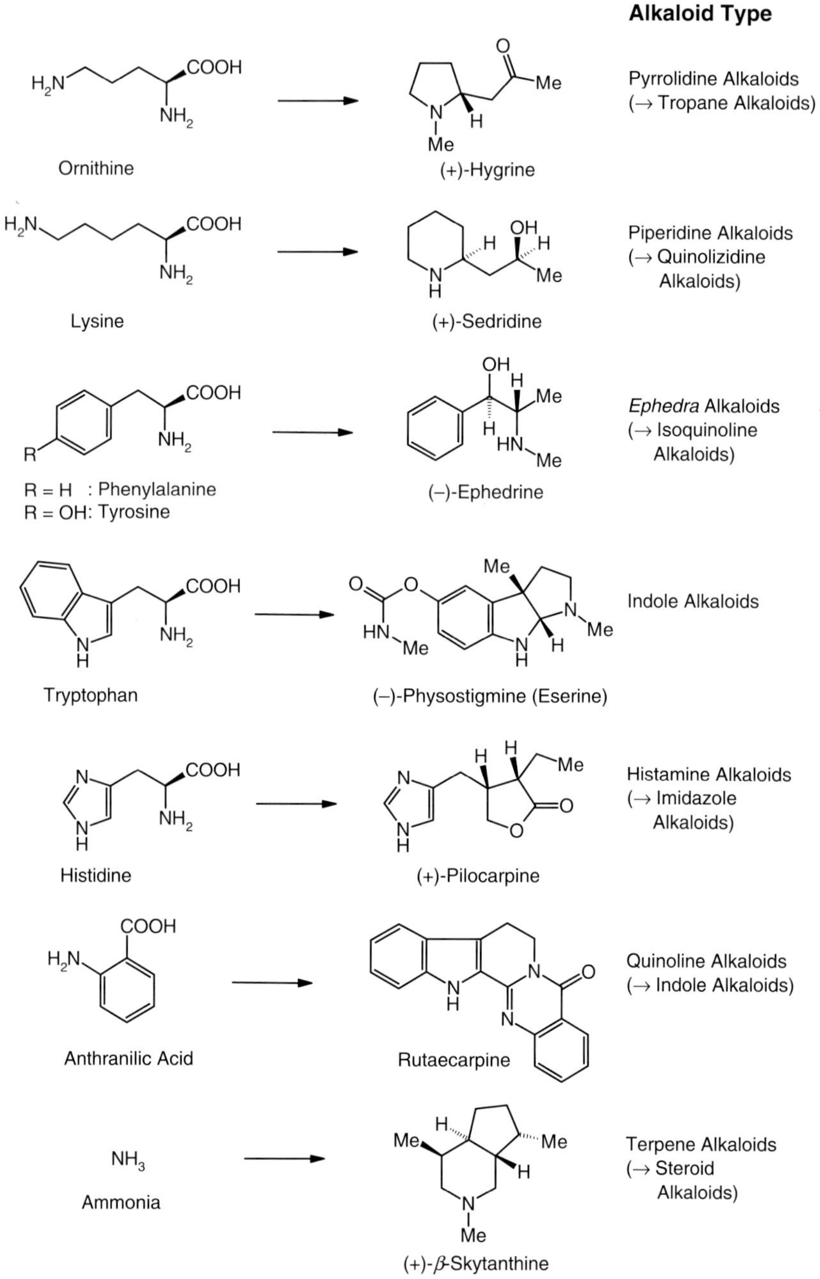

Alkaloid Type

Ornithine → (+)-Hygrine

Pyrrolidine Alkaloids
(→ Tropane Alkaloids)

Lysine → (+)-Sedridine

Piperidine Alkaloids
(→ Quinolizidine
Alkaloids)

R = H : Phenylalanine
R = OH: Tyrosine → (−)-Ephedrine

Ephedra Alkaloids
(→ Isoquinoline
Alkaloids)

Tryptophan → (−)-Physostigmine (Eserine)

Indole Alkaloids

Histidine → (+)-Pilocarpine

Histamine Alkaloids
(→ Imidazole
Alkaloids)

Anthranilic Acid → Rutaecarpine

Quinoline Alkaloids
(→ Indole Alkaloids)

NH₃
Ammonia → (+)-β-Skytanthine

Terpene Alkaloids
(→ Steroid
Alkaloids)

cessfully be used for this, as demonstrated in the following section. Only when structural relationships within a given alkaloid family are evident, together with basic aspects of their biogenesis, specific feeding experiments with suitably radiolabeled precursors can be carried out. Statistical methods, though, should be applied first.

8.2. 1-Benzylisoquinoline Alkaloids

8.2.1. Botanical Origin

1-Benzylisoquinoline alkaloids and their natural derivatives are found in many plant families, as listed below (in parentheses: skeleton types present, *cf. Scheme 8.3*):

Alangiaceae (IX)   Lauraceae (I–III, VI, IX)
Annonaceae (I, II, IX)   Leguminosae (V, IX)
Araceae (II)   Magnoliaceae (I, II)
Aristolochiaceae (II)   Menispermaceae (I–IV, VI, IX)
Berberidaceae (I, II, VI, IX–XI)   Monimiaceae (I–III)
Combretaceae (I)   Nymphaeceae (I–III)
Convolvulaceae (IX)   Papaveraceae (I–IV, VI–XII)
Euphorbiaceae (II, III)   Ranunculaceae (I ,II, VI, VII, IX–XI)
Fumariaceae (II, IX–XII)   Rhamnaceae (I, II)
Hernandiaceae (I, II)   Rutaceae (I, IX, X, XII)
Hydrastidaceae (IX, XI)

As is evident from this inventory, it is common for several skeleton types to be found in one plant family. On closer examination, it becomes clear that many of these structures may similarly be encountered in certain individual plants, *e.g.*, *Papaver somniferum* (I, IV, IX–XI) and *P. orientale* (II–IV, IX, X). This kind of common structures in a single plant constitutes strong evidence for a biogenetic interrelationship between the individual alkaloids.

8.2.2. Statistics Relating to Oxidation Sites in Basic Skeletons

The class of 'isoquinoline alkaloids' is divided into various subgroups on the basis of structural criteria (*cf. Chapt. 2*). Here, attention should be paid to the fact that there are some subgroups with only a few members, and others with a great many. A statistical breakdown in terms of the location of oxygenated substituents on the ring skeletons of the most important subclasses is shown in *Scheme 8.3*. Subclasses with at least ten known members are viewed as 'important' here. To provide a uniform basis, the representatives listed in the tabular compilation[1] have been numbered, with bisisoquinoline alkaloids disregarded. These two assumptions underlying the statistics may seem arbitrary, since it would certainly have been possible to set the boundary conditions for qualifying alkaloids in a different way, while the restriction to only one literature source is also problematic. Both serve to ensure a statistical profile as 'unfalsified' as possible. Other alkaloids attributable to these classes have also been isolated in recent years. It is assumed, though, that the additional data (not included in *Scheme 8.3*) would hardly alter the overall picture.

Scheme 8.3. *Statistical Overview of the Most Common Subgroups of 1-Benzyl-1,2,3,4-tetrahydroisoquinoline Alkaloids in Terms of the Distribution of O-Containing Functional Groups Attached to the Ring Skeleton*[2]

Skeleton: C₁₆N

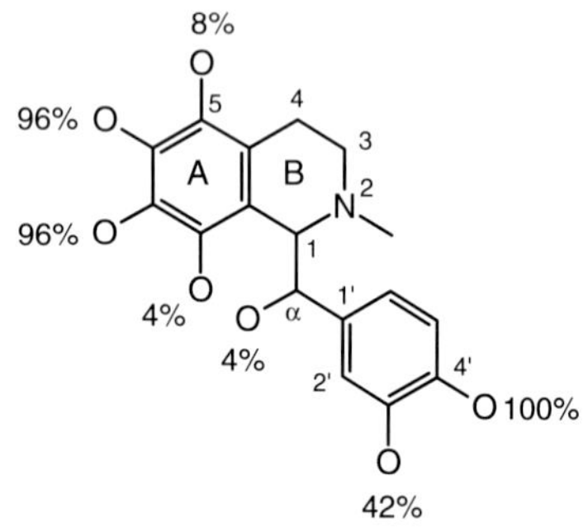

Type I. 1-Benzyl-1,2,3,4-tetrahydro-isoquinoline Alkaloids[3]

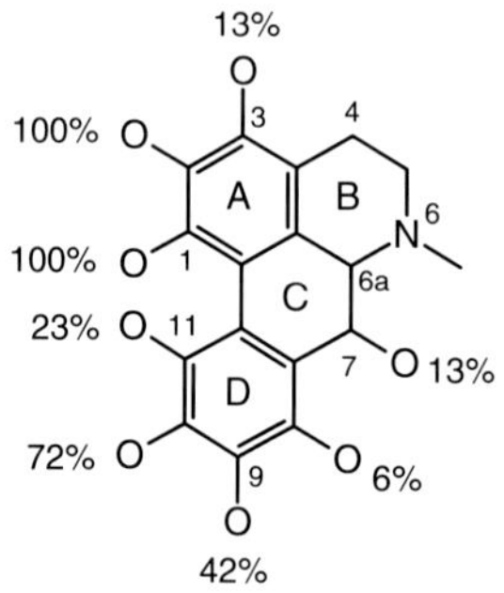

Type II. Aporphine Alkaloids[3]

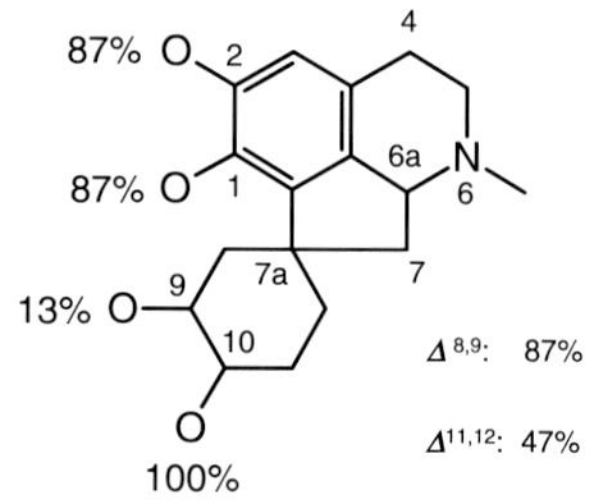

Type III. Proaporphine Alkaloids[4]

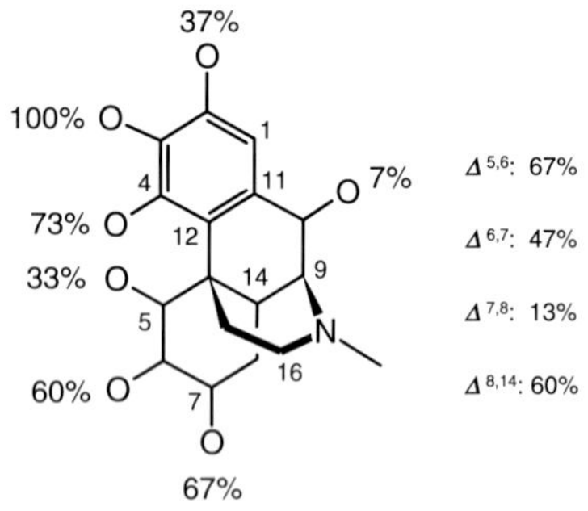

Type IV. Morphine Alkaloids[5]

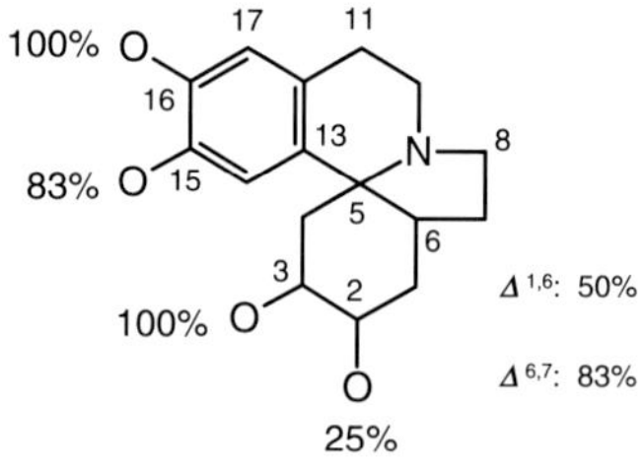

Type V. *Erythrina* Alkaloids

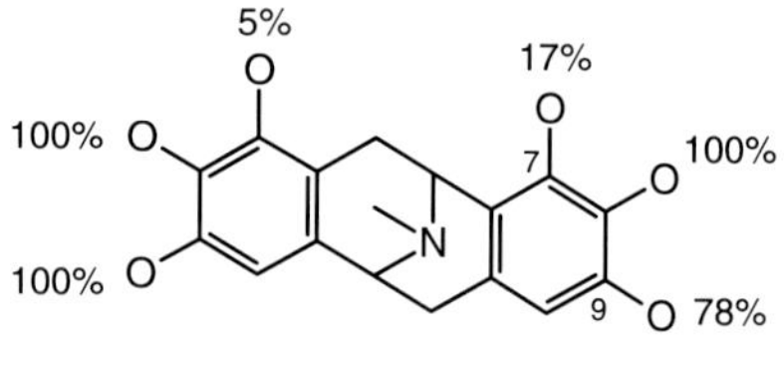

Type VI. Pavine Alkaloids[6]

Regarding the twelve skeleton types depicted in *Scheme 8.3*, the first striking feature is that only nine of them contain an isoquinoline chromophore. The remainder – namely the morphine, protopine, rhoeadine, and papaverrubine alkaloids – seem not to belong to this alkaloid class, if the relevant criterion (the chromophore) is used. In addition, it is conspicuous that the number of skeleton C-atoms may be either 16 or 17. Nevertheless, paying attention to the types of O-substitution, it is possible to recognize some definite family relationships among all of these skeleton types.

In some ways, the 1-benzyltetrahydroisoquinolines (Type I) seem to represent a fundamental subclass among the isoquinoline alkaloids. This class possesses both the smallest number of rings and C-atoms. Furthermore, all members bear an oxy group at C(4'), and almost as many have residues of

Scheme 8.3 (*cont.*)

Skeleton: C$_{17}$N

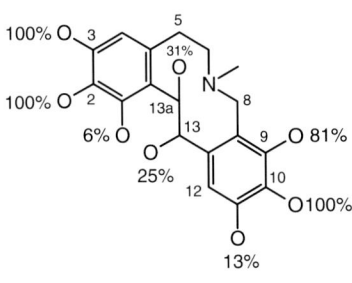

Type VII.   Isopavine Alkaloids

Type VIII.   Rhoeadine and Papaverrubine Alkaloids

Type IX.   Protoberberine Alkaloids[7,8]

Type X.   Protopine Alkaloids[8]

Type XI.   Phthalideisoquinoline Alkaloids

Type XII.   Benzophenanthridine Alkaloids[3]

this type at C(6) and C(7). Aside from these three favored substitution sites, only C(3′) stands out with a relative value of 42%. If we now compare the substitution pattern of the 1-benzyltetrahydroisoquinoline alkaloids with those of the other skeleton types presented in *Scheme 8.3*, we can sometimes find very good agreement.

This is most notable in the *proaporphine* alkaloids (oxy substituents mostly at C(1), C(2), C(10), and (more rarely) at C(9)), in the *Erythrina* alkaloids (C(15), C(16), C(3), and, occasionally, C(2)), and in the *protoberberine* and *protopine* alkaloids (C(2), C(3), C(10), and – more rarely – C(9)).

Among the *phthalideisoquinoline, pavine, isopavine, benzophenanthridine,* and *rhoeadine* alkaloids, we find four equally or almost equally often substituted C-atoms, the pattern of which fits the first-mentioned group above. Only the *aporphine* and *morphine* alkaloid groups differ conspicuously from the types named so far. The number of recorded members of the former of these two groups is particularly large, which implies that the deviation from the substituent pattern of Type I is no coincidence.

It should be emphasized that both the number and relative orientation of O-substituted ring C-atoms is highly conserved. Of the four C-atoms most commonly bound to oxygen, two are consistently found in a *vicinal* relationship. In the majority of cases, one C-atom in such a 'pair' carries a 2-aminoethyl residue in *para*-position (groups VIII and XII are exceptions here), and so the most common arrangement is given by the general formula **3**.

On the basis of these observations, it should be possible to put together some hypotheses about the transformations involved to explain the biogenetic interconversion of individual skeleton types in the plant.

The proaporphine alkaloids (Type III) differ from the 1-benzyl-1,2,3,4-tetra-hydroisoquinoline alkaloids (Type I) in having an additional bond between C(8) and C(1′), which results both in a spiro center (Type III) and in the fact that ring C can no longer be aromatic. In the lower part of the molecule an O-atom is found in the *para*-position to the spiro center, while, in the upper part, the *ortho*-position is affected. This implies that the formation of the new bond takes place in a radical manner.

The aporphine alkaloids (Type II) also differ from those of Type I solely by the presence of an additional bond, once more involving C(8), but this time joined with C(2′) of Type I. Both the rings A and C can, therefore, remain aromatic. Hence, attempts to explain the formation of this new bond in a similar manner (by means of a radical mechanism) run into difficulties. True, in

3

the upper part of the molecule the new bond is once more positioned *ortho* to an O-atom, but, in the lower part, neither the *ortho*- (C(11)) nor the *para*-position (C(9)) is especially frequently substituted. To make up for this, the *meta*-position does display a relatively high degree of substitution (72%). It would not be unreasonable to conclude that – as a result of the inferior radical stabilization supplied by OH groups in *meta*-positions – a radical mechanism can be ruled out in favor of another. This other mechanism should, however, also be capable of explaining the significantly lower percentage of O-substitution at C(10) in ring D, relative to alkaloids of Type I, and also of providing clues as to why substitution by O-containing functional groups is also observed at both C(9) and C(11) in compounds of Type II, rather than at only one of the two atoms, as seen in alkaloids of Type I and Type III.

One possible explanation could be the reaction sequence I → III → II.

The transition I → III would have to be regarded as a radical phenol oxidation between C(8) (in Type III) and C(1′) (in Type I). For the transition III → II, one can postulate an acid-catalyzed rearrangement under re-aromatization of ring D, as outlined in *Scheme 8.4*.

Since Type-III alkaloids are spiro compounds, two possible rearrangement mechanisms (by means of a dienone–phenol rearrangement) can be proposed (*Scheme 8.4*), leading to products that would differ in the O-substitution of the aporphine skeleton. In addition, compounds of Type III are not necessarily impelled to contain an oxo group at C(10); the rearrangement may also take place if a C(10)–OH group is present. In any case, this site is going to be removed in the course of the dienol–benzene rearrangement. These two modes of reaction, *i.e.,* two possible modes of rearrangement plus one means of removal of the C(10)–O function, fit well with the statistical findings (*cf. Scheme 8.3*).

Similar considerations also apply for the examination of biosynthetic options for the morphine alkaloids (Type IV), which could arise from alkaloids of Type I (*cf. Scheme 8.5*). The corresponding cyclization, once again most likely radical in character, is plausible on the same grounds as given for the transformations I → III → II. The upper ring of the morphine alkaloids, however, does not necessarily show signs of phenolic oxidation, if we take account of the statistical data, but is much more reminiscent of the substitution pattern of ring D in the aporphine alkaloids. On the basis of the arguments advanced there, an intermediate stage, corresponding to **4**, would be necessary. Type-IV alkaloids would then be formed *via* **4** by an acid-catalyzed rearrangement. If, though, we now examine the substitution patterns of the individual subclasses, then we observe that in the morphine alkaloids either

4

Scheme 8.4. *Rearrangement of 1-Benzyl-1,2,3,4-tetrahydroisoquinoline Alkaloids* (Type I) *by Phenol Oxidation to the Proaporphine Alkaloids* (Type III), *Which May Be Converted to Two Differently Substituted Aporphine Alkaloids* (Type II) *through a Dienone–Phenol Rearrangement*

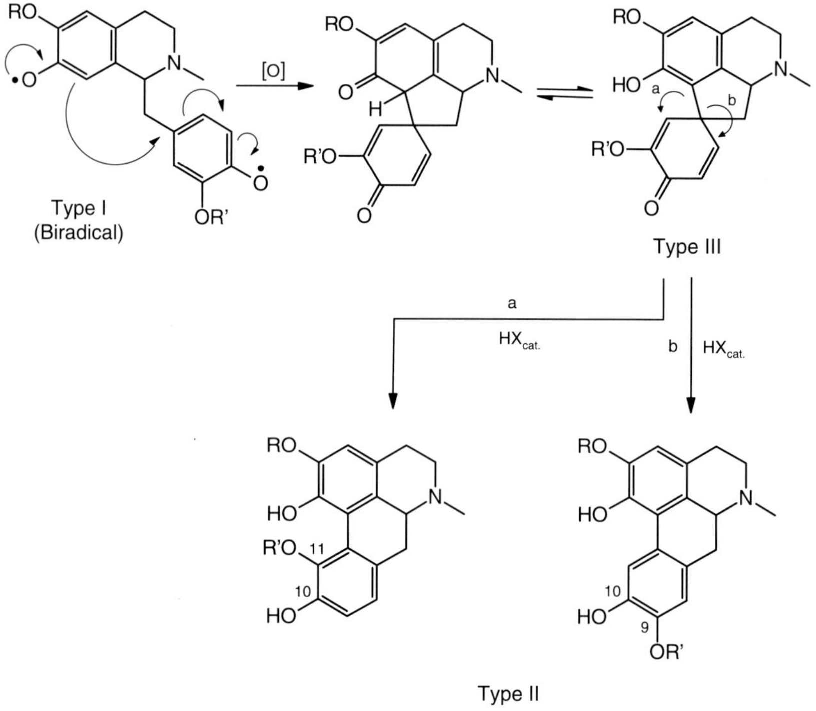

Type I
(Biradical)

Type III

Type II

Scheme 8.5. *Formation of Morphine Alkaloids* (Type IV) *from 1-Benzyl-1,2,3,4-tetrahydroisoquinoline Alkaloids* (Type I)

Type I

Type IV

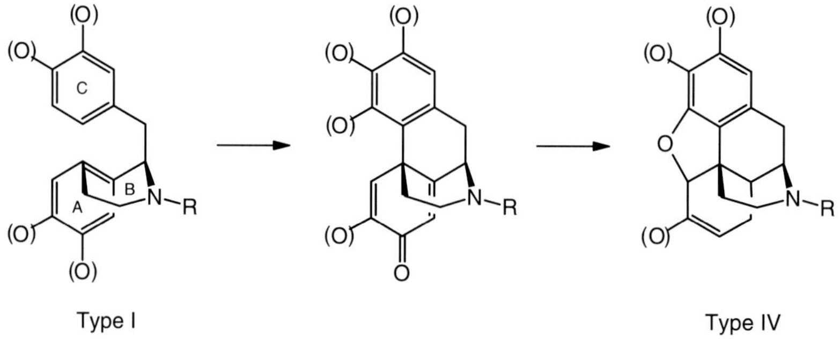

C(2) or C(4) (together with C(3)) are substituted. Consequently, direct coupling (*ortho* or *para*) is possible in any case, making **4** as a putative intermediate superfluous.

Conspicuous in the *Erythrina* alkaloids of Type V is the restriction of O-substituents to only four C-atoms. These seem to correlate with representatives of the 1-benzyl-1,2,3,4-tetrahydroisoquinoline alkaloids as follows: C(6) (Type I) and C(16) (Type V); C(7) and C(15); C(4′) and C(3), as well as C(3′) and C(2). Building on these assumptions, a biosynthetic pathway as shown in *Scheme 8.6* can be proposed.

A phenol oxidation with *para/para* coupling is necessary for the transformation of alkaloids of Type I into the intermediate **5**. The next two steps include a *retro-Mannich* reaction and a dienone–phenol rearrangement, leading to **6** and **7**, respectively. Finally, the conversion of **7** into *Erythrina* alkaloids of Type V may be regarded as an oxidative reaction of the lower benzene ring in a *Michael*-style ring closure. The reaction pathway shown then results only in one alkaloid, which fits the observed substitution pattern.

Scheme 8.6. *Formation of Erythrina Alkaloids* (Type V) *from 1-Benzyl-1,2,3,4-tetrahydroisoquinoline Alkaloids* (Type I)

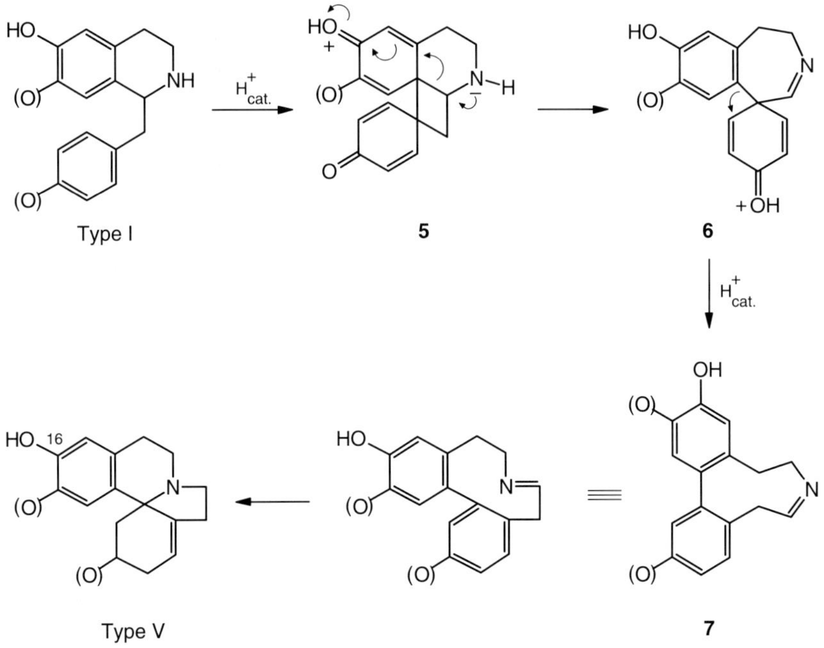

Scheme 8.7. *Transformation of 1-Benzyl-1,2,3,4,-tetrahydroisoquinoline Alkaloids* (I) *into Those of the Pavine* (VI) *and isopavine* (VII) *Types*

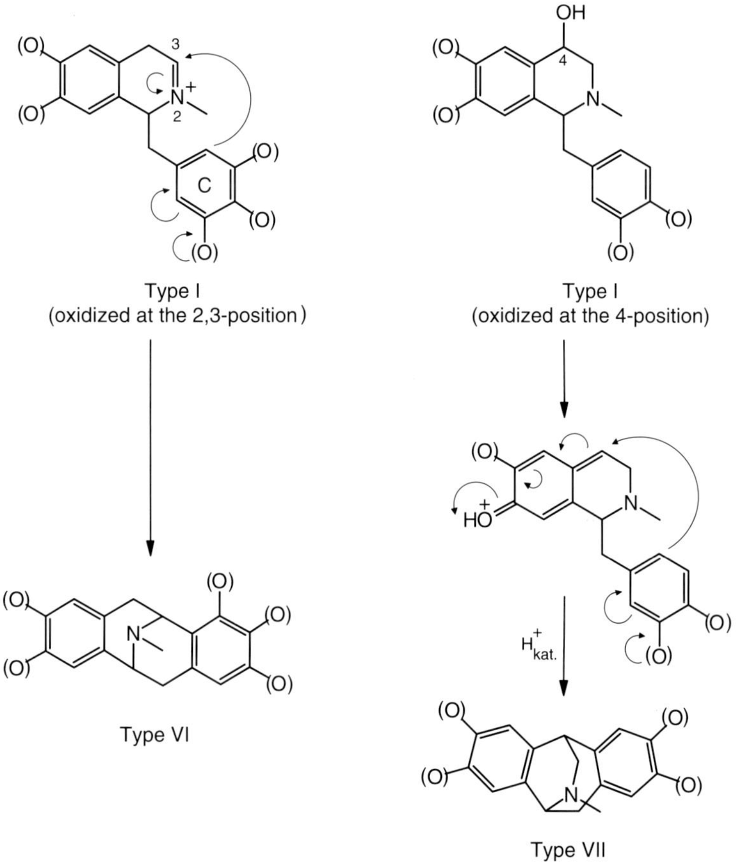

Type I
(oxidized at the 2,3-position)

Type I
(oxidized at the 4-position)

Type VI

Type VII

The pavine (Type VI) and isopavine (Type VII) alkaloids are clearly recognizable as cyclization products of 1-benzyl-1,2,3,4-tetrahydroisoquinoline (Type I) alkaloids. The formation of pavine alkaloids formally requires a dehydrogenation at both C(2) and C(3) in the skeleton of Type I, followed by ring closure to Type VI; the latter process is promoted by both the *para-* and *ortho*-O-substituents in ring C (*Scheme 8.7*). To undergo transformation into isopavine alkaloids of Type VI, however, compounds belonging to Type I must be oxidized at C(4). Here, the basicity of the N-atom is irrelevant. The selection principle already discussed is once more remarkable, and so only alkaloids of Type I with O-substituents at C(6), C(7), C(3'), and C(4') are converted into those of the isopavine type.

Interestingly, the isoquinoline alkaloids with C_{17} skeletons (*cf. Scheme 8.3*) also feature a substitution pattern that matches that of their C_{16} relatives. It is, therefore, once more plausible to assume Type-I alkaloids as progenitors. For instance, it is conspicuous that in the protoberberine (IX), protopine (X), phthalideisoquinoline (XI), and rhoeadine (VIII) alkaloids, additional C-atoms – relative to Type I – are located in *ortho*-positions to an O-atom. This gives grounds to assume that the same reaction might be responsible for the incorporation of the extra C-atom in all cases. As shown in *Scheme 8.8*, a Type-IX alkaloid may be formed from an immonium ion of a Type-I alkaloid by a *Mannich* reaction *via* **8**. The statistically appreciably higher O-content at C(3′) (or C(5′)) in alkaloids of Types VII–XI, as compared to alkaloids of Type I, fits the 'natural' selection criteria in that only those 1-benzyl-1,2,3,4-tetrahydroisoquinoline alkaloids that possess an O-function at C(3′) (or at C(5′)) enter into the reaction shown in *Scheme 8.8*. Oxidative ring opening of the C(13a)–N bond would produce the protopine skeleton (Type X). If the C(6)–N bond is oxidatively opened and a new bond created between C(6) and C(13), then the benzophenanthridine skeleton results with a substitution pattern observed in the natural alkaloids[9]. Of the two tetrahydroisoquinoline chromophores in the protoberberine type, only the lower one is preserved as such in benzophenanthridine bases.

From the substitution patterns in rings A and D, it is possible to recognize a relationship between rhoeadine (Type VIII) and phthalideisoquinoline alkaloids (Type XI).

The enlargement of the six-membered into a seven-membered ring may take place through an initial ring-opening in the phthalideisoquinoline alkaloid (Type XI), which results in an intermediary enol ether (*Scheme 8.9*). The latter is converted into the corresponding oxo acid, which cyclizes to yield a seven-membered ring (*Schiff* base). After reducing the carboxylic acid to an aldehyde and adding the latter to the nearby C=C bond, the rhoeadine skeleton (Type VIII) is, upon hydrolysis, finally produced.

It should be noted that the proposed pathways for the biosynthesis of complicated alkaloids from more simple compounds are based exclusively on the comparison of both skeleton types and the statistically observed abundance of O-substituted C-atoms.

The above approach neglects one point mentioned at the beginning of this section: different plants may synthesize the same substance in different ways. This viewpoint has been wholly disregarded in favor of mechanistic considerations leading to the assembly of the principal isoquinoline alkaloid skeletons.

Scheme 8.8. *Transformation of 1-Benzyl-1,2,3,4-tetrahydroisoquinoline Alkaloids* (Type I) *into Those of the Protoberberine Type* (Type IX)[11]

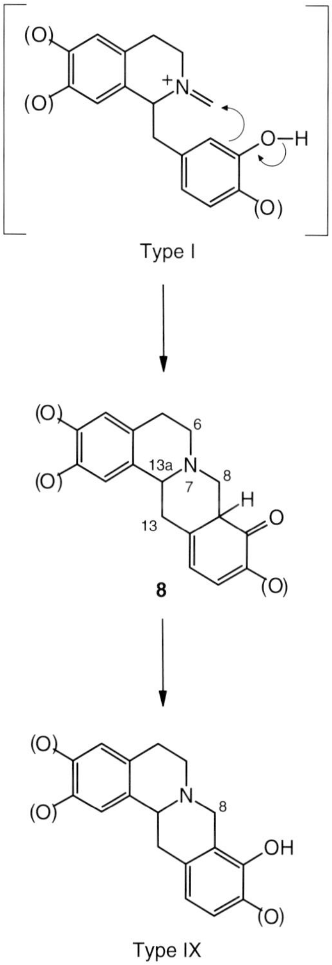

Type I

8

Type IX

Type XI

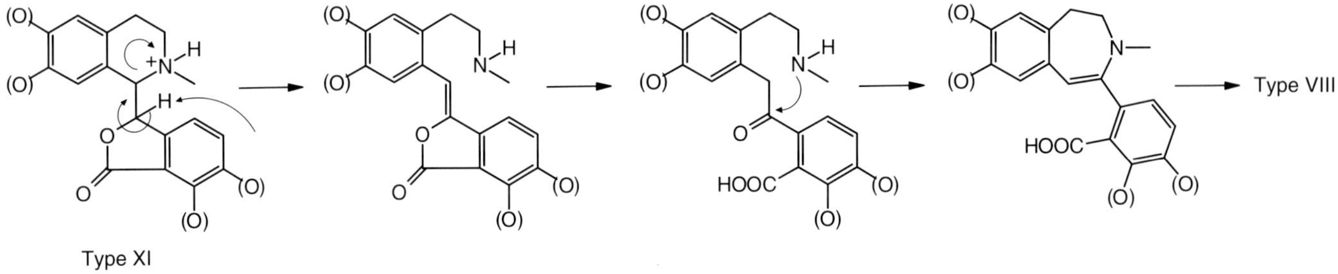

On one hand, statistical methods provide pointers to possible biosynthetic pathways. On the other, though, subtle nuances that might be able to point at a single alkaloid or a group of bases from the same plant get lost in the background noise and disappear.

It is possible to apply similar statistical analyses to a number of other isoquinoline-alkaloid structural types, from which analogous results are to be expected. Many of the statistically derived biosyntheses discussed above have been confirmed experimentally. Thus, statistical methods have shown themselves to be excellently suited for preliminary investigations of structurally related compounds, alkaloids as well as natural products in general.

8.2.3. Chemical Transformations Involved in the Rearrangements of Basic Skeletons [13]

The clearly close relationship between individual subgroups of 1-benzyl-1,2,3,4-tetrahydroisoquinoline alkaloids has acted as a spur to the identification of reactions allowing for chemical correlation between certain alkaloid groups. Some of these are dealt with below.

Rearrangement of 1-Benzyl-1,2,3,4-tetrahydroisoquinoline Alkaloids to Aporphine Alkaloids (Type I → Type II)

Orientaline (**9**) is an alkaloid that belongs to the 1-benzyl-1,2,3,4-tetrahydroisoquinoline alkaloid group (*Scheme 8.10*). In an aqueous solution of ammonium acetate and potassium hexacyanoferrate(III), **9** is being converted (in *ca.* 1% yield) into the proaporphine alkaloid orientalinone (**10**), which naturally occurs in a variety of *Papaver* species (*e.g.*, *P. orientale*). Reduction of **10** affords the diastereoisomeric dienol **11**, known as orientalinol. If the lat-

Scheme 8.10. *Synthesis of (+)-Isothebaine (12) from (+)-Orientaline (9)*

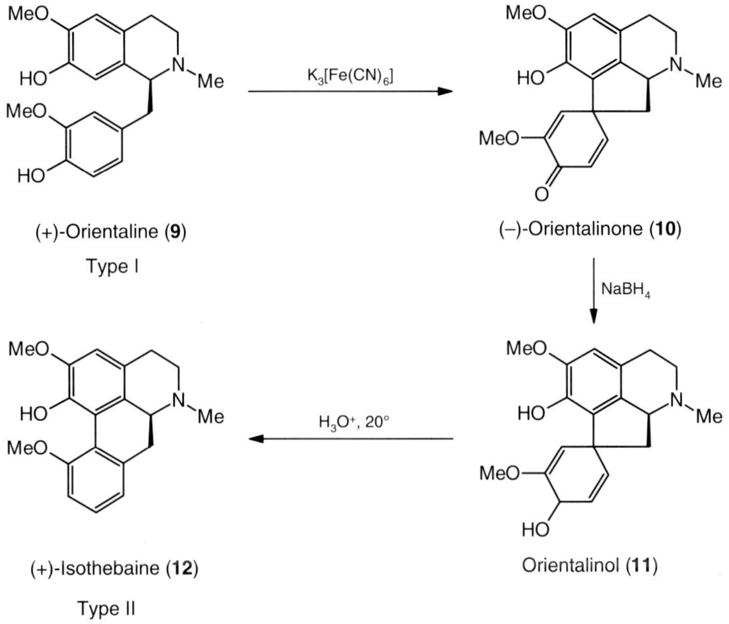

(+)-Orientaline (**9**)

Type I

(−)-Orientalinone (**10**)

K$_3$[Fe(CN)$_6$]

NaBH$_4$

(+)-Isothebaine (**12**)

Type II

H$_3$O$^+$, 20°

Orientalinol (**11**)

ter is allowed to rest at ambient temperature in aqueous/methanolic HCl solution overnight, then a dienol–benzene rearrangement takes place, in which the aporphine alkaloid isothebaine (**12**) is formed [14].

Even if the overall yield of this *in vitro* reaction is very modest, it still demonstrates that such transformations are generally possible even under very mild reaction conditions comparable to those *in vivo* (plant).

Rearrangement of 1-Benzyl-1,2,3,4-tetrahydroisoquinoline Alkaloids to Morphine Alkaloids (Type I → Type IV)

Reticuline (**13**), a Type-I alkaloid, can be oxidized with MnO$_2$ to the dienone **14** in a (pathetic) yield of 0.024% (*Scheme 8.11*). By analogy with the rearrangement of orientalinone (**10**) into isothebaine (**12**), **14** may similarly be converted into **15** with NaBH$_4$, and this (under mildly acidic conditions at room temperature) into thebaine (**16**) [15].

Detection of the very small quantities of thebaine was accomplished with the aid of radioisotopes. ^{14}C-Labeled reticuline (**13**) was added and the reaction

Scheme 8.11. *Synthesis of Thebaine* (**16**) *from the 1-Benzyl-1,2,3,4-tetrahydroisoquinoline Alkaloid Reticuline* (**13**)

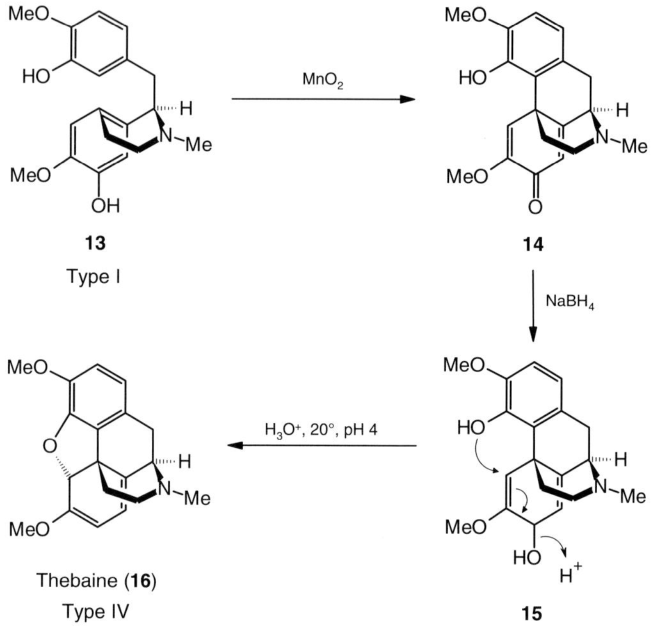

mixture, containing **16**, was diluted with unlabeled thebaine (dilution analysis). The yield was then determined on the basis of the activity of the isolated thebaine.

Rearrangement of 1-Benzyl-1,2,3,4-tetrahydroisoquinoline Alkaloids to Protoberberine Alkaloids (Type I → Type IX)

For the formation of protoberberine-type alkaloids from 1-benzyl-1,2,3,4-tetrahydroisoquinolines, it is necessary to incorporate a C-atom between the N-atom and the lower benzene ring. Formaldehyde is one possible C_1 source, and its incorporation might take place through a *Mannich*-type reaction. A condition for this, though, is that a free phenolic OH group has to be present in the lower ring, either in *ortho-* or in *para*-position to the new bond to be formed.

If (–)-norreticuline (**17**), a Type-I alkaloid, is allowed to stand for two days at pH 6.3 in MeOH containing an aqueous solution of $NaHCO_3$ and formalin, the result is a mixture from which it is possible to isolate the two protoberberine alkaloids (–)-scoulerine (**18**) and (–)-coreximine (**19**) in 43% and

Scheme 8.12. *Synthesis of the Protoberberine Alkaloids Scoulerine (**18**) and Coreximine (**19**) from the 1-Benzyl-1,2,3,4-tetrahydroisoquinoline Alkaloid Norreticuline (**17**)*

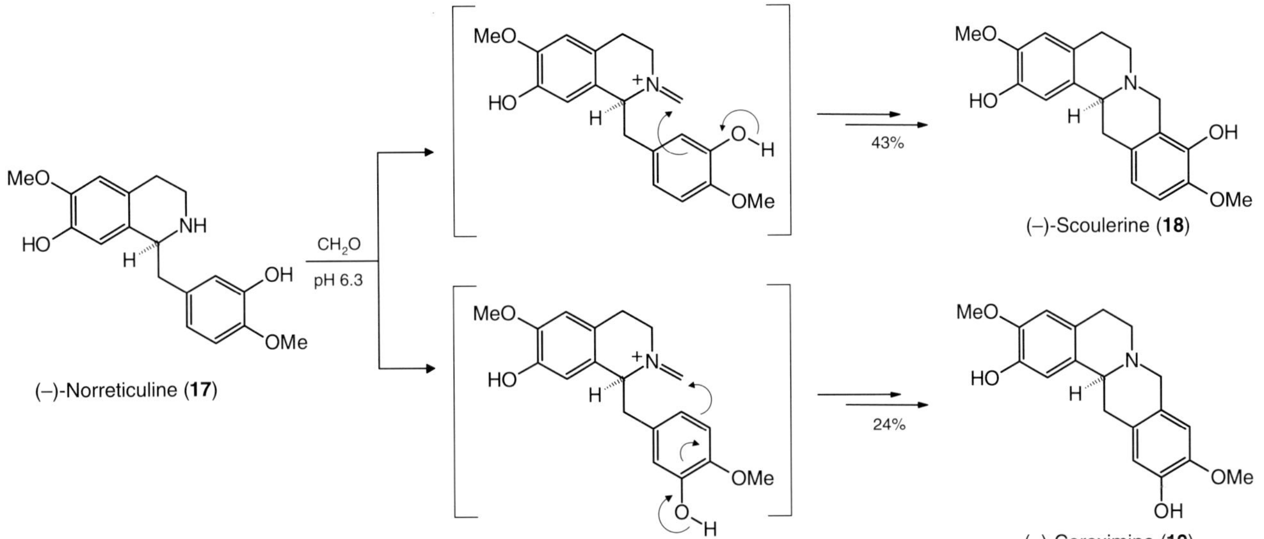

24% yields, respectively (*Scheme 8.12*) [16]. Also, the product ratio can be tuned by varying the pH.

Rearrangement of 1-Benzylisoquinoline Alkaloids to Pavine Alkaloids (Type VI)

Upon heating papaverine hydrochloride at 210–230° C, an achiral member of the 1-benzylisoquinoline alkaloids, a thermal transmethylation takes place, giving rise, among other compounds, to the betaine phenol **a** (*Scheme 8.13*). Compound **a** can be converted into **b** by alkylation with MeI. Acid-catalyzed reduction of **b** finally leads, by way of the immonium intermediate **c**, to the pavine alkaloid codamine [17].

Rearrangement of the 1-Benzyl-1,2-dihydroisoquinoline Alkaloids to the Isopavine Alkaloids (Type VII)

For a transformation of this kind, 1,2-dihydro-*N*-methylpapaverine (**d**) was converted to the desired 1,2,3,4-tetrahydro-4-hydroxy-*N*-methylpapaverine

Scheme 8.13. *Conversion of the 1-Benzylisoquinoline Alkaloid Papaverine to the Pavine-Type Alkaloid Codamine*

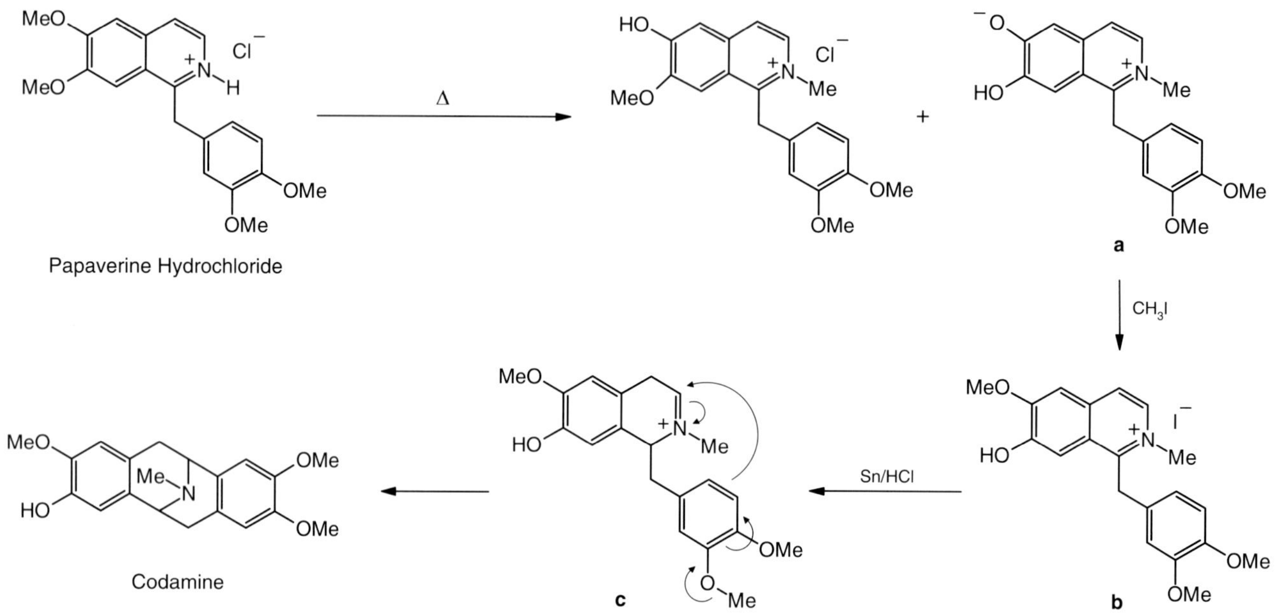

(**e**) by hydroboration (*Scheme 8.14*). In hydrochloric-acid solution, compound **e** spontaneously rearranges to the isopavine alkaloid *O*-methylthalisopavine (**f**)[18].

These kinds of reactions can be listed almost *ad infinitum*, but the examples given should suffice to demonstrate that such transformations of simple, natural isoquinoline derivatives into more complex natural products can also successfully be performed in the laboratory. Finally, these synthetic transformations also support the proposed biogenetic pathways derived statistically.

8.2.4. Alkaloid Biogenesis in *Papaver somniferum*

As mentioned before, the synthesis of a particular alkaloid in a number of plants does not have to proceed in the same manner in all cases. In the next section, we shall examine the biogenesis of alkaloids in just one plant: *Papaver somniferum* L. (Papaveraceae). This example has been chosen, because this particular plant has been intensively studied from a biogenetic

Scheme 8.14. *Formation of O-Methylthalisopavine from N-Methyl-1,2-dihydropapaverine*

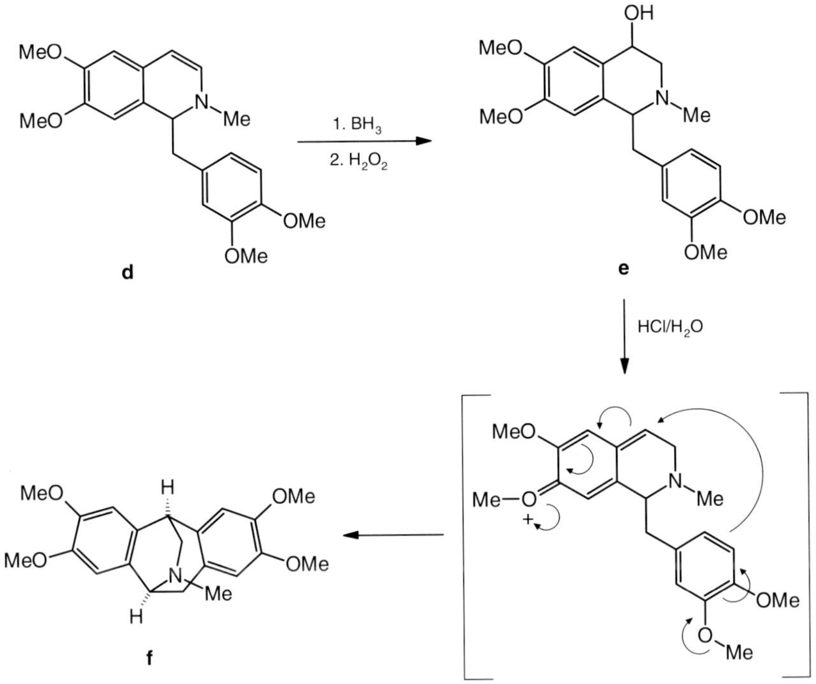

viewpoint; also, several members of the isoquinoline alkaloids subgroups have been detected in it.

Investigations, geared towards establishing the biogenesis of the morphine alkaloids (especially thebaine (**16**), codeine (**28**), and morphine (**29**)) in *P. somniferum*, have lasted over many years and required a great number of individual experiments[19–21]. The results of these studies are summarized in *Scheme 8.15*.

Initially, the question of the true precursors of the isoquinoline alkaloids produced by *P. somniferum* was a matter of considerable interest. After labeled tyrosine (**20** = [2-^{14}C]-tyrosine) had been fed to the plant, the following active alkaloids were isolated: papaverine (**23**; [1-^{14}C,3-^{14}C]-**23**), thebaine (**16**), codeine (**28**), and morphine (**29**; [9-^{14}C,16-^{14}C]-**29**). These findings clearly demonstrated that two molecules of tyrosine are consumed in the production of both papaverine (**23**) and morphine (**29**). Incorporation of phenyl-

Scheme 8.15. *Biogenesis of Morphine (29) in Papaver somniferum*

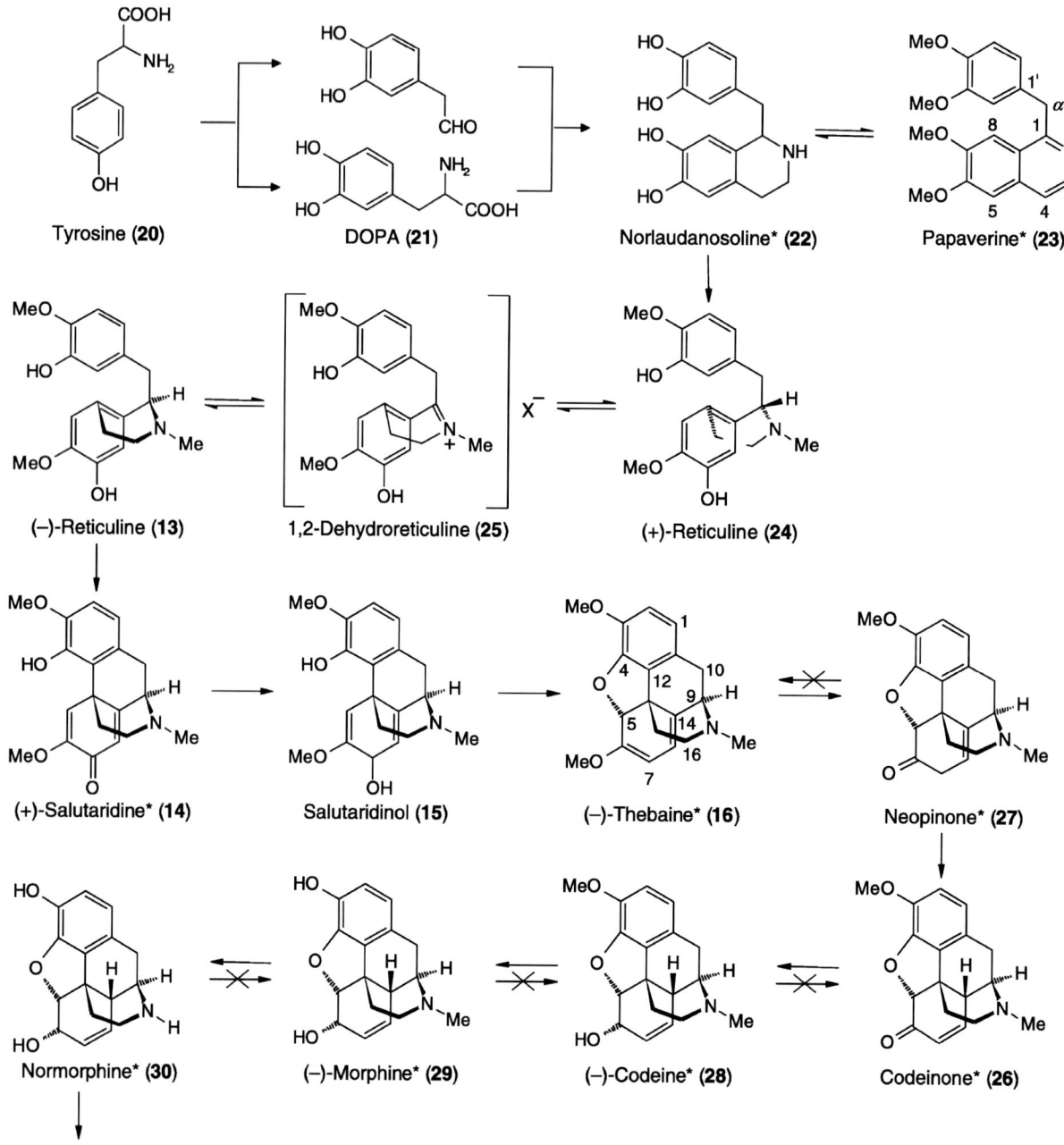

Tyrosine (**20**)

DOPA (**21**)

Norlaudanosoline* (**22**)

Papaverine* (**23**)

(−)-Reticuline (**13**)

1,2-Dehydroreticuline (**25**)

(+)-Reticuline (**24**)

(+)-Salutaridine* (**14**)

Salutaridinol (**15**)

(−)-Thebaine* (**16**)

Neopinone* (**27**)

Normorphine* (**30**)

(−)-Morphine* (**29**)

(−)-Codeine* (**28**)

Codeinone* (**26**)

Non-Alkaloid Metabolites

* Isolated from or detected in *P. somniferum*.

alanine into morphine did take place as well, but far less readily compared to tyrosine. It was also found that labeled norlaudanosoline (**22**) was aromatized and methylated in the plant, yielding papaverine (**23**).

If DOPA (**21**; [2-^{14}C]-**21**) was given to the plant in place of tyrosine, then the radioactive label was found exclusively at C(16) in thebaine, codeine, and morphine. This means that, while both parts of the 1-benzyl-1,2,3,4-tetrahydroisoquinoline alkaloids can certainly be produced from tyrosine, DOPA, in contrast, is only incorporated into the amino component. The plant thus does not convert DOPA into the corresponding arylbenzaldehyde.

With the aid of appropriate feeding experiments, using labeled alkaloid intermediates, it was possible to explain the biosynthesis of morphine (**29**) (*Scheme 8.15*):

1) Codeine (**28**) is demethylated to morphine; the reverse process, however, is not observed.

2) The rate of incorporation of $^{14}CO_2$ and labeled tyrosine (**20**; [2-^{14}C]-**20**) is initially greater for thebaine (**16**) than for both codeine (**28**) and morphine (**29**). After a few days, though, morphine begins to exhibit the highest rate of incorporation, which indicates that thebaine is formed first during biosynthesis, followed by codeine, and then by morphine.

3) Labeled thebaine is converted to both codeine and morphine, while neither codeine nor morphine are transformed into thebaine.

4) Labeled codeinone (**26**) is converted specifically to both codeine (**28**) and morphine (**29**).

5) Neopinone (**27**) is transformed into codeine (**28**) in the plant.

6) Labeled salutaridine (**14**) is incorporated at a high rate into the morphinan alkaloids thebaine, codeine, and morphine. Moreover, the two C(7)-epimers of salutaridinol (**15**) also serve as precursors for **16**, **28**, and **29**.

7) The *O*- and *N*-Me groups preferentially originate from methionine, whereas formates are obviously less well suited for C_1 incorporation.

8) The nonmethylated 1-benzyl-1,2,3,4-tetrahydrosoquinoline derivative norlaudanosoline (**22**) is incorporated into papaverine (**23**), thebaine (**16**), codeine (**28**), and morphine (**29**) at a greater rate compared to tyrosine. This implies that **22** is 'closer' to the mentioned alkaloids than tyrosine. *O,O,O,O*-Tetramethylnorlaudanosoline, in contrast, is only incorporated into morphine alkaloids to an extremely small extent. Obviously, the additional Me groups block the reaction. The *O*-Me substitution pattern of thebaine (**16**) matches neither that of **22** nor that of its tetramethyl ether. Reticuline (**24** and **13**), however, possesses the correct number of Me groups in the same positions as thebaine does. Reticuline, logically, is therefore incorporated into thebaine without demethylation.

9) *P. somniferum* degrades morphine (**29**) into normorphine (**30**). The plant is not able, however, to methylate the latter to produce **29**. Normorphine, for its part, seems to be degraded further to give non-alkaloidal metabolites.

10) Finally, a very interesting stereochemical aspect of this biosynthesis was examined more closely. The absolute configurations of the alkaloids (–)-thebaine, (–)-codeine, and (–)-morphine are shown in *Scheme 8.15*. The true precursor of the morphine alkaloids would, according to this, be (–)-reticuline (**13**) rather than (+)-reticuline (**24**). It might, therefore, be anticipated that only **13** would be taken up into the biosynthesis pathway, while **24** would not. However, both (+)- and (–)-reticuline are incorporated into these alkaloids. In the light of this, the tritium-labeled compounds $[1\text{-}^3H_1]$-**24** and $[1\text{-}^3H_1]$-**13** were used in feeding experiments. It was found that tritiated (+)-reticuline produced tritium-free morphine, while tritiated (–)-reticuline, matching morphine in its absolute configuration, was converted into partly labeled morphine featuring 60% of the original $^3H_1$ content. At this point, it was concluded that there must be some intermediate between **24** and **13** in the biogenetic scheme, existing in equilibrium with both epimers. Hence, if (+)-reticuline (**24**) is used in the feeding experiment, it has to be converted first into **13** *via* the dehydrogenated intermediate **25** (1,2-dehydroreticuline). (–)-Reticuline (**13**), however, is transformed directly to **14** by phenolic oxidation. The partial loss of $^3H_1$ in the feeding experiment with **13** then is a consequence of the equilibrium between **24** and **13**. Evidence that incorporation did indeed take place was provided by double labeling, in tandem with $[3\text{-}^3H_1]$ labeling[22].

These experimental findings are in agreement with the biosynthetic pathway depicted in *Scheme 8.15*. However, two more decisive factors for successful biogenetic experiments should be stressed here:

1) Clear results may only be expected from feeding experiments with precisely labeled precursors. Their chemical synthesis, therefore, takes on great importance. For reasons of space, though, it is not possible to go into detail here.

2) The chemical degradation of labeled metabolites is of no less importance. Total radioactivity is measured only rarely. The goal is usually to determine the position in a molecule at which incorporation of a radioactive element has taken place as compared to its precursor. To prevent false conclusions, the degradation reactions must proceed chemically unambiguously. Synthesis and degradation are frequently the time-determining factors in biogenesis-related experimentation. In view of this, it must also be borne in mind that, in plants, the degree of incorporation is subject to seasonal variation.

In the case at hand, the two atoms C(9) and C(16) in morphine (**29**) are particularly significant, since they can be isolated specifically by chemical degradation. In *Schemes 8.16* and *8.17* the corresponding degradation reactions are shown and will not have to be further commented. Fundamental works on biogenesis have been carried out by *Battersby* (*Fig. 8.1.*) and co-workers.

8.2.5. Papaver orientale [24]

Papaver orientale L. is particularly well-suited to study the biosynthesis of aporphine alkaloids. The following compounds have been isolated from the plant: the aporphine alkaloid isothebaine (**12**), the proaporphines orientalinone (**10**) and dihydroorientalinone, and the morphine alkaloid thebaine (**16**). The latter is of significance insofar as it had been assumed that isothebaine would be synthesized from thebaine. This assumption, though, was not substantiated by the labeling experiments discussed below.

The essential results from biogenetic investigations are summarized in *Scheme 8.18*.

1) When labeled orientaline (**9**), a 1-benzyl-1,2,3,4-tetrahydroisoquinoline alkaloid (Type I), was fed to *P. orientale*, both radioactive isothebaine (**12**) and unlabeled thebaine (**16**) were isolated. Thereby, specifically labeled MeO groups in orientaline were preserved as such in isothebaine. Hence, no demethylation/re-methylation took place during the transformation of orientaline (**9**) to isothebaine (**12**).

2) Also, feeding (+)-orientaline to the plant established that its incorporation was *ca.* 28 times more efficient compared to the corresponding (−)-enantiomer.

Fig. 8.1. *Sir Alan Rushton Battersby* (born March 4, 1925). Studied chemistry in Manchester; Ph.D. in 1959 in Dundee; reader at Bristol, Liverpool, and Cambridge. Natural products chemist, particularly well-known for the structure elucidation and biosynthesis of alkaloids.

Scheme 8.16. *Specific Degradation of Isotopically Labeled Morphine (29) for the Isolation of the C(9)-Atom (present in carbon dioxide)*[23]

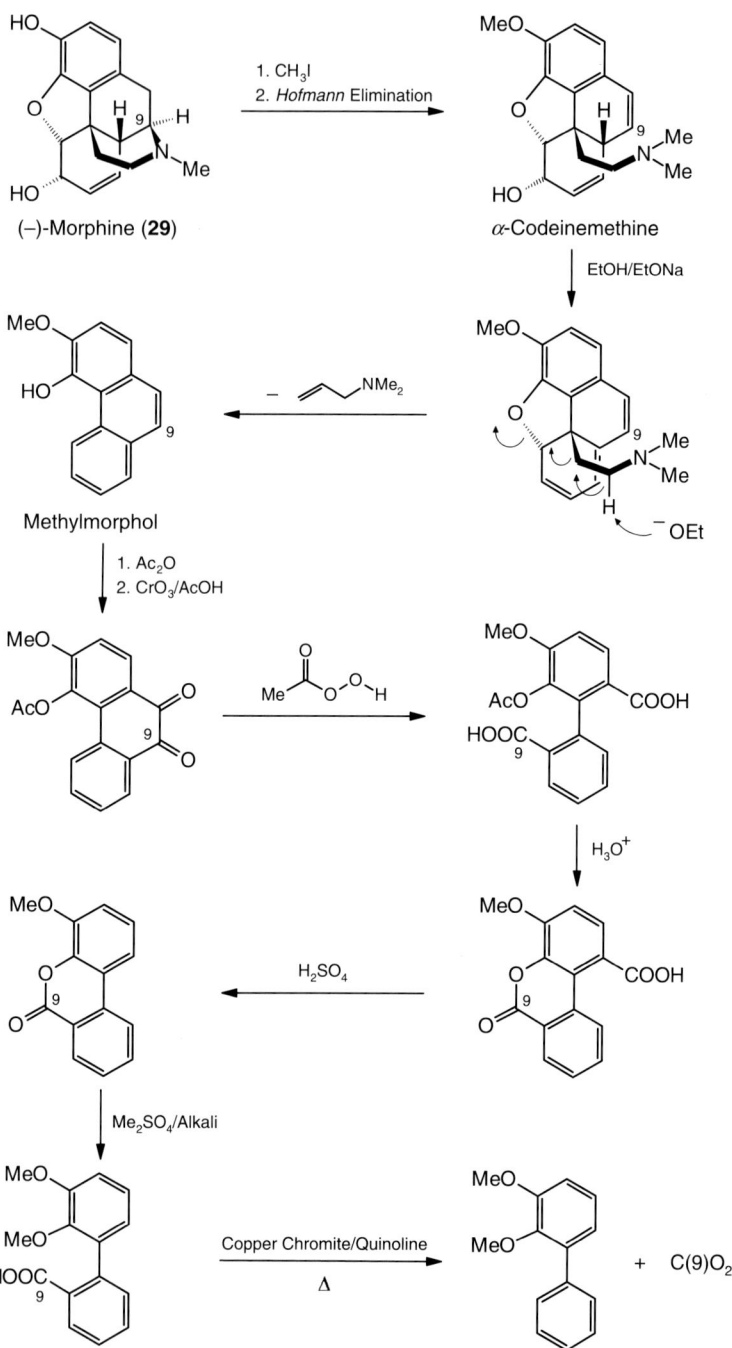

Scheme 8.17. *Specific Degradation of Isotopically Labeled Morphine* (**29**) *for the Isolation of the C(16)-Atom* (present in formaldehyde)[24]

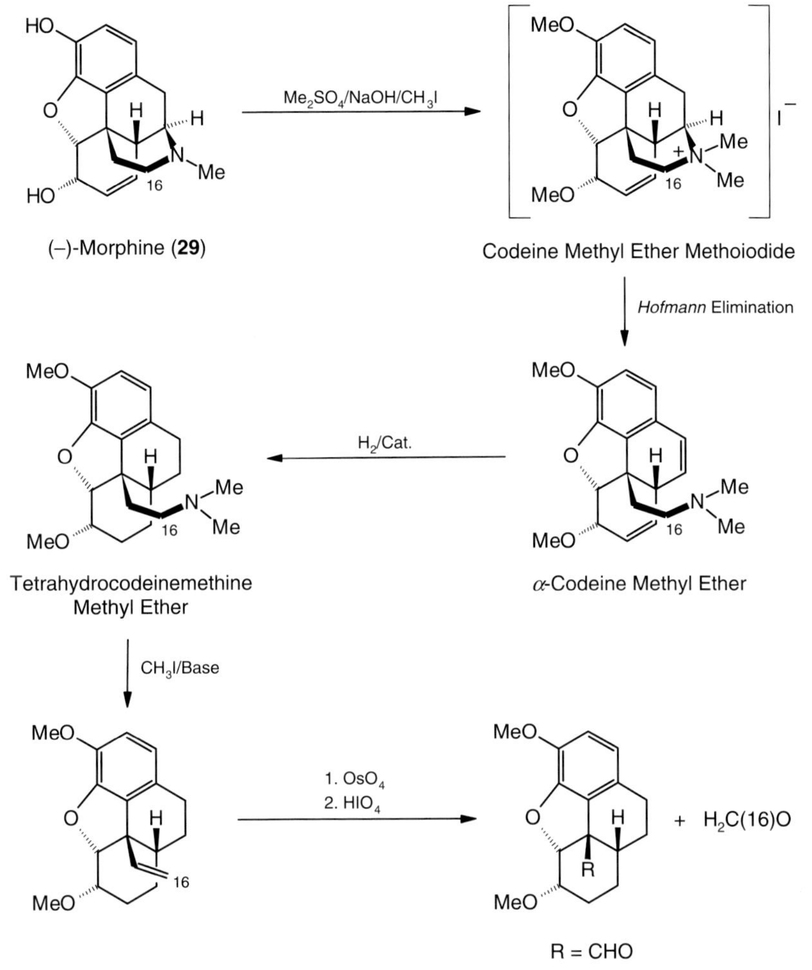
HO
HO
O
H
H
N
Me
16
(–)-Morphine (29)
Me2SO4/NaOH/CH3I
MeO
MeO
Me
Me
I–
Codeine Methyl Ether Methoiodide
Hofmann Elimination
α-Codeine Methyl Ether
H2/Cat.
Tetrahydrocodeinemethine Methyl Ether
CH3I/Base
1. OsO4
2. HIO4
R
+ H2C(16)O
R = CHO

3) Interestingly, labeled reticuline (**24**) was incorporated only into thebaine (**16**), and not into isothebaine.

4) It was also shown that orientalinone (**10**) could similarly be incorporated specifically into isothebaine.

5) The subsequent reactions of the two alkaloids reticuline (**24**) and orientaline (**9**) appeared particularly informative (*cf. Points 3* and *4*). These

Scheme 8.18. *Aporphine Biogenesis in Papaver orientale*

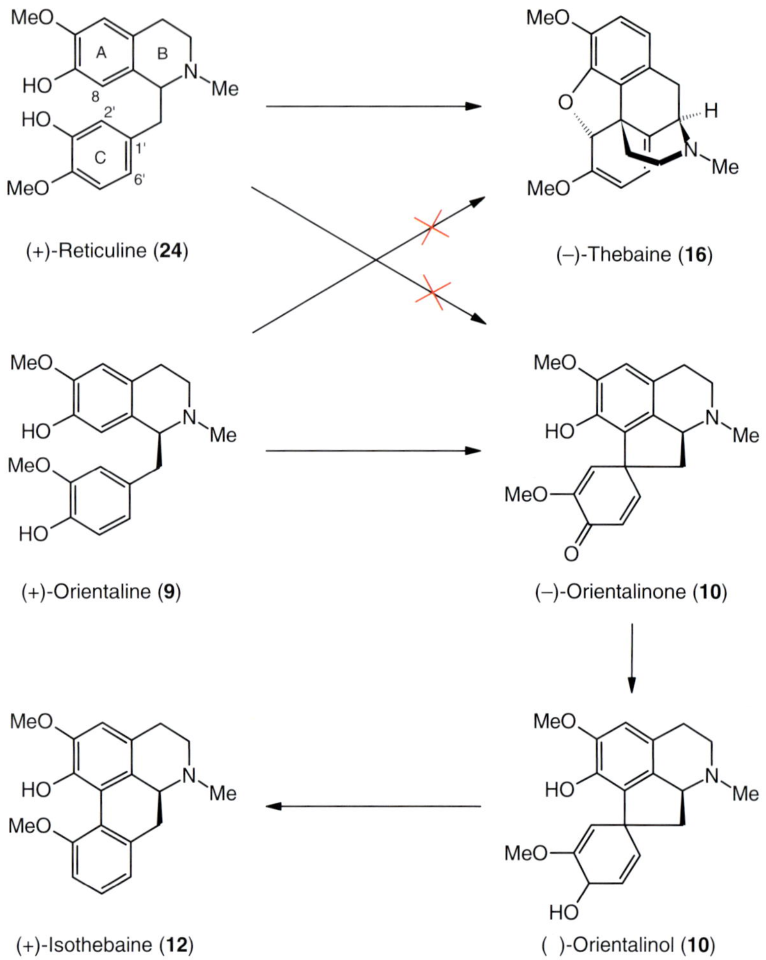

(+)-Reticuline (**24**)

(−)-Thebaine (**16**)

(+)-Orientaline (**9**)

(−)-Orientalinone (**10**)

(+)-Isothebaine (**12**)

( )-Orientalinol (**10**)

compounds differ only in their substitution pattern in ring C: the MeO and OH groups are found in exchanged positions. In ring C of orientaline, the phenolic OH group occupies the *para*-position. In the orientaline → orientalinone transformation, a phenol oxidation is taking place. In this reaction, only a phenolic OH group located *ortho* to the new formed C–C bond in orientalinone is capable of supporting the reaction. Since the plant is clearly not able to perform transmethylations at the stages of orientaline (**9**) or reticuline (**24**), as has been shown experimentally, reticuline cannot be directly converted into orientalinone. An alternative process that would also result in an aporphine ring system – namely, a phenol coupling between C(2′) and C(8), or C(6′) and C(8) – has yet to be detected in *P. orientale*. Whether this pathway is indeed operative is not known. The subsequent reaction of orientalinone (**10**) to isothebaine (**12**) has already been described (*vide supra*).

The biosynthesis of the aporphine alkaloid isothebaine (**12**), shown in *Scheme 8.18*, has been corroborated by investigations into other aporphines. Examples of this include the formation of stephanine (**31**) and aristolochic acid I (**32**) from orientaline (**9**), and that of roemerine (**33**) from coclaurine (**34**) (*Scheme 8.19*).

The above survey of two biosynthetic pathways was meant to demonstrate how investigations of this kind are undertaken. It can clearly be seen, though, that none of the presented schemes should in any way be viewed as truly complete. Some questions always remain open, *e.g.*, is isothebaine degraded to N-free products in a similar way to morphine? Is there a particular order for the methylation of phenolic OH groups in norlaudanosoline (**22**)? Might there still be alternative biosyntheses to aporphine alkaloids?

Scheme 8.19

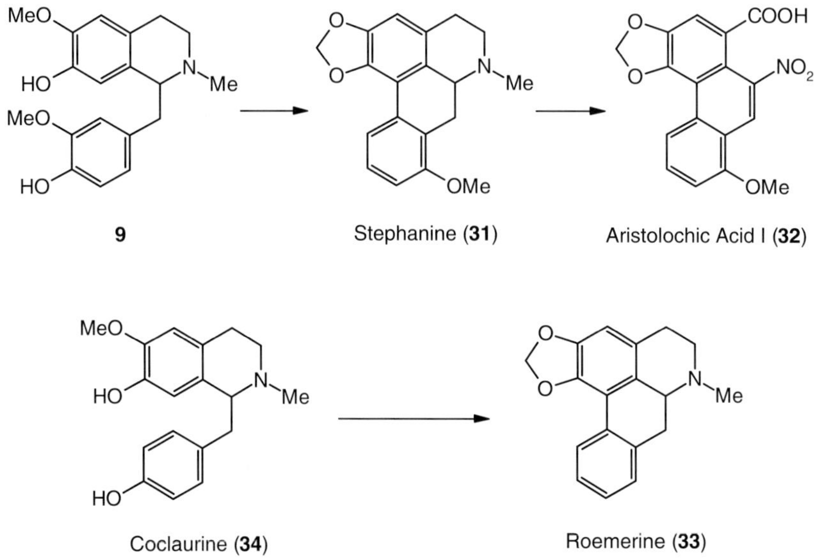

9   Stephanine (**31**)   Aristolochic Acid I (**32**)

Coclaurine (**34**)   Roemerine (**33**)

References and Notes

1. R. F. Raffauf, *A Handbook of Alkaloids and Alkaloid-Containing Plants*, John Wiley & Sons, New York, 1970.

2. In all cases, no distinctions were made as to the nature of the O-containing functional group (OH, MeO, –O–CH_2–O–, C=O), nor as to that of substitution at the N-atom (H, Me, Me_2, oxide). The percentages given relate to all recorded representatives of the particular alkaloid type.

3. Representatives in which ring B is aromatic are also included here.

4. Some 80% of proaporphine alkaloids possess an oxo group at C(10).

5. Some 60% of representatives of morphine alkaloids possess an oxo group at C(7).

6. In virtually all cases, either C(7) or C(9) is O-substituted.

7. Representatives in which rings B and C are aromatic are also included here.

8. Alkaloids with C-substituents at C(13) were included, but are not accentuated specially.

9. The synthetic pathway outlined for the formation of the benzophenanthridine alkaloids has been confirmed by means of feeding experiments[10] on *Chelidonium majus* L.

10. E. Leete, '*Biosynthesis of the Alkaloids of Chelidonium majus. I. The Incorporation of Tyrosine into Chelidonine*', *J. Am. Chem. Soc.* **1963**, *85*, 473.

11. The reaction pathway presented in *Scheme 8.8* agrees with the findings from feeding experiments carried out on *Hydrastis canadensis*[12,22].

12. I. Monkovič, I. D. Spenser, '*Biosynthesis of Berberastine*', *J. Am. Chem. Soc.* **1965**, *87*, 1136.

13. S. Tobinaga. '*A Review: Synthesis of Alkaloids by Oxidative Phenol and Nonphenol Coupling Reactions*', *Bioorg. Chem.* **1975**, *4*, 110.

14. A. R. Battersby, T. H. Brown, J. H. Clements, '*Syntheses along Biosynthetic Pathways, Part I. Synthesis of (+)-Isothebaine*', *J. Chem. Soc.* **1965**, 4550.

15. D. H. R. Barton, G. W. Kirby, W. Steglich, G. W. Thomas, '*The Biosynthesis and Synthesis of Morphine Alkaloids*', *Proc. Chem. Soc.* **1963**, 203.

16. A. R. Battersby, R. Southgate, J. Staunton, M. Hirst, '*Synthesis of (+)- and (–)-Scoulerine and (+)- and (–)-Coreximine*', *J. Chem. Soc. C* **1966**, 1052

17. F. R. Stermitz, J. N. Seiber, '*Alkaloids of the Papaveraceae. III. The Synthesis and Structure of Norargemonine*', *Tetrahedron Lett.* **1966**, 1177.

18. S. F. Dyke, A. C. Ellis, '*The Synthesis of Isopavine Alkaloids*', *Tetrahedron* **1971**, *27*, 3803.

19. R. J. Miller, C. Jolles, H. Rapoport, '*Morphine Metabolism and Normorphine in Papaver somniferum*', *Phytochemistry* **1973**, *12*, 597.

20. H. I. Parker, G. Blaschke, H. Rapoport, '*Biosynthetic Conversion of Thebaine to Codeine*', *J. Am. Chem. Soc.* **1972**, *94*, 1276.

21. H. R. Schütte, '*Isochinolin Alkaloide*', in *Biosynthese der Alkaloide*, Eds. K. Mothes and H. R. Schütte, VEB Deutscher Verlag der Wissenschaften, Berlin, 1969.

22. D. H. R. Barton, R. H. Hesse, G. W. Kirby, '*Phenol Oxidation and Biosynthesis. Part VIII. Investigations on the Biosynthesis of Berberine and Protopine*', *J. Chem. Soc.* **1965**, 6379.

23. A. R. Battersby, B. J. T. Harper, '*Biogenesis of Morphine*', *Chem. Ind. (London)* **1958**, 364.

24. A. R. Battersby, R. Binks, D. J. Le Count, '*Biosynthesis of Morphine*', *Proc. Chem. Soc.* **1960**, 287.

9. Biological Significance of Alkaloids

9.1. General

Plants, animals, and microorganisms certainly do not produce alkaloids for the benefit of humans. It is much more likely that all these life forms have specific reasons to produce or store such a wealth of compounds.

As late as 1969, *Kurt Mothes* (*Fig. 9.1*), one of the great alkaloid chemists of the 20th century, was still able to write '*According to all investigations carried out so far, one can say that alkaloids are without any vital physiological significance for those species that synthesize them*'. He qualified this, though: '*It would be entirely possible that, in individual cases, the formation of alkaloids might have some physiological significance [...] in the sense of a detoxification process that makes metabolically active substances less reactive by stabilization, and hence enables damagingly high concentrations of these substances to be avoided*'. One supporting pillar for his opinion was that: '*It cannot be overlooked that there is no known alkaloid-producing plant that is resistant to animal, fungal, and bacterial pests*'[1].

It is highly unlikely, though, that living organisms should produce substances such as alkaloids in large amounts and then store them in bark or roots, or in skin and other organs, without actually having any particular requirement for them. Production of substances of this kind not only demands starting materials, but also, regardless of their structures, consumes greater or lesser quantities of energy and auxiliary materials. Often, a whole army of the most diverse enzymes is needed for all the biochemical transformations lying between the starting material and the alkaloid, and these certainly cannot be maintained *only* for production of useless materials. Only *Homo sapiens* allows itself the luxury of squandering raw materials and energy. 'Lower' organisms, for technical reasons of survival, possess 'well-planned', 'rational', and 'economical' systems for managing their resources.

If we base our examination of the significance of alkaloids on these premises, then the principal question ('why do organisms produce alkaloids?') might be answered by looking at their actions. There are some difficulties in connection with this, especially because there are so many alkaloid-producing plants. To determine both their alkaloid content and the corresponding chemical structures, the plants are usually examined under 'denatured' conditions, *i.e.*, in a dried state and removed from their natural environment. What the natural 'friends and foes' of a plant might be is largely unknown

Fig. 9.1. *Kurt Albin Mothes* (November 3, 1900 – February 12, 1983). Studied chemistry, pharmacy, and botany in Leipzig; Ph.D. in 1925 (*W. Ruhland*); 1935 Professor of botany and pharmacognosy in Kaliningrad (Königsberg); 1958 Director of the Institute for Plant Biochemistry, Halle/Saale. Worked particularly in the fields of plant biochemistry and alkaloid biosynthesis.

and very seldom analyzed. To address this question properly requires close cooperation between scientists of such diverse disciplines as organic chemistry, microbiology, soil science, botany, zoology *etc.*

Since, in the confines of this book, it is virtually impossible to attempt a detailed study of the biological effects of alkaloids, we shall refer here to some specialist literature reviews in which some of the following aspects are examined more closely[2–5].

9.2. Antifungal Activity

The aporphine alkaloid liriodenine (**1**), isolated from the tulip tree *Liriodendron tulipifera* (Magnoliaceae), displays a broad activity spectrum against fungi such as the yeast *Candida albicans*[6], (MIC[7] value: 6.2 mg/ml), *Trichophyton mentagrophytes*, *Aspergillus niger*, *Syncephalastrum racemosum*, and *Mucor griseocyanus*. Particularly noteworthy is its high activity against phytopathogenic fungi, which provides a plausible reason for the synthesis of the compound in terms of the survival strategy of the plant. Thus, liriodenine (**1**) is active against such pests as *Helminthosporium teres* (Net Blotch of barley), *Botrytis fabae* (a fungus that causes brown spots in *Vicia faba*, the faba bean), *Plasmospora viticola* (grape downy mildew), and *Pericularia oryzae* (blast of rice in *Oryza sativa*).

Other oxoaporphine alkaloids, structurally similar to liriodenine (**1**), *e.g.*, lysicamine (**2**), oxonantenine (**3**), and cassameridine (**4**), display only moderate activity, compared with **1**, against *Trichophyton mentagrophytes* and *Saccharomyces cerevisiae*. Of the great number of possible phytopathogenic fungi that might be found in the natural environment of the 'alkaloid-producer', it would be entirely feasible, extrapolating on the basis of the established antifungal activities, that liriodenine (**1**) might be an even stronger toxin to non-species-specific fungi; however, those native to the natural environment of the plant have not been tested yet for their sensitivity to liriodenine[2].

Another example of alkaloidal antifungal activity is represented by juliprosopine (**5**; from *Prosopis juliflora* (Leguminosae))[8]. This was tested not against phytopathogenic fungi, but against several dermatophytic fungi of various species of *Trichophyton*, *Microsporum*, and *Epidermophyton floccosum*, in the course of which MIC values of as little as 1.5 mg/ml were determined[9], thus making its *in vitro* antifungal activity comparable to that of griseofulvine[10]. Since, at *ca.* 0.2%, the alkaloid content of roots of *P. juliflora* is very high, it seems reasonable to assume that juliprosopine (**5**) and relat-

Liriodenine (**1**)

Lysicamine (**2**)

Oxonantenine (**3**)

Cassameridine (**4**)

ed alkaloidal compounds found in the plant are not just random derivatives or metabolic degradation products, but serve to protect the plant from phytopathogenic fungi.

Juliprosopine (**5**)

In further, extensive studies, very large differences in activities have been reported for alkaloids tested against certain microorganisms; a finding that fits expectations for reasons of species diversity.

Higher plants occasionally co-exist with fungi, often in a symbiotic relationship. These fungi may be found within the plant itself, or they occur externally, in the region of the roots, for instance. A well-known example of the latter is provided by the nodule bacteria (*Rhizobium*) of Leguminosae, which reduce atmospheric nitrogen (N_2) to ammonia (NH_3) or to simple amines, needed by the plant; the latter, in return, shares other substances with the bacteria. It is certain that alkaloids are involved in such symbiotic relationships. However, little is yet known about their structure and function.

9.3. Alkaloids as Protectors against UV Irradiation

In humans and other vertebrates, copulation involves the transfer into the female reproductive tract not only of sperm cells but also of polyamines, found in the seminal fluid. It has long been known that these polyamines include spermidine (**6**) and spermine (**7**)[11]. These amines, also known as biogenic amines, are water-soluble, strongly basic compounds, the functions of which appear to include, among others, pH regulation (sperm: pH 7.2; vaginal environment: pH 4) and the stabilization of the acidic phosphoric acid residues in ribonucleic (RNA) and deoxyribonucleic acids (DNA) (*Fig. 9.2*)[12]. In flowering plants, essentially the same process takes place; pollen is transferred from the male flower organs (anthers) to the female flower organs (stigmas, pistils), ultimately resulting in fertilization. In the following discussion, it is irrelevant whether the plants are monoecious or dioecious (*cf. Corylus avellana, Fig. 9.3*), or whether the pollen is trans-

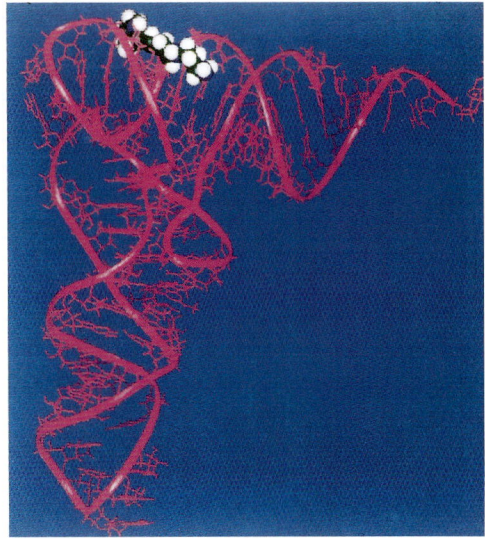

Fig. 9.2. *Localization of a specific binding region for polyamines in tRNA, assisting better understanding of the roll of this molecule, ubiquitous in living organisms* (picture B. Frydman)

Fig. 9.3. *Hazel, Corylus avellana (Corylaceae), which releases its pollen in early spring, contains, like the pollen of many other plants, p-cumaroylspermidines*

ferred by the wind or by insects. In chemical investigations of the pollen of a variety of plants (*e.g., Aphelandra chamissoniana, A. tetragona* (Acanthaceae)), there have been found a number of cinnamic acid derivatives of polyamines including N^1,N^5- and N^5,N^{10}-di(*p*-cumaroyl)spermidine, together with N^1,N^5,N^{10}-tri(*p*-cumaroyl)spermidine. Dicumaroylspermidines have also been isolated from or detected in the pollen of Arales, Fagales, Juglandales, and Myricales.

Spermidine (**6**)

Spermine (**7**)

(*E,E*)-Isomer

(*Z,Z*)-Isomer

(*Z,E*)-Isomer

((*E,Z*)-Isomer not depicted)

Isomers of N^1,N^5-Di(*p*-cumaroyl)spermidine

The dicumaroylspermidines are derivatives of polyamines and thus belong to the alkaloids. Their biological functions may coincide with those of polyamines in other living organisms. The additional substitution, though, appears to protect the transferred genetic material from UV light[13]. For the $(E) \rightarrow (Z)$ isomerization of *p*-hydroxycinnamic acid or *p*-methoxycinnamic acid, UV light is required, whereas the $(Z) \rightarrow (E)$ isomerization takes place thermally in daylight. With the aid of this cycle, dangerous UV irradiation is quenched and damage to the (photolabile) genetic material prevented. Were this protection not available, then, presumably, the consistency of transmission of genetic information would be seriously compromised. The HPLC diagram in *Fig. 9.4* shows a mixture of isomers produced on irradiation of (E,E)-N^1,N^5-di(p-cumaroyl)spermidine.

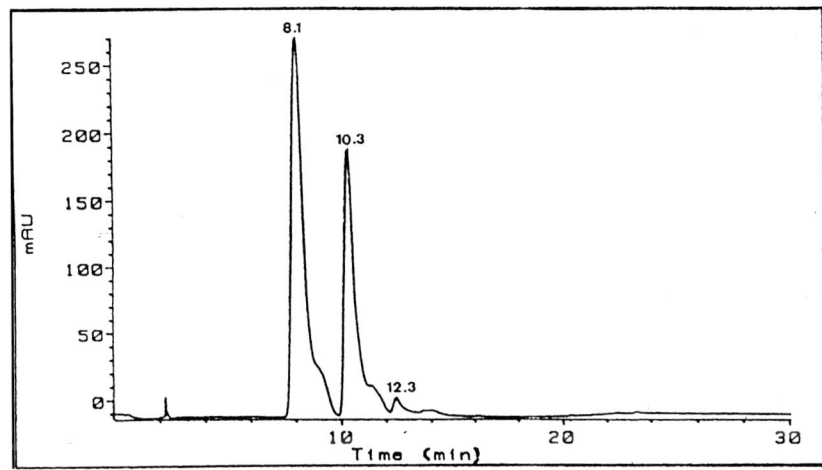

Fig. 9.4. *HPLC diagram* (reversed phase, *Nucleosil-C-8* column (5μm, 200 × 4 mm i.d.), flow rate 1 ml/min, 2-step linear gradient, H_3PO_4, AcOH, MeCN, H_2O) *of* N^1,N^5-*di(p-cumaroyl)spermidine after irradiation of the (E,E)-isomer with UV light* (quartz vessel, low-pressure Hg-vapor lamp, 366 nm, 50 min, MeOH). Identification: 8.1 min: (*Z,Z*)-isomer, 10.3 min: (*E,Z*)- and (*Z,E*)-isomers, 12.3 min: (*E,E*)-isomer[14].

9.4. Hydroxamic Acids as Insecticides, Herbicides, Fungicides, and Siderophores

The hydroxamic acids and their rearrangement products, the benzoxazol-2(3*H*)-ones, similarly belong to the alkaloids. Biogenetically, they are derived from anthranilic acid and ribose-5-phosphate, or, according to more recent investigation, from indole (*cf. Chapt. 2*). In plants, these compounds mainly exist in the form of their 2β-*O*-D-glucopyranosides (*cf.* **8** or **9**). If the plant cells are damaged, β-glucosidase is released, which enzymatically liberates highly reactive species such as DIBOA (3,4-dihydro-2,4-dihydroxy-1,4-benzoxazin-3(2*H*)-one) from the corresponding glycosides (*Scheme 9.1*). DIBOA decomposes into formic acid (HCOOH) and benzoxazol-2(3*H*)-one (**12**; BOA). The half-life of the methoxy-substituted hydroxamic acid DIMBOA (**11**) at pH 5.6 and 28° C is *ca.* 24 h.

Scheme 9.1

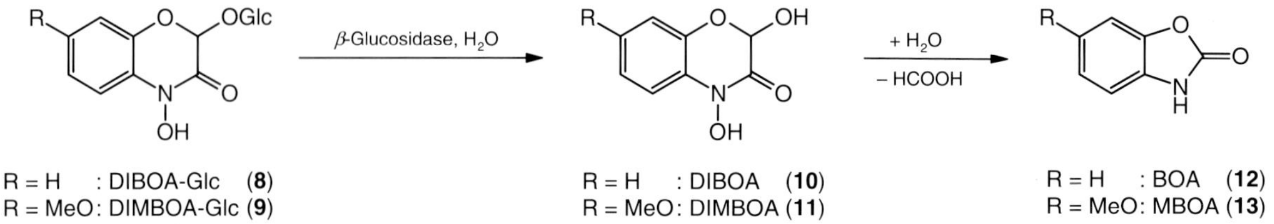
β-Glucosidase, H_2O
$+ H_2O$
$- HCOOH$
R = H : DIBOA-Glc (**8**)
R = MeO: DIMBOA-Glc (**9**)
R = H : DIBOA (**10**)
R = MeO: DIMBOA (**11**)
R = H : BOA (**12**)
R = MeO: MBOA (**13**)

Three different reaction mechanisms for this conversion have been proposed, the most likely of which is shown in *Scheme 9.2*. It comprises the elimination of the C(2)-atom as HCOOH, with formation of the reactive isocyanate[15].

Scheme 9.2. *Conversion of 3,4-Dihydro-2,4-dihydroxy-1,4-benzoxazin-3(2*H*)-one* (DIBOA; **10**) *into Benzoxazol-2(3*H*)-one* (BOA; **12**) *with Liberation of Formic Acid*

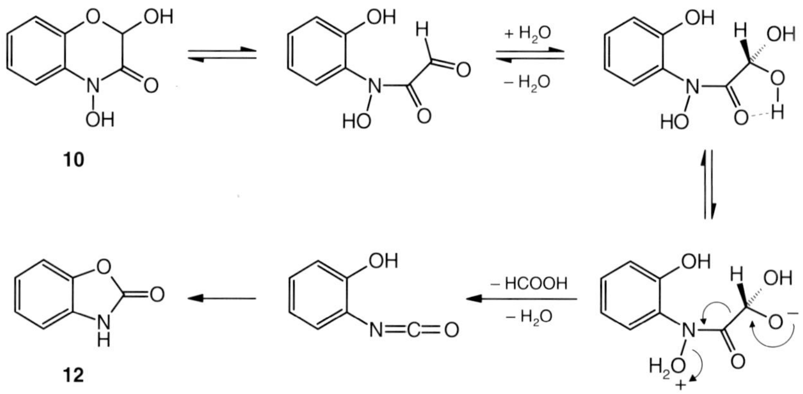
10
$+ H_2O$
$- H_2O$
$- HCOOH$
$- H_2O$
12

HCOOH is both a preservative and disinfectant, which is known to generally inhibit fungal infestation.

Hydroxamic acids of type **8** have been isolated from various Graminae genera, including *Aegilops*, *Arundo*, *Chusquea*, *Coix*, *Elymus*, *Secale*, *Sorghum*, *Tripsacum*, *Triticale*, and *Zea* (*cf. Fig. 9.5*), from *Aphelandra* and *Acanthus* (Acanthaceae), *Consolida* (Ranunculaceae), as well as from *Scoparia* and *Blepharis* (Scrophulariaceae).

To date, the properties of hydroxamic acids in defending against insect pests of all types have been most closely studied in the Graminae. They are active against the European and Asian Corn Borers (*Ostrinia nubilalis, O. furunacalis*), as well as the Southwestern corn borer (*Dietraea grandiosella*). In addition, aphids (*Schizaphis graminum, Sitobion avenae, Metopolophium dirhodum, Rhopalosiphum maidis*) and other insects have been successfully controlled with the aid of hydroxamic acids.

Hydroxamic acids also inhibit the germination of fungal spores (*Helminthosporium turcicum*). Levels of stem rot in grasses (*Diplodia maydis, Fusarium moniliforme, Gibberella zeae, Cephalosporium maydis*) and of Black Rust in grain (*Puccinia graminis*) are inversely proportional to the hydroxamic acid content of the grasses. These compounds also make plants more resistant to *Erwinia* species, causes of soft rot.

The physiological modes of action of hydroxamic acids are diverse. They may act directly on enzymes and enzyme systems, causing inhibition of certain reactions. One known effect, for example, consists of the oxidation of cysteine (and other thiols) to cystine derivatives within peptide chains. This becomes especially relevant if sucking or biting insects or fungi attack a plant. If the concentration of hydroxamic acids in the plant is high enough, the parasitic organisms literally burn their 'tongues'.

The phytotoxic alkaloids DIBOA (**10**) and BOA (**12**) also display another interesting property; they can cause allelopathic action[16]. Alkaloids **10** and **12** both influence the germination capability of seeds and reduce root growth. When these compounds are secreted by a plant, they may inhibit the growth of other plant species (*e.g.*, weeds) and so act as natural herbicides[17].

Another important function of hydroxamic acids in plants lies in controlling iron metabolism. The alkaloid DIMBOA (**11**), for instance, is a siderophore secreted by an organism, which complexes Fe^{III} ions in the soil. Only with the aid of **11** is it possible for the plant to take up the biologically important Fe^{III} ions and have them available for metabolism.

Electrospray-ionization mass spectrometry (ESI-MS) has provided insight into the structure of this complex. It has been established that (in aqueous methanolic solution) each Fe^{III} is bound by three hydroxamic acid molecules (*Fig. 9.6*). These results also corroborated previously determined complexation constants in solution[18].

Benzoxazol-2(3*H*)-one (BOA; **12**) and its MeO-substituted derivatives, derived from the corresponding hydroxamic acids, are, as mentioned, allelo-

Fig. 9.5. *Zea mays* L., *one of the first plants in which hydroxamic acids of the DIBOA type were found* (young corn plants, July, Switzerland)

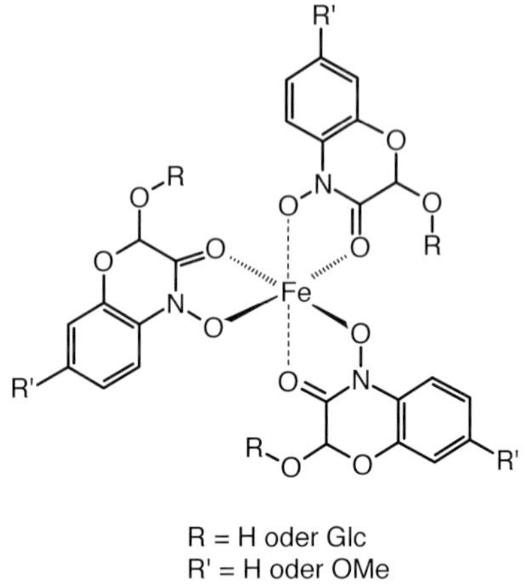

R = H oder Glc
R' = H oder OMe

Fig. 9.6. *Complex of Fe^{III} ions and hydroxamic acids in a 1:3 ratio in H$_2$O/MeOH solution. The geometric representation of the isomers is arbitrary.*

pathically active. It has been shown that grasses such as oats (*Avena sativa*) and wheat (*Triticum aestivum*), and also beans (*Vicia faba*), take up BOA (**12**) through their roots. These plants, however, have developed detoxification mechanisms through which they are able to eliminate these substances when no longer needed. Thus, oats and wheat produce different metabolites out of BOA, namely 6-hydroxybenzoxazol-2(3*H*)-one (6-OH-BOA) and its glycoside 6-*O*-Glc-BOA. These two compounds, though, are still toxic to the plants. It is thought that they are stored in vacuoles in the roots, where they act as feeding deterrents. A true detoxification product of BOA (**12**) is its *N*-glucoside *N*-Glc-BOA, which can likewise be detected in these grasses.

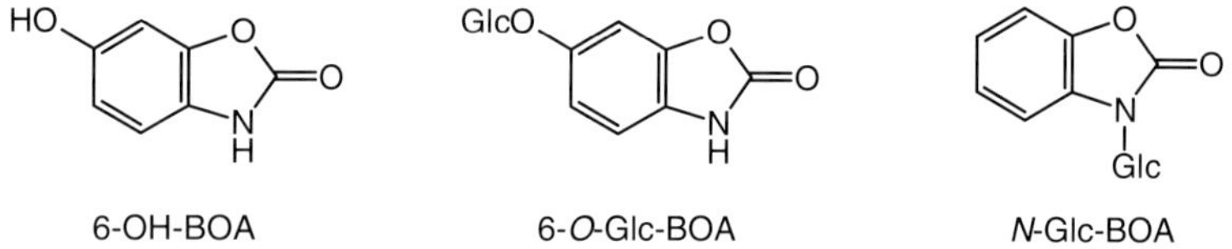

6-OH-BOA   6-*O*-Glc-BOA   *N*-Glc-BOA

Detoxification reactions are also a consequence of the interplay between grasses and pathogenic fungi such as *Fusarium culmorum*, *F. moniliforme*, or *Gaeumannomyces graminis*. These fungi, especially *F. culmorum* and *G. graminis*, appear to be responsible for the transformation of BOA (**12**) into *N*-(2-hydroxyphenyl)malonic acid monoamide (**15**) (and its MeO-substituted derivatives) and into 2-amino-3*H*-phenoxazin-3-one (**14**) (*Scheme 9.3*)[19].

It cannot be denied that the hydroxamic acids, thanks to their agricultural significance, constitute a particularly thoroughly studied group of alkaloids. They may be summarized as comprising a group of universal insecticides,

Scheme 9.3

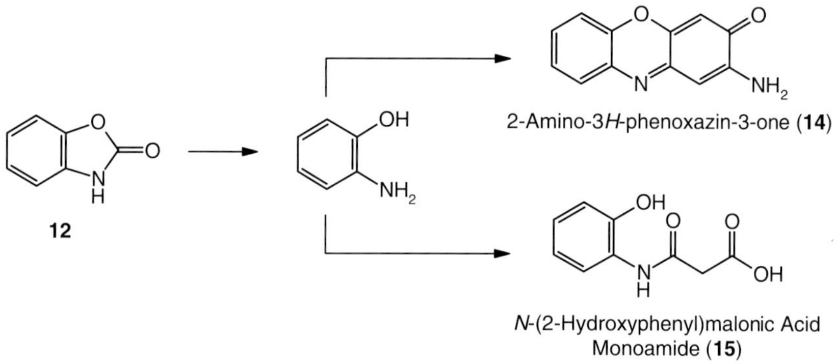

2-Amino-3*H*-phenoxazin-3-one (**14**)

N-(2-Hydroxyphenyl)malonic Acid Monoamide (**15**)

herbicides, and fungicides of natural origin, which can also function as side-rophores. Many alkaloids are less multifaceted, but, in compensation, more specific in their action. It is, however, possible that other alkaloids also possess similarly broad spectra of activities, but this must remain speculation for now, since further investigations have yet to be carried out.

From this example, it clearly emerges that alkaloids in plants fulfill important tasks by means of protecting their 'producers' from a range of external factors.

9.5. Alkaloids as Feeding Deterrents

Many insect species are plant eaters (herbivores). Some favor a large range of host plants (polyphagous herbivores), while others are specialized towards one or a very few plants (monophages and oligophages). The polyphagous insects change plant species frequently to avoid death from plant poisons. Many of these insects have also developed detoxification mechanisms, which allow them rapidly to rid their bodies of the harmful substances.

Monophagous and oligophagous species, on the other hand, sometimes specially pick out the plants with the 'toxic' contents. A convincing example of this can be found in the alkaloid-containing plant *Lupinus polyphyllus* (Leguminosae), endemic in North America and introduced to Europe and cultivated *ca.* 300 years ago. In its homeland, the plant is being used as a food source by a number of herbivores. In Europe, for a long time, it was largely unmolested by herbivores. This was to change when, in the 1980s, the American Lupin Aphid *Macrosiphum albifrons* was (accidentally) brought to Europe, where it spread widely.

Pyrrolizidine alkaloids are highly effective means of defense for the plants that produce them, as well as for some insects, which ingest and put them to use themselves. This process involves storage of the alkaloids in their *N*-oxide forms, in which they are nontoxic. In this manner, pyrrolizidine alkaloids can not only be stored safely, but also rapidly transformed into their toxic forms by enzymatic degradation. Senecionine (**16**; isolated, *inter alia*, from *Senecio*, *Adenostyles*, *Brachyglottis*, *Emilia*, *Erechtites*, *Gynura*, and *Ligularia* (all Compositae)) and its *N*-oxide are examples of such compounds.

(–)-Senecionine (**16**)
(abs. config.)

Vertebrates, for instance, refuse to consume food that has been mixed with certain pyrrolizidine alkaloids. A similar state of affairs applies for some insects, notably butterfly larvae. There are some butterflies, on the other hand (*e.g.*, *Utetheisa*, *Hyalurga*, and *Ithomiinae*), that protect themselves by ingesting pyrrolizidine alkaloids and, in turn, are not eaten by the tropical web-building spider *Nephila clavipes*. The spider will immediately cut the butterfly out of its web when it gets caught. If, however, the spider is offered *Utetheisa*, fed as larvae with a diet free of pyrrolizidine alkaloids, so that they are demonstrably alkaloid-free, it will eat them. A number of similar examples are known[20].

Investigations have shown that herbivores generally prefer a plant with a low alkaloid content to one with a high alkaloid content. The monophagous insects, though, represent an exception, 'consciously' favoring plants containing particular alkaloids as food sources.

A host of other alkaloids have also been tested for their biological properties such as general (*e.g.*, anabasine) or specific (*e.g.*, nicotine against bees or *Spodoptera*) insecticidal action, feeding deterrence (*e.g.*, nicotine against *Locusta*), larva toxicity, and similar characteristics.

In summary, it can be regarded as certain that alkaloids are not just some waste products of nature; they seem to be highly significant for the survival of plants, even if far from all their functions are known and, by no means, all alkaloid classes have been tested. Plants ensure their survival against microorganisms, insects, and plant-eaters, and also against other plants, by means of allelopathically active chemicals. In addition, they have 'learned' how to cope with damaging irradiation. Here, we can gain some impression of the fantastic world that has developed over millions of years to what we see today, and about which we know only a fraction. In general, two fundamentally different approaches have emerged in this 'struggle of the fittest'. First, there is the strategy of somehow removing opponents, competitors, and rivals for food and light from the scene, or at least to render them harmless. The second strategy is based on deterrence and/or repulsion with physical and/or chemical methods.

The physical method *de facto* results in the annihilation of the opponent; the victim, as a rule, is either eaten or buried. Here, both energy and raw materials (resources) are only partially utilized – a particular specialty of humans. When plants are involved, the physical method also means that they are used as donors of food and energy, or as construction materials *etc*. The chemical method, on the other hand, is much more refined, with chemicals (not only alkaloids) used to keep the opponent at a safe distance, or, if it does not understand this language, to give it a sharp lesson. This process makes use of especially aggressive chemicals, which can variously enfeeble the opponent (*e.g.*, spider toxins) or inhibit it from consuming 'chemically protected' prey (*e.g.*, caterpillars containing cyanohydrins), or, particularly unpleasant, enemies are occasionally exposed to powerful oxidizing agents such as 1,4-benzoquinone or similar compounds (*e.g.*, bombardier beetles of the *Brachinus* genus). These deterrence scenarios not only protect the individuals in question, but also contribute to the survival prospects of whole species.

Chemical weapons, ingestion poisons, and deterrents used by prey creatures, though, have been overcome by some selected specialists that have learnt how to render the toxic substances harmless by rapid disposal or chemical manipulation, or, in some cases, even to use them for their own purpose. These creatures, in the course of the evolutionary process, have thus learnt to adapt to these particular aspects of their environment.

9.6. Alkaloids from Animals

Many animal species, especially insects and amphibians, synthesize alkaloids that, as far as we know, fulfill or may fulfill a number of different functions. These alkaloids include pheromones, poisons for paralyzing victims (to be stored as food resources), and spray reagents, which serve to deter hunters. The structural diversity of these compounds matches that of their vegetable counterparts in every way. Once again, only a few examples have been picked for illustrative purposes.

9.6.1. Poisons from Spiders, Wasps, and Sponges

The poisons of spiders and wasps are complex mixtures of free amino acids, peptide toxins of high molecular weight, single purine bases, and smaller polyamine toxins, with molecular weights scarcely ever exceeding 1000 Da (*Table 9.1*)[5]. These polyamines can also be regarded as alkaloids, since, in structural terms, they are very similar to plant alkaloids. Of the *ca.* 100,000 species of spider, only *ca.* 30 have been studied to date. All the following details are, therefore, based on these few investigations. To obtain the toxins in reasonable quantities, special methods (*e.g.*, regular 'milking') have been developed. The separation of the low molecular weight components (up to 30 different polyamines in varying amounts) and their structural elucidation are not unproblematic, even with the most modern equipment. Chiefly responsible for this, apart from the small quantities of compounds (submicro scale), is the structural similarity of many of these compounds to one another (*cf. Table 9.1*).

Table 9.1. *Examples of Polyamine Toxins from Spiders, Wasps, and Sponges*

Structures	Molecular Weight	Trivial Name and Synonyms	Occurrence
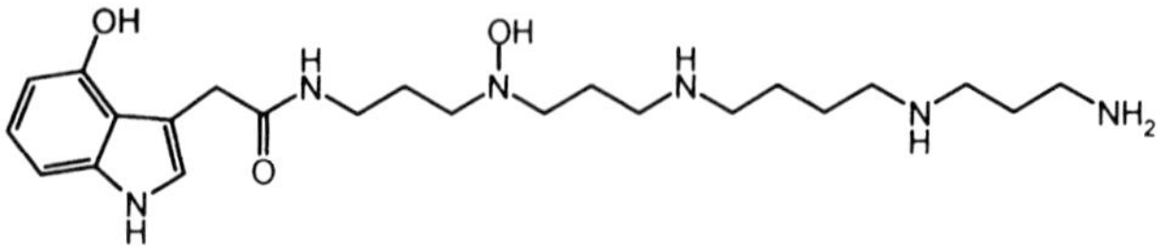	448	Agel 448 AG 448	*Agelenopsis aperta* (spider)
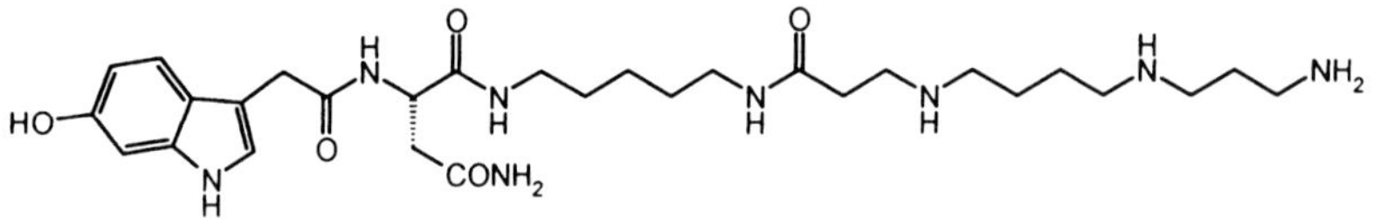	588	NPTX-1	*Nephila clavata* (spider)
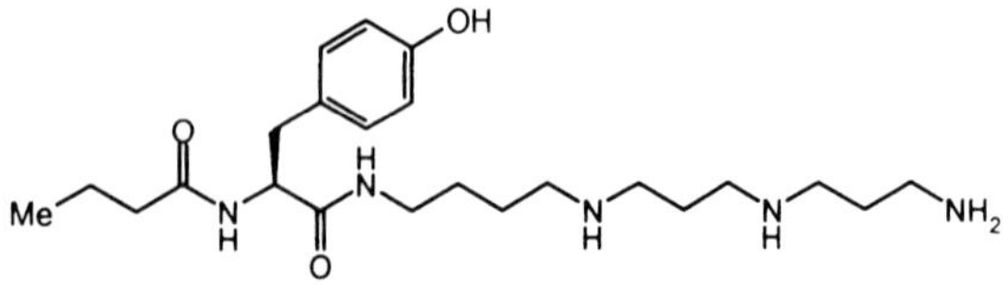	435	δ-PhTX δ-Philanthotoxin Philanthotoxin 433 PhTX-433 PTX-433	*Philanthus triangulum* (wasp; *Fig. 9.7*)
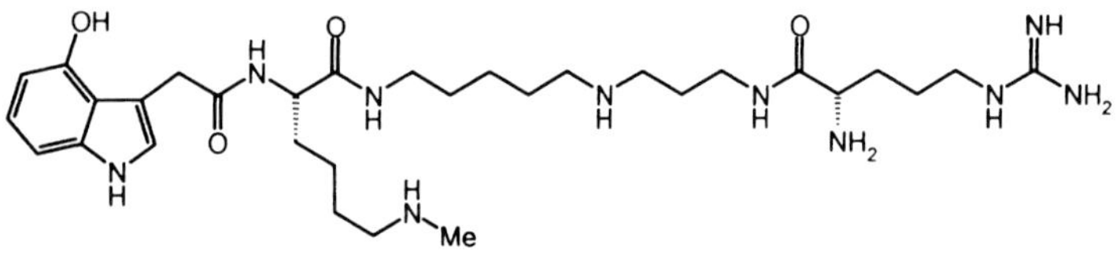	630	Arg 630 Argiopinine IV	*Argiope lobata* (spider; *Fig. 9.8*)
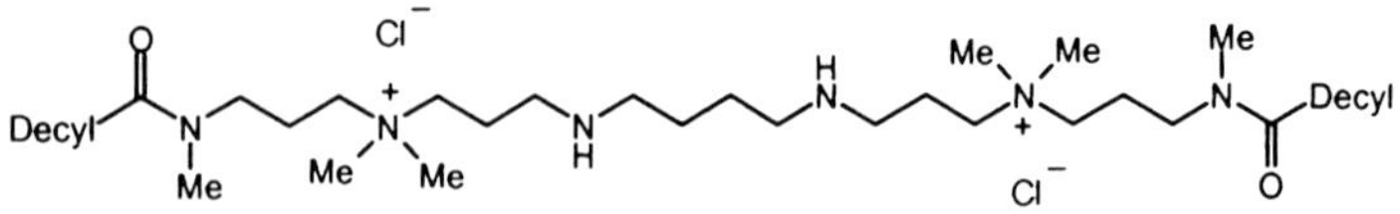	773	Penaramide A	*Penares aff. incrustans*[21] (sponges; *Fig. 9.9*)

Spiders use their poison to bite and paralyze their prey, so that they may later consume the victims. Some spider toxins are capable of irreversibly blocking neuromuscular transmission (which results in rapid death), while others produce only temporary paralysis.

The biosynthesis of these polyamine toxins is presumed to take place in the spider itself, since no similarly constituted polyamines have yet been detected in their food. No more is known at present, nor is any information available concerning the 'chemical' fate of the polyamine derivatives in the victims.

The structural similarity of the penaramides (such as penaramide A), isolated from the marine sponge *Penares aff. incrustans* (*cf. Table 9.1*), to the spider poisons is evident, but nothing is known at present about the biological significance of sponge toxins.

9.6.2. Poisons from Amphibians

Some amphibians also belong to the venomous animals, although, unlike spiders, they only use their poison to defend themselves. The toxins, mostly alkaloids, are stored in the skin and can often be found on the surface. Transport of the alkaloids takes place in conjunction with the mucus produced by special glands responsible for maintaining the moisture content of the animal skin. It has been possible to show that the toxins (mostly present as mixtures) inhibit the growth of microorganisms. This allows these continually wet creatures to survive in (tropical) wetlands. Amphibian alkaloids are also physiologically active against animals and humans. Extracts from amphibian skins have occasionally been described as sources of arrow (or blowpipe dart) poisons.

'In the Colombian coastal provinces of Buenaventura and Choco, there lives a small, green frog, which, when brought near a small fire, secretes from the skin on its back a liquid suitable for poisoning darts. The Indians dip their small darts in this liquid. The poison is so strong that if a jaguar should get some in its blood it immediately suffers convulsions and dies' (cited from a traveler's journal from *ca.* 1820)[22].

The frogs that act as sources of arrow poisons are of the Dendrobatidae family (colored frogs): namely *Phyllobates aurotaenia* (= *P. chocoensis*, *P. bicolor* var. *toxicaria*), *Dendrobates histrionicus*, *D. pumilio*, and *D. auratus* (Panama). The structures of some toxins, **17–20**, are shown. Batrachotoxin (**17**), clearly a 'modified' steroid, acts on the central nervous system, causing irreversible blocking of motor synapses.

With the exception of five *Phyllobates* species, batrachotoxin (**17**) has yet to be detected in other amphibians. Interestingly, though, an alkaloid of the batrachotoxin group, homobatrachotoxin (**18**), has been found in the bird species *Pitohui dichrous* (Pachycephallinae family), indigenous to New Guinea and known by the indigenous people to be poisonous. The highest

Fig. 9.7. *Philanthus triangulum* (Sphecidae), a predatory wasp, possessing in its venom toxins that display a great structural resemblance to spider toxins (*e.g.*, δ-philanthotoxin (δ-PhTX)) (photo H. Bellmann)

Fig. 9.8. A sample page from a classic among spider books: Carl Wilhelm Hahn, 'Monographie der Spinnen', Nürnberg, 1820–1836. a) Epeïra sericea OLIV.; now Argiope lobata PALLAS. Its poison contains, among other compounds, the polyamine Arg 630 (cf. Table 9.1). b) Epeïra angulata WALK, female; now Araneus angulatus CLERK. c) Epeïra Schreibersii HAHN, female; now Araneus circe AUDOUIN. Polyamine toxins have also been found in the venom of Araneus species.

Fig. 9.9. *The sponge Penares aff. incrustans from which a series of penaramides have been isolated* (photo *N. Fusetani*)

Fig. 9.10. *Phantasmal poison dart frog (Epipedobates tricolor)* (photo *G. Grall*)

concentrations of this compound were found in the skin and feathers. However, no batrachotoxin-producing frogs are known to live in New Guinea, and the bird exclusively feed on insects and seeds. Therefore, it is possible that a second, independent synthetic pathway to this alkaloid class exists[23]. Other steroid alkaloids have also been obtained from salamanders of both the genera *Salamandra* and *Triturus*. Samandarine (*cf. Chapt. 1*), for example, originates from the skin glands of the Fire Salamander (*Salamandra maculosa*).

Wild *Phyllobatis terribilis* frogs contain a total of *ca.*1140 µg of **17/18** per animal. When raised in a terrarium, however, even second-generation specimens are found to have an alkaloid content 10,000 times smaller than that of their wild parents. These findings pose a number of questions. How does the synthesis of these compounds take place? Are the substances introduced from outside, through food or contact? There are no convincing answers yet.

Dendrobates auratus and *D. pumilio* produce, among other compounds, the pumiliotoxins A (**19**) and C (**20**). Histrionicotoxin (**21**) has been isolated from the skin of *D. histrionicus*, representing a further structure type. It was later also found in several species of the genera *Dendrobates*, *Epipedobates* (*Fig. 9.10*), and *Phyllobates*. At present, *ca.* 20 representatives of the same basic structural type are known. To give some idea of the *in vivo* alkaloid content, it may be mentioned, in passing, that one frog of the *D. pumilio* type contains *ca.* 200 µg of toxins.

The knowledge of the biogenesis of these amphibian toxins is still very limited. Since, for example, batrachotoxin (**17**) has until now only been found in frogs, it is assumed that the animal organism produces it from naturally occurring steroids. There are a few clues regarding the origins of other amphibian alkaloids: the pyrrolizidine alkaloid *cis*-2-heptyl-8-methyl-1-azabicyclo[3.3.0]octane (= xenovenine, **22**), isolated from the poison glands of the red Fire Ant (*Solenopsis xenovenenum*) proved to be identical with a compound isolated from the frogs *Dendrobates auratus* and *Melanophryniscus stelzneri*. Compound **22** may also be a substance specifically active against microorganisms, used by the ant as a fungicide.

A direct relationship between the substances found in animals and those found in their food has been observed in *Dendrobates auratus* frogs, fed with Pharaoh's ants (*Monomorium pharaonis*). In its poison glands, the ant stores the two alkaloids monomorine I (**23**)[24] and *trans*-5-(hex-5-en-1-yl)-2-pentylpyrrolidine (**24**) in a ratio of *ca.* 1:2. In the skin of these frogs, **23** can indeed be found, but not **24** (not even in trace amounts). Hence, a significant enrichment of one substance, taken up in the food, has taken place. Wheth-

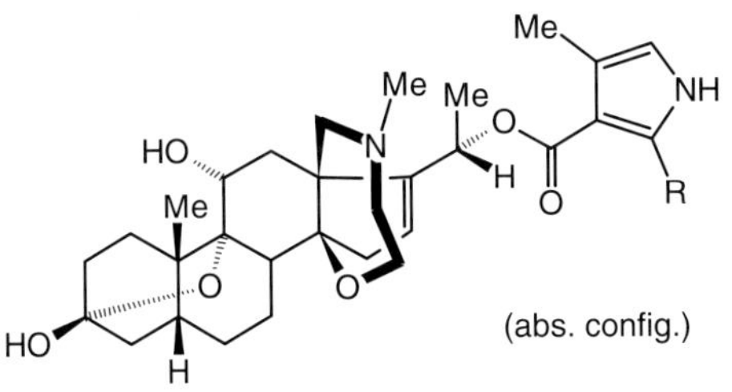

R = Me: Batrachotoxin (**17**) from *Phyllobates*

R = Et : Homobatrachotoxin (**18**)
from *Pitohui* and *Phyllobates*

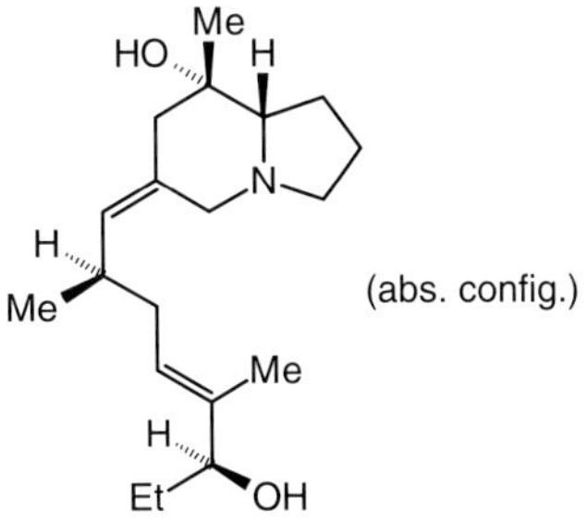

(abs. config.)

Pumiliotoxin A (**19**)
from *Dendrobates, Pseudophryne,
Mantella, Melanophryniscus*

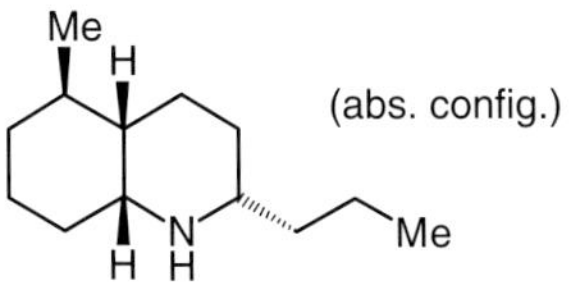

(abs. config.)

Pumiliotoxin C (**20**)
from *Dendrobates, Mantella,
Epipedobates, Phyllobates*

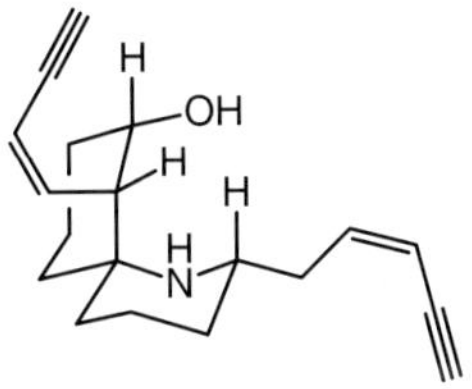

Histrionicotoxin (**21**)
from *Dendrobates*

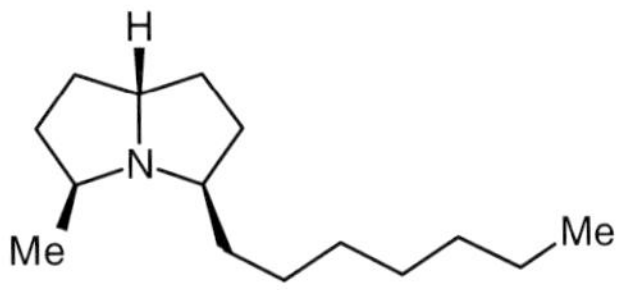

Xenovenine (**22**) from *Dendrobates,
Melanophryniscus*, and *Solenopsis*

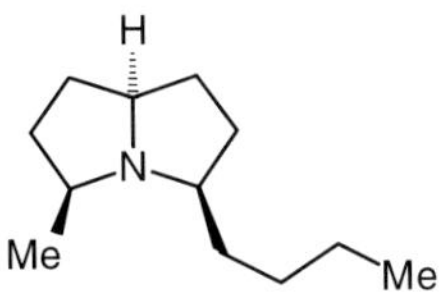

Monomorine I (**23**)
from *Monomorium*

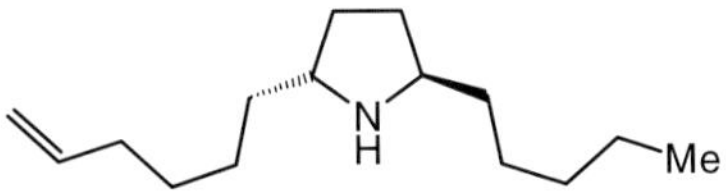

24
from *Monomorium*

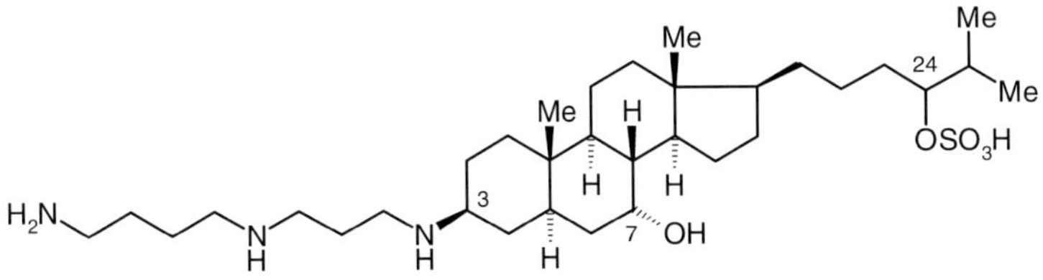

Squalamine (**25**)
from the stomach of *Squalus acanthias*

er the substance taken up through the food chain is ultimately produced by plants or animals, or both, is still an open question[4, 25].

Like their vegetable counterparts, alkaloids from the animal kingdom display multifaceted and fascinating activities. They are used, among other things, for defense in the widest sense, for attacking prey, and also for trail finding.

9.7. Alkaloid Content of Plants: Local Differences and Changes over Time

Uncaria tomentosa (Rubiaceae)

The alkaloids of a given plant generally display considerable variation in terms of composition, amount, and distribution between root, leaf, stem, and fruit. As an example, the recently published systematic examination of *Uncaria tomentosa* (Rubiaceae)[26] is discussed here. The alkaloid distribution in different parts of the plant is presented in *Table 9.2*.

The four oxindole alkaloids pteropodine, isopteropodine, speciophylline, and uncarine F differ from one another only in their configurations at C(3) and C(7) (*cf. Table 9.2*). The pteropodine/isopteropodine diastereoisomeric pair is present at significantly higher concentrations in fully grown leaves than it is in younger ones, while the exact opposite applies for the other pair. No rational explanation for this fact has yet been forthcoming.

The alkaloid content of a plant can also vary over the year. These continuous fluctuations may be considerable, depending not only on the plant but also on the alkaloid types and concentrations involved. The latter are usually given relative to the dry weight of the whole plant (or of relevant parts). The varying alkaloid content of a species can only be determined if a large number of plants under identical conditions are extracted and examined sequentially. The plants described below have been investigated in terms of their alkaloid contents in greater detail.

Chaenorhinum minus (Scrophulariaceae)

Chaenorhinum minus is an annual plant (or weed), germinating in Central Europe from March/April. It rapidly grows into a *ca.* 30 cm high bush, blooming continually (*cf. Fig. 2.37*). From about the middle of August, after the seeds have ripened, it dies off and remains standing as a dry, slightly

Table 9.2. *Distribution of Alkaloids in Various Parts of the Plant Uncaria tomentosa.* Concentrations are given in mg/g (alkaloid/plant material)

Alkaloid		Young Leaves	Adult Leaves	Branch Bark	Trunk Bark	Root
Pteropodine	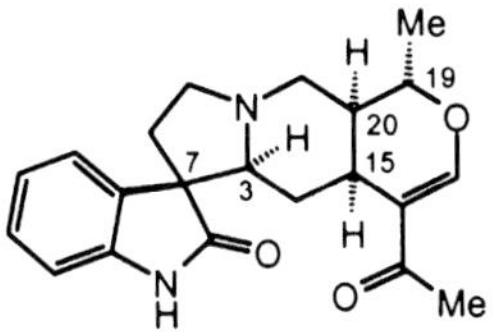	2.8	9.3	3	0.05	1.1
Isopteropodine	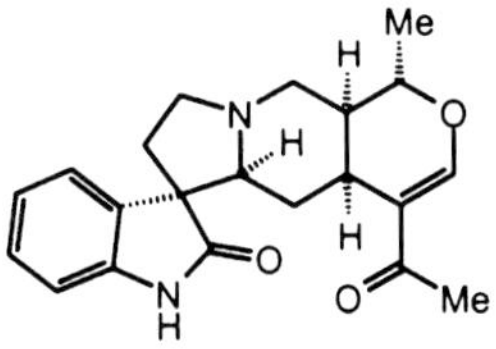	1.9	4.1	0.63	0.06	0.57
Speciophylline	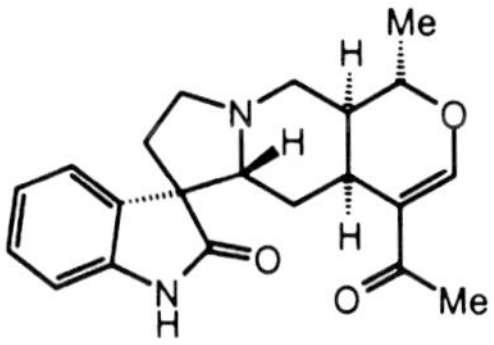	12.2	9.6	1.6	0.08	2.1
Uncarine F	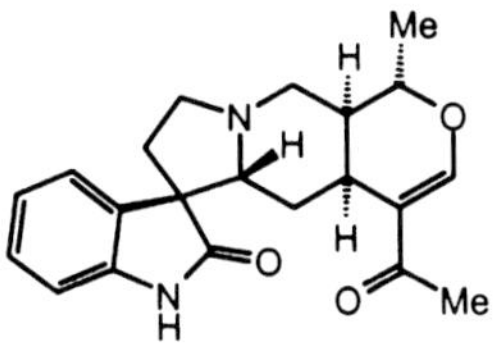	47.9	2.4	0.63	0.03	0.91

woody plant. While its content of the spermine alkaloid chaenorhine (*cf. Chapt. 2*) amounts to *ca.* 0.15% in the fresh green plant (around June), it decreases slowly in the yellowed state and approaches zero in the dead, dried plant (*i.e.*, chaenorhine is no longer detectable). The highest alkaloid content was naturally measured in the fully green plant. It is not exactly clear what happens to the chaenorhine; a hydroxylation *ortho* to the existing O-function probably takes place, in analogy to aphelandrine, which is hydroxylated enzymatically.

(+)-Aphelandrine (**26**)
(abs. config.)

Aphelandra tetragona (**Acanthaceae**)

The main alkaloid of this tropical plant (native to northern South America, *cf.* also *Fig. 9.11, Aphelandra squarrosa*), which was cultivated in a Swiss hothouse, is aphelandrine (**26**), found exclusively in the roots. In a large number of experiments, it was shown that the concentration of **26** increases over the year, reaching its maximum after *ca.* 26 weeks, when the plant starts to bloom (*cf. Fig. 9.12*). After that, the alkaloid concentration decreases again. The reason for this is not known. Aphelandrine might be catabolized by plant-specific enzymes. It is also conceivable that the numerous endophytes (endoparasites) of *A. tetragona* are responsible for the chemical modification of the alkaloid[27].

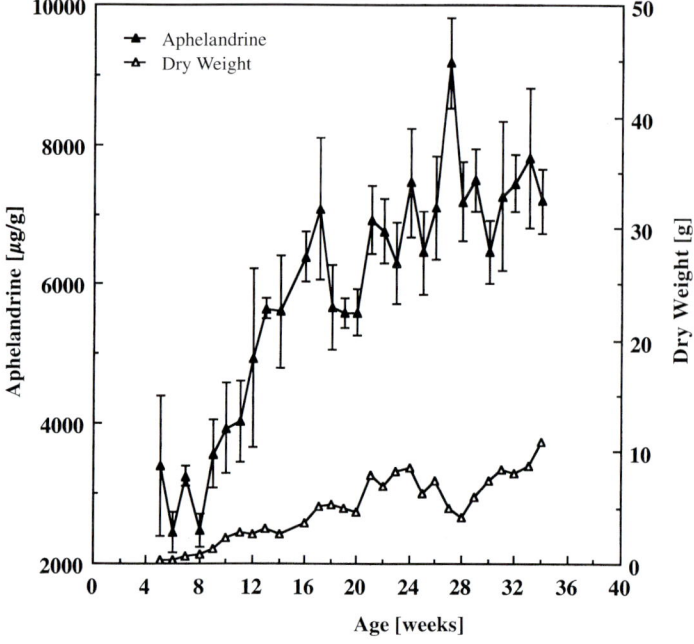

Fig. 9.11. *Aphelandra squarrosa* (Acanthaceae). The tropical Acanthacea, much prized as a houseplant, contains the alkaloid aphelandrine in its roots (photo *C. Werner*).

Fig. 9.12. *Aphelandrine content over time in the roots of Aphelandra tetragona as a function of dry root weight.* Each value represents the average of two measurements, each corresponding to samples prepared from the lyophilized roots of three different plants (six values)[27].

Conium maculatum and *Papaver somniferum*

Much more impressive is the change in the alkaloid content of a plant over the course of a day. The hemlock alkaloids coniine and γ-coniceine in the fruits of *C. maculatum* display inverse variations in concentration over the weekly and daily cycles (*Figs. 9.13* and *9.14*). It may be concluded from this that the two compounds exist in a close biogenetic relationship. Similar observations have been made concerning the morphine, thebaine, and codeine contents of *P. somniferum* (*Fig. 9.15*). Their biogenetic relationship can be read off directly. Morphine is present in higher concentration in the morning, while its successor, codeine, as well as its precursor, thebaine, are found in higher concentrations in the afternoon (*cf. Chapt. 8.2*).

Even when certain variations in concentrations are easily recognizable, in particular the increase in alkaloid content towards the flowering of the plant, the phenomenon, as a whole, is not yet fully understood. It is to be assumed, though, that such variations reflect different functions of alkaloids in the plant, in some form or another.

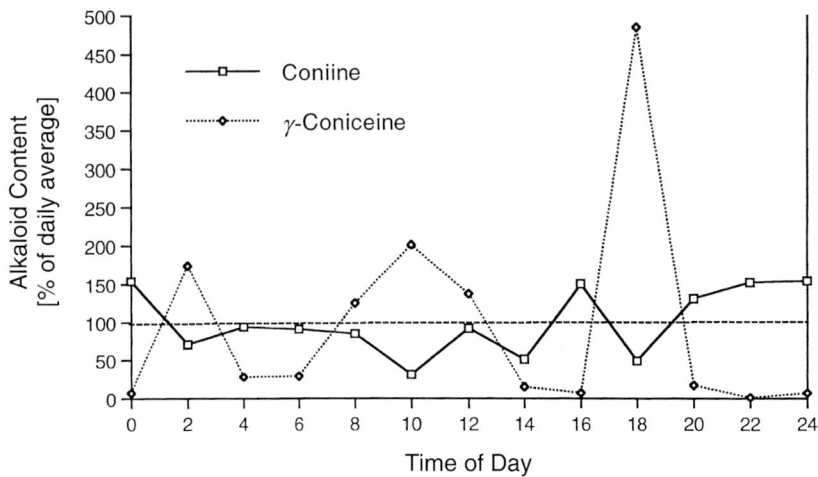

Fig. 9.13. *Variation in alkaloid content in hemlock over the course of a day* [28]

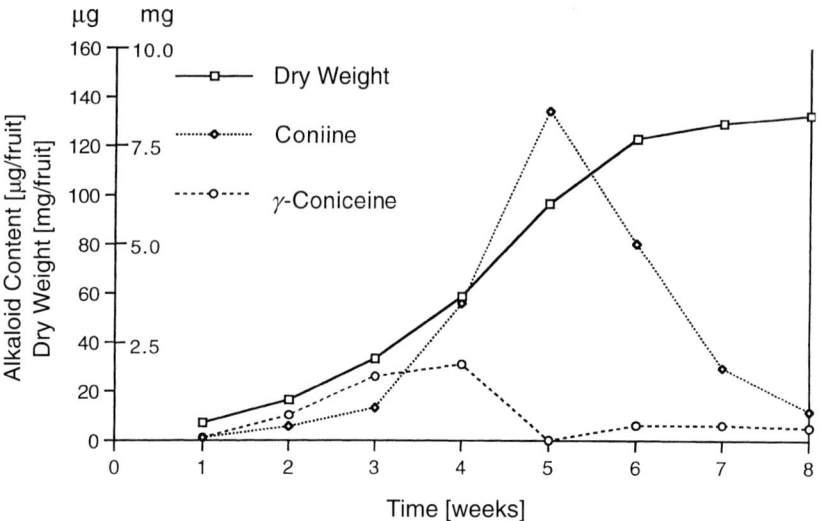

Fig. 9.14. *Variation in alkaloid content in hemlock over weeks* [28]

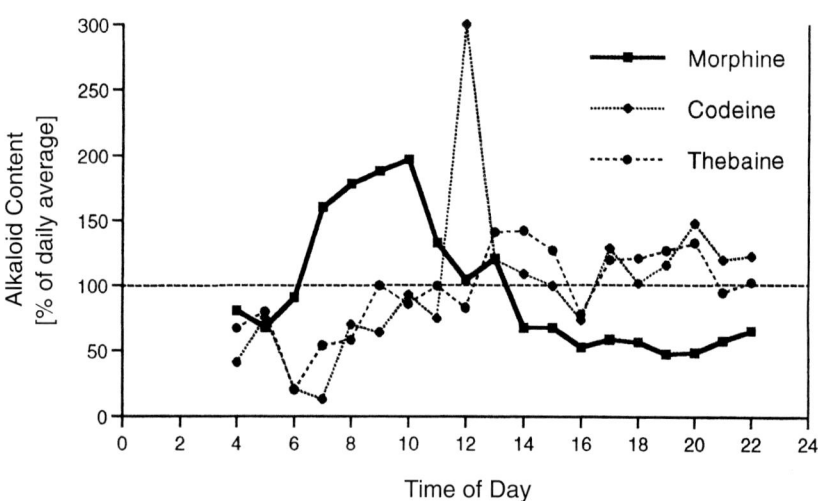

Fig. 9.15. *Circadian variation of concentrations of the alkaloids thebaine, codeine, and morphine (in developmental biological sequence) in Papaver somniferum (whole plant)* [29]

9.8. Alkaloids as Medicines

Mankind has always striven to cure diseases, or at least to ameliorate their course. Any kind of 'operations' on the human body, such as the removal of ulcers or amputation of limbs, should be made as painless as possible. Effective treatment of highly localized pain (such as toothache), which negatively affects the quality of life, is also sought after. Also, people have always tried to change their naturally programmed life rhythms to fit their (personal) desires, to stay awake when tired, or to be able to sleep when fully awake. If people are dispirited or clinically depressed, they yearn to be able to view the world optimistically. In brief, our demand for well-being and freedom from pain is so strong that (almost) any means of reaching this goal can be justified, despite of possible risks involved. If taking these risks then does indeed give rise to severe side effects or even to incurable disease (*e.g.*, lung cancer from tobacco), regret often comes too late for a return to the original condition.

Since this struggle seems to be basic to the human nature, it is not too surprising that even the most ancient records contain information concerning medicinally active vegetable substances. Hence, herbals and plants have from the earliest times frequently been combined with details of methods for curing various diseases. A selection is given in *Table 9.3*.

The bible and histories from that time also offer clues concerning the use of alkaloids and of possible harm arising from their use (*cf. Chapt. 11.4*, ergot alkaloids).

Chemistry, appearing on the scene at the start of the 19th century, also devoted itself principally to investigation of medicinal plants and drugs (*cf. Chapt. 10*). Attempts were made to isolate the active principles and, later, to interpret their structures, so as to be able to synthesize them if necessary[31]. The successes of extraction works slowly appeared, and many alkaloids were isolated from their natural sources and came onto the market. Some naturally occurring alkaloids are still isolated from plants in industrial quantities today, and are marketed as pharmaceuticals in pure form as either salts or derivatives.

According to the 1995 *International Drug Directory*[32], several alkaloids are still in use today. *Ajmaline* (an antiadrenergic drug with a sedative effect in higher doses) and *sparteine* are used in the treatment of arrhythmia. *Atropine*, *hyoscyamine*, and *scopolamine* are parasympatholytics, substances that inhibit acetyl cholinergic neurotransmission at parasympathetic nerve endings. *Atropine* is a mydriatic, a substance that induces pupil dilation. It is

Table 9.3. *Selection of Historical Medicinal Records Preferentially Based on Plants*

Year	Description
1900 – 400 B.C.	Assyrian plant medicine, recorded on *ca.* 400 clay tablets.
1500 B.C.	The *Ebers* papyrus[30] contains instructions for pharmaceutical preparations and techniques for treatment of diseases. As well as plant and animal parts, minerals are also listed as crude medicinal agents. Plants mentioned include *Ricinus communis* and *Papaver somniferum*.
250 – 20 B.C.	*Shen-nong Ben-cao Jing* ('Book of Domestic Plants', more generally known as 'The Divine Farmers Materia Medica'). This Chinese book contains more than 350 remedies for diseases. The plants mentioned include *Ephedra* spec., *Ricinus communis*, and *Papaver somniferum*.
ca. 50 A.D.	*De Materia Medica*, by *Pedanios Dioskorides*, a compilation in five volumes, deals particularly with medicinally useful plants. Used in Europe and Asia Minor in various translations until the 18th century.
ca. 70	*Naturalis historia* (37 volumes) by *Gaius Plinius Secundus* (*Pliny the Elder*) contains plant and animal remedies. This discourse had great influence in Antiquity and in the Medieval age.
ca. 190	*Claudius Galenus* (129 – 199) published a comprehensive work on all aspects of medicine (*cf. Fig. 9.16*).
ca. 630	*Paulos Aíginetes* (*Paul of Aígina*) compiled a medical encyclopedia that was to influence Arabic medicine particularly strongly.
659	*Li Ji, Xin-xiu Ben-cao* (a new edition of the *Shen-nong Ben-cao Jing*). Discussion of 844 therapeutic preparations and their application in Chinese medicine. As in the *Shen-nong*, healing plants are divided into three classes: 'good, intermediate, and less well suited'. Only fragments of this work have been preserved.
ca. 970	*Masawaih al-Mardini* (*Mesue the Younger*) compiled a very extensive Arabic medical book.
1080 – 1107	*Tang Shen-Wei* produced the '*Jing-Shi Zheng-lei Bei-ji Ben-cao*' ('*Materia Medica Arranged According to Pattern*'), a historical, detailed index of therapeutics in classified arrangement containing 1558 preparations (plants, animals, minerals). First attempt at a scientific classification of therapeutics.
1150 – 1170	*Physica* ('Natural History') and *Causae et curae* ('Medicine') by *Hildegard von Bingen* (1098 – 1179). Description of nature, including folk medicines and treatments, from a physician's point of view (*cf. Fig. 9.17*).
1484	The book '*Hortus sanitatis*' ('Garden of Well-Being'), collected by *Johann von Kube*, was published by *Peter Schöffer* in the house of *Gutenberg*, Mainz. It deals with 382 plants and plant products, 25 animals and animal products, and 28 minerals, all of which may be used as therapeutic agents. Because of its index of ailments and their treatment with the medicines described, the work was of excellent and lasting practical value. It underwent many reprints and much unauthorized copying.
1557	Herbal named '*Kreutterbuch von allem Erdtgewächs*', by *Adamus Lonicerus*, Frankfurt am Main. Based on older works (including those of *Eucharius Rösslin* (Rhodion)) and enlarged in its later editions, this became a standard work in medicinal healing on the basis of plants, animals, metals, and stones. First description of the ergot and tobacco drugs.
1578	*Li Shi-zhen* (1518 – 1593; *Fig. 9.18*) produced the *Ben-cao gang mu* ('*Compendium of Materia Medica*'), in which almost 1900 Chinese therapeutics are discussed. An extension and improvement of the classification system of *Tang Shen-Wei* (*cf. Figs. 9.19* and *9.20*).
1616	The book '*Arzneigarten, von Kreutern so in den Gärten gemeinlichen wachsen*', by *Anthonius Mizaldus*, Basel, was a popular, non-illustrated edition, provided with an index of all treatable ailments (*Figs. 9.21* and *9.22*, *cf.* also *Fig. 9.23*).

Fig. 9.16. *Galen* (Lat. *Claudius Galenus*), *a Roman physician of Greek Origin (130 – 200)*. Created a comprehensive system of medicine (building on the works of *Hippocrates* and *Aristotle*), which dominated for centuries (from E. Heilbronner, F. A. Miller, *A Philatelic Ramble through Chemistry*, Verlag Helvetica Chimica Acta, Basel, 1998).

Commemorative Issue

»Hildegard von Bingen«

Fig. 9.17. *Hildegard von Bingen* (1098–1179) *at the entrance to the Rupertsberg Benedictine monastery at Bingen am Rhein.* Stamp commemorating the 800th anniversary of her death, August 9, 1979. In her '*Causae et curae*', she presents, *inter alia*, a list of folk remedies and treatments from a physician's perspective (*Michel* catalog, Germany, No. 1018).

also used as an antidote to morphine, hydrogen cyanide, and alkyl phosphate poisoning. *Combopen*® autoinjectors (containing 2 mg atropine plus 150 mg obidoxime) are modern, easily operable automatic 'syringes' with which all known nerve gases can be 'neutralized'.

Caffeine, a psychostimulant, is presumably largely obtained from the decaffeination of *Coffea* species. *Codeine* is used as an antitussive (agent that suppresses the coughing reflex), offered by several firms. It also possesses opioid properties, although these are significantly less powerful than those of morphine. *Cocaine* is still in use as a local anesthetic (*e.g.*, in eye surgery). Other possible applications (including dentistry) have been discontinued be-

Fig. 9.18. *Li Shi-zhen* (1518–1593), *Chinese physician and pharmacologist.* He compiled the Compendium of Materia Medica, containing almost 1900 Chinese therapeutic agents (from E. Heilbronner, F. A. Miller, *A Philatelic Ramble through Chemistry*, Verlag Helvetica Chimica Acta, Basel, 1998).

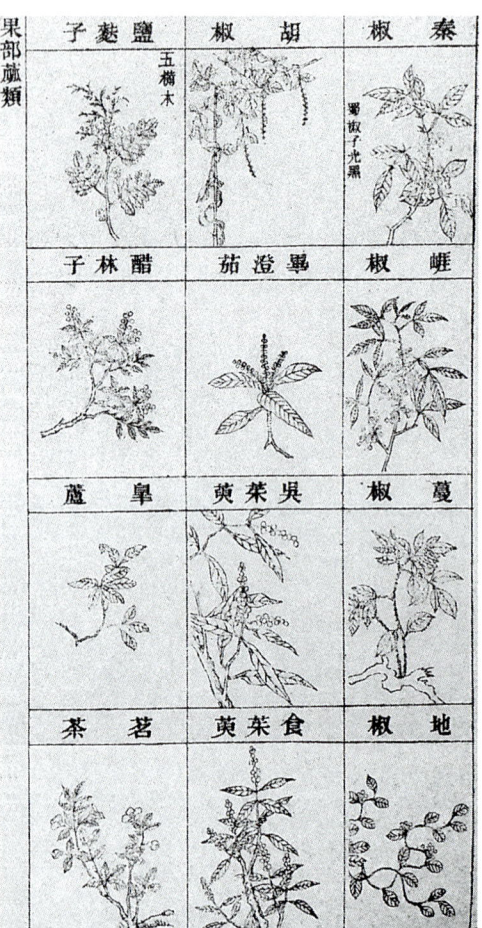

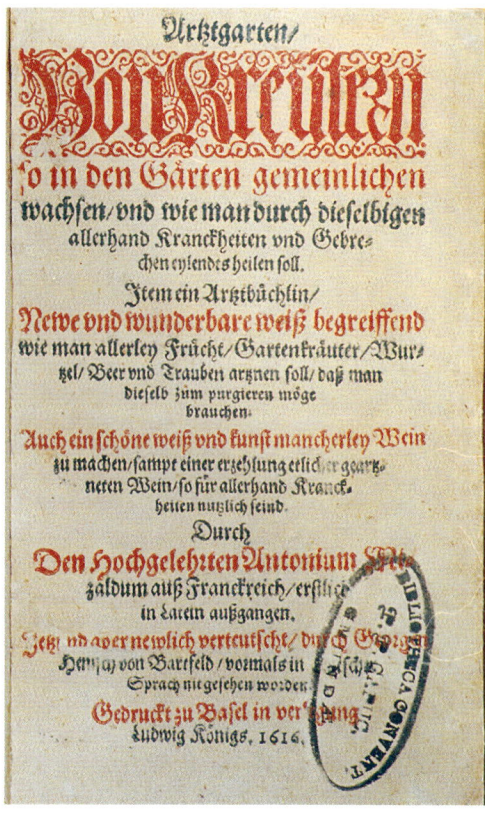

Fig. 9.19. *Title page of the 'Compendium of Materia Medica' by Li Shi-zhen (1578)*

Fig. 9.20. *Page of illustrations from Li Shi-zhen's 'Compendium of Materia Medica'*. The Chinese plant descriptions were kindly translated by Prof. *Y. Kao*, Wuhan and by *W. Hu*, Zürich. First row, from the left: 1. *Rhus chinensis*; 2. *Piper nigrum, fructus* (piperidine alkaloids); 3. *Zanthoxylum bungeanum* (β-carboline alkaloids). Second row: 4. *Prunus* sp.; 5. *Piper cubeba* (piperidine alkaloids); 6. *Zanthoxylum* sp. (canthinone). Third row: 7. *Camellia sinensis var. macrophylla* (caffeine); 8. *Evodia rutaecarpa* (evodiamine); 9. unclear. Fourth row: 10. *Camellia sinensis* (caffeine); 11. *Zanthoxylum ailanthoides* (β-carboline alkaloids); 12. *Thymus mongolicus* (or *Th. quinquecostatus*).

Fig. 9.21. *Title page from the book 'Arzneigarten, von Kreutern so in den Gärten gemeinlichen wachsen' by A. Mizaldus, Basel, 1616*

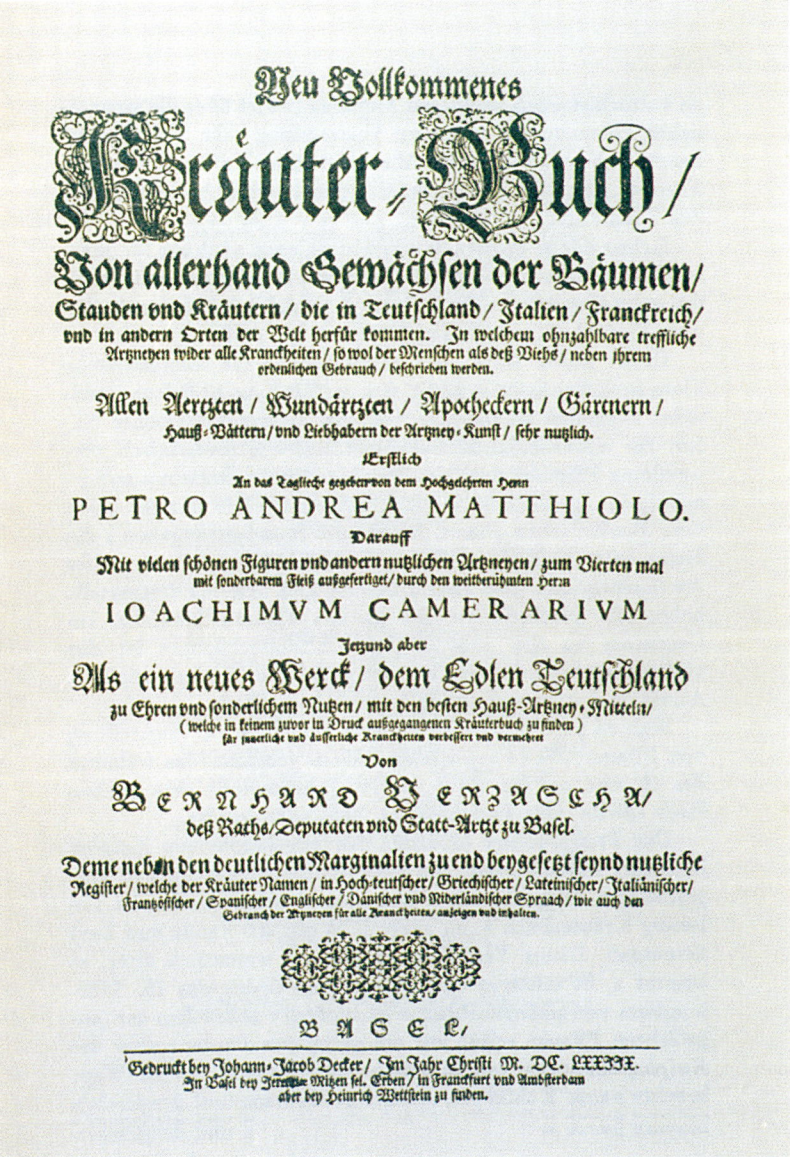

Fig. 9.22. *Title page of Verzascha's edition of the Metthioli translation of J. Camerarius' 'Kräuterbuch',
Basel, 1678.* Numerous other herbals with strong emphasis on botanical or medical aspects were published
in Europe after 1550. Works of earlier authors were often taken directly or newly revised, thanks to which new
aspects and preparations were to make their impression (A. Schmid, *Über alte Kräuterbücher*, Haupt-Verlag,
Bern, 1939).

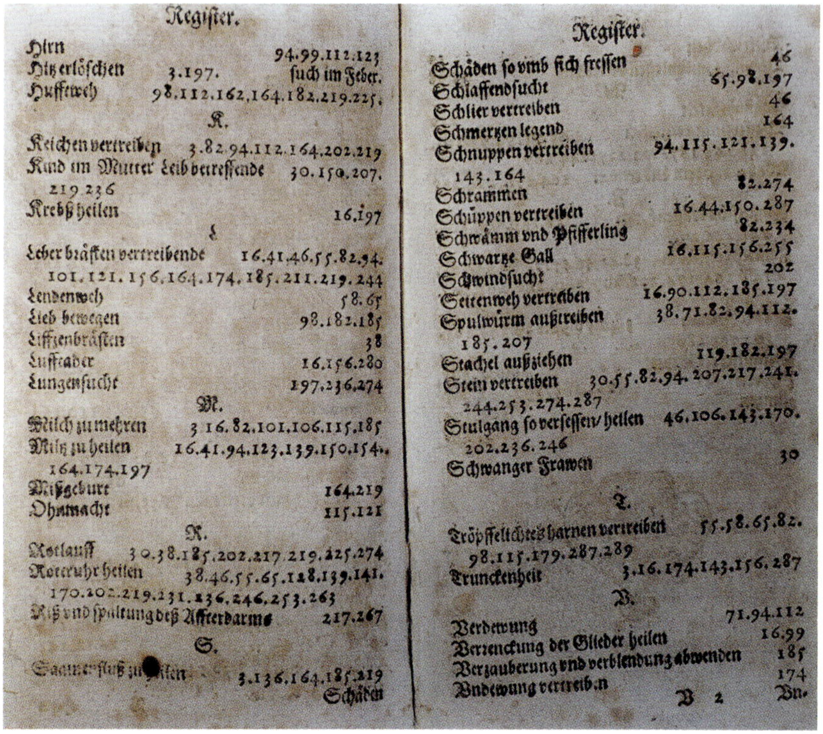

Fig. 9.23. *Index pages of the book 'Arzneigarten, von Kreutern so in den Gärten gemeinlichen wachsen' by A. Mizaldus, Basel, 1616*

cause of the great addiction risk. *Colchicine* is used in treatment of gout. The *Ipecacuanha* alkaloid *dihydroemetine* is still used for its original purpose as an antiprotozoic and amoebacide. The *Rauwolfia* alkaloids *deserpidine, rescinnamine, reserpiline,* and *reserpine* are used as antihypertensives (blood-pressure depressants). The number of active substances from *Claviceps purpurea*, the ergot alkaloids, currently in use as therapeutic agents, is especially large. Examples include the antimigraine treatments *dihydroergotamine* and *ergotamine*, the oxytocic agent *ergometrine* (inducing muscular contraction of the uterus), and the vasodilator (and hence blood-pressure depressant) *dihydroergocristine*. α-*Dihydroergocryptine* is used against *Parkinson*'s disease. *Tubocurarine dichloride* and *dimethyltubocurarinium dichloride*, two bis(benzyl-1,2,3,4-tetrahydroisoquinoline) alkaloids, are, like *alcuronium chloride* (=4,4'-di(prop-2-enyl)-4,4'-bis(nortoxiferine) dichloride, *Alloferin*®) effective neuromuscular receptor blockers, which inhibit the transmission of neurosignals to the muscles at the synapses. Their physiological effect can be overcome by the use of *curare* antagonists. *Physostigmine* and its oxidation product *eseridine* (*geneserine*) are used as choline-esterase inhibitors and as parasympathomimetics (*i.e.*, substances that directly stimulate the cholinergic system), a class to which *arecoline* and *muscarine* also belong. *Quinidine* and its dihydro derivative *hydroquinidine* find application as antiarrhythmics. *Morphine* is still an indispensable analgesic (painkiller), used in particular for treatment of severe pain. There is still always the greatest danger of addiction associated with its use, though. The antitussive *noscapine* contains an *8-methoxyhydrastine*, thus belonging to the phthalideisoquinoline alkaloids. *Quinine* remains on the market as an antipyretic (fever suppressant), antiprotozoic, and, with increasing tendency, as an antimalarial medicine, although its earlier dominance has been eroded by the application of synthetic drugs. Antimalarial chemoprophylaxis and therapy is increasingly dogged by resistance problems, however. In some malarial areas, 90% of parasite strains are already resistant to *chloroquine*, the most widely used drug for treatment of and protection against malaria from the 1950s onwards. A similar situation applies for other medicines, and so the natural product *quinine*, the first antimalarial agent of all, is consequently enjoying a comeback. The *Rauwolfia* alkaloid *raubasine* is in use as a vasodilator. *Theophylline*, an antiasthmatic with cardiostimulatory and diuretic properties, is offered by almost all pharmaceutical firms. An important cerebral vasodilator, which is used in cases of interrupted blood supply to the brain, is *vincamine*, accessible both by extraction from *Vinca* species and by total synthesis. *Eburnamonine* (= *vinburnine*) is also used to the same end.

Of course, many synthetic derivatives of natural alkaloids have also been manufactured, serving as medicines themselves, sometimes with similar or altered physiological effects. *Lysergic acid*, *tropine* derivatives, and *morphine* analogues in particular are licensed as semisynthetic or fully synthetic medicines. Alkaloids, with their various and important pharmacological effects, have in the past often played the role of godparents in the planning of new pharmaceuticals.

Alkaloids that find application in the treatment of cancer have not yet been mentioned. *Catharanthus roseus* (Apocynaceae), long known as *Vinca rosea*, occurs in the tropics, although it seems to have originated in Madagascar. It is a perpetual flowering, evergreen plant, up to 60 cm high, grown ubiquitously as a garden plant in warm regions. Based on earlier experiments, researchers from *Eli Lilly & Co.* (Indianapolis, Indiana) performed extractions on the plant and observed oncolytic activity against mouse P-1534 leukemia in certain fractions. Tremendous activity in isolation and structural elucidation then ensued. By 1963, after *ca.* six years of research, 41 different alkaloids had already been isolated. Antileukemic activity was observed in four

R¹ = Me
R² = MeOOC
R³ = Ac
R⁴ = MeO
} Vinblastine (**27**)

R¹ = CHO
R² = MeOOC
R³ = Ac
R⁴ = MeO
} Vincristine (**28**)

R¹ = Me
R² = MeOOC
R³ = H
R⁴ = NH₂
} Vindesine (**29**)

Vinorelbine (**30**)

bisindole alkaloids: *vinblastine* (**27**), *vincristine* (**28**), *leurosidine*, and *leurosine*. These important compounds are used today in the treatment of *Hodgkin*'s disease, lymphoblastic leukemia, and testicular cancer. *Vincristine* (**28**) is excellently suited to treat childhood leukemia (remission rates of 60–70%). The following four alkaloids are used today therapeutically as antineoplastics (substances inhibiting tumor growth): *vinblastine* (= *vincaleukoblastine*, VLB; **27**), *vincristine* (= *22-oxovincaleukoblastine*, *leurocristine*, VCR, N-formyl-N-demethylvincaleukoblastine; **28**), *vindesine* (= *3-aminocarbonyl-O⁴-deacetyl-3-(demethoxycarbonyl)vincaleukoblastine*; **29**), and *vinorelbine* (**30**).

Since all of these substances from *Catharanthus rosea* are synthesized *in vivo* only in tiny amounts (*ca.* 1 mg of the alkaloid per 1 kg of dried plant), it is easy enough to picture the amount of work required to cultivate and process the plants, as well as to purify all the component substances.

In summary, it can be seen that a number of natural alkaloids are still isolated from plants for direct medicinal application today. If an active substance possesses particularly rare properties, then great efforts are made to isolate the compound, unless cheaper treatment for the particular ailment is available.

Therapeutically important alkaloids sometimes include remarkably toxic substances (*e.g.*, *colchicine*, *atropine*, or *curare* alkaloids). This does not mean, though, that some medical application can also be found for every toxic base. The hemlock alkaloid *coniine* is unlikely ever to be used for purposes other than killing[33].

Probably of even greater significance than the direct application of natural alkaloids in medicine is their value as 'lead compounds'. Today, innumerable structural modifications are carried out with the goal of removing or mitigating undesired side effects of drugs. A great many biologically active heterocycles can be viewed as spin-off products of earlier alkaloid research, and so it is hard not to conclude that the direct and indirect significance of alkaloids for medicine can scarcely be overstated.

References and Notes

1. K. Mothes, H. R. Schütte, *Biosynthese der Alkaloide,* VEB Deutscher Verlag der Wissenschaften, Berlin, 1969.
2. A. M. Clark, C. D. Hufford, '*Antifungal Alkaloids*', The Alkaloids **1992**, *42*, 117.
3. M. Wink, '*Allelochemical Properties or the Raison d'Etre of Alkaloids*', The Alkaloids **1993**, *43*, 1.

4. J. W. Daly, H. M. Garraffo, T. F. Spande, 'Amphibian Alkaloids', The Alkaloids 1993, 43, 185.
5. A. Schäfer, H. Benz, W. Fiedler, A. Guggisberg, S. Bienz, M. Hesse, 'Polyamine Toxins from Spiders and Wasps', The Alkaloids 1994, 45, 1; S. Chesnov, L. Bigler, M. Hesse, 'The Spider Paracoelotes birulai: Detection and Structure Elucidation of New Acylpolyamines by On-Line Coupled HPLC-APCI-MS and HPLC-APCI-MS/MS', Helv. Chim. Acta 2000, 83, 3295.
6. Candida is a convenient organism for studies of this kind.
7. MIC = 'minimum inhibitory concentration'.
8. P. Dätwyler, R. Ott-Longoni, E. Schöpp, M. Hesse, 'Über Juliprosin, ein weiteres Alkaloid aus Prosopis juliflora A. DC.', Helv. Chim. Acta 1981, 64, 1959.
9. A. K. Khursheed, H. F. Arshad, A. Viqaruddin, Q. Sabiha, A. R. Sheikh, S. H. Tahir, 'In Vitro Studies of Antidermatophytic Activity of Juliflorine and its Screening as Carcinogen in Salmonella/Microsome Test System', Arzneimittelforsch. 1986, 36, 17.
10. Griseofulvine is an antibiotic produced by Penicillium griseofulvum.
11. S. S. Cohen, Introduction to Polyamines, Prentice-Hall, Englewood Cliffs, N. J., 1971; S. S. Cohen, A Guide to the Polyamines, Oxford University Press, Oxford, 1998.
12. B. Frydman, W. M. Westler, K. Samejima, 'Spermine Binds in Solution to the $T\psi C$ Loop of tRNA^{Phe}: Evidence from a 750 MHz ^{1}H-NMR Analysis', J. Org. Chem. 1996, 61, 2588.
13. Sunshield creams contain methoxycinnamic acid derivatives for the same reason.
14. C. Werner, W. Hu, A. Lorenzi-Riatsch, M. Hesse, 'Di-coumaroylspermidines and Tri-coumaroylspermidine in Anthers of Different Species of the Genus Aphelandra', Phytochemistry 1995, 40, 461.
15. H. J. Grambow, J. Lückge, A. Klausener, E. Müller, 'Occurrence of 2-(2-Hydroxy-4,7-dimethoxy-2H-1,4-benzoxazin-3-one)-β-D-glucopyranoside in Triticum aestivum Leaves and its Conversion into 6-Methoxybenzoxazolinone', Z. Naturforsch. 1986, 41c, 684.
16. Allelopathy means the direct influence of one plant species on another by means of chemical substances.
17. H. M. Niemeyer, 'Hydroxamic Acids (4-Hydroxy-1,4-benzoxazin-3-ones), Defense Chemicals in the Gramineae', Phytochemistry 1988, 27, 3349.
18. L. Bigler, A. Baumeler, C. Werner, M. Hesse, 'Detection of Noncovalent Complexes of Hydroxamic-Acid Derivatives by Means of Electrospray Mass Spectrometry', Helv. Chim. Acta 1996, 79, 1701; D. Sicker, M. Frey, M. Schulz, A. Gierl, 'Role of Natural Benzoxazinones in the Survival Strategy of Plants', Int. Rev. Cytology 2000, 198, 319; D. Sicker, H. Hartenstein, M. Kluge, 'Natural Benzoxazinoids – Synthesis of Acetal Glucosides, Aglucones, and Analogues', Recent Res. Dev. Phytochem. 1997, 1, 203; A. Baumeler, M. Hesse, C. Werner, 'Benzoxazinoids – Cyclic Hydroxamic Acids, Lactams and Their Corresponding Glucosides in the Genus Aphelandra (Acanthaceae)', Phytochemistry 2000, 53, 213.
19. A. Friebe, I. Wieland, M. Schulz, 'Tolerance of Avena sativa to the Allelochemical Benzoxazolinone. Degradation of BOA by Root-colonizing Bacteria', Angew. Bot. 1996, 70, 150.
20. T. Hartmann, L. Witte, 'Chemistry, Biology, and Chemoecology of the Pyrrolizidine Alkaloids', in Alkaloids: Chemical and Biological Perspectives, Ed. S. W. Pelletier, 1995, 9, 155.
21. N. Ushio-Sata, S. Matsunaga, N. Fusetani, K. Honda, K. Yasumuro, 'Penaramides, which Inhibit Binding of ω-Conotoxin GVIA to N-type Ca^{2+}-Channels, from the Marine Sponge Penares aff. incrustans', Tetrahedron Lett. 1996, 37, 225.
22. L. Lewin, Die Pfeilgifte, J. A. Barth-Verlag, Leipzig, 1923.

23. P. Dumbacher, B. M. Beehler, T. F. Spande, H. M. Garraffo, J. W. Daly, '*Homoba-trachotoxin in the Genus Pitohui: Chemical Defense in Birds*', Science **1992**, *258*, 799.

24. Monomorine I (**23**) serves the ant as a trail pheromone.

25. G. G. Habermehl, *Gift-Tiere und ihre Waffen*, 5th edn., Springer-Verlag, Berlin, 1994; A. Numata, T. Ibuka, '*Alkaloids from Ants and Other Insects*', The Alkaloids **1987**, *31*, 194; K. S. Brown, J. R. Trigo, '*The Ecological Activity of Alkaloids*', The Alkaloids **1995**, *47*, 227.

26. G. Laus, D. Brössner, K. Keplinger, '*Alkaloids of Peruvian Uncaria tomentosa*', Phytochemistry **1997**, *45*, 855.

27. C. Werner, C. Hedberg, A. Lorenzi-Riatsch, M. Hesse, '*Accumulation and Metabolism of the Spermine Alkaloid Aphelandrine in Roots of Aphelandra tetragona*', Phytochemistry **1993**, *33*, 1033.

28. J. W. Fairbairn, P. N. Suwal, '*The Alkaloids of Hemlock (Conium maculatum L.) – II. Evidence for a Rapid Turnover of the Major Alkaloids*', Phytochemistry **1961**, *1*, 38.

29. J. W. Fairbairn, G. Wassel, '*The Alkaloids of Papaver somniferum L. – I. Evidence for a Rapid Turnover of the Major Alkaloids*', Phytochemistry **1964**, *3*, 253.

30. *Papyrus Ebers* is named after the Egyptologist *G. M. Ebers* (1837–1898), who discovered the scroll, 21 m long and 30 cm wide, in a tomb in the necropolis of the ancient Egyptian town of Thebes.

31. The first synthetic medicinal agent, not an alkaloid, though, but still based on a plant-derived substance (salicylic acid from the willow *Salix alba*), was acetylsalicylic acid, first marketed as 'aspirin' by the *Farbenfabrik vormals Friedrich Bayer und Co.*, Elberfeld (today known as *Bayer AG*, Leverkusen). It is still today the most widely sold medicine worldwide, as it was as early as 1914.

32. *Index Nominum, International Drug Directory*, Medpharm, Stuttgart, 1995.

33. Formulae not explicitly shown in this section can be found elsewhere in the book.

10. Historical Aspects of Alkaloid Chemistry

The history of alkaloid chemistry, in structural terms, begins in 1804, when the Paderborn apothecary *Sertürner* (*Fig. 10.1*) discovered the so-called *principium somniferum* (also 'soporific principle' or 'soporific substance') in opium[1,2], which he reported the following year in the *Journal der Pharmacie* (*Fig. 10.2*)[3,4]. The attention of scientists, however, was only aroused twelve years later by a publication appearing in the *Annalen der Physik* (*Fig. 10.3*)[5,6]. There, *Sertürner* named his *principium somniferum* for the first time 'morphium' (after *Morpheus*, the son or servant of sleep and creator of dream states in *Ovid*; altered to 'morphinium' by the French physicist *Gay-Lussac*). In his publication[5], he described 'morphium' as '*one of the most peculiar substances, which appears first of all to be associated with ammonia*'. Thus, he had isolated an organic base containing '*oxygen, carbon, and hydrogen, perhaps also nitrogen*' from a plant. As was customary at that period, the products of a researcher's own chemical endeavors were tasted and its effects examined. *Sertürner* thus tested '*morphium*' on himself; he wrote[5]: '*We may also expect the likelihood of a number of healing properties against infirmity from the different morphium salts. From my own experience, I can testify that a very intense toothache, not eased by application of opium, was immediately relieved by a tincture of the morphium in alcohol, although it was not strongly concentrated*'. *Sertürner*'s (second) work caused a great sensation, since the discovery of a basic substance in a plant at that time was very surprising, especially as chemists' interests were directed more towards acidic plant substances after the important work by *Scheele*[7]. However it may be, the closing comments by *Gilbert* (the editor of *Annalen der Physik*) were taken by a number of researchers as a call to arms directed towards investigation of plants, which was duly taken up with enthusiasm (*Table 10.1*). At that period, however, it was frequently the case that newly isolated alkaloids were not immediately recognized as such.

In his 1837 *Lehrbuch der Chemie* (Vol. 6, p. 362), *Berzelius* (*Fig. 10.4*) carried out a summary of the '*well-studied vegetable salt bases*' known at that time. The alkaloids were even assigned symbols, like those of the chemical elements. The chemical compositions, measured by combustion analysis, were, notwithstanding, still far removed from the correct values (*Fig. 10.5*).

Often, a very great deal of time would pass between the isolation of an alkaloid and the determination of both its structure and absolute configuration

Fig. 10.1. *Friedrich Wilhelm Adam Sertürner* (June 19, 1783 – February 20, 1841). Apothecary in Paderborn, Einbeck, Hameln; awarded an honorary doctorate by the University of Jena in 1817 (with kind permission of *Petra Klein*, Museum of Hameln).

III.

Darstellung
der reinen Mohnſäure *) (Opiumſäure)

nebſt einer

chemiſchen Unterſuchung des Opiums

mit

vorzüglicher Hinſicht auf einen darin neu entdeckten
Stoff und die dahin gehörigen Bemerkungen.

Vom

Herrn Sertürner in Paderborn.

Im Journale der Pharmazie 13ten Bandes
machte ich einige Bemerkungen über die
beſondern Eigenſchaften des im Handel vor-
kommenden Opiums, welche mir nach den bis
jetzt bekannten Beſtandtheilen deſſelben uner-
klärbar waren; auch äußerte ich zugleich, daß
jene

*) Dieſes ſcheint mir der angemeſſenſte Name zu
ſeyn, weil ich ſie bis jetzt in keinem andern Vege-
tabil als dem Mohne gefunden habe.

Fig. 10.2. *Title page of Journal der Pharmacie* 1805, *14*, 47. With this original work concerning the isolation of morphine from opium, *Sertürner* ushered in the age of alkaloid chemistry.

III.

*Ueber das Morphium, eine neue ſalzfähige
Grundlage, und die Mekonſäure, als
Hauptbeſtandtheile des Opiums,*

von

SERTUERNER,
Pharmac. zu Eimbeck im Königr. Hannover.

„ Vor ungefähr 14 Jahren hat Herr Derosne, Pharma-
ceut zu Paris, beinahe gleichzeitig mit mir eine Analyſe des
Opiums unternommen, und ſie in den *Annales de Chimie* t. 45.
Jahrg. 1803 bekannt gemacht; unſere Reſultate waren aber ſo
verſchieden und widerſprechend, daß dieſer Gegenſtand ſo gut
wie im Dunkel blieb. Meine Abhandlung insbeſondere hat man
nur wenig berückſichtigt; ſie war flüchtig geſchrieben, die Men-
gen, mit denen ich gearbeitet hatte, waren nur klein, und Einige wollten mehrere meiner Verſuche nicht mit glücklichem Erfol-
ge wiederholt haben. Von der Richtigkeit derſelben im Allgemeinen
überzeugt, ob ich ſie gleich in einem frühen Alter unternommen
hatte, glaubte ich dieſes Mislingen in ihrem Verfahren ſuchen
zu müſſen. Um daher dieſe Widerſprüche zu heben und die
früheren Arbeiten über das Opium zu berichtigen, ſchritt ich zu
einer zweiten Analyſe dieſes merkwürdigen Pflanzenkörpers, und
habe das Vergnügen beinahe alle meine frühern Beobachtungen
in ihrem ganzen Umfange beſtätigt und mich im Beſitze neuer

Fig. 10.3. *Title page of Annalen der Physik* 1817, *55* (new system *25*), 56. General recognition was only to come with this second publication about the isolation of morphine.

Berzelius
Gezeichnet von Prof. Krüger
im Jahre 1827

Fig. 10.4. *Jöns Jakob Freiherr von Berzelius* (August 20, 1779 – August 7, 1848). Swedish scientist, studied chemistry and medicine in Uppsala, professor of medicine and pharmacology in Stockholm in 1807. Founder of elemental analysis; introduced the chemical element symbols still in use today as well as the term 'organic chemistry'.

Namen der Pflanzenbasen.	Symbol.	Atom des Ammoniaks.	Atome von			Atomgewicht.	Procente von			
			Kohlenstoff.	Wasserstoff.	Sauerstoff.		Kohlenstoff.	Wasserstoff.	Stickstoff.	Sauerstoff.
Morphin	M̊o	1	34	30	6	3600,00	72,30	6,24	4,92	16,66
Codeïn	C̊d	1	32	32	5	3366,00	72,66	7,23	5,26	14,86
Narcotin	N̊a	1	40	34	12	4684,11	65,27	5,32	3,78	25,63
Thebaïn	T̊h	1	25	21	4	2656,37	71,94	6,34	6,66	15,06
Strychnin	S̊t	1	30	26	3	3034,00	77,16	6,72	5,95	10,11
Brucin	B̊r	1	32	30	6	3447,67	70,96	6,50	5,14	17,40
Cinchonin	C̊i	1	20	18	1	1955,55	78,67	7,06	9,11	5,16
Chinin	Q̊u	1	20	18	2	2055,55	74,39	7,25	8,62	9,74
Aricin	År	1	20	18	3	2155,55	70,93	6,95	8,21	13,96
Veratrin	V̊e	1	34	37	6	3644,25	71,25	7,57	4,85	16,39
Solanin	S̊o	1	84	140	28	10308,67	62,29	8,84	1,72	27,15
Atropin	Åt	1	34	40	6	3662,95	70,98	7,83	4,83	16,36
Coniin	C̊n	1	13	22	1	1369,00	67.00	12,77	12,93	7,30

Fig. 10.5. *State of knowledge concerning the best studied 'vegetable salt bases' in 1837* (from J. J. Berzelius, *'Lehrbuch der Chemie'* (translated into German by F. Wöhler), Vol. 6, Arnoldische Buchhandlung, 3rd edn., Dresden, 1835 – 1841)

(*Table 10.1*). In the case of strychnine, 138 years passed by, and for morphine 150 years! Today, it is usual to determine the structure of a substance in the year of its isolation, especially when it seems to possess pharmacological properties as promising as those of strychnine and morphine.

The physical methods of structure determination routinely in use today were to conquer organic chemistry laboratories relatively late. The first commercial infrared (IR) spectrometers became available around 1950, ^{1}H-NMR spectrometers in 1961, and mass spectrometers in 1962, with ^{13}C-NMR spectrometers completing the list in 1973. The development of 'modern' separation techniques had begun only a few years earlier: the advent of distri-

Table 10.1. *Historical Data Concerning Some Well-Known Alkaloids*

Alkaloid	Original Source	Isolation of the Pure Compound	Correct Structure Determination	Absolute Configuration	Synthesis
Morphine	Opium	1805 *Sertürner*	1925 *Gulland & Robinson*	1955 *Mackay & Hodgkin*	1952 *Tschudi & Gates*
Xanthine	Bladder Stones	1817 *Marcet*	1882 *E. Fischer*	–	1882 *E. Fischer (Fig. 10.6)*
Emetine	Emetic Root (*Psychotria ipecacuanha, P. granadensis*)	1817 *Pelletier[a] & Magendie*	1948 *Robinson (Fig. 10.7)*	1959 *Battersby & Garrat*	1950 *Preobrazhenski et al.*
Strychnine	*Strychnos colubrina, S. ignatii, S. nux-vomica*	1818 *Pelletier & Caventou (Fig. 10.8)*	1947 *Woodward, Brehm & Nelson*	1956 *Peerdeman*	1954 *Woodward et al. (Fig. 10.9)*
Piperine	*Piper nigrum*	1819 *Oersted*	1874 *Fittig & Mielch*	–	1882 *Rügheimer*
Atropine	*Atropa belladonna*	1819 *Runge[b]*	1883 *Ladenburg (Fig. 6.1)*	1959 *Fodor & Csepreghy*	1902 *Willstätter (Fig. 10.10)*
Quinine (Cinchonine)	Bark of *Cinchona*	1820 *Pelletier & Caventou[c] (Fig. 10.8)*	1908 *Rabe*	1944 *Prelog & Zalán (Fig. 10.11)*	1944 *Woodward & Doering (Fig. 10.9)*
Caffeine	*Coffea arabica*	1820 *Runge*	1882 *E. Fischer (Fig. 10.6)*	–	1895 *E. Fischer & Ach*
Solanine	*Solanum nigrum, S. dulcamara*	1822 *Desfosses*	1954 *Kuhn & Löw*	1955 *Kuhn*	1964 *Schreiber & Rönsch*
Chelidonine	*Chelidonium majus*	1824 *Godefroy*	1931 *v. Bruchhausen, Bersch, Späth & Kuffner*	1979 *Takao et al.*	1971 *Oppolzer & Keller (Fig. 6.3)*
Coniine[d]	*Conium maculatum*	1827 *Gisecke*	1885 *Hofmann (Fig. 3.2)*	1932 *Leithe*	1886 *Ladenburg (Fig. 6.1)*
Nicotine[d]	*Nicotiana tabacum*	1828 *Posselt & Reimann*	1893 *Pinner*	1972 *Dagne & Castagnoli*	1904 *Pictet & Rotschy (Fig. 10.12)*
Aconitine	*Aconitum napellus*	1821 *Reimann & Peschier*	1963 *Wiesner et al.*	1971 *Wiesner et al.*	1969 *Wiesner et al.*
Colchicine	*Colchicum autumnale*	1833 *Geiger & Hesse*	1955 *Müller & Velluz*	1955 *Corrodi & Hardegger*	1961 *Eschenmoser et al.*
Sparteine	*Spartium scoparium*	1851 *Stenhouse*	1931 *Clemo et al.*	1961 *Okuda*	1960 *v. Tamelen & Foltz*
Cocaine	*Erythroxylum coca*	1860 *Niemann & Wöhler*	1898 *Willstätter*	1955 *Hardegger & Ott; Fodor*	1898 *Willstätter (Fig. 10.10)*

bution chromatography came in 1941, with paper chromatography following in 1943, and thin layer chromatography in 1960.

Natural compounds largely occur in complex mixtures of similar substances. At the beginning of the study of alkaloids – in *Sertürner*'s time – there was heavy reliance on the fact that alkaloids crystallized well and could be purified by dissolution/precipitation techniques using various solvents or, if possible, by fractionated recrystallization. It was, therefore, a great advantage when only one main alkaloid was present. It can easily be seen that instrumental analysis and chromatography had only marginal influences on the structure elucidation of most of the alkaloids listed in *Table 10.1*. Other factors were appreciably more important; especially that, at the beginning of the 19th century, organic chemistry was just starting to be accepted as a discipline on its own. The techniques necessary for structure determination had to be established first, and reagents for particular transformations hat to be found or developed. Many compounds were discovered for the first time as a result of degradation reactions of alkaloids and other natural products, as attested by some parent names immortalized in *IUPAC* nomenclature. One example, among many others, is quinoline, which *Gerhardt* (1816–1856) obtained in 1842 on distilling quinine (or cinchonine or strychnine) from molten KOH.

Fig. 10.6. *Emil Hermann Fischer* (October 9, 1852 – July 15, 1919). Studied chemistry in Bonn and Strasbourg, where he received his Ph.D. (*A. von Bayer*). Professor in Munich (1879–82), Erlangen (1882–85), Würzburg (1885–92), and Berlin (1892); *Nobel* Prize 1902. One of the most important natural products chemists, noted, among other things, for the structure determination and synthesis of purine bases.

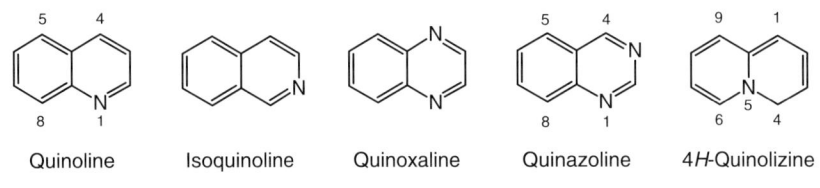

| Quinoline | Isoquinoline | Quinoxaline | Quinazoline | 4*H*-Quinolizine |

[a] The basic character of emetine was not recognized until 1823 (by *Pelletier* and *Dumas*).

[b] The alkaloidal nature of atropine was discovered by *Mein, Geiger*, and *Hesse* in 1833.

[c] *Vauquelin, Duncan, Reuss, and Gomez* had isolated a curious, crystallizable component from the bark of the Cinchona tree as early as 1811. *Gomez* named it cinchonine, and it was later crystallized in pure form by *Lambert*. This cinchonine was recognized by *Pelletier* and *Caventou* as (variously) pure quinine or a mixture of quinine and cinchonine.

[d] Until 1827, only solid alkaloids were known, and so the isolation of the volatile coniine from *Conium maculatum* caused great excitement and the establishment of an investigatory commission by the Parisian association of apothecaries to clarify the basic character of coniine, which confirmed the original findings in 1835.

Fig. 10.7. *Sir Robert Robinson* (September 13, 1886 – February 8, 1975). Studied chemistry in Manchester (*W. H. Perkin Jr.*) 1912 –1915. Professor in Sydney, then in Liverpool, Saint Andrews, Manchester, London, and, from 1930, in Oxford. Made fundamental contributions concerning the chemistry of biologically important alkaloids (*e.g.*, emetine, morphine). *Nobel* Prize 1947.

Quinine was consequently to play godfather at the naming of other heterocycles. The quinuclidine ring, for example, represents the non-aromatic, bicyclic part of the quinine ring-system. Piperine, the alkaloid of pepper (isolated, *e.g.*, from the fruits of *Piper nigrum* L.), affords piperidine upon hydrolysis. This is also a *IUPAC*-approved name, while its influence can also be seen in the naming of hexahydropyrazine as piperazine.

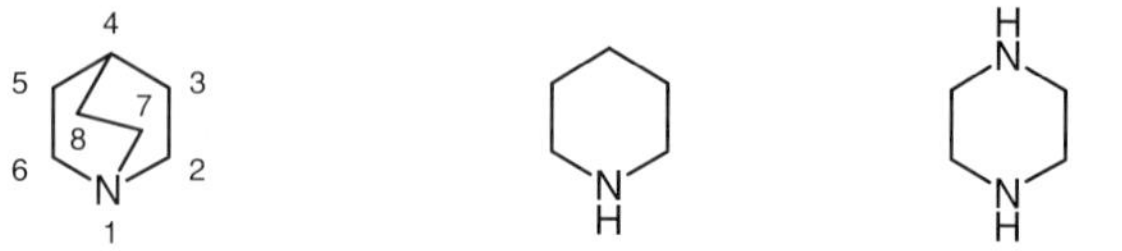

Quinuclidine
(= 1-Azabicyclo[2.2.2]octane)   Piperidine   Piperazine

The term 'indole' is derived from 'indigo', an Indian vat dye[8] from East Asian *Indigofera* species (*e.g.*, *I. tinctoria*; *I. leptostachya* (Leguminosae)), from woad (*Isatis tinctoria* (Cruciferaceae)), and from the Japanese indigo plant 'ai' (*Polygonum tinctorium*), in all of which the indole moiety is found as an indoxyl glucoside, called indican. After cleavage of the glucoside bond, there follows an oxidative dimerization of indoxyl (= 3-hydroxyindole) to indigo. Indigo was once the most important organic dye, since – in particular when applied on wool and cotton – it is resistant to light, washing, alkali, and acids. The dye itself is completely insoluble in water and alcohol. Both its structure determination and (later) its synthesis by *von Baeyer* (1870) was a major challenge to the chemistry of the 19th century and took place in parallel with the ongoing structure elucidation of alkaloids (*Scheme 10.1*). On performing a zinc-dust distillation[9] of a degradation product of indigo, oxindole, *Bayer* obtained 'ind-ol' (with the suffix 'ol' standing for 'oil'; Lat. *oleum*). Indole is also the archetype for the names indene, indolizine, indazole *etc.*

Scheme 10.1. *Degradation Reactions Performed on the Vat Dye Indigo, which Lead to Oxindole and Indole*

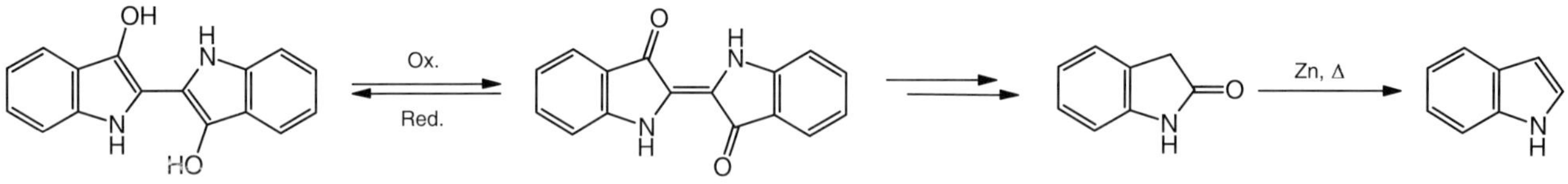

Indigo White
water-soluble
leuco base

Indigo (Indigotine)
insoluble in water
dye

Oxindole

Indole

The term 'morpholine' has its origin in the name of the alkaloid morphine, not because it can be obtained from this alkaloid, but because *Knorr* (1889) erroneously assumed that morphine contained (as its core component) an oxazine ring[10]:

'The oxazines have been objects of notably increased interest, since my investigations into morphine suggest that this important base, and presumably other bases closely related to morphine, must be viewed as oxazines. The interpretation of morphine printed in the formula

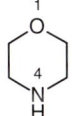

which appears to me to provide the best explanation for the facts known to date, presents morphine as a derivative of the hypothetical base

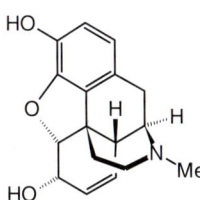

[...], *which I refer to by the name morpholine'.*

While the term 'morpholine' has endured to the present day, the assumed structural elements present in morphine turned out to be wrong.

Alkaloids or alkaloid-degradation products containing pyrrole or pyrrolidine rings produce a fiery red color (Greek πυρρος) on heating with spruce or beech shavings soaked in HCl. In 1834, *Runge* was able to make a connection between this characteristic and the correct structure of the 'oil' pyrrole, isolated from hard coal tar. The names 'pyrrolidine' or 'pyrroline' for its unsaturated analogs naturally followed.

These examples clearly testify to the decisive role played by the structure determination of alkaloids in the development of both heterocyclic and general organic chemistry, which were rapidly and continually developed.

Fig. 10.8. *Pierre Joseph Pelletier* (March 22, 1788–July 19, 1842). French apothecary and chemist, professor at the 'Ecole de pharmacie', Paris. *Jean Bienaimé Caventou* (1795–1877). Discoverer of numerous alkaloids, *e.g.*, brucine, quinine, and strychnine (from E. Heilbronner, F. A. Miller, *A Philatelic Ramble through Chemistry*, Verlag Helvetica Chimica Acta, Basel, 1998).

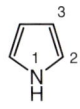

Morpholine

(–)-Morphine
(final structure, abs. config.)

Pyrrole   Pyrrolidine   2-Pyrroline

Fig. 10.9. *Robert Burns Woodward* (April 10, 1917 – July 8, 1979). Studied chemistry at the Massachusetts Institute of Technology, Cambridge (MA), Ph.D. 1937 (at age 20!), Professor at Harvard University, Cambridge (MA). Synthesized numerous natural products, notably the alkaloids quinine and strychnine. *Nobel* Prize 1965.

Fig. 10.10. *Richard Willstätter* (August 13, 1872 – August 3, 1942). Studied chemistry in Munich, Ph.D. 1894 (*A. von Bayer*). 1902 professor of chemistry in Munich, then in Zürich and Berlin. Director of the 'Kaiser Wilhelm Institute', Berlin. Important natural products chemist. Synthesized atropine and cocaine. *Nobel* Prize 1915.

Research into alkaloid structures was cross-fertilized by important advances in organic chemistry, among them such developments as *Kekulé*'s postulation of the tetravalent nature of carbon (1857) and his proposal of the formula of benzene (1865), *Le Bel*'s and *Van't Hoff*'s theory of the tetrahedral grouping of C-substituents (1874), and *Wislicenius*'s (1873) and *von Bayer*'s (1888) explanation of geometric isomerism.

Description of further structure determinations from the classical period (*cf.* coniine, *Chapt. 3.3*) seems excessive here, although a truly thrilling organic chemical discourse – almost like a crime novel – could be looked forward to. For a scientific perspective, *Boit*'s alkaloid book[11] is recommended.

Today, structure determinations of alkaloids are undertaken mainly with the aid of physical methods: NMR spectroscopy, mass spectrometry, and X-ray crystal-structure analysis. Chemical degradation reactions now play only a very subordinate role in the elucidation process.

Fig. 10.11. *Vladimir Prelog* (July 23, 1906 – January 7, 1998). Ph.D. 1929 with *Votoček*, Prague; 1935 lecturer at the Technical University of Zagreb, 1941 lecturer and 1950 – 1976 full professor in organic chemistry at the Swiss Federal Institute of Technology (ETH), Zürich. Many structural and stereochemically-oriented contributions, particularly well-known in the field of alkaloids (strychnine, steroid alkaloids *etc.*). *Nobel* Prize 1975.

Fig. 10.12. *Amé Pictet* (July 12, 1857 – March 11, 1937). Studied medicine and chemistry in Geneva, chemistry in Dresden, Bonn, and Geneva; Ph.D. 1881 in Geneva; postdoctoral study in Paris (*Wurtz*); professor in Geneva. Natural products chemist, specializing in alkaloids, who performed the first synthesis of nicotine. Other syntheses (laudanosine, papaverine) followed together with the *Pictet–Spengler* isoquinoline synthesis. He was an exponent of the view that alkaloids were degradation products of non-excretable, poisonous plant substances.

The advent of spectroscopic and chromatographic methods in organic chemistry manifests itself particularly in the increasing number of alkaloid publications appearing in the leading journal *Helvetica Chimica Acta* (founded 1917; *cf. Fig. 10.13*). A significant increase in the number of publications, largely attributable to new and improved analytical procedures, can be seen from the 1950s onwards[12].

The structure elucidation of a complex alkaloid (villalstonine, established in 1965) by chemical and spectroscopic means, particularly by mass spectrometry, is specially dealt with elsewhere in this book (*Chapt. 3*). Although more than 30 years ago, the 'logic' of structure elucidation remains vividly illustrated in the form of this example.

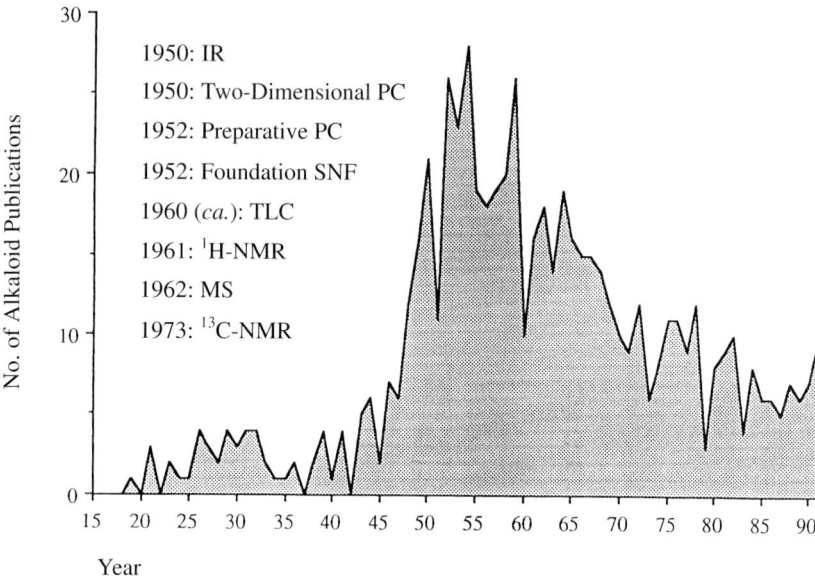

1950: IR
1950: Two-Dimensional PC
1952: Preparative PC
1952: Foundation SNF
1960 (*ca.*): TLC
1961: ^{1}H-NMR
1962: MS
1973: ^{13}C-NMR

Fig. 10.13. *Introduction of different physical methods into alkaloid chemistry and their impact in the form of publications in Helvetica Chimica Acta*[12]. *TLC = thin-layer chromatography, PC = paper chromatography, SNF = Swiss National Science Foundation.*

References and Notes

1. *J. B. Trommsdorff*, still unaware of the discovery of morphine, but aptly expressing the *spirit of those times*, wrote in his *Handbuch der gesammten Chemie*[2], Vol. 2, p. 534:
   '*Narcotic plant substances. § 1831. The narcotic effects of different plants and their constituent parts on the living organism has been attributed to a curious substance, named 'the intoxicating' or 'the narcotic principle' (Principium narcoticum). In fact, though, such a substance has merely been assumed hypothetically, since it has not been possible neither to prepare it, nor to identify any reagent with which to detect the same. I am unable to convince myself of the existence of the substance, and I rather assume that the narcotic property of many plant materials, e.g., the poppy juice, the leaves of the cherry laurel (Prunus lauro-cerasus), the deadly nightshade (Atropa belladonna), the thorn apple or Jimson weed (Datura stramonium), the henbane (Hyoscyamus niger), the tobacco plant (Nicotiana tabacum), and many others, is a property of the previously examined intimate components of the plant kingdom and depends on their particular admixture*'.
2. J. B. Trommsdorff, *Handbuch der gesammten Chemie*, 8 Vols., 2nd edn., Henning'- sche Buchhandlung, Erfurt, 1805–1818.
3. At the end of the report[4], *J. B. Trommsdorff*, the editor of the journal, put the following afterword (which was common practice at that time):

'*The author's publications contain some very interesting opinions, for which the chemical community is greatly in his debt. Numerous as investigations into opium now are, it is not in any way possible to view the files as closed, and it is much more to be desired that this topic should be examined further to shed light on some still enduring dark areas. I would especially wish that the experiment be repeated with somewhat larger quantities*'.

4. Sertürner, '*Darstellung der reinen Mohnsäure (Opiumsäure) nebst einer chemischen Untersuchung des Opiums*', *Journal der Pharmacie (Trommsdorff)* **1805**, *14*, 47.

5. Sertürner, '*Ueber das Morphium, eine neue salzfähige Grundlage, und die Mekonsäure, als Hauptbestandtheile des Opiums*', *Annalen der Physik* **1817**, 55 (new series *25*), 56.

Meconic Acid: Open *vs.* Closed Form

3-Hydroxy-2,4,6-trioxo-
heptanedioic Acid

2-Oxo-2-(3,4,5,6-tetrahydro-3,5,6-
trioxo-2*H*-pyran-2-yl)ethanoic Acid

6. R. Schmitz, '*Friedrich Wilhelm A. Sertürner und die Morphinentdeckung*', *Pharm. Zeitung* **1983**, *128*, 1350.

7. E. Winterstein, G. Trier, *Die Alkaloide*, Verlag Bornträger, Berlin, 1910.

8. In vat dyeing, the dye is initially reduced (in a vat), whereupon a colorless, so-called 'leuco compound' is produced. The latter, being water-soluble, has an enhanced affinity for fibers. Atmospheric oxidation finally converts the leuco compound back to the water-insoluble dye, which is, therefore, fixed onto the fiber.

9. In a zinc-dust distillation, the organic test substance is mixed with powdered zinc and dry distilled. The products (or product mixtures), separating in the air condenser, consist of smaller cyclic compounds (aromatics, heterocycles), which may sometimes provide valuable clues about the structure of the original test substance. If, however, (thermal) skeletal rearrangements take place during this process (*e.g.*, the formation of quinoline from indole alkaloids), structure determination becomes significantly more difficult.

10. L. Knorr, '*Synthesen in der 'Oxazinreihe*'', *Ber. Dtsch. Chem. Ges.* **1889**, *22*, 2081.

11. H.-G. Boit, '*Ergebnisse der Alkaloidchemie bis 1960*', Akademie-Verlag, Berlin, 1961.

12. A. Guggisberg, M. Hesse, '*Meilensteine der Alkaloid-Forschung in den Helvetica Chimica Acta*', *Helv. Chim. Acta* **1992**, *75*, 647.

11. Active Principles from Selected Alkaloid Sources and Their Cultural and Historical Significance

11.1. General

Many alkaloids display pronounced physiological activity. As a result, they have been feared and craved, proscribed by churches, and outlawed by states. Certain alkaloid-producing plant crops get burned or otherwise destroyed in police actions; but people, the consumers of these alkaloid-containing drugs, are always ready to break the law and offer certain alkaloids to other people for cash. Sometimes, the banned substance may even be tolerated in certain borderline cases, such as pain relief in illness. There would, in principle, be nothing to object to in the occasional taking of many of these alkaloids in small quantities – their use in this way may even be recommendable in seemingly hopeless situations – were it not for the tendency to addiction. For it is addictiveness that is ultimately responsible for all the known, ghastly side effects; whether for the consumers themselves, the suppliers, the producers, the community, or the organs of state.

'Addiction' is a state that can be brought on by repeated use of drugs (narcotics, medicinal drugs, alcohol)[1]. It is characterized by psychological (overpowering desire or compulsion to take more of the effective substance) and physical (potentially fatal symptoms on withdrawal) dependence on the 'substance' in question. The addiction is largely associated with a development of tolerance, which, in turn, produces a need for higher doses to obtain the same effect with continued use.

'Drugs' were originally natural products, plant components used medicinally or for technical purposes (hence 'drugstore'). In a narrower and more modern sense, the word can mean a substance that may induce dependency (narcotic drugs).

This chapter offers some examples of the use of alkaloids (either alkaloid-containing drugs or pure substances) that stand out by virtue of their pharmacological effects or their economic significance. The word 'drug' is used here in the general sense of 'natural product of plant origin'.

11.2.   Hemlock (*Conium maculatum* L.)

The poison hemlock (*Fig. 11.1*), also known as 'spotted parsley' (*Conium maculatum*), commonly encountered in pastures and meadows, grows up to *ca.* 2.5 m in height and is found in North and South America, as well as from Europe up to Siberia. The plant blooms in July and August, with small, white blossoms. In certain parts of southern Germany, though, its eradication has been so thorough that consideration has been given to making it a protected species.

The main alkaloid in all parts of the plant, but especially in the ripe fruits, is (+)-coniine (for structure elucidation and synthesis, *cf. Chapt. 3.3* and *Chapt. 6.2*, respectively).

Generally, (+)-coniine is accompanied by *N*-methylconiine, conydrine, and γ-coniceine, *etc.* In any case, *C. maculatum* is an extremely poisonous plant, and instances of unintended poisoning are still known even today. The most common mishap is to confuse the foliage with parsley or chervil, or the root with parsnip (or parsley root or white beet). A report from earlier literature dwells on that theme[2]:

> '*In Spain, in the year 1812, two soldiers ate the roots of hemlock, which they had mistaken for parsnip, in a soup. Soon afterwards, they fell into a narcotic state, experiencing headaches and nausea. One of them, who had eaten more, lay down and went to sleep. After an hour and a half, however, his breathing became difficult and a doctor was sent for. This gentleman found the patient in an unconscious state, with a pulse rate of only thirty beats per minute against a normal figure of seventy. The patient's limbs were cold, his face blue, engorged with blood, like that of a hanged man. He was given the appropriate treatment. Half an hour after taking the emetic, flatulence ensued, without any vomiting, though. His condition became steadily worse, and he gave up the ghost three hours after the repast. – The other man was saved*'.

In ancient Athens, the juice of *Conium maculatum* was used as a punishment for common criminals. It is reliably known that the Thirty Tyrants introduced (404–403 B.C.) this judicial means of death and made fulsome use of it against 'unpopular' people.

In the year 399 B.C., *Socrates* (*Fig. 11.2*) was condemned to death by drinking hemlock[3]:

(+)-Coniine
(abs. config.)

N-Methylconiine
(abs. config.)

Fig. 11.1. *Conium maculatum* L., *the poison hemlock or spotted parsley.* Contains the alkaloid coniine, among other compounds (J. Pecirka, *Giftgewächse*, Verlag K. André, Prague, 1859).

Fig. 11.2. *Socrates, surrounded by his students, takes the cup of hemlock* (woodcut, 19th century; Bildarchiv Stiftung Preussischer Schlösser und Gärten, Berlin-Brandenburg)

'The charge that supposedly justified his death sentence went: 'Socrates sins and commits foolishness in that he studies infernal and heavenly things and turns injustice into justice and also teaches this to others' or, according to the prosecutors Anitos and Meletos: 'Socrates is godless and corrupts the young'. And so, after his condemnation, he was turned over to the eleven men whose duty was to supervise his execution. In prison, where he did not remain for long, the draught, already prepared, was offered to him in a cup by the executioner. He took it, 'without any trembling, paling, or changing his expression'. He had previously let the poison-maker inform him as to the nature of the course of the poisoning. What this person told him, and what Plato reported, is a good outline of

the nature of action of hemlock as recognized by modern toxicological analysis, which, by virtue of its content of coniine, conydrine etc., causes paralysis of the motor and sensory synapses, together with paralysis of spinal cord and brain. The respiratory centers in the spinal marrow are the first to give way. The victim suffocates, while consciousness persists to the end. Sometimes, slight spasms precede death. 'Once you have drunk', said the poison-maker, 'you have to do no more than walk around until your legs feel heavy and then lie down'. As Socrates commented that his legs were getting heavy, he lay down on his back. After that, they would touch him from time to time, and examine his feet and legs. Then, they would press hard on his foot and inquire whether he could feel it, to which he said no. And then the knee, and it went on ever higher, and those waiting by him saw how he became cold and rigid. Finally, he twitched. When he was uncovered, he was dead'.

11.3. Fly Agaric (*Amanita muscaria*)

In north-eastern Asia, in Siberia between the *Ob* river and the *Bering* Sea, live the Samoyed, Ostyak, Tungus, Yakut, Yukagir, Chukchi, Koryak, and Kamchadal tribes. They make use of the poisonous fly agaric toadstool (*Fig. 11.3*) to achieve a state of intoxication. Whether alkaloids are solely responsible for this effect or merely involved in combination with other substances, has yet to be completely resolved. Compounds isolated from fly agaric – muscimol, ibotenic acid, and muscazone – inhibit motor function, while muscazone, in particular, is a psychotropic agent. Since muscimol is a decarboxylation product of ibotenic acid, it is possible that it might have been produced from the latter during the isolation process. Photochemical investigation has shown that muscazone is produced from ibotenic acid on UV irradiation.

The alkaloid muscarine, isolated from the toadstool, is at least not purely responsible for the attainment of this state of intoxication, which produces arousal, confusion, wild laughter, and hallucination, and also maniacal rage in bad cases (without permanent damage, however). After a few hours, normality reasserts itself.

'To bring about the desired effect, one large mushroom or two or three small ones, dried in air or by smoking, are sufficient for one day. Infusions of the toadstools in either water or milk are taken warm or cold. Koryaks and Chukchi have been seen taking out small, round, raffia containers made from birch, or leather holders, used to keep small pieces of dried fly agaric. From time to time they would put one of these in their

Fig. 11.3. *Amanita muscaria L., the mysterious Fly Agaric – poison mushroom or narcotic drug (photo M. Hesse)*

Muscimol

Ibotenic Acid

Muscazone

(+)-Muscarine
(abs. config.)

mouths and keep it there for a long time, without swallowing it. Among the Koryaks, it seems that its use is in such a way that the women chew the dried toadstools and roll the chewed mass between their hands to make small pellets, which the men swallow.

Among the great number of mysteries connected with the use of this mushroom, not the least is the fact that Koryaks, Kamchadals, and others have found the urine of those intoxicated by fly agaric to possess equally intoxicating properties. As soon as Koryaks notice that their state of intoxication is wearing off, and if they have no more of the toadstool or wish to conserve their supplies, they drink their own urine. The Koryak women pass intoxicated people a metal vessel, specially kept for this purpose, into which they urinate in the presence of all. The urine, often still warm, is drunk by people on awakening from sleep and, after a few minutes, once more has its effect, which can apparently be renewed many times in this way. It seems unlikely that – as was maintained previously – the intoxicating principle is still to be found in effective quantities in the urine after passage through four or five persons. It cannot always be parsimony or poverty that leads to the use of the urine, though, since even shamans of the Yukagiri, Tungus, and Lamut always take some toadstool urine of this kind before entering their ecstatic states[4].

11.4. Ergot

Ergot (also known as *Secale cornutum*; *Figs. 11.4* and *11.5*) is the product of a plant disease. A cylindrical, rather bent, hard body, dark violet on the outside and white inside, grows between the cornhusks, especially those of rye and barley. This is the sclerotium of the ergot fungus *Claviceps purpurea* Tul. (Ascomycetes). It is necessary to remove this ergot before grinding the grain, as it will otherwise poison the flour. If this is not done properly, then poisoning may occur, especially with prolonged ingestion of foodstuffs prepared from the contaminated flour. Epidemic-like outbreaks of mass poisoning have been known in western, central, and eastern Europe.

Literary evidence suggests that people have long been familiar with fungal parasites on grain (*cf. Table 11.1*). A magic spell found in a small temple in Ischaly (ancient Neribtum) in Mesopotamia was dated to *ca.* 1900–1700 B.C. There, the term *mehru* was used for abnormally infested grain. The Assyrians were already able to distinguish between different grain diseases. Sumerian clay tablets, dated to *ca.* 1700 B.C., describe the reddening of damp grain, which is called *samona*. There are also some references concerning grain diseases in the Old Testament (850–550 B.C.). The *Hearst* medical papyrus

(Secale cornutum)
(Mutterkorn)

Fig. 11.4. *Secale cornutum, ergot* (J. Pecirka, *Giftgewächse*, Verlag K. André, Prague, 1859)

331

Fig. 11.5. *Overwintering resting structures* (sclerotia) *of the fungus Claviceps purpurea on rye* (Secale cereale L.) (Photo S. Johne)

(*ca.* 550 B.C., Egypt) describes a preparation (No. 145) in which a mixture of ergot, oil, and honey is recommended as a treatment for hair growth, although the translation of the word *ergot* is controversial[5].

For centuries, the afflictions caused by *C. purpurea* raged through Europe, being given such names as the 'holy' or 'infernal fire', or *St. Anthony*'s fire (*Fig. 11.6*). These names for the condition were widespread between the 9th and the 13th century; afterwards, expressions such as ergotism are found more commonly. One particular form of the condition, convulsive ergotism (*Ergotismus convulsivus*), particularly common in Germany, initially manifested itself in damage to the nervous system. Its onset was marked by convulsion of the limbs, followed by spasmodic, epilepsy-like muscle contractions, resulting sometimes in irreversible contortion of the limbs. The inhabitants of France suffered chiefly from its second manifestation, gangrenous ergotism (*Ergotismus gangraenosus*). Damage to peripheral blood vessels resulted in the dying of individual limbs – commonly hands and feet, but also facial features, genitals, and breasts; these became cold and blackened, and the flesh fell from the bones. The few survivors presented a horrific sight. Not uncommonly, it would particularly affect young, unmarried women who made use of a drug mixture (prescribed by midwife folk her-

Table 11.1. *Historical Mentions of Fungal Infections* (rust) *of Grains and Grasses*

Time (B.C.)	Source	Mention
1900 – 1700	Charm in Neribtum (Mesopotamia)	Use of the expression *mehru* for abnormal grain afflicted with fungal infections
1700	Sumerian clay tablets (Mesopotamia)	The expression *samona* is used to describe the reddening of wet grain
850 – 550	Old Testament	Examples of grain diseases mentioned: Genesis **41**, 27; Amos **4**, 9; Deuteronomy **28**, 20; Haggai **2**, 17; I. Kings **8**, 37
700	Roman feast of Robigalia	Introduced under King *Numa Pompilius*, celebrated each April 24 (the day of the start of wheat ripening; prayers and offerings to the corn rustgoddess *Robigo* and/or the corn rust god *Robigus*)
550	*Hearst* medical papyrus (Egypt)	Ergot is described as an ingredient in a treatment for hair growth (preparation no. 145), first mention of a medicine from corn fungi
460 – 370	*Hippocrates,* Greek physician	Description of corn blight
384 – 322	*Aristotle,* Greek philosopher	He assumes that grain rust is caused by warm vapors
371 – 286	*Theophrastus,* Greek philosopher	Concludes from observations on fungal diseases of grain that barley is more susceptible than wheat, and that windy fields have less rust than damp, shady low-lying ones

balists) containing a high, indeterminate proportion of ergot to bring about abortion.

A contemporary account gives a precise description of the course of the condition[2]:

'... *On February 5, 1855, the same siblings were once again affected by the ergotism in the most severe degrees, after having eaten hot dumplings made from the remains of the ergot-containing flour for their midday meal. Immediately after the meal, the boy complained of nausea, giddiness, weakness of vision, a feeling as of ants crawling over his limbs, and fell into the most terrible convulsions with subsequent lockjaw; he lost*

Fig. 11.6. *Sections from Pieter Bruegel the Elder's 'Fight between Carnival and Lent'* (1559; Kunsthistorisches Museum, Vienna). People with misshapen bodies and missing extremities can be seen in the left and central parts of the picture (*cf.* enlargement). The cause of these mutilations is ergotism, which is in turn caused by consumption of food contaminated with ergot.

the ability to speak and became unconscious. At this, repeated vomiting set in, followed after six hours by death. The older of the girls, at the sight of her dying brother, fell into the most severe convulsions, and although these ceased after a few days, her strength disappeared, vision and hearing faded away almost completely, and a drowsy, semiconscious state set in. Searingly painful boils developed on the torso; these were encircled in red and became gangrenous. The gangrene then attacked the whole torso and the upper legs, the surface of the skin peeled off, and a corpse-

like smell filled the room. The abdomen finally became distended, intestinal cramps, diarrhea, delirium, sobbing, fainting, and unconsciousness ensued, and on March 9, 1855, the poor thing breathed her last. – The younger girl recuperated after many repeated attacks, once she was completely removed from the parental home, and recovered fully after two months'.

The malady was named after *St. Anthony* (*Fig. 11.7*), because the afflicted sought to be cured by pilgrimage to the saint's church of St. Didier-la-Mothe (Dauphiné, south-eastern France) and by invocation.

The ideal conditions for the fungus to grow are damp and marshy regions in combination with cold, wet years. In these times, harvest failures would occur, and the resulting hunger of the poor, who tended to favor a rye-based diet, encouraged less scrupulous practices in ensuring the quality of the rye grains, after which fate would duly take its course. A first key turning point came with the discovery of the relationship between ergot and ergotism in 1717 thanks to the work of *C. N. Langen* (*Fig. 11.8*). The final turning point came in 1918 with the isolation of the ergot alkaloid ergotamine by *A. Stoll* (*Fig. 11.9*).

During the course of intensive pharmacological and clinical studies at *Sandoz* (Basel, Switzerland), *Stoll* and *Hofmann* identified ergotamine (*cf. Chapt. 2*) as the primary agent of ergot's physiological and therapeutic properties. Today, the compound is used with great success not only in obstetrics, but also, thanks to its sympatholytic action, in other areas of medicine. Its most outstanding property is induction of uterine contraction, as a result of which the ergot alkaloids find widespread application in gynecology.

During the course of investigations into the chemistry of lysergic acid, a new and surprising aspect of ergot effects was stumbled upon, laying the foundation stone for the development of novel psychopharmaceutics. The synthesis of lysergic acid diethyl amide (LSD) was published in 1955, although *Hofmann* (*Fig. 11.10*) had voluntarily intoxicated himself already in April 1943. He later described this as follows[6]:

'...in the spring of 1943, I thus repeated the synthesis of LSD-25. As in the first synthesis, this involved the production of only a few centigrams of the compound.

In the final stage of the synthesis, the purification and crystallization of lysergic acid diethyl amide in the form of a tartrate (tartaric acid salt), I was interrupted in my work by unusual sensations. The following de-

Fig. 11.7. *Saint Anthony the Hermit, surrounded by sufferers of ergotism (Swabia, 1440–1450; Staatliche graphische Sammlung, Munich)*

Carl Niclaus Langen
Phil & Med. D. Acad. Leopoldino- Carolinæ
Societ. Reg. Prussicæ. & Physio- Crit. Sen.
wie auch deß Raths einer Hochl. Cant. Luzern/
und würckl. Landvogten zu Knutwyl.

Beschreibung
Deß bis dahin bey uns niemahl erhörten/
und zu Zeiten sehr schädlichen Genuß

Der Korn = Zapffen
In dem Brot/
Und deß darauff folgenden unversehenen
Kalten Brandts/
Darin
seine innerliche und äusserliche Ursachen
sambt den erforderlichen Mittlen und Weiß
deß Auswachs und Vergifftung der Korn-Zapffen
begriffen seynd.
Worbey weitläuffig von dem grossen Nutzen
deß natürlichen Taus
Und hergegen von dem entsetzlichen Schaden
deß vergifften Mühltaus/
Darauß auch öffters der Vich-Presten entstehet
gehandlet wird.
Neben einem kleinen Anhang
Etwelcher seltzsammen und zu der Artzney
sehr nutzlichen Observationen.
LUCERN/

PERMISSU SUPERIORUM.
Bey Heinrich Rennward Wyssing Statt-Bucht. 1717.

Fig. 11.8. *Title page of Carl Niclaus Langen's 1771 book about grain sclerotia. Langen first recognized the connection between ingestion of ergot or ergot-contaminated flour and the onset of ergotism* (private collection)

Fig. 11.9. *Arthur Stoll* (January 8, 1887–January 13, 1971). Studied botany, geology, and chemistry at the Swiss Federal Institute of Technology (ETH) in Zurich; Ph. D. in 1911 (*R. Willstätter*); director and president of *Sandoz AG*, Basel, Switzerland. In 1918, *Stoll* isolated the first ergot alkaloid, ergotamine, which was the beginning of an extremely fruitful collaboration with *A. Hofmann* regarding the ergot alkaloids (= lysergic acid derivatives) (photo *P. Heman*).

scription of this incident comes from the report that I sent at the time to Professor Stoll.

Last Friday, April 16,1943, I was forced to interrupt my work in the laboratory in the middle of the afternoon and to go home, since I was affected by a remarkable restlessness combined with a slight dizziness. At home, I lay down and sank into a not unpleasant state, as though intoxicated, characterized by a highly active imagination. In a drowsy state, with eyes closed (I found the daylight to be unpleasantly glaring), I perceived an uninterrupted stream of fantastic pictures, extraordinary shapes with intense, kaleidoscopic play of colors. After some two hours, this condition faded away.

To get to the bottom of this, I decided on a self-experiment. I wanted to be cautious and so began the planned series of experiments with the smallest quantity that might be expected to produce some effect, in view of the activity of the ergot alkaloids known at the time: namely with 0.25 mg (mg = milligram = one thousandth of a gram) of lysergic acid diethyl amide tartrate.

The last words I was able to write only with great effort. By now, it was already clear to me that lysergic acid diethyl amide had been the cause of the remarkable experience of the previous Friday, for the altered sensations and experiences were of the same type as before, only much more intense. I could only speak intelligibly with intense efforts and asked my laboratory assistant, who was aware of the self-experiment, to accompany me home. On the way home, by bicycle – no automobile was available then, since during the war they were reserved only for a few privileged persons – my condition began to assume threatening forms. Everything in my field of vision wavered and was distorted as if seen in a curved mirror. I also had the sensation of being unable to move with the bicycle from the spot. Nevertheless, my assistant later told me that we had traveled very rapidly. Finally, arriving home safe and sound, I was just about still capable of asking my companion to call our family doctor and request milk from the neighbors. ...'

Fig. 11.10. *Albert Hofmann* (January 11, 1906). Studied chemistry at the University of Zurich; Ph. D. in 1929 (*P. Karrer*); head of the Natural Product Section at *Sandoz AG*, Basel. Famous among other things for his work in the field of ergot alkaloids.

LSD, at times a drug highly sought-after, is less *en vogue* today, perhaps because accessing the semi-synthetic lysergic acid is more difficult nowadays.

Fig. 11.11. *Papaver somniferum* L. Scoured, unripe seed heads exuding latex for opium production (photo *S. Johne*)

(−)-Morphine

11.5. Opium [7]

Opium (Greek *οπιου* = poppy juice) is a latex exuded from appropriately damaged, unripe seed heads of the opium poppy (*Papaver somniferum* L.), which subsequently dries out. A few days after the shedding of the petals, surface incisions are made in the seed capsules, allowing the latex to exude overnight (*Fig. 11.11*). The next morning, it is scraped off with a knife, collected on a poppy leaf and kneaded into a cake. One seed head delivers *ca.* 20 mg of opium. Commercially, it used to be available in balls of *ca.* 1 kg in weight, for which a good 50,000 poppy seed heads would have been required (estimates vary).

The oldest evidence for the use of the poppy comes from the time of the pile dwelling culture (Neolithic, *ca.* 3000–2500 B.C.). In Switzerland, *e.g.*, poppy seeds and capsules have been found near various lakes in the canton Zurich, Thurgau, and Berne. What they were used for at the time is unclear, but they probably served as a foodstuff (added to breads and cakes or as a source of oil). Even today, poppy seeds remain a popular baking ingredient, especially in central and eastern Europe (poppyseed cake, poppyseed rolls, *etc.*; *Fig. 11.12*). The opiate content in ripe poppy seeds is exceedingly small. In gas chromatographic/mass spectrometric (GC/MS) measurements on Indian poppy seeds, only 1.7×10^{-4} g morphine were detected per g seeds.

Both written and pictographic documents from the eastern Mediterranean region testify to the very early knowledge of opium, its production, and applications.

Homer, in the *Odyssey*, refers to an opium-containing, wine-based drink that he calls *Nepenthes* (Greek *νηπενϑης*; *ne*: not, *penthos*: grief, sorrow → thus a drink to bring about forgetfulness). It was drunk by warriors before battle, to stop feelings of fear. It also served to deaden feelings towards horrific incidents in battle.

Minoan culture venerated the poppy in the form of a goddess with a crown of poppy seedpods (*Fig. 11.13*). Later, the center of poppy cultivation was to be found at Mecone[8], a town east of Corinth. Across the Aegean, in Asia Minor, the center of poppy cultivation was the town of Afion (modern-day Afyon, Turkey). The poppy, opium, and its Turkish (*afyon*) and Arabic names (*afion*) spread eastwards along the caravan routes, becoming *afium* (Persia), *aphuka* or *ahiphena* (India), *á phiên* (colloquial Vietnamese), *nha phiên* (standard Vietnamese), *ya pian* (Chinese, *Fig. 11.14*), or *ahen* (Japanese). The Chinese expression at the same time means 'dark pane', which, in the sense conveyed, may be translated as 'disastrous'.

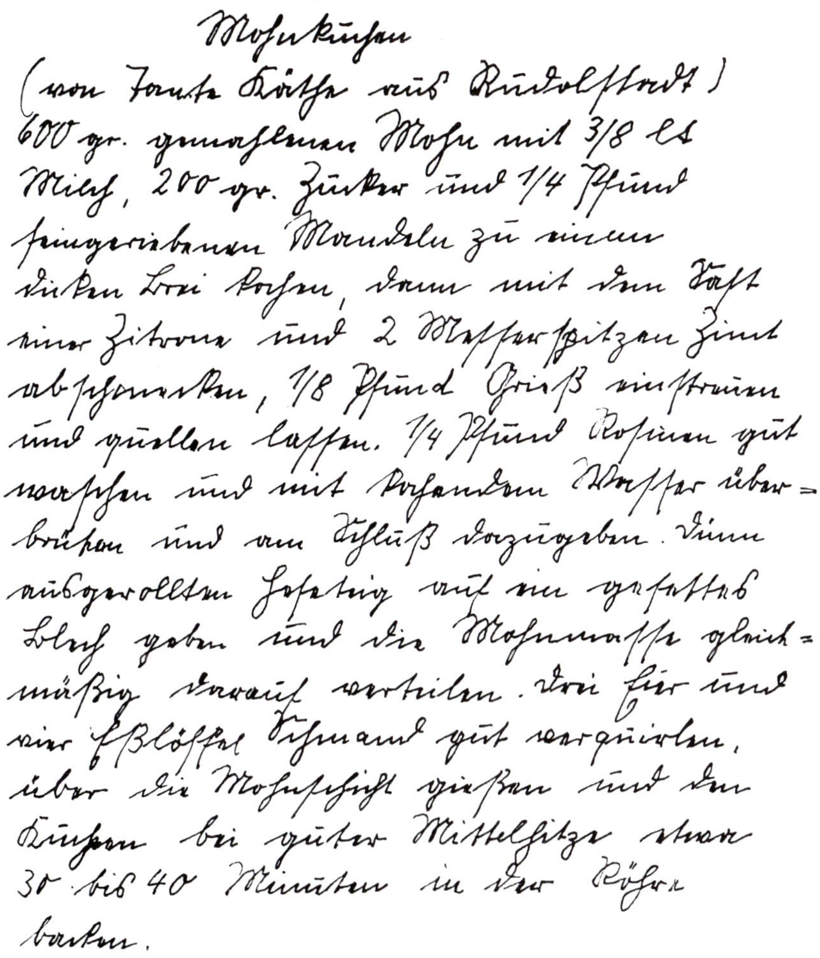

Mohnkuchen

(von Tante Käthe aus Rudolstadt)
600 gr. gemahlenen Mohn mit 3/8 lt.
Milch, 200 gr. Zucker und 1/4 Pfund
feingeriebenen Mandeln zu einem
dicken Brei kochen, dann mit dem Saft
einer Zitrone und 2 Messerspitzen Zimt
abschmecken, 1/8 Pfund Grieß einstreuen
und quellen lassen. 1/4 Pfund Rosinen gut
waschen und mit kochendem Wasser über-
brühen und am Schluß dazugeben. Dann
ausgerollten Hefeteig auf ein gefettetes
Blech geben und die Mohnmasse gleich-
mäßig darauf verteilen. Drei Eier und
vier Eßlöffel Schmand gut verquirlen,
über die Mohnschicht ziehen und den
Kuchen bei guter Mittelhitze etwa
30 bis 40 Minuten in der Röhre
backen.

Fig. 11.13. *Poppy Goddess, Crete, ca. 1400 B.C.* (M. Seefelder, *Opium*, Landsberg, 1996)

Fig. 11.12. *Thuringian poppyseed cake* (Mohnkuchen nach Thüringer Art). Recipe from 'aunt Käthe' of Rudolstadt (Thuringia): Heat 600 g ground poppy seeds with 3/8 liter milk, 200 g sugar, and 1/4 'pfund' (1 pfund = 500 g) finely grated almonds to make a thick paste, then flavor with the juice of a lemon and two pinches cinnamon, mix in 1/8 pfund semolina and allow to swell. Wash 1/4 pfund raisins thoroughly, steep in boiling water, and add. Grease a baking tray and place thinly rolled out yeast dough on it, then spread the poppy seed mixture evenly over this. Thoroughly mix three eggs and four dessertspoons sour cream, pour over the poppy layer, and bake the cake in a moderate to hot oven for 30 to 40 minutes.

Until the beginning of the modern era (about the middle of the 17th century), opium was used in all cultures (*Figs. 11.15* and *11.16*) almost exclusively as a medicine for pain relief, in the broadest sense, rather than as a euphoric drug (*Fig. 11.17*).

Fig. 11.14. *Chinese characters for opium* (Ya Pian) (drawn by *W. Hu*)

In India, poppy cultivation was first practiced in the central countryside of Malwa, while the regions around Bihar and, later, Patna followed. At that time, China was importing a lot of opium from India for medical purposes. Smoking of opium, however, came to China only in the second half of the 17th century. The British *East India Company* cultivated the poppy for opium production in Bengal, establishing a monopoly, and, from 1773 onwards, exported opium to China in ever greater quantities[9]. Opium became the major commodity traded in exchange for silk and tea. In 1820, the Chinese government outlawed the importation of opium, which resulted in an organized black market of undreamed size. In 1839, the Chinese government allowed 20,000 cases of opium, with a value of *ca.* four million pound sterling, to be destroyed. London viewed this act as a provocation and introduced punitive measures against China. British warships bombarded the southeastern Chinese coast in 1840, and China was finally forced to capitulate in the Opium War of 1842. In the peace treaty of Nanking, Hong Kong was handed over to the British Empire, five Chinese ports were opened to European trade, and, in addition to this, China was force to pay war reparations to London. In 1858, with the Treaty of Tiëntsin – again under the pressure from London – the Chinese government once more legalized the opium trade (*Fig. 11.18*)[10,11].

Fig. 11.15. *Greek bronze coin from Ankyra, Phrygia* (second century A.D.; obverse: a poppy head between two ears of grain [reverse: anchor]; Winterthur Coin Museum, Switzerland, Inventory No. G4038)

Fig. 11.16. *Roman bronze coin* (81–96 A.D. [obverse: IMP.DOMIT.AVG.GERM.COSXI. Half-length portrait of Emperor's wife *Domitia* as *Ceres*, bounded to the right with ears of grain]; reverse: bundle of two ears of spelt wheat (*Triticum spelta* L.) and three poppy heads; Winterthur Coin Museum).The poppy (*Papaver somniferum* L.), thanks to its anesthetizing properties, was known to the ancients even in the earliest periods, and its juice was a popular medicine. Cultivation of the poppies was already practiced during Antiquity, with opium production mainly distributed in Asia Minor (modern Turkey). Distinction was made between Opos (᾽οποσ), the dried latex from the unripe seedpod, and the less effective extract of the entire plant, called *Mekonium* (μεχωυιου).

In 19th century England, use of opium as a recreational drug was widespread, being as easily available as aspirin is today. In 1821, the English essayist and critic *Thomas de Quincey* (1785 – 1859) published his autobiographical piece *Confessions of an English Opium-Eater*, in which the effects of the drug were vividly described. The publisher *Alethea Hayter* wrote:

Fig. 11.17. *Allegoric illustration in the Chapter 'Der Tabak und die übrigen narkotischen Genussmittel' [Tobacco and Other Recreational Narcotics] (W. v. Hamm, T. Schwartze, H. Wagner, J. Zöllner, 'Die Chemie des täglichen Lebens', in Das neue Buch der Erfindungen, Gewerbe und Industrien, 6th edn., Vol. 5, O. Spamer, Leipzig, 1873)*

'De Quincey wrote his famous Confessions at a time when opium was as easily available as aspirin today, and almost as frequently used, and when its dangers were not understood. Though something of a fugitive from respectable society, he shared his addiction with some of the most distinguished men of his age. But the Confessions are not about drug addiction. 'They are a meditation on the mechanism of the imagination, an exploration of the interior life of an altogether exceptional being'. Brilliantly gifted and charming, de Quincey suffered from what he himself called a 'chronic passion of anxiety' which led him from the security and success he might have enjoyed into the direst poverty, and into the experiences which form the subjects of the terrible, drug-induced dreams he describes so superbly'.

Fig. 11.18. *Importing opium to China.* The balls are of crude opium, exported from Bengal (*The Illustrated London News*, December 8, 1883)

An extract from the chapter '*The Pleasures of Opium*':

'The next morning, as I need hardly say, I awoke with excruciating rheumatic pains of the head and face, from which I had hardly any respite for about twenty days. On the twenty-first day, I think it was, and on a Sunday, that I went out into the streets; rather to run away, if possible, from my torments, than with any distinct purpose. By accident I met a college acquaintance who recommended opium. Opium! Dread agent of unimaginable pleasure and pain! [...]. My road homewards led through Oxford

street; and near 'the stately Pantheon' I saw a druggist's shop. The druggist, unconscious minister of celestial pleasures! – as if in sympathy with the rainy Sunday, looked dull and stupid, just as any mortal druggist might be expected to look on a Sunday: and, when I asked for the tincture of opium[12], he gave it to me as any other man might do: and furthermore, out of my shilling, returned to me what seemed to real copper halfpence, taken out of a real wooden drawer. […].

Arrived at my lodgings, it may be supposed that I lost not a moment in taking the quantity prescribed. I was necessarily ignorant of the whole art and mystery of opium-taking: and, what I took, I took under every disadvantage. But I took it: – and in an hour, oh! heavens! what a revulsion! what an upheaving from its lowest depths, of the inner spirit! what an apocalypse of the world within me! That my pains had vanished, was now a trifle in my eyes: – this negative effect was swallowed up in the immensity of those positive effects which had opened before me – in the abyss of divine enjoyment thus suddenly revealed'.

In medicine, opium served as an antispasmodic agent for neuralgia or colic, as a sleeping drug, as a painkiller, and also for the easing of death throes. Use of opium as a euphoric (by means of eating, smoking, or taking tinctures of opium) leads to addiction and thus to dangerous dependency (*Fig. 11.19*)[13,15].

As a medical doctor, *Lewin* characterized the morphinists as follows[4]:

'The individual stages of continuous morphine use, which cannot be sharply delineated but are nevertheless present, have their particular characters. The beginning sees the morphinist in a euphoric state, founded on self-deception in the estimation of their skill, ability, and joie de vivre. The ego falsely overvalues itself, both in itself and in its relationship with the outside world – but, however this altered state may have come to be – the individual feels it, work appears to progress more easily, the small blows that crude reality deals out are not felt, or not as badly as before, and this elevated vitality, of six to eight hours' duration, is the result of a single dose of morphine.

This introductory, seductive stage, perhaps lasting for months, leads with increasing dosages into the still more positive second episode, which is filled with contentment with life, happiness without desires, absolute self-sufficiency, peace of mind, unshakable by anything. […].

The waves of life's calamities beat against the morphinized brain leaving no impression or consequences. No unpleasant physical state is found

Burmas Kampf gegen Drogen

ku. Die burmesischen Militärmachthaber der Slorc-Junta (State Law and Order Restoration Council) haben im Frühjahr dieses Jahres eine Kampagne gegen die Drogen lanciert. Zur Propagierung hat die Post der Union of Myanmar, wie das Land offiziell seit 1989 heisst, einen Einzelwert zu 2 Kyat herausgegeben. Eine mit roter Farbe durchkreuzte Kapselfrucht eines Schlafmohngewächses soll darauf

hinweisen, dass der Handel mit Drogen bei harten Strafen verboten ist. Interessanterweise hat die Militärjunta mit dem «Opium-König» Khun Sa, der im sogenannten Goldenen Dreieck zwischen Burma, Laos und Thailand seit Jahren als der mächtigste Drogenhändler galt, ein Arrangement getroffen. Nun wohnt Khun Sa, der in seinem Reich eine eigene «Armee des San-Staates» befehligte, in einem Gästehaus der Regierung am Inya-See in Rangun.

Fig. 11.19. *Extract from the Swiss Newspaper 'Neue Zürcher Zeitung'* (No. 279, November 29, 1996). The text reads: 'Burma's war against drugs. The Burmese military rulers of the Slorc junta (State Law and Order Restoration Council) have in the spring of this year launched a campaign against drugs. For propaganda purposes, the postal service of the Union of Myanmar, as the country has officially been known since 1989, has produced a special issue 2-kyat stamp. An opium poppy seedpod, crossed through in red, puts over the message that dealing in drugs is illegal, with severe penalties. Interestingly, the military junta has come to an arrangement with the 'Opium King' *Khun Sa*, for years regarded as the most powerful drug dealer in the so-called Golden Triangle between Burma, Laos, and Thailand. *Khun Sa*, who once commanded his own 'Army of the San State' in his kingdom, now resides in a government guest house by lake Inya in Rangoon'.

discomforting, sorrow and worries scarcely touch the soul, and lesser agitations like frustration and annoyance dissipate without impression. Freed from everything that ties people to the earth, free even of the feeling of possessing a body, the individual consciously lives a kind of wak-

ing dreamlife in the daily routine of social occupation. This life is pure-ly a me-life, though, merely a life of the moment. Thoughts are not direct-ed to the future, just to the day and its need for opium or morphine. In this way, the higher sensitivities soon become defective. Heart and soul suffer. The confinement of the world to the self produces moral dullness and indifference to partner and children. Concern for these comes either far behind that for morphine or does not figure at all. [...].

The effective duration of a dose, which has already increased to 0.2 to 0.5 g, becomes shorter. The drug must be injected more often and in larger quantities, if it is still to be as agreeable as in the beginning, the slave's chain becomes ever shorter and pulls at the morphinist. The cred-itors, the brain cells, knock, make demands, shout, and take revenge by causing pain, if they are not satisfied promptly. If money to secure the substance is lacking, then stealing and cheating are resorted to. Reputa-ble women are said to have turned to prostitution in order to be able to buy morphine. If, at the start of the passion for morphine, one pleasure was superceded by another, still more intense, a state now sets in, in which the craving brain, after a large enough dose of morphine, will still act as it used to, but in the period between two doses, it will start to act disagreeably as soon as the full effect begins to wear off.

The passing of time, amid severe suffering, leads to the last stage: the awak-ening of the recognition of the morphinist's abject surrender, body and soul, to the drug. Willpower is totally incapacitated. The resolve needed for the slightest effort is absent, and the continual battle between the need to take action and the inability to do so now results in the feeling of inner wretch-edness accompanied by terrible suffering. Even in dreams, the mental or-deal goes on, because the happy, carefree past stands in such cruel contrast to the hopeless present. Professional duties can only be fulfilled with the aid of huge doses. It is possible, though, for a 'morphinist surgeon' – I saw this once with one of the best in this field, as though divinely inspired to perform work of permanent value – to use these means to calm previously trembling hands, and restore clouded vision and unclear judgement, for a rider on the racetrack to achieve victory, the judge to make a correct ruling, but the will-power needed for this subsides quickly. [...].

In a gradual development, physical disorders now manifest as a conse-quence of the disturbed state of the brain. The brain, as the director of so many physical functions, flags in its regulatory tasks. Diet suffers, ap-pearance deteriorates, weight loss sets in, and capacity for work dimin-ishes conspicuously. Only in toxic doses does morphine now allow en-forcement of physical effort of whatever kind.

The morphinist is now usually just skin, bone, and quivering nerves. Cold sweats occur, especially at night, either over the whole body or confined to the head. Appearance and personal hygiene are neglected. From time to time, there are outbreaks of feverishness of several hours duration, with shivering, headache, feelings of oppression. Itching of the skin, also in combination with a rash, torments the sufferer. These are joined by: stomach pains, colic, diarrhea with rectal inflammation after evacuation – presumably due to some unknown morphine metabolite – sporadic urinary disorders, and also conjunctivitis, lachrymation, disorders of accommodation in the eye, and weak vision. [...].

Thus afflicted, the morphinist looks for help. [...]. The individual, harrowed by the long, terrifyingly overwhelming work of morphine, is now no more than a wreck, whose collapse into a heap of rubble is only rarely still preventable. If withdrawal is eventually to be successful, it makes no difference whether it is carried through in one go or in stages. In the first case, the suffering produced is severe, with arousal in the previously disregarded sexual sphere, agitation, morbid craving for morphine, explosions of rage, violence, and fear, all of which set the scene for delirium or suicide attempts. At the same time, there are sensations of pain in various nerve centers, vomiting, diarrhea, angina, succeeded by cardiac collapse etc., lasting for several days. Slow withdrawal from the drug, after each lessening of the dose, results in a renewed cry of the brain cells for the full dose to which they have become habituated. In both cases, the morphinist may break free of the immediate craving for the drug, but that is all. Some 80–90% of these unfortunates – possibly even more – suffer relapses. This figure also includes those who have temporarily been freed of their habit, not in a medical establishment, but by forced internment in prison. [...].

The authorities in all civilized countries are conscious of the danger implicit in the increasing use of morphine, cocaine, and other narcotics. A deluge of ordinances has been the result. These are almost all regulations formulated at the green tables of welfare officials with no experience of the matter. Many of these rules proved useless decades ago and have been discarded. Their goal is simply to restrict drug supply to purely medical purposes, to make access to supplies possible only through a central authority, and to subject pharmacists to strict controls and monitoring. The [German] 'Opiumgesetz' obliges pharmacists to keep records of all prescriptions of opium, morphine, cocaine, and heroin, even if the prescribed medicine is not to be repeated without renewed written instruction. [...].

All measures attempted to suppress the problem can be and are circumvented. Their imposition is necessary, but it is unrealistic to hope for total compliance. The needs of the morphinists and the profit motive of the dealers, even when the latter are states, overcome all defenses. This is the final conclusion arrived at by all those with an interest in the matter. […].

A blessing and a curse are united in morphine. The blessing comes from the godlike power of the substance, which only the hand of the physician should dispense. Those who spend sleepless nights in their beds in excruciating pain, those who, because of some incurable condition racking body and soul, see only blackness and despair in each next day and on into the future, those who curse life because death will not come, those who live lives unworthy of the name, because destructive powers, blind forces of nature, are mercilessly and unceasingly at work, with the prospect of death certain – the physician should come to all of these as a bringer of relief, as an easer of suffering, and as a facilitator of death. Not as a hastener of death, though, although life and death do merge into one another under morphine's shadow. The physician has no right to do that. Morphine should be used only in those circumstances in which the causing of morphinism may be viewed as inconsequential in comparison to the suffering it relieves. But it should not be dispensed indiscriminately as an analgesic. That would be to produce more morphinists, with the associated stigma, should they find the substance to give them pleasure beyond the temporarily necessary relief of pain and so continue its use purely for hedonistic purposes. They do not have any right to be judged leniently, but it must nonetheless be conceded that they are subject to the compulsion produced by the ever morphine-hungry cerebral cells, which are able to overcome less resolute wills. Only those whose lives have become a state of martyrdom, into which morphine enters as a miraculous relief, are due sympathy'.

Fig. 11.20. *Philippus Aureolus Theophrastus Paracelsus* (*Theophrastus Bombastus von Hohenheim*; 1493 – September 24, 1541). Ph. D. in 1515 from Ferrara; since 1527 town physician in Basel and lecturer at the University of Basel; founder of pharmaceutical chemistry (iatrochemistry) (Swiss postage stamp, 1993, from E. Heilbronner, F. A. Miller, *A Philatelic Ramble through Chemistry*, Verlag Helvetica Chimica Acta, Basel, 1998).

Of the opium-containing medicines, *laudanum* should be mentioned in particular. To physicians of the Middle Ages, the term stood for (especially opium-containing) sedatives in general and also for every preparation in which they believed the effect to be caused by a particular substance. Laudanum is frequently mentioned in the works of *Theophrastus von Hohenheim*, more usually known as *Paracelsus* (*Fig. 11.20*), but not always in the same sense or as the same preparation or composition. Until the beginning of the 20th century, *Laudanum Sydenhams* (*Laudanum liquidum Sydenhami*) was used. This was a wine extract of opium and saffron, to which cloves and cinnamon were added, termed *Tinctura opii crocata officinell*[16].

As a drug, opium today no longer plays any significant role in the industrialized world, having been usurped by its active principle, morphine, and its acetylated derivative, heroin. In Asia (Iran, Iraq, India, China), however, opium is even now a drug to be taken very seriously.

Informations relating to police seizures of heroin in Germany[17] are presented in *Fig. 11.21*. Similar figures probably apply for other countries. It is gen-

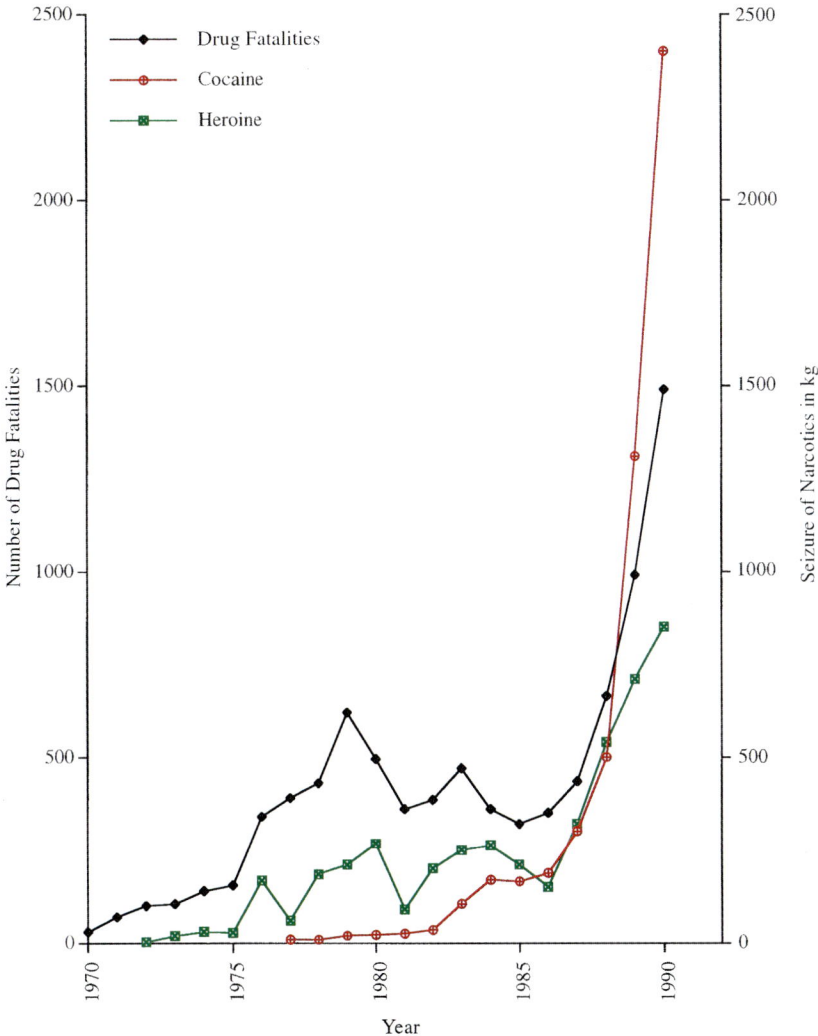

Fig. 11.21. *Comparison of quantities of cocaine and heroin* (in kg) *seized in Germany from 1970–1990.* The drug fatality figures run almost in parallel[17].

Fig. 11.22. *The Coca plant (Erythroxylum coca) (Meyers Konversations-Lexikon, 1898)*

COOMe

Me–N

H

H

O

(–)-Cocaine (abs. config.)

erally assumed that the quantities actually imported are at least four times as high.

11.6. The Coca Plant (*Erythroxylum coca*)

According to a saying of the Andean peoples, the coca leaf (*Erythroxylum coca* (Erythroxylaceae), a 0.5–1.5 m high shrub; *Fig. 11.22*), which '*nourished the hungry, gave new strength to the tired and exhausted, and made the unfortunate forget their sorrows*', was a gift of the sun god after the founding of *Tawantinsuyu*, the 'Inca Empire'. The plant was used as an emblem of kingship and employed by priests at a variety of different ceremonies. One of their queens was even called *Mama Cuca*. With time, use also spread to the populace. When *Francisco Pizarro* marched his troops into the Peruvian highlands in 1533 and overthrew the empire, the plant was already cultivated everywhere and used as a euphoric. Conquistadors as well as plantation and mine owners forced the indigenous peoples into servitude, paying them in coca leaves. The forced labor and abuse was outlawed by the government of the Spanish viceroyalty of Peru in 1560. In 1569, coca was declared a state monopoly, and after it had failed, it was lifted again. Finally, the trade with coca leaves was privatized towards the end of the 18th century.

The dried coca leaves are chewed in a wad, mixed with chalk and plant ash to liberate the alkaloid cocaine. The drug is still taken in this form by indigenous Andean peoples today, apparently without causing addiction. The requirement per day is *ca*. 50 g of leaves; thereby, the cocaine content varies between 1% (Peru) and 1.6% (Java) of dry weight.

Since it blocks peripheral pain sensation, cocaine, to some extent, is still used today as a local anesthetic in surgery.

About 100 years ago, cocaine became a fashionable euphoric[18], especially in England, France, and Germany, resulting in increasing exports from the producing countries (*ca*. 8 tons in 1877; 3,000 tons in 1906; 2,200 tons in 1920). The daily requirement of a habituated cocaine user typically amounts to 1–4 g (in extreme cases as much as 30 g of 'snow')[19]. Current global use of the drug certainly vastly exceeds that in 1920 (*cf. Fig. 11.21*; police seizures of pure cocaine powder in 1990, Germany: 2.4 tons).

Persistent intake of cocaine (cocainism), after a brief euphoric beginning, turns a healthy person into a mental and physical wreck in the course of a few months. The physician and toxicologist *Lewin* described this as follows[4]:

'As far as the craving for cocaine is concerned, I have seen appalling things in recent years. They all believe that it is the ascent through the portal of pleasure into the temple of happiness, but they pay for their momentary delight with their bodies and souls, soon passing on through the portal of unhappiness into the night of nothingness'.

Interestingly, there appeared a report in 1992 that nicotine, cocaine, and tetrahydrocannabinol (one of the active principles of hashish) had been found in the hair, soft parts, and bones of nine Egyptian mummies, *ca.* 3,000 years old[20]. A reasonable explanation for this phenomenon is still awaited. Today, we know of only one cocaine-containing plant (*Erythroxylum coca*) found exclusively in South America. Hashish, from *Cannabis sativa*, indigenous to India, might well have been known to the ancient Egyptians, while nicotine is found in many plants of a variety of families. In other words, the only puzzling thing is the detection of 'mumified' cocaine. Possible explanations include some sort of microbial transformation of atropine into cocaine, transatlantic contact between Egypt and South America, alternative cocaine-producing plants that have since died out, and others. Also unanswered is the question of whether the substances found had been used as drugs, painkillers, sleeping preparations, or ingredients in the mummification process.

11.7. Deadly Nightshade (*Atropa belladonna*)[21]

Belonging to the Solanaceae, *Atropa belladonna* L. (*Fig. 11.23*) takes its name from *Atropos* (ατροπος), one of the three (female) Fates of Greek mythology, also called *Moerae*.

In Renaissance Italy, deadly nightshade, thanks to its pupil-dilating properties, was used as a cosmetic (giving rise to 'doe-eyed' beauties), hence *bella donna* (Italian for 'beautiful woman'). There is certain regularity of instances of poisoning occurring in Summer and early Fall, both in children and in adults who have been unable to resist the lure of the lustrous, black berries of the deadly nightshade[22]. Depending on the quantity of fruit consumed (the roots and leaves possess similarly properties, though), the symptoms consist of a dry mouth, dilated pupils, facial flushing, arousal, confusion, hallucinations, unconsciousness, and convulsions. The poisoning is occasionally fatal.

The Deadly Nightshade is a perennial, herbaceous plant with a robust, branching rootstock. The stalks, up to 1.5 m in height and branching, bear oval leaves. The bell-like, purplish violet flowers produce dark, cherry-like berries. The poisonous fruits are palatable and not bitter, which assists the con-

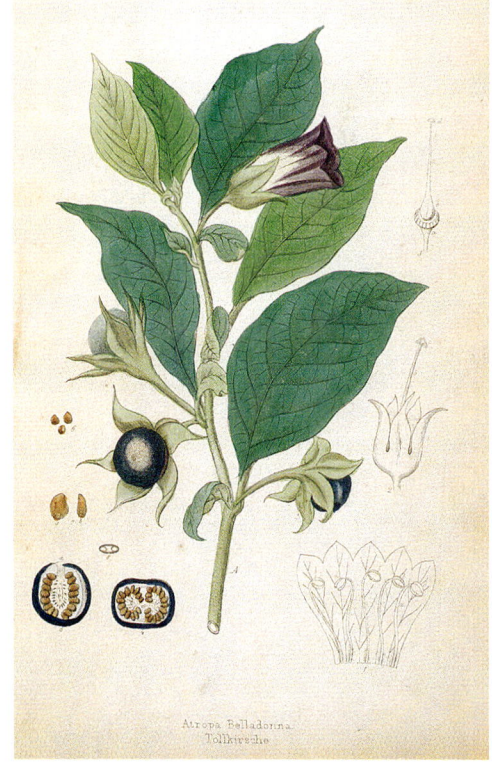

Fig. 11.23. *Atropa belladonna* L. (Deadly Nightshade). The lustrous, black fruits are occasionally confused with cherries and eaten, which may have severe consequences, as they contain atropine and hyoscyamine (J. Pecirka, *Giftgewächse*, Verlag K. André, Prague, 1859).

(–)-Hyoscyamine

Atropine
(racemic)

fusion. The Deadly Nightshade is indigenous to southern and central Europe and is cultivated for pharmaceutical purposes in the Balkan, Russia, and Iran. The roots contain up to 0.3% of atropine and hyoscyamine, the two main alkaloids of *A. belladonna*. Many stories have grown up about the deadly nightshade, one of the most important poisonous plants north of the Alps.

Thus, at the end of the 18th century, an old woman in an inn in the Rheinland Palatinate village of Kirkel was said to boil soft *Atropa* roots in water to make a poison that she put in wine, which she gave to her guests. This would befuddle the guests for a time. When they woke up, their property would be missing.

The anesthetizing effect of the deadly nightshade is also said to have been exploited in war, as attested in the following 11th century account. A successful *mass poisoning* with the aid of *Atropa belladonna* is attributed to King *Duncan I* of Scotland (stroke dead in 1040), a grandson of *Malcolm II* and cousin of *Macbeth*[23].

'Duncan, a mild and good-natured man, perhaps precisely because of these aspects of his character, was to experience an outbreak of rebellion in his lands, instigated by Macdugald. Malcolm, sent out against the rebels, was defeated, taken prisoner, and beheaded. To avenge this disgrace, Duncan sent Banquo, Thane of Loch Abyr, and Macbeth against Macdugald. Their victory was complete. Scarcely was Duncan free of this worry, though, there came a new one. Sweyn, King of Denmark, who also owned England, landed with his Northmen in Fifeshire to conquer Scotland as well. Macbeth was tasked with collecting an army and awaiting further orders. King Duncan himself moved against Sweyn, but he was defeated at Culross and forced to retreat to Perth. The Danes immediately laid siege. Duncan, feeling completely secure, sent an order to Macbeth to remain where he was. He himself trusted in an advised stratagem. He sent ambassadors to Sweyn, under the guise of negotiating surrender. The Scots demanded freedom to withdraw. Sweyn took this offer as an act of utter weakness and so demanded unconditional surrender. At the same time, the negotiators had let it be known that their king, during the course of further negotiations, wished to relieve the supply problems of the victors. That was gratefully accepted, but it also appeared to be a concession smacking of desperation. The Scots now brought large amounts of bread and wine, and a strong, spirituous drink made from barley. To this, they had added the juice of a poisonous plant, abundant in certain areas in Scotland, named the sleeping nightshade, 'Solanum somniferum', the blacks fruits, roots and seeds of which produces sleep and, taken in larger quantities, madness. The unsuspecting Danes de-*

voured this unexpected gift, emptying great tankards of it. Sweyn also did so, as a sign of his goodwill. Duncan, who knew very well when the effects of this brew would set in, had meanwhile secretly arranged for Macbeth to come with his troops and to enter the town by a particular route. As soon as the scouts reported that the Danes were in deep sleep in their camp, the king sent Banquo, who was informed of the plan, into the enemy camp with troops, keeping his own forces and those of Macbeth in reserve. The attacking Scots raised a warcry. Only a few Danes ran around, as though mentally confused, and were killed. Most passed straight from their poisoned sleep into death. Sweyn himself staggered as though half dead, but it was his good fortune to be taken to a ship by a few of his men who were not so badly afflicted, and made his escape'.

Deadly Nightshade has a special significance as a component of witch's salve (*Fig. 11.24*). Recipes, called *Receptarium diabolicum*, dating from as early as the 15th century, are known. Two examples are given here:

In 1591, *Gödelmann* (1559–1611) wrote[16]:

'*Paracelsus [...] tells of such a witch's salve made from the flesh of a young infant, cooked to a broth, and sleep-bringing herbs, such as poppy or oil [Papaver], nightshade or Jew's cherry [Solanum], hemlock or pig hemlock, and Cicuta virosa etc*'.

Fig. 11.24. *Hans Baldung* (known as *Grien*), *'Die Hexen beim Einsalben'* [the witches at their anointing] (1514, woodcut; Städelsches Kunstinstitut, Frankfurt a. M.)

Gianbattista della Porta (1535–1615)[16]:

Sium	*Greater water parsnip (Sium latifolium L.)*
Acorum vulgare	*Common myrtle*
Pentaphyllon	*Devil's bit scabious*
Vespertilionis sanguinem	*Blood of bat*
Solanum somniferum et oleum...	*Nightshade and oil...*

Many other recipes also recommend other nightshades (*e.g.*, henbane (*Hyoscyamus niger*), thorn apple (*Datura stramonium, Fig. 11.25*), Angel's Trumpet (*Datura suaveolens, Fig. 11.26*), or mandrake (*Mandragora officinarum*)) for the preparation of the salves. These were rubbed into the mucous membranes in the genital region and under the armpits. This gave rise to hallucinations involving erotic excesses and the '*congress with the Devil*', and was later to form the basis of the Inquisition's witch trials. In interrogation of suspected witches, these salves were sometimes used to confound the accused.

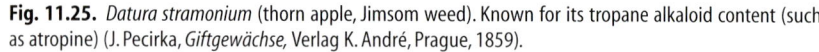

Fig. 11.26. *Datura suaveolens* (Angel's Trumpet). Modern ornamental plant popular in European gardens. All plant parts contain atropine, hyoscyamine, and scopolamine (photo *M. Hesse*).

Datura Stramonium L
Stechapfel

Fig. 11.25. *Datura stramonium* (thorn apple, Jimsom weed). Known for its tropane alkaloid content (such as atropine) (J. Pecirka, *Giftgewächse*, Verlag K. André, Prague, 1859).

During festive gatherings or in medieval bathhouses (*Fig. 11.27*), powders (made from *Mandragora*, *Datura*, *Hyocyamus*, and *Atropa*) would be sprinkled onto ovens, much like incense burning, whereupon the volatile ingredients (alkaloids and other substances) would evaporate and produce euphoric effects, with delight being taken in people's uncontrolled behavior.

The Spanish writer *Miguel de Cervantes Saavedra* (1547–1616), in his *Novelas ejemplares*, has an old witch telling[16]:

'I was the best in preparing the salves that we witches would rub ourselves with, nor can anyone among the younger ones, who practice our art today, surpass me in this skill. […]. Some people believe we were just imagining it, that we would visit these banquets, and that the Devil made us see pictures of all the things that we would later say happened to us. Others, though, maintain that this is not so, but that we were really there, in body and soul. For my part, I think that there is something true in both views, since we ourselves don't know how many of these experiences are real. Everything that our imagination shows us is so clear and vivid that it is not possible to tell it from a real experience. The inquisitors, if they caught one of us, used to collect some testimonies about it, and I believe they would endorse my words. This salve that we witches would rub ourselves with, is made from ice-cold juice of different herbs, and not, as people say, from the blood of children that we have strangled. […]. [The devil] does, no doubt, let us kill small children as well, to cause pain to the parents, the most awful pain that can be imagined. Probably, though, it is still more important to him [the devil] that we commit the cruelest and most abominable sins wherever we go. But God allows all this for the sake of our sins; as without his leave, as I know from personal experience, the devil cannot harm an ant. […]. The misfortune and ill that is called 'sin', however, that comes from ourselves'.

[This is all relayed through a conversation between two dogs at the Hospital of the Resurrection at Valladolid. One of the dogs now thinks:]

'Why then, if she knows so much, does she not stop her witchery and return to God? She must know that He would rather forgive sins than allow them to happen'!

Fig. 11.27. *Albrecht Dürer, 'Die vier Hexen (Discordia)'* [the four witches] (1497, copperplate engraving; Staatliche Graphische Sammlung, Munich)

[Whereupon the old woman, guessing his thoughts, continues:] '*The habit of the vice becomes second nature and the witch's calling passes into a person's flesh and blood. In the burning heat that it produces, it shifts the soul into such an icy cold that it cools and freezes even in belief. […]. I see and understand everything, but because pleasure has my will in fetters, I am wicked and will also always remain wicked. I assure you, they [the salves] are so cold that our senses depart from us when we rub them in, and we stay lying naked on the ground. Then, even though we may live out all that we are thinking of in our imagination, as people say, it really happens to us. Sometimes it also happens that we change our form afterwards; we then transform into cockerels, owls, or ravens, and fly to the place where our master awaits us. […]. My salve has given me the most wonderful hours. Seventy-five years old am I now. […]. Now, even if the joy that the devil has to give us is just appearance and illusion, it still is joy, and sensuality is always much greater in the imagination than in reality, although, for true joy, the exact opposite should be the case*'.

[Then, while softly muttering to herself and applying the salve from head to toe, she sinks to the ground and lies here stretched out and as if dead. The dog of the tale tries in vain to wake her, by biting and dragging her back and forth. He finally resumes:] '*Who made this wicked old woman so cunning and evil? […]. How does she recognize the difference between a misfortune that comes through harm and one that arises through our fault? How is it possible that she knows so much about God and can talk about Him when she carries out the orders of the devil? How can she sin so spitefully and knowingly, although not being excused by ignorance?*'

The pharmacological application of *Atropa* has been known since the 16th century[24]. Thus, in the above extract, we probably have one of the first reported instances of drug dependency.

Narcotic agents from *Atropa belladonna* were very easy to prepare and the source of a lot of harm. In Montpellier, France (*ca.* in 1400), *e.g.*, delinquents condemned to death were not executed as such. They would much more commonly be handed over to the university's physicians, who 'pumped them full' with narcotic plant juices and then freely performed operations on them (vivisection), without the subjects sensing anything.

As well as this, four Dominican friars are said to have performed a particularly 'real' passion play in Berne in 1509, giving a lay-brother a narcotic drink and then nailing him to a cross. They were condemned '*on the last day of May 1509 to be burned to powder and ashes, as they well deserved*'.

11.8. Betel

Betel (or betel nut, *Fig. 11.28*) is a mixture made from the leaves of the betel pepper vine *Piper betle* (Piperaceae), parts of the *Areca* nut (seeds of the betel palm *Areca catechu* (Palmae)), and burnt lime (CaO). Sometimes, spices such as cardamom, tamarind, and clove are added, with small, bite-size pieces (betel nuts) being made out of the mixture and chewed. Contact with saliva, mainly by the action of basic $Ca(OH)_2$, frees the alkaloid arecoline, which is swallowed together with the bright red stained saliva. Betel possesses mild stimulatory properties and is supposed to produce a feeling of euphoria, but this is not solely attributable to arecoline.

The craving for chewing betel is similar strong as any other psychoactive agent. The betel chewer, however, is probably more persistent in this regard than any other users of psychopharmacologically active agents. The nuts are chewed by people of both sexes, all ages and standings, in free time and at

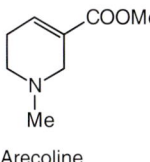

Arecoline

Fig. 11.28. *Ingredients for the preparation of betel* (betel nuts). Leaves of the betel pepper vine *Piper betle* with (from left to right) quicklime, portions of Areca nut, fruit of the Betel palm *Areca catechu* plus spices such as cardamom, tamarind, clove *etc.* (photo *B. Häsler*).

work. Its consumption does not appear to have any evident adverse health consequences, not even when taken excessively, *i.e.,* more than the usual 24 leaves per day.

With habitual use, a tartar of chalk ($CaCO_3$) forms, said to be evident as a shining black 'tooth growth'. Betel chewing is widely distributed: through East Africa, Madagascar, India, Indonesia, Polynesia, and South-east Asia. The betel chewing habit is very ancient, appearing to date back to prehistoric times, as attested by the existence of the corresponding words in ancient languages, such as the Sanskrit *guváka*, or the Chinese *pinlang*. The actual word 'betel', however, comes from the Malay. Descriptions of the use of betel by *Theophrastus* (340 B.C.) and *Marco Polo* (13th century) are also known.

11.9. Tobacco (*Nicotiana tabacum*)

Smoking of tobacco in *'cylindrical rolls'* has been customary among indigenous Americans since ancient times. The Spanish obtained knowledge of this after the discovery of the New World in 1492, and the first tobacco plant arrived in Spain in 1511. In 1560, the French ambassador in Portugal, *Jean Nicot de Villemain* (1530–1600, who gave his name to the *Linné*an generic descriptor of the plant), brought the seeds of some tobacco plants (*Nicotiana tabacum*)[25], he had grown himself, to Paris. The rest of Europe and the whole world were to 'profit' from this. In subsequent years, a full-blown craze for tobacco consumption took place (*Fig. 11.29*). The way tobacco was taken followed either the American style, through the lungs (pipes, cigars), or, more European-like, through mucous membranes (snuff, or chewing, *Figs. 11.30–11.32*). State and church sought to combat this new luxury product. Some particular events from this time are notable:

During the 16th century, among Spanish priests, snuff-taking and smoking of tobacco had become so habitual that its use was not interrupted even for mass or communion. Pope *Urban VIII* (1568–1644) ended this excess by decree of excommunication for any priest who, in future, took tobacco in Spain's churches.

In Russia, on the other hand, Tsar *Mikhail Feodorovich* (1613–1645) ordered in 1634 the death penalty for those of his serfs who were found in possession of, smoked, or dealt in tobacco. Their goods were sold and the proceeds delivered to *'Your Majesty's Treasury'*. The condemned could be chastened further by corporal punishment and mutilation (cutting off the nose).

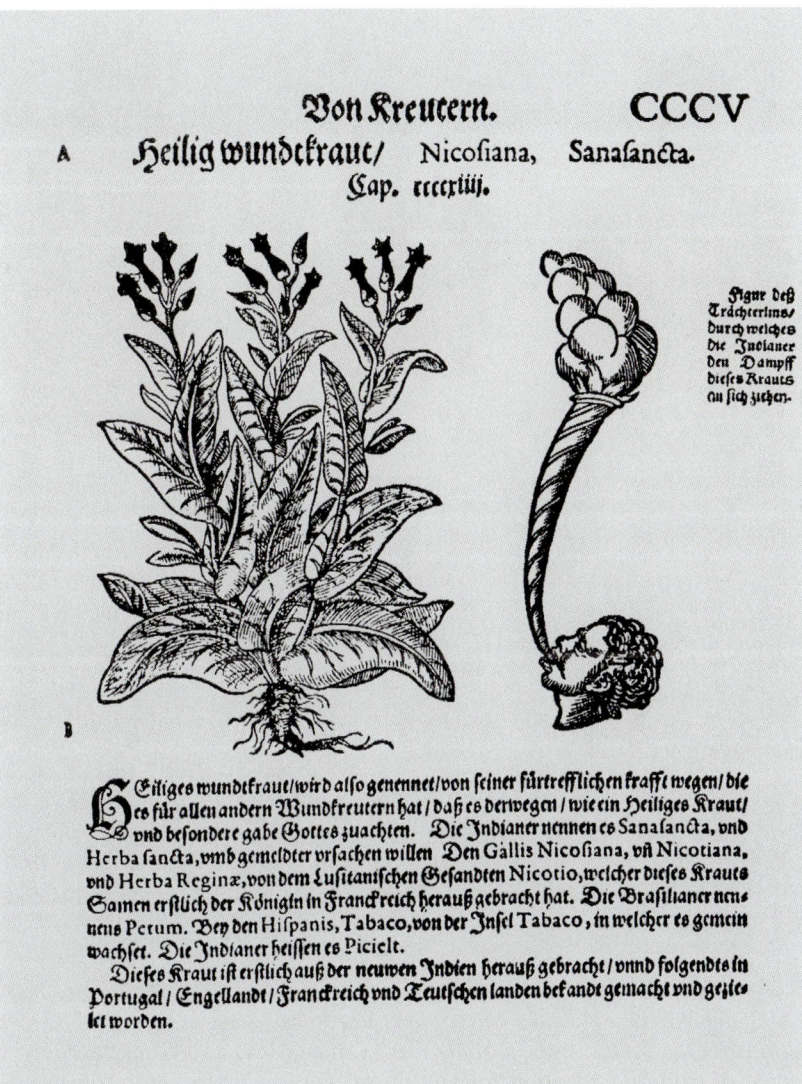

Fig. 11.29. *Nicotiana tabacum and a very early depiction of smoking* (from Adam Lonicerus, *Kreuterbuch*, 6th German edn., 1587)

Fig. 11.30. *Grater for snuff tobacco* (Dieppe, France, *ca.* 1695; length 20 cm). Front: portrait in ivory; back: steel grater (photo R. J. Foster)

Fig. 11.31. *German porcelain tabatière* (1897, Berlin; porcelain with brass settings, *ca*. 9 cm long) (private collection)

Fig. 11.32. *Snuff flasks from China* (from left to right: coral (*ca*. 1810, height *ca*. 8 cm); brown and yellow amber (*ca*. 1800, vase on recumbent deer); enamel flask (18th century) with a European scene (Victoria & Albert Museum, London; photo *A. Hornak* and *K. Hoddle*)

Shah *Safi I* of Persia, in 1634, would punish smokers by having molten lead poured down their throats.

In Switzerland, in the canton of Berne, tobacco smoking was listed as a serious crime in 1660.

In 1590, the Medical Faculty in Holland warned against smoking because *'the brain becomes black from it'*.

These interdictions were lifted when tobacco tax was discovered to be a valuable supplementary source of state revenue, introduced in England in 1652 (*Figs. 11.33 – 11.36*).

Nowadays, prohibition and restrictions on tobacco smoking have less to do with irrational reaction and more with medical opinion. Seen globally, unfortunately, these restrictions have had no evident success, since worldwide tobacco production[26] continues to increase more or less in line with world population (*Fig. 11.37*).

It has to be mentioned at this point that nicotine is not used as a medicine but has acquired a certain degree of importance as a pest control agent.

Fig. 11.33. *'Tobacco harvest on Cuba'* (contemporary depiction: *W. v. Hamm, T. Schwartze, H. Wagner, J. Zöllner, 'Die Chemie des täglichen Lebens',* in *Das neue Buch der Erfindungen, Gewerbe und Industrien,* 6th edn., Vol. 5, O. Spamer, Leipzig, 1873)

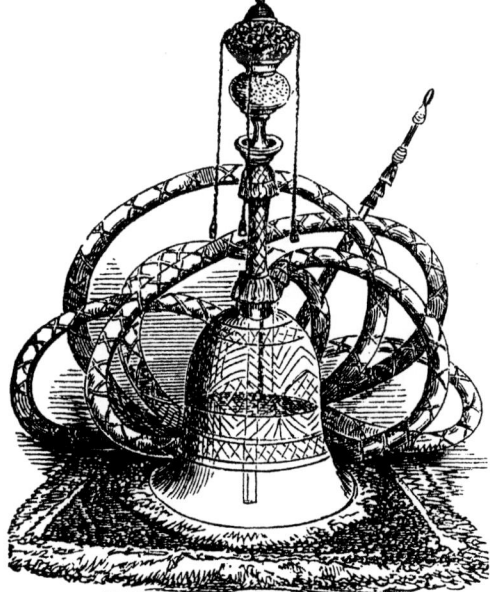

Fig. 11.34. *'The Persian Huka'* (waterpipe; *W. v. Hamm, T. Schwartze, H. Wagner, J. Zöllner, 'Die Chemie des täglichen Lebens',* in *Das neue Buch der Erfindungen, Gewerbe und Industrien,* 6th edn., Vol. 5, O. Spamer, Leipzig, 1873)

Fig. 11.35. *Georg Lisiewski, 'Das Tabakskollegium König Friedrich Wilhelm I. von Preussen'* (1737/38; oil on canvas (130 × 175 cm)). The 'Tabakskollegium' ['tobacco club'] was a more or less daily evening gathering of a selected circle. It was convened by Prussian king *Friedrich Wilhelm I.* in Berlin, Potsdam, or Königswusterhausen for the purpose of enjoying tobacco and informal political discussions (Potsdam, Sanssouci, Neues Palais).

Fig. 11.36. *Carl Spitzweg, 'Oriental, Smoking on a Divan'* (section) (1856; oil on board (30.6 × 24.8 cm), from S. Wichmann, *Spitzweg*, Schuler Verlagsgesellschaft, Herrsching, 1985)

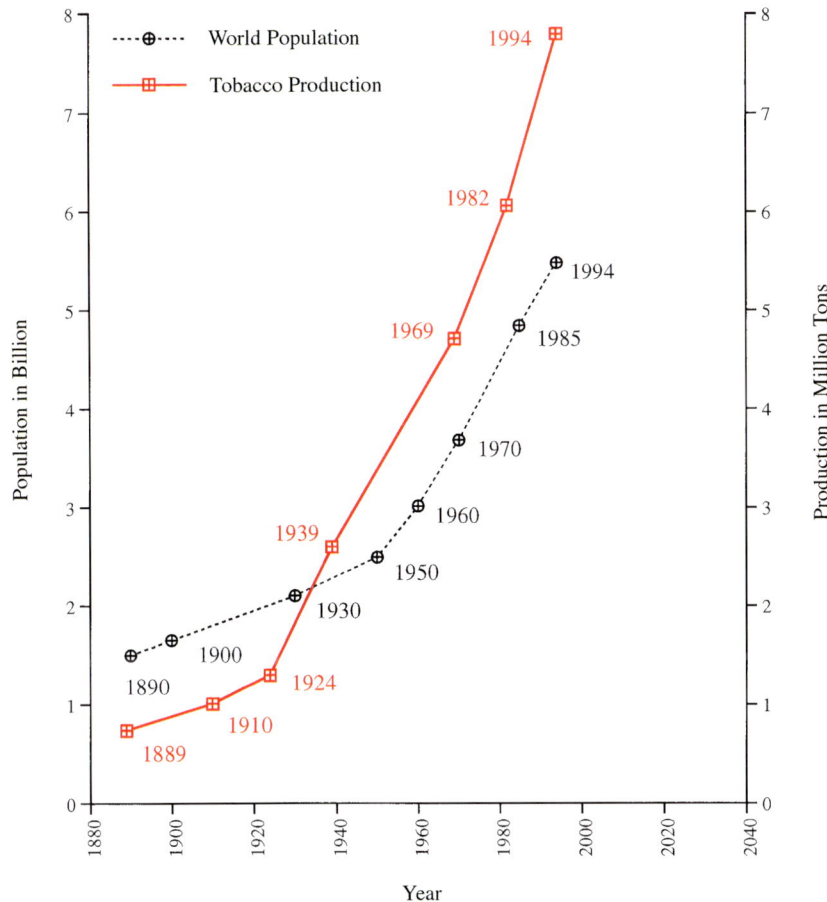

Fig. 11.37. *Tobacco production vs. world population.* Lung cancer rates would probably parallel both curves (data from various sources).

Nicotine is a powerful poison. Depending on the dose, after initial stimulation of the central nervous system and the peripheral autonomous ganglia, vital centers in the medulla and the mid-brain are put out of action by rapid onset of paralysis. After very high doses, death by respiratory paralysis occurs quickly. In humans, administration of 50–100 mg of nicotine (about the content of half a cigar) can be lethal. Interestingly, naturally occurring (–)-nicotine has the same effect as synthetic (+)-nicotine, a rather rare case of enantiomer behavior (*cf. Chapt. 5.6*).

(–)-(*S*)-Nicotine

(+)-(*R*)-Nicotine

Fig. 11.38. *Emancipation in tobacco advertising.* The oriental cigarette as a masculine status symbol (*H. Leupin*, 1943). Modern, self-confident woman with cigar (photo *Annabelle*, 1997). The premium cigar enjoys increased popularity (sales increases in the USA: 15% in 1994, 30% in 1995, and 60% in 1996 corresponding to a total of 3 billion cigars of all types).

Tobacco smoke contains over 1,300 different chemical substances. This complex mixture acts in a variety of ways – independently of the effects of nicotine as the main alkaloid – especially on the mouth, pharynx, larynx, and lungs. It is a source of heart and circulatory diseases, causing peripheral vascular disease and lung and bronchial cancers, even when consuming 'light' tobacco products (*Fig. 11.38*).

Nicotine does not lead people into actual physical addiction. Tobacco withdrawal thus is possible without serious symptoms, except for psychological ones.

The historian *H. Spode* wrote[27]:

> '*The history of tobacco in Europe shows a different characteristic in each century. The 17th century is the epoch of the hotly contested introduction of modern drug culture, its emblem is the pipe; the 18th century is the age of the feudal snuffbox; the 19th that of the bourgeois cigar; the*

20th that of the quick, democratic cigarette – and the 21st? Looking at America, it is imaginable that the epoch of nicotine chewing-gum is now dawning'.

11.10. Mandrake (*Mandragora officinarum*)

The main alkaloid in mandrake[28] (*Mandragora officinarum* L. (Solanaceae), *Figs. 11.39* and *11.40*), which grows especially widely in the Mediterranean region, is scopolamine. The roots of this plant frequently grow in a form resembling a human figure, thanks to which the plant has had magical or healing properties attributed to it from the earliest times. In Antiquity, patients would be administered with the narcotic and anesthetic mandrake root before difficult operations. The fruits (many-seeded berries) are supposed to stimulate sensuality (and so are also known as 'Love Apples') and ensure fertility, and so have commonly been used since Antiquity in the preparation of love potions. The roots are said to bless women with fertility and easy

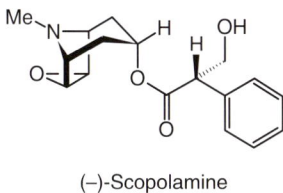

(–)-Scopolamine

Fig. 11.39. *Mandrake (Mandragora officinarum), an illustration from the work of Johannes de Cuba: 'Hortis Sanitatis' (Johann von Kube, Gart der Gesundheit,* Mainz, Germany, 1485)

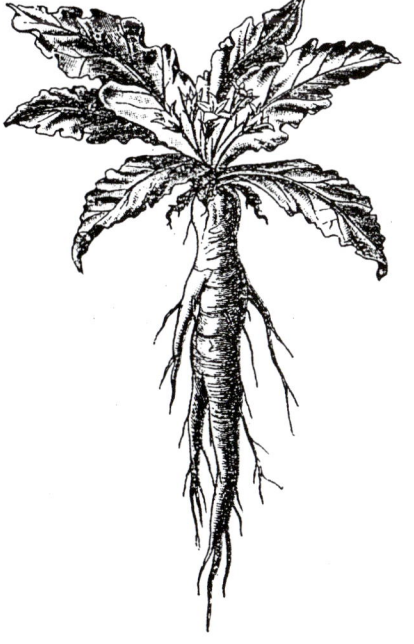

Fig. 11.40. *Mandragora officinarum (Brockhaus-*Lexikon, 1893)

births. They have been used as talismans, as good luck charms, as amulets, and the plant is mentioned in the Old Testament (Genesis **30**, 14, 15):

> '*And Reuben went in the days of wheat harvest, and found mandrakes in the field, and brought them unto his mother Leah. Then Rachel said to Leah, Give me, I pray thee, of thy son's mandrakes. And she said unto her, Is it a small matter that thou hast taken my husband? and wouldest thou take away my son's mandrakes also? And Rachel said, Therefore he shall lie with thee to night for thy son's mandrakes*'.

11.11. *Cinchona* Bark (Peruvian, Fever, or Jesuits' Bark)

This term covers the stem, twig, and root barks of numerous species of the *Cinchona* genus. All of them contain the alkaloid quinine, with *C. officinalis*, *C. pubescens* (Fig. 11.41), *C. condaminea* (Fig. 11.42), *C. ledgeriana*, and *C. calisaya* being particularly noteworthy. The last of these can have a total alkaloid content of up to 13%, or a quinine content of up to 11.6%.

(–)-Quinine (abs. config.)

In view of these figures, these barks may be counted among those plant parts with the highest of all known alkaloid contents. The word *Quina* is of Quechua origin (*kina*) and means *bark*. It appears that earliest knowledge of fever bark was confined to the region of Loja, the southernmost province of Ecuador. There, in 1630, the Spanish corregidor of Loja, *Don Juan Lopez de Cañizares*, is supposed to have been cured of malaria by the bark. When, in 1638, the Countess of Chinchón, consort of the viceroy of Peru, took sick with a fever, the corregidor sent fever bark to the personal physician of the vice-king, *Juan de Vega*, who successfully cured the Countess, from which the 'bark remedy' acquired the name 'polvo de la condesa' (Countess's Powder) and its *Linné*an generic descriptor *Cinchona*. Thanks to *Vega*'s work, the bark came to Europe in 1639, where it found wide distribution. As early as 1890, though, the primary producers of fever bark had become Ceylon (7,600 t) and Java (4,300 t), followed by South America (1,100 t). Demand for fever bark had increased enormously in the 19th century, generating

Fig. 11.41. *Branches and fruits of Cinchona pubescens* VAHL. (photo *J. Valverde*)

large profits. In the South American rainforests, collectors carried out mass bark-stripping projects, which resulted in the deaths of the trees: the beginning of the rape of the tropical rainforests.

Quinine used to provide great service in the treatment of malaria as an antipyretic, in the induction of uterine contraction in labor, and also as a treatment for infectious diseases. It was later largely superceded by synthetic substances (*Fig. 11.43*). Since malarial agents are ever more commonly developing resistance, recourse to quinine, which remains effective, is increasing again. This alkaloid, by the way, is also being used as an additive in soft drinks, *e.g.*, in *Quinine Water* or in *Indian Tonic Water* (*Fig. 11.44*), which contain *ca.* 0.007% of quinine hydrosulfate.

11.12. Coffee (*Coffea arabica*)

In commercial terms, coffee is probably the most important semi-luxury item in the world; and even when decaffeinated – this too is exceptional – it still remains a prized product.

The *Coffea* genus (Rubiaceae) occurs in many varieties (such as *C. arabica*, *C. liberica*, *C. canephora*). These are shrubs and trees with white clusters of flowers and cherry-like fruits, mostly containing two kernels in which the actual coffee beans are found (*Fig. 11.45*). Roasting the beans produces coffee containing 1–1.5% of caffeine that can be efficiently removed with the aid of pressurized CO_2 (destraction).

Probably originating in the southwestern Ethiopian highland province of *Kaffa*, coffee was known by the 13th century in Aden, Yemen, and further afield in Arabia, both as a medical agent and a luxury item. News of coffee came to Europe in 1582, in a literary description by the Augsburg physician *L. Rauwolfen* of his journey to the Orient[4]:

'*Among other things, they have a pleasant beverage that they value highly, called chaube, almost as black as ink and highly efficacious for many ailments, particularly of the stomach. They drink in the early morning, in public in front of everyone, without any stigma attached, from small, deep earthenware or porcelain cups, as hot as they can bear it. They take frequent small sips, passing the drink on to their neighbor as they sit in a circle. They add a fruit that the natives call bunnu to water; this fruit superficially resembles a laurel berry both in size and color. The drink is very popular with everyone, and so it is not uncommon around the bazars to find either sellers of the drink itself or shopkeepers selling the fruits*'.

Fig. 11.42. *'Sprig of Cinchona condaminea'* (from *W. v. Hamm, T. Schwartze, H. Wagner, J. Zöllner, 'Die Chemie des täglichen Lebens'*, in *Das neue Buch der Erfindungen, Gewerbe und Industrien*, 6th edn., Vol. 5, O. Spamer, Leipzig, 1873)

Caffeine

Fig. 11.43. *Synthetic antimalarial agents 'chloroquinine' and 'primaquine'.* The *Cinchona* plant (Cuban postage stamp, issued December 14, 1962; from E. Heilbronner, F. A. Miller, *A Philatelic Ramble through Chemistry*, Verlag Helvetica Chimica Acta, Basel, 1998).

Fig. 11.44. *Quinine-containing soft drinks* (1996). The British colonists in India introduced the drinking at sundown of gin and tonic, which contains quinine. It was effective as a preventive against malaria. Quinine water remains a popular drink, or mixer, today. 'Quinine Water' or 'Indian Tonic Water' has a quinine content of 70 mg/l.

Fig. 11.45. *'Sprig of the coffee plant'* (*Coffea arabica* L.). (from *W. v. Hamm, T. Schwartze, H. Wagner, J. Zöllner, 'Die Chemie des täglichen Lebens',* in *Das neue Buch der Erfindungen, Gewerbe und Industrien,* 6th edn., Vol. 5, O. Spamer, Leipzig, 1873)

The Venetians would bring coffee to Europe in significant quantities from 1642 onwards. A coffee house was opened in Venice in 1647, after the first one in Europe had been opened in Paris in 1643 (*Fig. 11.46*). Coffee drinking became popular in Europe very rapidly, although the practice was not tolerated everywhere. In 1777, for example, the Prince-Bishop of Paderborn declared coffee drinking a prerogative of nobles, the priesthood, and senior civil service officials: it was strictly forbidden to anyone else. But, as with tobacco, it was to turn out that a high rate of taxation and the consequent enhancement of rulers' revenues made prohibition untenable, and since then, these revenues have constituted an important part of the budget in many countries (*e.g.*, coffee-tax revenues in Germany in 1990: 1.93 billion DM) (*Figs. 11.47 – 11.50*).

The main coffee production zones today are in Central and South America, Central Africa, and Indonesia (*cf. Table 11.2*). World production has increased almost tenfold in the last 100 years, during which time some of the main production centers have also shifted. The USA leads the world in coffee consumption (*Table 11.3*), although in terms of 'consumption per head', Switzerland enjoys a significant lead. The rather low levels of coffee consumption in Britain are compensated for by tea-drinking.

Table 11.2. *Coffee Production* (green beans, in 1000 t)

Country	1989[a,b]	1889[c]	(Ranking)
Brazil	1556	384	(1)[d]
Columbia	720	ca. 100	(9)[d]
Indonesia (Java)	455	71	(2)[a]
Mexico	382	9	(10)[d]
Ivory Coast	332	ca. 2	(17)[d]
India (British East India)	228	12	(7)[d]
Guatemala	218	32	(3)[a]
Ethiopia	204	ca. 4	(15)[d]
Uganda	180	–	–
El Salvador	137	10	(8)[d]
Worldwide	5437	649	

[a] Harvest. [b] *cf. Ref. 31.* [c] *cf. Ref. 29.* [d] Export.

Fig. 11.46. *Oriental motive (sculpture) above the entrance of the oldest coffee house in Germany: 'Zum Arabischen Coffe Baum' (built 1570, restored 1997: Leipzig, Kleine Fleischergasse; photo D. Sicker)*

Table 11.3. *Coffee Imports and Annual Consumption* (A: imports in 1000 t; B: annual per-head consumption in kg)

	1885 – 1889 (annual average)[a]		1990[b]	
	A	B	A	B
USA	227	3.79	1,260	5.1
Germany	114	2.38	820	10.3
United Kingdom	14	0.37	174	3.1
Switzerland	8	2.79	70	10.8

[a] *Cf. Ref. 29.* [b] Source: *Cafe, Cacao, Thé* **1993**, *37*, 88.

Fig. 11.47. *'Scarcely had I enjoyed your wondrous fragrance…'* (*J. Delisle*; from *W. v. Hamm, T. Schwartze, H. Wagner, J. Zöllner, 'Die Chemie des täglichen Lebens'*, in *Das neue Buch der Erfindungen, Gewerbe und Industrien*, 6th edn., Vol. 5, O. Spamer, Leipzig, 1873)

Fig. 11.49. *Traditional East African pot for mokka, prepared with cinnamon and cardamom.* Glowing charcoal is placed in a perforated, removable base (height 38 cm). There are pieces up to 80 cm high in use (photo *Ch. Hesse*).

Fig. 11.48. *'Serving coffee in the desert'* (*W. v. Hamm, T. Schwartze, H. Wagner, J. Zöllner, 'Die Chemie des täglichen Lebens'*, in *Das neue Buch der Erfindungen, Gewerbe und Industrien*, 6th edn., Vol. 5, O. Spamer, Leipzig, 1873)

Fig. 11.50. *Examples of caffeine-containing soft drinks* (photo *Ch. Hesse*)

Fig. 11.51. *Coffee pot* (Christofle, Paris; height 22 cm) *and two coffee cups* (oval opening) *with translucent pattern* (Königlich-Preussische Porzellanmanufactur, Berlin, *ca.* 1830) (photo *Ch. Hesse*)

Fig. 11.52. *Mokka cup* (Muscovite porcelain) *in silver filigree stand* (height 6 cm) (photo *Ch. Hesse*)

Fig. 11.53. *Mokka cup* (Kütahya, Turkey, hand-painted with Ottoman motifs) *with chocolatière* (silver, *ca.* 1840, Paris) *on Hereke silk carpet* (photo *Ch. Hesse*)

A story from the beginning of the 16th century tells of how sheik *Ali ibn Omar al-Shadhili* of the port of Mokha (Mokka, today Mocha, Arab Republic of Yemen on the Red Sea) set a strange, black brew before some passing Portuguese merchants, telling them that he himself would prepare it from roasted, powdered beans, boiled in water, and that he was careful to drink it daily. It was described as kavak, kaweh, or kahwa. Enthused by the drink, the Portuguese took several sacks of the green beans aboard. Mokha consequently developed into the most important coffee-exporting port, until the eventual rise of Brazil as the world's biggest coffee producer (*Figs. 11.51–11.53*).

11.13. Tea (*Camellia sinensis*)

The home of the tea plant (*Thea sinensis* L. = *Camellia sinensis* L.; *Fig. 11.54*) is said to be Assam, where it still grows wild on the southern slopes of the Himalayas. From there, probably in prehistoric times, it made its way to China, which has been largely responsible for spreading tea in the historical period.

The tea plant is an evergreen shrub or small tree, growing 1–1.5 m high and flourishing in moderately hot, damp climates not subject to periods of drought, up to a latitude of 45°N. The harvested young leaves are subjected immediately after collection to an extremely elaborate process of drying, heating, rolling, and fermenting. Green tea is not fermented, but immediately steamed under exclusion of air to preserve its color. Addition of fragrances, special processing techniques, and the use of especially young leaves are some of the reasons behind the immense number of teas available on the world market.

The discovery of tea is recounted in a legend attributed to the pious *Bodhidharma*, an Indian ascetic and third son of a southern Indian king, who was spreading the teachings of Zen in China at the beginning of the 6th century. This holy Buddhist is said, in pious zeal, to have made a vow to meditate upon *Buddha*'s teachings for nine years, without sleeping. After three years, however, he did fall asleep. When he awoke, he was so distressed that he cut off his eyelids in atonement and threw them on the ground. In the place where they lay, however, the stimulatory tea plant sprouted. After a further five years, he became sleepy again. He took leaves from the plant and chewed them, whereupon the tiredness went away, and he was able to fulfill his task.

It is known that tea was already in medical use in China in the 8th century, when Chinese bonzes brought the plant to Japan, where it enjoyed a dissem-

Fig. 11.54. *'Branch from the tea plant'* (from *W. v. Hamm, T. Schwartze, H. Wagner, J. Zöllner, 'Die Chemie des täglichen Lebens', in Das neue Buch der Erfindungen, Gewerbe und Industrien, 6th edn., Vol. 5, O. Spamer, Leipzig, 1873)*

ination as great as in China. *Marco Polo* reported the dismissal of a Chinese finance minister because of an unauthorized increase in the tea tax in *ca.* 1285. In the 15th century, the custom of tea drinking spread through Asia. Rumors of this new beverage reached Europe in the 16th century, thanks to the travelers *G. B. Ramusion* (1559), *Almeida* (1576), *Maffeno* (1588), and others[30]. In 1610, the Dutch merchants brought tea, acquired from Chinese merchants in Indonesia, onto the European market. By 1635, tea was available in Paris. By the land route, tea arrived in Russia in 1637, when the Russian emissary to the court of *Altyn Khan* received it as a gift for the Tsar. In London, tea was offered in the coffee houses as a luxurious drink as early as 1660. Wider distribution was impeded not only by the disapproval of many public figures, but also by the high tea tax. This tax was also, formally, responsible for the American War of Independence, generally regarded as beginning with the 'Boston Tea Party' on December 16, 1773.

The still increasing production of tea today is attested to by statistics from the largest producer countries (*Table 11.4*) and reflects increasing consumption, mainly in the Western countries (*cf. Table 11.5*).

The infusion of tea in water contains caffeine (occasionally and unscientifically also referred to as *theine*) and tannins, the levels of which increase with longer brewing time. Today, tea is used medicinally, for pleasure and refreshment (*cf. Fig. 11.55*). The formal Japanese tea ceremony (*chanoyu*) is performed in especially simple rooms called *chashitsu* (tea pavilion) with the use of exquisite instruments, *e.g.*, *chavan* (tea bowl), *chaire* (tea caddy), and *chashaku* (bamboo spoon). Following strict rules, the tea is prepared from a

Fig. 11.55. *Russian samovar* (Tula, 1865; brass, height 59 cm) (photo *Ch. Hesse*)

Table 11.4. *Tea Production* (in 1000 t)

Country	1886[a]	1935[b]	1976[c]	1998[d]
India[e]	16	179	512	810
China[f]	*ca.* 113	11[g]	325	648
Sri Lanka[h]	0.2	96	197	270
Russia[i]	–	–	92	34[j]
Japan	21[k]	46	100	90
Turkey	–	*ca.* 3	60	120
Kenya	–	–	–	280
Worldwide			1,633	2,840

[a] *Meyers* Konversations-Lexikon, 1897. [b] Statistisches Jahrbuch des Völkerbundes. [c] DTV-*Brockhaus* Lexikon, Wiesbaden, 1984. [d] Fischer Weltalmanach 2000, Fischer Taschenbuch Verlag, Frankfurt a. M., 1999. [e] British India, India. [f] Formosa, People's Republic of China, China. [g] Formosa. [h] Ceylon, Sri Lanka. [i] UDSSR, Russia. [j] Georgia. [k] Export.

Table 11.5. *Tea Consumption* (in kg per head of population)

Country	1989[a]	1935[b]	1987/89 (average)
United Kingdom	2.24	4.22	2.81
USA	0.63	0.29	0.34
Germany	0.04	0.07	0.24
Switzerland	0.05	–	0.24

[a] *Meyers* Konversations-Lexikon, 1897. [b] Statistisches Jahrbuch des Völkerbundes. [c] Annual Bulletin of Statistics, International Tee Committee, 1991. Countries with the highest *per-capita* consumptions: Ireland (3.00) and Qatar (2.87); countries with the lowest *per-capita* consumptions: Thailand (0.01) and Italy (0.06).

powder of green tea leaves and stirred in hot water with a special, fine whisk (*chasan*) made from split bamboo (*Fig. 11.56*). After the tea, which, to the novice, tastes very bitter as well as refreshing, a piece of sugar is offered (*Fig. 11.57*).

Fig. 11.56. *Chasan, the Japanese bamboo whisk, an important instrument for the tea ceremony* (photo *Ch. Hesse*)

Fig. 11.57. *Japanese tea bowl* (porcelain, hand-painted, height 8 cm) (photo *Ch. Hesse*)

Fig. 11.58. *Tea brick*. Style of dried tea traditional in Mongolia, China, and Tibet, mixed with rice water and pressed into a cake for better keeping qualities (P. Oppliger, *Der Grüne Tee*, Midena-Verlag, Küttigen, Switzerland, 1997).

Fig. 11.59. *Tea caddy* (Thuringia, *ca.* 1850; Volkstedt porcelain, height 15 cm) (photo *Ch. Hesse*)

Fig. 11.60. *Tea service in Empire style by Odiot* (Paris, early 19th century; gilded silver). From right to left: teapot, water jug, samovar, sugar bowl, tea caddy (photo *S. J. Phillips*).

Chinesisches Haus

Erbaut 1754 - 1757
nach einer Entwurfsskizze Friedrichs II.
unter Leitung des Architekten J.G. Büring.
Die vergoldeten Sandsteinfiguren schufen
J.P. Benckert und M.C. Heymüller.
Die Chinesenfigur auf dem Dach
ist ein Werk des Bildhauers B. Giese
und des Kupferschmiedes F. Jury.

GEÖFFNET

Mai-Oktober: 10.00 - 17.00 Uhr

Mittagspause: 12.30 - 13.00 Uhr

Schließtag jeden Freitag

Fig. 11.61. *The Chinese House* (formerly Japanese House) *at Sanssouci, Potsdam* (built 1754 – 1757). Gilded sandstone figures representing Chinese tea-drinkers (photo *G. Grosskopf*, 1997).

Fig. 11.62. *Chinese characters representing 'tea'.* Pronunciation: *tiä* (Fujian province) or *cha* (rest of China) (drawn by *W. Hu*).

This tea ceremony developed during the Chinese *Sung* dynasty (960–1278) and reached Japan, where it remained preserved. Originally, the Chinese style of tea drinking had been completely different: '*The leaves were steamed, crushed with a pestle and mortar, made into a cake, and boiled together with rice, ginger, salt, orange peel, spices, milk, and sometimes onions as well*'[30]. This practice has been preserved by the Tibetans and by many Mongolian peoples (*Fig. 11.58*).

Europe similarly learned the custom of making tea from the Chinese by brewing the leaves in hot water. However, at that time, the classical style of Chinese tea-making had already changed after the invasions of the Mongols in the 13th century (*Figs. 11.59–11.61*).

The word 'tea' comes from the Chinese '*tiä*', a dialect spoken in the province of Fujian. This expression has found its way into other languages, *e.g.*, *tea* (English, Hungarian), *thé* (French); *Tee* or *tee* (German, Hebrew, Finnish), *té* (Spanish), *tè* (Italian). There is, by the way, a second Chinese term for the word *tea*: '*cha*'. The latter has, in this or in modified forms, also passed into other languages: *tschai* (Slavic), *shài* (Arabian), *cai* (Persian), *çay* (Turkish), *chá* (Portuguese), *o-cha* (Japanese), *trá* (Vietnamese). Interestingly, the same Chinese characters are used for both expressions, '*tjä*' and '*cha*' (*Fig. 11.62*). Presumably, migration and trade (*via* sea and land routes) brought the product and its names from China to the West and East. The use of both terms in modern Greek is compelling: common people drink τσαι (*tsai*), while the 'upper classes' enjoy τειον (*teïon*)!

11.14. Cocoa (*Theobroma cacao*)

The home of the cocoa tree (*Theobroma cacao* L. (Sterculiaceae)) is tropical America. The word *Theobroma*, like *cacao*, means 'food of the gods' (from the Greek δεοσ = god, βρωμα = food). The plants (*ca.* 20 species are known) grow to heights of 6–12 m. In Mexico, the cocoa tree has been cultivated since the 12th century, but the use of cocoa as a luxury commodity in Central America is more ancient. Around 1528, *Hernando Cortez* introduced cocoa (*cacahuatl* in Aztec) and chocolate (*chocolatl*) to Spain, from where it began its spread into the rest of Europe from 1650 (*Figs. 11.63–11.65*).

In catholic regions, calorie-rich drinks, which were permissible during strict periods of fasting, were extremely popular. Consequently, the custom of drinking the pleasant-tasting, rich, and nutritious chocolate spread from the catholic Mediterranean states. The custom of coffee drinking, on the other hand, developed more quickly in protestant communities.

Modern areas of cocoa cultivation lie in the hottest and rainiest parts of the tropics, mostly in the Ivory Coast (1.12 million t, 38.9%), Ghana (13.2%), Indonesia (12.8%), Brazil (9.4%), Nigeria (5.4%), and Cameroon (4.3%)[32]. A cocoa fruit contains up to 50 cocoa beans. Fermentation, drying, roasting, and crushing produces the cocoa mass. Cocoa beans contain 1–3% theobromine and smaller quantities of caffeine. Both substances act as stimulants, although the effects of theobromine on humans are significantly weaker than those of caffeine. Interestingly, both caffeine and theobromine are listed as illegal substances on the horse doping-list of the *Fédération Equestre Internationale* (*FEI*, April 1995: threshold value: 2 µg/ml urine), although recommended energy feedstuffs for horses contain a high proportion of theobromine- and caffeine-rich cocoa husks.

The world cocoa harvest in 1998 was 2.88 million tons[30].

Typical compositions of cocoa and chocolate are given in *Table 11.6*.

The annual (*per capita*) variation in consumption of chocolate and chocolate-based products (in kg per person) in different countries is also interest-

Fig. 11.63. *Copperplate engraving from 1689: Coffee, tea, and chocolate; the new hot beverages.* The Arab as a coffee drinker, the Chinese as a tea drinker, and the native American as a chocolate drinker (Cliché Bibliothèque Nationale de France, Paris).

Table 11.6. *Composition of Cocoa* versus *Chocolate* (in %)[a]

	Cocoa Beans	Cocoa Butter	Dark Chocolate
Proteins	18.0	6.4	6.0
Lipids	56.0	54.0	27.0
Carbohydrates	13.5	28.0	54.0
Water	3.0	2.0	1.0
Theobromine	1.45	1.10	0.5
Caffeine	0.05	0.50	0.07
Theophylline	–	–	0.001

[a] These average values are subject to considerable variations[31].

R[1] = Me, R[2] = H : Theophylline

R[1] = H, R[2] = Me : Theobromine

R[1] = R[2] = Me : Caffeine

ing (source: *International Committee for Harmonization of Confectionery Statistics*; 1992):

Switzerland	10.0	USA	4.8
Austria	7.9	France	4.8
UK	7.4	Italy	1.9
Germany	6.5	Japan	1.6
Australia	5.5	Spain	1.5

As well as for recreational and nutritional purposes, chocolate for medicinal purposes has also been commercially available on occasion. It was 'invented' in France *ca.* in 1800, and a wealth of such products were on the market in the 19th century[31]. A few sample recipes are reproduced below (figures in percentages by weight):

Anthelmintic chocolate (for the preparation of small pastilles (2 g)):

Soft cocoa butter	66.2
Sugar	20.7
Croton oil	1.4
Cinnamon	1.4
Calomel (Hg_2Cl_2)	10.3

Chocolate remedy for sexually transmitted diseases:

Ordinary chocolate	84.7
Sugar	12.1
Peru balsam	3.0
Sublimate ($HgCl_2$)	0.2

Fig. 11.64. *Jean-Etienne Liotard, 'The Chocolate-Girl'* (1744/45; pastels on parchment; Staatliche Kunstsammlung Dresden, Gemäldegalerie Alte Meister)

Fig. 11.65. *Straight-handled pot for hot chocolate, with matching cup; top of pot fitted with a brass handle and a special device for stirring (Meissen, Germany, ca. 1775) (G. Sterba, Meissener Tafelgeschirr, Edition Leipzig, 1998)*

Chocolate for health and trinity (homeopathic, 'cocoa-free' chocolate):

Cocoa butter	0.0
Carolina rice	2.2
Chicory root (powdered)	30.4
Mocha coffee	14.0
Rhizome of iris of Florence (powdered)	1.0
Lactose powder	1.0
Olive oil	1.4

The practice of administering medicines with a chocolate coating has lasted to this day.

References and Notes

1. dtv Lexikon in 20 Volumes, Deutscher Taschenbuch-Verlag, Munich, 1992, Vol. 17.
2. J. Pecirka, *Die Giftgewächse des österreichischen Kaiserstaates und Deutschlands*, Verlag K. André, Prague, 1859.
3. L. Lewin, *Die Gifte in der Weltgeschichte*, 2nd edn., Springer-Verlag, Berlin, 1920.
4. L. Lewin, *Phantastica*, Verlag G. Stilke, Berlin, 1927.
5. S. Aaronson, '*Fungal Parasites of Grasses and Cereals: their Role as Food or Medicine, Now and in the Past*', Antiquity **1989**, *63*, 247.
6. A. Hofmann, *LSD – Mein Sorgenkind*, Klett-Cotta, Stuttgart, 1979; published in English under the titel: *LSD – My Problem Child*.
7. M. Seefelder, *Opium, eine Kulturgeschichte*, ecomed Verlagsgesellschaft, Landsberg, 1996.
8. This ancient town gave its name to the word 'meconium', which is another term for 'opium' (Chambers Dictionary). In other documents, meconium is described as an extract of *Papaver somniferum* (whole plant), supposed to be less effective than opium itself. *Mecon*: Greek, poppy.
9. Some figures[7]: Imports to China: 2,000 cases in 1762; 4,000–5,000 cases in 1819; 18,000 cases in 1828; 40,000 cases in 1839; 81,000 cases (or *ca.* 5,000 t) in 1884. Opium production in 1906: 500 t in Mediterranean countries; 850 t in Persia; 7,000 t in India; 35,000 t in China. With the beginning of international conferences targeted at the use of opium exclusively as a narcotic drug (1909 Shanghai, 1909 den Haag, 1914 den Haag), leading to the international agreement on opium abuse (Geneva Convention, July 13, 1931), official production figures are no longer available. However, from the annual amounts of confiscated drugs, it is possible to guess at the magnitude of opium produced and smuggled today. Thereby, the center of modern opium production is the so-called 'Golden Triangle' between Burma, Laos, and Thailand, where 70% of the crude opium is harvested for worldwide heroin production.
10. K. Fairbank (Ed.), *The Cambridge History of China*, 1800–1911, Vol. 10, Cambridge University Press, Cambridge, 1978.
11. F. Abegg, *Vom Reich der Mitte zu Mao Tse-Tung*, Verlag Bucher AG, Luzern, 1966.
12. *De Quincey*, like most addicts in the early 19th century, almost always took opium in the form of *laudanum*, a tincture of opium in alcohol.

13. At variance with the modern view, the pharmacologist and toxicologist *F. Hauschild* wrote in his renowned textbook[14] as late as 1956:

    '*Society does a disservice if the morphinist is, for instance, a gifted artist or an important figure in commercial or scientific life. It is not possible to discount the fact that detoxification therapies are only rarely successful and, even if they are, that 'another personality' tends to come out of them, to make no mention of the fact that the therapy itself removes the subject from their social functioning for a prolonged period of time.*

    It is wrong to brand all morphinists or addicts as damnable and dangerous criminals, deserving of punishment!

    If we consider the damage done to society at present by opiates and on the other hand by alcohol abuse (traffic accidents, violence, absence from work), the discrepancy becomes obvious. The beneficial effects of opiates, the ability to remove physical and mental pain, make them among the most valuable medicines. To demonize them as 'drugs' is a sign of primitive thinking; their role in weakening whole populations (opium war) is scandalous.

    It is unquestionably necessary to regulate traffic in these substances. The 'drug laws' constitute a necessary measure, which may, on sensible interpretation, be useful, but which may also, through narrow and bureaucratic application […], be damaging'.

14. F. Hauschild, *Pharmakologie und Grundlagen der Toxikologie*, VEB Thieme Verlag, Leipzig, 1956.

15. The lethal opium doses are *ca.* 10 mg for children and 0.5–2 g for adults. Addicts can tolerate up to 7.5 g/day, according to the degree of habituation; these values, though, are only approximate, since the alkaloid content of opium varies, depending on its origin[14].

16. F.-J. Kuhlen, *Zur Geschichte der Schmerz-, Schlaf-, und Betäubungsmittel im Mittelalter und früher Neuzeit*, Deutscher Apotheker Verlag, Stuttgart, 1983.

17. *Brockhaus* Enzyklopädie, 24 Vols., 19th edn., Mannheim, 1986–1994.

18. Drinks containing cocaine became commercially available around 1900, *e.g.*, 'Coca des Incas, vin tonique & digestiv, Paris' or 'Vino Mariani', likewise of French origin. 'Coca Cola' (Coke), on the US market since 1886, also contained cocaine until 1903.

19. The lethal dose in a non-habituated adult is *ca.* 1 g.

20. S. Balabanova, F. Parsche, W. Pirsig, '*First Identification of Drugs in Egyptian Mummies*', *Naturwissenschaften* **1992**, *79*, 358.

21. Also known as belladonna, black nightshade, nightshade, or sleeping nightshade.

22. Between 1973 and 1996, 14–27 cases of poisoning from *Atropa belladonna* were registered annually by the Swiss Toxicological Information Center in Zurich, 10% of which leading to serious or even fatal consequences. (Source: annual reports).

23. G. Buchanan, *Rerum Scoticarum Historia*, Francofurti, 1624.

24. Until the beginning of the 19th century, the following agents (either individually or in mixtures) were used for narcotic purposes: hashish, opium, poppy (*Papaver*), *Atropa*, *Hyoscyamus*, and theriac. From 1818 onwards, diethyl ether, popularized by *Faraday*, came into use, together with chloroform after 1853 (Queen *Victoria*, *e.g.*, gave birth under chloroform narcosis in 1857). Theriac was an antique, universal panacea, described by *Galen* in his *De antidotis*. It was composed out of 70 separate components, opium among them.

25. The genus *Nicotiana* (Solanaceae) comprises *ca.* 40 species, most of them annual, rather clammy-leafed plants, which are herbaceous (1–2 m high) or, more rarely, tree-like (*e.g.*, *N. fruticosa* L.), and are grown worldwide, preferentially in temperate and subtropical zones. The nicotine content in the leaves of *N. tabacum* varies between 0.05–10%.

26. Worldwide production of tobacco (excluding home production), given in million tons (per year): 0.743 (1889); 1.034 (1910); 1.310 (1924); 2.580 (1939); 5.780 (1978); 6.908 (1998); 6.854 (1999). World population, in billions: 1.480 (1890); 1.650 (1900); 2.100 (1930); 3.020 (1960); 4.124 (1977); 5.930 (1998). Revenue from tobacco tax in Germany in 1998: 21.65 billion DM. (Source: Der *Fischer* Weltalmanach, Fischer Taschenbuch Verlag, Frankfurt a.M., 1979, 1994, 1999, 2000).

27. H. Spode in '*Feuer, bitte!*', *NZZ-Folio* **1996**, *11*, 18.

28. The German word 'Alraune' (Old High German *alruna*, Gothic *rûna*) means 'secret' or 'mystery'.

29. *Meyers* Konversations-Lexikon, 17 Vols., 5th edn., Leipzig, 1897.

30. K. Okakura, German translation by H. Hammitzsch, *Das Buch vom Tee*, Insel-Verlag, 1995.

31. M. Montignac, *Gesund mit Schokolade*, Artulen-Verlag, Offenburg, 1996.

32. Der *Fischer* Weltalmanach 2000, Fischer Taschenbuch Verlag, Frankfurt a.M., 1999.

Index